Flame Retardant Materials

Manufacturing, Applications, and Environmental Perspectives

Edited by

Inamuddin[1], Arwa Alrooqi[2], Tariq Altalhi[3], Ayham Safklou[4]

[1]Advanced Functional Materials Laboratory, Department of Applied Chemistry, Zakir Husain College of Engineering and Technology, Faculty of Engineering and Technology, Aligarh Muslim University, India

[2]Department of Chemistry, Faculty of Science, Al-Baha University, Al Baha, Saudi Arabia

[3]Department of Chemistry, College of Science, Taif University, 21944 Taif, Kingdom of Saudi Arabia

[4]Sustainable Environmental Works Trading Company, Jeddah, Saudi Arabia

Published by **Materials Research Forum LLC**
Millersville, PA 17551, USA

Published as part of the book series
Materials Research Foundations
Volume 190 (2026)
ISSN 2471-8890 (Print)
ISSN 2471-8904 (Online)

Print ISBN 978-1-64490-406-0
eBook ISBN 978-1-64490-407-7

Distributed worldwide by

Materials Research Forum LLC
105 Springdale Lane
Millersville, PA 17551
USA
https://mrforum.com

Manufactured in the United States of America
10 9 8 7 6 5 4 3 2 1

Table of Contents

Preface

In the era of modern science and technology, the continuous development of new technologies has become essential for progress and modernization. These advancements require materials that are resistant to fire to ensure safety and reliability. Such materials, known as flame-retardant materials, are increasingly incorporated into products across modern society to enhance safety and sustainability. A wide variety of these materials are used in plastics, textiles, sensors, batteries, and construction materials to inhibit, suppress, or delay the spread of fire.

This book presents a comprehensive and systematic exploration of flame-retardant materials, focusing on their principles, sources, properties, and industrial significance. It covers a broad spectrum of retardants, including biomass-derived, synthetic, mineral-based, and nanotechnology-enabled systems, along with their evolving roles in modern materials science.

Special emphasis is placed on advanced materials such as MXenes, metal–organic frameworks (MOFs), carbon based, and electrospun networks, highlighting their potential in enhancing fire safety performance. The book also addresses the growing need for sustainability by examining bio-based and eco-friendly flame retardants as viable alternatives to conventional toxic materials.

Applications across key industrial sectors—including textiles, paints, polymers, electronics, sensors, batteries and transportation—are discussed in detail, with a focus on meeting stringent fire safety standards. The content is designed to bridge fundamental understanding with practical implementation, offering insights into current challenges and future directions in the field.

This volume is intended for researchers, scientists, engineers, and industry professionals engaged in materials development and safety engineering. It also serves as a valuable academic resource for undergraduate and postgraduate students in engineering, materials science, chemistry, nanotechnology, and related disciplines.

Overall, the book aims to provide a balanced integration of foundational concepts and emerging advancements, supporting the development of safe, sustainable, and high-performance flame-retardant materials. This book contains 14 chapters, which are summarized below:

Chapter 1 reviews sustainable and green flame retardants, covering bio-based, mineral-based, and nano-enhanced systems. Their mechanisms and limitations are described, including char formation in bio-derived additives, water-vapour release and barrier formation in mineral systems, and nanoscale barrier effects in advanced fillers. Applications in construction, electronics, textiles, transport, packaging, and furniture are summarized, along with current challenges related to stability, compatibility, and large-scale use.

Chapter 2 introduces phase change materials (PCMs) as sustainable thermal energy storage media and highlights their use in building insulation, transportation, and space applications. Despite their advantages, many PCMs pose safety concerns due to volatility and flammability. The chapter focuses on strategies to reduce flammability and improve the safety and performance of PCM-based systems.

Chapter 3 reviews biomass-based flame retardants as sustainable, non-toxic alternatives to halogenated systems. It summarizes key flame-retardant mechanisms, major biomass sources, modification strategies, applications in polymers and textiles, and emphasizes their environmental benefits and fire-safety performance.

Chapter 4 reviews carbon-based nanomaterials as advanced flame retardants in polymers, highlighting their mechanisms for fire risk reduction and performance at low loadings. Applications, benefits, challenges, and future trends toward hybrid, functionalized, and halogen-free systems are discussed.

Chapter 5 provides an overview of carbon-based nanomaterials used to improve flame retardancy in polymers. It describes how graphene, CNTs, fullerenes, carbon black, and expandable graphite enhance fire resistance by strengthening char layers, trapping radicals, and forming protective barriers. Key testing methods and basic synthesis approaches are briefly outlined, along with environmental considerations and challenges such as dispersion, large-scale production, and certification. Future directions for efficient, multifunctional, and sustainable flame-retardant systems are also highlighted.

Chapter 6 discusses the sources, structures, physical and chemical properties, preparation methods, and performance of lignin reported in the literature. The development prospects and challenges of using lignin as a natural flame retardant are also described.

Chapter 7 reviews the use of metal–organic frameworks (MOFs) and their hybrids as flame retardants in polymeric materials, discussing synthesis strategies, flame-retardant mechanisms, performance evaluation, current challenges, and future prospects.

Chapter 8 describes various base materials used in flame-retardant formulations, including compounds based on ammonia, melamine, phosphazenes, phospham, phosphoroxynitride, and nitrogen-based radical generators. Their action mechanisms, advantages, and common applications are outlined.

Chapter 9 discusses nitrogen-based flame retardants, including ammonium and melamine derivatives, cyanuric acid-related molecules, azoalkanes, phosphazenes, phospham, and N-alkoxy hindered amines, with a focus on their flame-retardant mechanisms and fire behavior.

Chapter 10 reviews advances in incorporating flame-retardant additives into electrospun polymeric matrices, discussing preparation techniques, nanofibrous membrane structures, mechanisms for enhanced flame resistance, performance in various applications, challenges, and future directions.

Chapter 11 examines flame-retardant strategies for mitigating thermal runaway in lithium-ion and solid-state batteries. Halogenated, halogen-free, and phosphorus-based materials, including non-flammable electrolytes and functional separators, are analyzed. The chapter also explores the relationship between safety and electrochemical performance and reviews international safety standards.

Chapter 12 introduces flame-retardant dielectric material substrates for microwave sensors, discussing material types, properties, interactions with microwave frequencies, applications, and the benefits and limitations of these substrates.

Chapter 13 reviews inorganic flame retardants and minerals used to enhance polymer fire resistance, covering mechanisms, key types, synergistic systems, applications, testing methods, and challenges. Emphasis is placed on halogen-free, eco-friendly, and nanoscale-enhanced solutions.

Chapter 14 describes fire safety standards for textiles, furniture, electronics, and building materials, focusing on BIS, NBC, ASTM, ISO, IEC, BS, and EN guidelines. It covers ignition behavior, heat release, smoke generation, and charring, as well as challenges in compliance and testing. Emerging trends, including sustainable flame retardants and AI-assisted material designs, are highlighted, along with relevant case studies.

Flame Retardant Materials | Materials Research Forum LLC
Materials Research Foundations 190 (2026) 1-15 | https://doi.org/10.21741/9781644904077-1

Chapter 1

Sustainable and Green Approaches in Flame Retardant Development

Charles Michael Albert[1], Liew Kang Chiang[1]*

[1]Faculty of Tropical Forestry, Universiti Malaysia Sabah, Jalan UMS, 88400, Kota Kinabalu, Sabah, Malaysia

liewkc@ums.edu.my

Abstract

Sustainable and green flame retardants are increasingly developed as alternatives to halogen-based chemicals that have raised environmental and health concerns. Their lower toxicity and better compatibility with recycling processes have increased their use across various material classes. This chapter reviews the mechanisms of action, integration methods, and industrial applications of bio-based, mineral-based inorganic, and nano-enhanced flame retardants. Phytic acid, lignin, and tannin-based formulations improve charring behaviour but still face limitations in heat resistance and moisture stability. Aluminium hydroxide and zinc borate are among the most widely applied mineral-based systems and provide stable, halogen-free protection, although high loading levels can reduce strength and processability in polymers. Nano-enhanced additives such as clays, layered double hydroxides, and graphene materials improve flame resistance at lower concentrations, but uniform dispersion and production cost remain practical challenges. These materials are now applied in construction, textiles, electronics, and packaging sectors to comply with strict fire safety standards. Research continues to focus on improving their long-term durability, scalability, and reliability during standard fire testing.

Keywords

Green Flame Retardant, Sustainable Materials, Fire Protection, Halogen-Free, Fire Safety

Contents

Materials Research Foundations 190 (2026) 1-15 https://doi.org/10.21741/9781644904077-1

1. Introduction

Flame retardants improve the fire resistance of combustible materials by suppressing ignition and reducing fire spread [1]. They are widely used in polymer-based products, textiles, structural components in construction, transportation, and electronic sectors [2]. The integration of these materials allowed the products to meet fire classifications such as UL-94, ASTM E84, EN 13501-1, and other region-specific codes [3,4]. Conventional flame retardants, especially halogenated types—polybrominated diphenyl ethers (PBDEs) and hexabromocyclododecane (HBCD)—have raised environmental and health concerns [5]. These compounds are resistant to degradation, bioaccumulative, and have been linked to endocrine disruption, developmental toxicity, and neurotoxicity [6].

Their degradation during combustion can promote the release of corrosive gases and toxic byproducts, such as brominated dioxins and furans [6]. The presence of these substances has prompted restrictions and bans under the EU REACH regulation and the Stockholm Convention [7]. In addition, these flame retardants are difficult to remove during recycling, and often remain in the plastic even after reprocessed [5]. This raises long-term environmental concerns, as contaminated materials may be reproduced into consumer products and disrupt safe waste management [1,5].

To address the limitations of the traditional flame retardants, research and development of sustainable alternatives that are both functionally effective and environmentally friendly are continuously being carried out [1]. Sustainable flame retardants are generally referred to as materials that reduce flammability without introducing hazardous substances throughout their life cycle [8]. These materials are derived from renewable feedstocks and non-halogenated additives that are designed for recyclability or biodegradation [8,9]. In fact, the presence of non-toxic additives in flame retardants improves occupational safety and reduces health risks during fire scenarios [2]. Sustainable flame retardants are required to achieve similar or better fire performance than conventional ones, especially when tested under standardized conditions [8,10]. They must also avoid increased smoke density, thermal degradation of the substrate, or incompatibility with recycling processes [3,11].

Halogenated flame retardants were originally adopted due to their capability to meet established fire classifications [10]. They enabled products to pass vertical burn tests, surface flame spread tests, and smoke emission limits [1]. Although many of these compounds are now restricted because of their toxicity, materials must still meet the same fire safety standards [8]. Therefore, sustainable flame retardants must not only exclude hazardous elements, but also achieve the minimum requirements under recognized standards such as ASTM E84, EN45545, and ISO 5660 [1,11]. As a result, new formulations are not only assessed for their chemical safety, but also for their flammability properties, such as smoke release, char formation, and thermal degradation [12]. This balance remains a key priority in creating environmentally safer flame retardants [9]. This chapter presents the overview of sustainable and green flame retardant development, by focusing on various flame retardant types, along with their mechanisms of actions, advantages, and drawbacks. In addition, the industrial applications are discussed in relation to material compatibility, and industrial scalability.

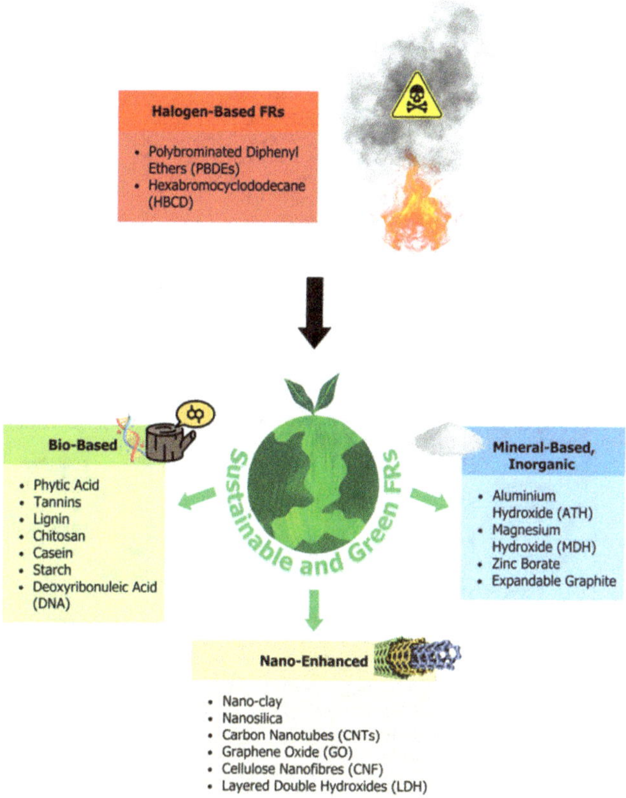

Figure 1. The shift from conventional halogen-based flame-retardants (FRs) to sustainable and green FRs.

2. Types of Green Flame Retardant

2.1 Bio-Based Flame Retardants

Bio-based flame retardants are derived from renewable sources, such as, agricultural waste, marine biomass, and lignocellulosic by-products [13]. These materials are increasingly being adopted as greener substitutes for halogen-based compounds, attributed to reduced toxicity and improved biodegradability [14]. Table 1 summarises main flame retardants from renewable sources and their use in different materials, methods of incorporation, mechanism of actions, advantages, and drawbacks.

Table 1. Comparative Properties of Flame Retardants Developed from Bio-Derived Materials

Flame Retardant Material	Target Material	Incorporation Technique	Flame Retardant Mechanism	Advantages	Disadvantages	Refs.
Phytic acid	Cotton fabric Polylactic acid Starch-based polymers	Impregnation Blending Layer-by-layer coating	Promotes early char formation Dilute combustible volatiles	Renewable Non-toxic	Moisture-sensitive Prone to leaching Limited life span	[1,15]
Tannins	Cotton, Epoxy resin Lignocellulosic boards	Impregnation Blending with phosphorus	Reduce flammable gases release Form dense and thermally resistant barrier layers	Bio-based Improve char yield	High thermal behaviour variation Low effectiveness alone	[16,17]
Lignin	Polylactic acid Epoxy Polyurethane Wood panels	Blending Phosphorylation	Produces a stable char layer Limits heat transfer and volatile release	Biodegradable Abundant High char yield	Incompatible with many polymers without chemical modifications	[4,18]
Chitosan	Cotton Polylactic acid Polyesters	Layer-by-layer coating Solution casting	Dilute combustible volatiles Forms intumescent char	Non-toxic Biodegradable	Limited thermal resistance Need phosphorus-based additives for better fire resistance	[19,20]
Casein	Cotton Polylactic acid Cellulose fibres	Surface coating Melt blending	Promotes early char formation Formed stable char layer	Waste-derived Rich in nitrogen and phosphorus	Low processability Less efficient standalone performance	[21,22]
Starch	Polylactic acid Polyester Cotton	Surface treatment Physical blending	Dehydrates and expands to form carbonaceous foam	Low cost Renewable Widely available	Hydrophilic Prone to leaching Weak thermal resistance	[13,23]
Deoxyribonucleic acid (DNA)	Cotton Polyamides Biopolymers	Layer-by-layer deposition	Dilute combustible volatiles Generate polyphosphoric acid that expands char	Rich in nitrogen and phosphorus	High cost Poor scalability Moisture-sensitive	[24,25]

Bio-based flame retardants generally operate in the condensed phase by promoting early char formation, dehydration, or thermal shielding [14]. During thermal exposure, these processes occur through acid-catalysed dehydration and carbonisation at the polymer surface [13]. Phosphorus-containing compounds produce phosphoric or polyphosphoric acid, which accelerates dehydration of hydroxyl groups in the substrate [1]. This reaction forms a carbon-rich char that acts as an insulating layer and restricts heat transfer [12]. Nitrogen-based compounds release ammonia and other non-flammable gases, lowering oxygen concentration near the flame and suppressing radical chain reactions in the gas phase [11].

Compounds such as phytic acid, DNA, and casein contain high levels of phosphorus and nitrogen, which facilitate acid-catalysed dehydration and the emission of non-combustible gases that suppress flame propagation [13]. Although tannins and lignin are lacking in direct gas-phase activity, they function as aromatic carbon sources that stabilise the char layer and slow down thermal decomposition [17,18]. Most of the flame retardants are more effective when applied to natural fibre-based composites or cellulosic textiles, especially through surface-based methods such as impregnation or layer-by-layer coating [21]. Surface-oriented applications allow more efficient use of bio-based additives because their activity is concentrated at the ignition zone [25]. These coatings or impregnations preserve the internal mechanical integrity of the substrate while improving surface protection [25]. Layer-by-layer deposition also enables precise control of thickness and chemical composition, which is useful for adjusting flame retardancy in fabrics, veneers, and biodegradable polymer films [26]. This allows flame retardants to perform without significantly altering the bulk properties of the host material [1].

Some bio-based flame retardants have shown reduced stability when exposed to elevated temperatures [14]. In comparison with halogenated compounds, they often begin to decompose before ignition occurs, which reduces their effectiveness in delaying combustion [15]. Their hydrophilicity—especially in starch, chitosan, and casein—can cause them to retain moisture [19,22]. This occurrence leads to leaching or functional loss after exposure to humidity or repeated cleaning cycles [18]. Additionally, inconsistencies in the molecular structure of lignin and tannins make it difficult to maintain uniform flame-retardant behaviour [13,16]. Without chemical treatment, their compatibility with low-polarity polymers such as polyethylene remains limited due to poor dispersion during processing [9]. To overcome this limitation, chemical modification strategies such as phosphorylation, esterification, and silane coupling have been applied to improve bonding with non-polar matrices [9,13]. These treatments enhance dispersion and reduce additive migration during thermal cycling [1]. Improved interfacial adhesion also contributes to better char stability and long-term retention of flame-retardant properties [27]. However, each additional processing step must be assessed for cost, environmental compliance, and potential solvent use [27].

Various formulation strategies have been studied, such as surface modification through grafting, incorporation of phosphorus groups, and blending with inorganic additives or nanoscale particles [28,29]. These approaches can improve thermal behaviour, enhance char layer stability, and increase compatibility with synthetic polymers [28]. However, the complexity of these hybrid formulations may affect recyclability or cost efficiency [27]. While various studies have investigated strategies to improve the performance of bio-based flame retardants, many of them still face challenges in meeting standardised fire safety criteria without high loading levels or hybrid formulations [1,28]. As a result, their current use remains limited to sectors where environmental criteria are prioritized over strict fire performance demands [13]. These include short-life or biodegradable products, cellulose-rich textiles, and surface-coated composites [1].

Achieving a balance between fire performance, material compatibility, and environmental safety remains the main challenge for bio-based flame retardants [28]. Their renewability and low toxicity make them attractive substitutes, but their thermal endurance is generally lower than mineral or nano-scale additives [27]. These systems are best suited for products where moderate heat resistance is acceptable, such as textiles, coatings, and biodegradable packaging [14,29]. Recent research trends have focused on hybrid approaches that combine bio-based phosphorus donors with mineral fillers or nano-clays to enhance barrier stability and reduce additive loading [28]. Therefore, further research and development is needed to improve scalability, consistency, and long-term stability under varied operating conditions [10].

2.2 Mineral-Based Inorganic Flame Retardants

Mineral-based inorganic flame retardants are widely applied in polymeric and fibre-reinforced composites as alternatives to restricted halogen-based compounds [30]. These materials are either naturally derived from minerals or synthesised from non-carbon inorganic compounds, such as metal hydroxides, borates, and silicates [31]. Although not renewable, they are considered environmentally preferable due to their low toxicity, non-halogenated chemistry, and stable behaviour during combustion and disposal [1,32]. Their mechanism of action is primarily physical—most operate in the condensed phase by releasing inert gases, promoting endothermic cooling, or forming a protective char or glassy layer [31]. These flame retardants are generally known for their compatibility with fire safety regulations and recyclability [9]. However, they often require increased loading levels to meet performance demands, which may negatively impact the structural and processing properties of the base material [31]. Table 2 summarises main mineral-based inorganic flame retardants, their application areas, incorporation methods, mechanisms of action, advantages, and limitations.

Table 2. Comparative Properties of Mineral-Based Inorganic Flame Retardants

Flame Retardant Material	Target Material	Incorporation Technique	Flame Retardant Mechanism	Advantages	Disadvantages	Refs.
Aluminium hydroxide (ATH)	Polyethylene Polypropylene	Direct blending Melt compounding	Release water vapour to dilute flammable gases	Low-toxicity Halogen-free Smoke suppression	Requires high-loading Reduces mechanical strength	[33,34]
Magnesium hydroxide (MDH)	Cable insulation Biocomposites	Extrusion Bulk blending	Release water vapour at high temperature	Thermally stable Halogen-free Low smoke generation	High loading needed Affect dispersion in some polymers	[33,35]
Zinc borate	Wood composites Textiles	Co-additive blending Surface coating	Forms glassy boron layer Promotes more char formation	Smoke suppression Enhance hybrid flame retardant	Limited standalone effect Limited thermal stability	[36,37]
Expandable graphite	Polystyrene Intumescent coating	Physical blending Compression moulding	Expands under heat to form intumescent insulating barrier	Non-toxic Low smoke generation	Reduce surface finish quality Prone to agglomeration	[38,39]

Materials Research Foundations 190 (2026) 1-15 https://doi.org/10.21741/9781644904077-1

The flame retardants listed in Table 2 operate primarily through condensed-phase mechanisms, specifically heat absorption, inert gas release, and physical barrier formation [37]. These mechanisms reduce the temperature of the combustion zone and slow the release of flammable volatiles, delaying ignition and limiting flame spread [12]. Aluminium and magnesium hydroxides suppress ignition through endothermic decomposition, releasing water vapour that cools the surface and dilutes flammable gases near the combustion zone [33]. The released water also promotes the formation of a thin oxide layer that limits heat penetration into the substrate and stabilises the surface against further decomposition [30,33].

Zinc borate forms a glassy residue that helps to retain structural integrity, delay combustion, and reduce smoke emission [36]. This boron-rich glass seals surface pores and improves the continuity of the protective barrier, further restricting oxygen diffusion and heat feedback to the underlying polymer [36,37]. In contrast, expandable graphite increases in volume when exposed to heat, creating a dense carbon barrier that restricts rtoxygen access and reduces thermal conductivity at the surface [38]. The expansion begins rapidly after heating, and the intumescent layer acts as both a thermal shield and a gas-diffusion barrier that slows degradation of the polymer below [12,38].

Despite their low toxicity and halogen-free composition, mineral-based flame retardants often require high loading levels—typically exceeding 40–50 wt%—to meet regulatory fire performance thresholds such as UL-94 or EN 13501-1 [1,37]. Such high loadings are necessary because these additives act mainly through physical processes rather than chemical interference with flame radicals, so a larger quantity is needed to achieve sufficient cooling and gas dilution [2,3]. These high filler contents can weaken mechanical strength, increase material density, and make extrusion or moulding processes difficult [39]. In addition, dispersion issues are common, especially with materials like magnesium hydroxide or expandable graphite, which may settle or aggregate during mixing [37,38]. This occurrence affects both processing and final flame retardant efficiency, especially in polymers with low surface polarity [37]. Surface treatment or the use of coupling agents is often applied to improve filler dispersion and interfacial adhesion, which helps maintain mechanical strength and processing uniformity [33,38,40].

Although these materials are not renewable, they are often included in low-toxicity flame retardant systems due to their clean combustion and compatibility with recycling processes [11]. They are used in applications such as cable insulation, polymer foams, and fire-rated components where halogen-free performance is required [2]. Their inert behaviour and low smoke generation also make them suitable for interior, construction, and transport applications where air-quality regulations are strict [9,10]. To reduce the quantity needed and minimise processing constraints, mineral-based flame retardants are sometimes blended with surface-treated additives or combined with dispersants [8]. Recent work has investigated hybrid formulations that combine mineral fillers with small proportions of bio-based or nano-scale additives to maintain fire performance while reducing total loading [38,40,41].

2.3 Nano-Enhanced Flame Retardants

Nano-scale additives are increasingly applied in flame-retardant systems to improve thermal stability and reduce flammable gas release [10]. They act as barriers or contribute to char formation when exposed to heat [39]. These additives are generally effective at lower concentrations compared to conventional fillers, which helps maintain the base material's mechanical strength [9]. Table 3 lists the nano-enhanced flame retardants and their properties.

Table 3. Different Types of Nano-Enhanced Flame Retardants and Their Properties

Flame Retardant Material	Target Material	Incorporation Technique	Flame Retardant Mechanism	Advantages	Disadvantages	Refs.
Nano-clay	Polylactic acid Polyamide 11	Melt compounding Coating	Forms a maze-like path that slows gas diffusion	Low loading needed Improves char stability	Risk of agglomeration	[41,42]
Nanosilica (SiO_2)	Polyester Polyolefins	Sol gel Surface coating	Acts as a thermal barrier Stabilises the char	Good thermal stability Compatible with bio-based polymers	May cause brittleness Prone to sedimentation	[43,44]
Carbon nanotubes (CNTs)	Thermoplastic Epoxy	Solution blending Melt mixing	Promotes char formation Delays heat transfer	Improves mechanical strength and fire resistance	High cost Inhalation or handling risk in raw form	[45,46]
Graphene oxide (GO)	Polylactic acid Polyurethane	Solution casting In-situ growth	Forms a barrier layer that slows heat and gas transfer	Improves char integrity	Expensive	[40,47]
Cellulose nanofibres (CNF)	Polylactic acid Biopolymers	Surface modification Blending	Promotes char formation	Renewable Biodegradable	Moisture-sensitive Limited thermal stability	[48,49]
Layered double hydroxides (LDH)	Polyolefins Ethylene Vinyl Acetate	Melt blending	Release water to delay ignition Form a solid, ceramic-like barrier	Halogen-free Reduce peak heat release rate	Requires surface treatment for effective dispersion	[26,50]

The flame retardants listed in Table 3 improve fire resistance by forming physical barriers, altering thermal degradation pathways, or delaying the release of flammable gases [43]. At the nanoscale, barrier efficiency depends on the geometry and surface behaviour of the additive [41]. Plate-like particles such as nano-clays create a long diffusion path for heat and volatiles, while fibre- or sheet-shaped fillers such as carbon nanotubes and graphene oxide improve the cohesion of the developing char [40,41,46]. Layered double hydroxides further assist by releasing water during decomposition, which cools the surface and supports char consolidation [26,50]. Their nanoscale dimensions enable effective performance at relatively low concentrations, which helps retain the mechanical and processing characteristics of the host polymer [41].

Achieving stable dispersion is essential for reliable performance [10]. Surface functionalisation, compatibilisers, or coupling agents are often used to enhance wetting and limit particle aggregation [10]. In-situ polymerisation or masterbatch techniques can also distribute nanofillers more evenly during melt processing, while controlled shear or ultrasonic mixing improves interfacial contact without damaging the polymer matrix [10,43]. Nano-clays reduce gas and heat diffusion by creating a tortuous (maze-like) path through the matrix, while layered double hydroxides form a ceramic-like layer that slows heat propagation [42,50]. Carbon nanotubes, cellulose nanofibres, and graphene oxide promote char formation by forming compact residue layers that reduce heat transfer [40,46,49]. Sometimes, they are blended with phosphorus-containing compounds to enhance fire resistance at lower loading levels [41].

Despite these advantages, many nano-additives have processing difficulties, especially those that are prone to agglomeration during mixing [43]. Inconsistent distribution can reduce flame retardant efficiency and weaken the physical properties of the final composite form [8]. The use of carbon nanotubes and graphene oxide may require additional safety measures during handling, especially in dry powder form [10]. Cellulose nanofibres are less toxic and renewable, but their performance can decline under high temperatures or prolonged moisture exposure [1]. Handling precautions such as sealed-feed systems and dust-suppression measures are required when processing nanopowders [41]. Long-term exposure to humidity or ultraviolet radiation can change the surface chemistry of nanofillers and reduce their effectiveness [43]. Durability assessments through accelerated ageing and migration testing are therefore recommended for product qualification [10,43].

Nano-enhanced flame retardants are useful in systems where traditional fillers are either incompatible or would reduce product quality due to high loading [41]. They are commonly integrated into coatings, biopolymers, and electronic housing components—where both fire safety and material performance are important [43]. In some applications, they are used in combination with conventional or bio-based additives to create multifunctional flame retardant systems [47,48].

3. Applications of Sustainable and Green Flame Retardants

Sustainable flame retardants have been integrated across various sectors to meet both fire safety standards and environmental requirements [9]. These materials are selected based on their compatibility with substrates, functional effectiveness under specific fire conditions, and compliance with regional fire classifications and chemical safety regulations [8]. Table 4 summarises the applications of green flame retardants, which categorised by sector, and the example of applications.

Table 4. Applications of Sustainable and Green Flame Retardants in Different Sectors

Sector	Type of Flame Retardants	Example of Applications	References
Construction	Phytic acid Expandable graphite Aluminium hydroxide	Cellulose insulation boards Intumescent wood coatings	[15,30,38]
Electronics	Nano-clay Lignin	Circuit board encapsulants	[18,41]
Textiles	Chitosan DNA Boric acid	Flame retardant cotton fabrics Flame retardant linen fabrics	[3,19,25]
Transport	Zinc borate Layered double hydroxides Expandable graphite	Polypropylene door panels Under-hood thermoplastics	[36,38,50]
Packaging	Starch-based Cellulose nanofibres	Flame-resistant bioplastic food trays	[23,49]
Furniture	Casein Magnesium hydroxide	Upholstery foams Biopolyester-based padding	[22,35]

Materials Research Forum LLC
https://doi.org/10.21741/9781644904077-1

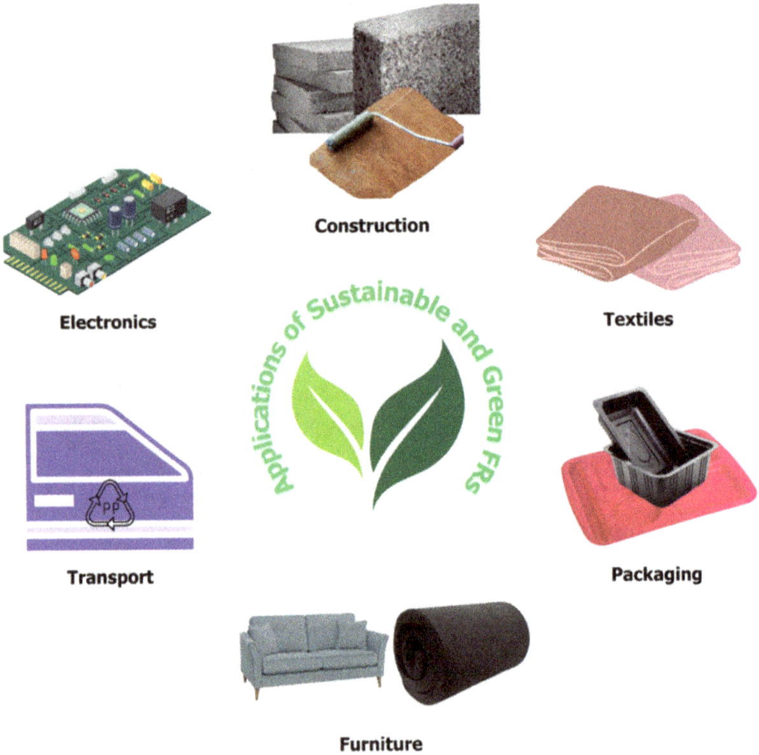

Figure 2. Applications of sustainable and green flame retardants across various sectors.

The application of green flame retardants varies based on substrate properties, fire risk exposure, and regulatory compliance [21]. In the construction sector, compounds such as expandable graphite and phytic acid are used in timber cladding and insulation boards due to their char-forming capabilities and stable combustion behaviour [1,4]. These materials are often preferred in non-load-bearing components where eco-labelling and indoor air quality criteria are implemented [4]. For electronics, low-loading nano-additives such as nano-clays or phosphorus-enriched bio-derived retardants provide thermal protection without reducing dielectric strength or moulding performance [10,41]. These additives comply with the RoHS directive and meet requirements for halogen-free electronics [41].

In textile applications, flame retardants derived from natural sources—such as chitosan and DNA—are often applied to cellulosic fabrics as surface treatments [20,25]. These materials help delay ignition while preserving tactile qualities like softness and air permeability [25]. However, their resistance to repeated washing is often limited unless stabilised through chemical bonding or crosslinking [19]. In the automotive sector, formulations containing layered double hydroxides (LDH) and zinc borate are added to thermoplastics for components located under the bonnet and within vehicle interiors [37,50]. These systems are designed to meet FMVSS 302 flammability criteria [50]. In some cases, they are combined with recycled polymers to align with end-of-life vehicle recovery regulations [26].

In packaging applications, starch-based additives and cellulose-derived nanomaterials are integrated into biodegradable trays and polymer films to reduce flammability [9]. Even though their performance under elevated temperatures can be limited, their usage complies with compostability standards, such as EN 13432 [23]. These materials are preferred where low toxicity and reduced environmental longevity are mandatory [1]. As for the furniture sector, casein and magnesium hydroxide have been applied to upholstery foams and biopolyester padding to improve fire resistance [14]. Their low emission levels and chemical stability make them suitable for indoor use, especially in residential and hospitality settings where air quality regulations are prioritised [8]. Across all sectors, the selection of flame retardants is guided by fire standards such as UL 94 (plastics), EN 13501 (construction), and ASTM E84 (surface burning characteristics) [2].

4. Conclusions

Different types of sustainable and green flame retardants were covered in this chapter, by emphasizing their mechanism of action, advantages, limitations, and industrial applications. Each flame retardant has unique advantages and practical limitations. Bio-derived compounds enhance char formation but may degrade under heat or absorb moisture. Mineral-based flame retardants improve thermal stability, although high filler content can affect mechanical strength. Nano-scale additives, on the other hand, have excellent performance, but are more expensive.

Despite these advancements, more studies are needed to develop formulations that adhere to both fire-safety classifications and environmental certification schemes. Future work should focus on hybrid systems that combine renewable and inorganic components to balance flame resistance, durability, and recyclability. Strengthening compatibility with existing manufacturing processes and maintaining compliance with international regulations are essential for broader industrial adoption of sustainable flame-retardant technologies.

References

[1] J.J. Andrew, M. Sain, S. Ramakrishna, M. Jawaid, H.N. Dhakal, Environmentally friendly fire retardant natural fibre composites: A review, Int. Mater. Rev. 69 (2024) 267-308. https://doi.org/10.1177/09506608241266302

[2] I. Soares, J.L. Ferreira, H. Silva, M.P. Rodrigues, Fire-retardant and fire-resistant coatings: From industry to the potential use on cultural heritage, J. Cult. Herit. 68 (2024) 316-327. https://doi.org/10.1016/j.culher.2024.06.014

[3] J.H. Troitzsch, Fire performance durability of flame retardants in polymers and coatings, Adv. Ind. Eng. Polym. Res. 7 (2024) 263-272. https://doi.org/10.1016/j.aiepr.2023.05.002

[4] L.E. Hasburgh, G.T. Kirker, K.M. Ohno, Flame-retardant design and protection for wood and wood products, in: Fire Retardancy of Polymeric Materials, CRC Press., 2024, pp. 627-647. https://doi.org/10.1201/9781003380689-26

[5] M.L. Marques, E. Cairrao, Occurrence and health effects of hexabromocyclododecane: An updated review, Toxics 11 (2023) 409. https://doi.org/10.3390/toxics11050409

[6] D. Zacs, I. Perkons, E. Abdulajeva, E. Pasecnaja, E. Bartkiene, V. Bartkevics, Polybrominated diphenyl ethers (PBDEs), hexabromocyclododecanes (HBCDD), dechlorane-related compounds (DRCs), and emerging brominated flame retardants (EBFRs) in foods: The levels, profiles, and dietary intake in Latvia, Sci. Total Environ. 752 (2021) 141996. https://doi.org/10.1016/j.scitotenv.2020.141996

[7] C. Olisah, L. Melymuk, O. Audy, P. Kukucka, P. Pribylova, M. Boudot, Extremely high levels of PBDEs in children's toys from European markets: Causes and implications for the circular economy, Environ. Sci. Eur. 36 (2024) 183. https://doi.org/10.1186/s12302-024-00999-2

[8] J. Feng, L. Liu, Y. Zhang, Q. Wang, H. Liang, H. Wang, P. Song, Rethinking the pathway to sustainable fire retardants, Exploration 3 (2023) 20220088. https://doi.org/10.1002/EXP.20220088

[9] R.A. Mensah, V. Shanmugam, S. Narayanan, J.S. Renner, K. Babu, R.E. Neisiany, M. Försth, G. Sas, O. Das, A review of sustainable and environment-friendly flame retardants used in plastics, Polym. Test. 108 (2022) 107511. https://doi.org/10.1016/j.polymertesting.2022.107511

[10] G. Vahidi, D.S. Bajwa, J. Shojaeiarani, N. Stark, A. Darabi, Advancements in traditional and nanosized flame retardants for polymers-A review, J. Appl. Polym. Sci. 138 (2021) 50050. https://doi.org/10.1002/app.50050

[11] H. Feuchter, F. Poutch, A. Beard, The impact of halogen free phosphorus, inorganic and nitrogen flame retardants on the toxicity and density of smoke from 10 common polymers, Fire. Mater. 47 (2023) 1003-1023. https://doi.org/10.1002/fam.3145

[12] C.M. Albert, K.C. Liew, Recent development and challenges in enhancing fire performance on wood and wood-based composites: A 10-year review from 2012 to 2021, J. Bioresour. Bioprod. 9 (2024) 27-42. https://doi.org/10.1016/j.jobab.2023.10.004

[13] M. Wang, G.Z. Yin, Y. Yang, W. Fu, J.L. Díaz Palencia, J. Zhao, N. Wang, Y. Jiang, D.Y. Wang, Bio-based flame retardants to polymers: A review, Adv. Ind. Eng. Polym. Res. 6 (2023) 132-155. https://doi.org/10.1016/j.aiepr.2022.07.003

[14] L. Costes, F. Laoutid, S. Brohez, P. Dubois, Bio-based flame retardants When nature meets fire protection, Mater. Sci. Eng. R Rep. 117 (2017) 1-25. https://doi.org/10.1016/j.mser.2017.04.001

[15] Y. Liu, A. Zhang, Y. Cheng, M. Li, Y. Cui, Z. Li, Recent advances in biomass phytic acid flame retardants, Polym. Test. 124 (2023) 108100. https://doi.org/10.1016/j.polymertesting.2023.108100

[16] X. Chen, J. Li, H. Essawy, A. Pizzi, E. Fredon, C. Gerardin, G. Du, X. Zhou, Flame-retardant and thermally-insulating tannin and soybean protein isolate (SPI) based foams for potential applications in building materials, Constr. Build. Mater. 315 (2022) 125711. https://doi.org/10.1016/j.conbuildmat.2021.125711

[17] T.T. Yang, J.P. Guan, R.C. Tang, G. Chen, Condensed tannin from Dioscorea cirrhosa tuber as an eco-friendly and durable flame retardant for silk textile, Ind. Crops. Prod. 115 (2018) 16-25. https://doi.org/10.1016/j.indcrop.2018.02.018

[18] N.N. Solihat, A.F. Hidayat, M.N.A.M. Taib, M.H. Hussin, S.H. Lee, M.A.A. Ghani, S.S.A.O. Al Edrus, H. Vahabi, W. Fatriasari, Recent developments in flame-retardant lignin-based biocomposite Manufacturing, and characterization, J. Polym. Environ. 30 (2022) 4517-4537. https://doi.org/10.1007/s10924-022-02494-2

[19] G. Malucelli, Flame-retardant systems based on chitosan and its derivatives: State of the art and perspectives, Molecules 25 (2020) 4046. https://doi.org/10.3390/molecules25184046

[20] P. Li, C. Liu, Y.J. Xu, Z.M. Jiang, Y. Liu, P. Zhu, Novel and eco-friendly flame-retardant cotton fabrics with lignosulfonate and chitosan through LbL: Flame retardancy, smoke suppression and flame-retardant mechanism, Polym. Degrad. Stab. 181 (2020) 109302. https://doi.org/10.1016/j.polymdegradstab.2020.109302

[21] S.H. Jeong, C.H. Park, H. Song, J.H. Heo, J.H. Lee, Biomolecules as green flame retardants: Recent progress, challenges, and opportunities, J. Clean. Prod. 368 (2022) 133241. https://doi.org/10.1016/j.jclepro.2022.133241

[22] S. Faheem, V. Baheti, N. Nahid, M. Tunak, J. Wiener, J. Militky, Flame retardancy, physiological comfort and durability of casein treated cotton fabrics, Fibers Polym. 20 (2019) 1011-1020. https://doi.org/10.1007/s12221-019-8826-y

[23] B.O. Carvalho, L.P.C. Gonçalves, P. V. Mendonça, J.P. Pereira, A.C. Serra, J.F.J. Coelho, Replacing harmful flame retardants with biodegradable starch-based materials in polyethylene formulations, Polym. 15 (2023) 4078. https://doi.org/10.3390/polym15204078

[24] J. Koedel, S. Seibt, C. Callsen, F. Puchtler, M. Weise, A. Weidinger, V. Altstaedt, H. Ruckdäschel, R. Schobert, B. Biersack, DNA as a natural flame retardant for cellulose acetate polymer mixtures, ChemistrySelect 6 (2021) 3605-3609. https://doi.org/10.1002/slct.202004493

[25] T. Suryaprabha, M.G. Sethuraman, Fabrication of a superhydrophobic and flame-retardant cotton fabric using a DNA-based coating, J. Mater. Sci. 55 (2020) 11959-11969. https://doi.org/10.1007/s10853-020-04911-0

[26] H. Oosthuizen, L. Jones, S. Naseem, F.J.W.J. Labuschagne, A. Leuteritz, Tailoring materials for their need: Sustainable layered double hydroxide polymer composites, J. Polym. Sci. 61 (2023) 1749-1777. https://doi.org/10.1002/pol.20230025

[27] M.N.A.M. Taib, P. Antov, V. Savov, W. Fatriasari, E.W. Madyaratri, R. Wirawan, L.M. Osvaldová, L.S. Hua, M.A.A. Ghani, S.S.A.O. Al Edrus, L.W. Chen, D. Trache, M.H. Hussin, Current progress of biopolymer-based flame retardant, Polym. Degrad. Stab. 205 (2022) 110153. https://doi.org/10.1016/j.polymdegradstab.2022.110153

[28] M. Chen, Q. Guo, Y. Yuan, A. Li, B. Lin, Y. Xiao, L. Xu, W. Wang, Recent advancements of bio-derived flame retardants for polymeric materials, Polym. 17 (2025) 249. https://doi.org/10.3390/polym17020249

[29] C.E. Hobbs, Recent advances in bio-based flame retardant additives for synthetic polymeric materials, Polym. 11 (2019) 224. https://doi.org/10.3390/polym11020224

[30] M.A. Iqbal, M.A. Iqbal, M. Fedel, Fire retardancy of aluminum hydroxide reinforced flame retardant modified epoxy resin composite, Russ. J. Appl. Chem. 91 (2018) 680-686. https://doi.org/10.1134/S1070427218040225

[31] M.J. Mochane, T.H. Mokhothu, T.C. Mokhena, Synthesis, mechanical, and flammability properties of metal hydroxide reinforced polymer composites A review, Polym. Eng. Sci. 62 (2022) 44-65. https://doi.org/10.1002/pen.25847

[32] J. Xiao, J. Hobson, A. Ghosh, M. Haranczyk, D.Y. Wang, Flame retardant properties of metal hydroxide-based polymer composites: A machine learning approach, Compos. Commun. 40 (2023) 101593. https://doi.org/10.1016/j.coco.2023.101593

[33] G. Scionti, E. Piperopoulos, M. Atria, L. Calabrese, E. Proverbio, Effect of magnesium hydroxide and aluminum hydroxide as thermal barriers on the flame-retardant behavior of acrylic-based coating, Coatings 13 (2023) 1517. https://doi.org/10.3390/coatings13091517

[34] J. Vaari, A. Paajanen, Evaluation of the reactive molecular dynamics method for Research on flame retardants ATH-filled polyethylene, Comput. Mater. Sci. 153 (2018) 103-112. https://doi.org/10.1016/j.commatsci.2018.06.032

[35] K. Zhou, Y. Hu, K. Zhao, Y. Hou, W. Hu, L. Song, F. Chu, Y. Hu, Transformation of magnesium hydroxide whiskers into high-performance multifunctional flame retardant synergists by interfacial engineering modulation, Constr. Build. Mater. 473 (2025) 141009. https://doi.org/10.1016/j.conbuildmat.2025.141009

[36] Q.F. Gillani, F. Ahmad, M.I. Abdul Mutalib, P.S.M. Megat-Yusoff, S. Ullah, P.J. Messet, M. Zia-ul-Mustafa, Thermal degradation and pyrolysis analysis of zinc borate reinforced intumescent fire retardant coatings, Prog. Org. Coat. 123 (2018) 82-98. https://doi.org/10.1016/j.porgcoat.2018.05.007

[37] R. Gupta, M.K. Singh, S.M. Rangappa, S. Siengchin, H.N. Dhakal, S. Zafar, Recent progress in additive inorganic flame retardants polymer composites Degradation mechanisms, modeling and applications, Heliyon 10 (2024) e39662. https://doi.org/10.1016/j.heliyon.2024.e39662

[38] B. Mazela, A. Batista, W. Grześkowiak, Expandable graphite as a fire retardant for cellulosic materials-A review, Forests 11 (2020) 755. https://doi.org/10.3390/f11070755

[39] L.T. Temane, S.S. Ray, J.T. Orasugh, Review on processing, flame-retardant properties, and applications of polyethylene composites with graphene-based nanomaterials, Macromol. Mater. Eng. 309 (2024) 2400104. https://doi.org/10.1002/mame.202400104

[40] K. Li, A. Fina, D. Marrè, F. Carosio, O. Monticelli, Graphite oxide nanocoatings as a sustainable route to extend the applicability of biopolymer-based film, Appl. Surf. Sci. 522 (2020) 146471. https://doi.org/10.1016/j.apsusc.2020.146471

[41] J. Rodrigues, N.G. Shimpi, Nanostructured flame retardants: An overview, Nano-Struct. Nano-Objects 39 (2024) 101253. https://doi.org/10.1016/j.nanoso.2024.101253

[42] M. Rajaei, N.K. Kim, S. Bickerton, D. Bhattacharyya, A comparative study on effects of natural and synthesised nano-clays on the fire and mechanical properties of epoxy composites, Compos. B. Eng. 165 (2019) 65-74. https://doi.org/10.1016/j.compositesb.2018.11.089

[43] A. Rabajczyk, M. Zielecka, T. Popielarczyk, T. Sowa, Nanotechnology in fire protection-application and requirements, Materials 14 (2021) 7849. https://doi.org/10.3390/ma14247849

[44] Y. Wang, B. Liu, R. Chen, Y. Wang, Z. Han, C. Wang, L. Weng, Synergistic effect of nano-silica and intumescent flame retardant on the fire reaction properties of polypropylene composites, Materials 16 (2023) 4759. https://doi.org/10.3390/ma16134759

[45] X. Ji, D. Chen, Q. Wang, J. Shen, S. Guo, Synergistic effect of flame retardants and carbon nanotubes on flame retarding and electromagnetic shielding properties of thermoplastic polyurethane, Compos. Sci. Technol. 163 (2018) 49-55. https://doi.org/10.1016/j.compscitech.2018.05.007

[46] G.Q. Chai, G.Q. Zhu, Y. Gao, J. Zhou, S. Gao, Flame retardancy of carbon nanotubes reinforced carbon fiber/epoxy resin composites, Appl. Sci. (Switz.) 9 (2019) 3275. https://doi.org/10.3390/app9163275

[47] M. Zhang, X. Ding, Y. Zhan, Y. Wang, X. Wang, Improving the flame retardancy of poly(lactic acid) using an efficient ternary hybrid flame retardant by dual modification of graphene oxide with phenylphosphinic acid and nano MOFs, J. Hazard. Mater. 384 (2020) 121260. https://doi.org/10.1016/j.jhazmat.2019.121260

[48] W. Yin, L. Chen, F. Lu, P. Song, J. Dai, L. Meng, Mechanically robust, flame-retardant poly (lactic acid) biocomposites via combining cellulose nanofibers and ammonium polyphosphate, ACS Omega 3 (2018) 5615-5626. https://doi.org/10.1021/acsomega.8b00540

[49] D. Wang, H. Peng, B. Yu, K. Zhou, H. Pan, L. Zhang, M. Li, M. Liu, A. Tian, S. Fu, Biomimetic structural cellulose nanofiber aerogels with exceptional mechanical, flame-retardant and thermal-insulating properties, Chem. Eng. J. 389 (2020) 124449. https://doi.org/10.1016/j.cej.2020.124449

[50] J. Kameliya, A. Verma, P. Dutta, C. Arora, S. Vyas, R.S. Varma, Layered double hydroxide materials: A review on their preparation, characterization, and applications, Inorganics 11 (2023) 121. https://doi.org/10.3390/inorganics11030121

 Materials Research Forum LLC
https://doi.org/10.21741/9781644904077-2

Chapter 2

Flame Retardancy of Phase Change Materials

Hikmet Yusuf Şahin[1], Melike Gürkan[1], Bilge Coşkuner Filiz[1,2]*

[1]Yıldız Technical University, Metallurgy and Materials Engineering Department, Istanbul, Türkiye

[2]Yıldız Technical University, Clean Energy Technologies Institution, Istanbul, Türkiye

coskuner@yildiz.edu.tr; bilgecoskuner@gmail.com

Abstract

Phase change materials (PCMs) are promising for thermal energy storage by utilizing latent heat, offering a green and sustainable solution for thermal management in various systems. Their application as insulation in buildings, in vehicles of different capacities, or in space technologies requires careful attention to safety measures. The volatility and flammability of PCMs must be controlled according to relevant standards. Therefore, additional research is needed to develop safer, more efficient materials. This chapter first provides an overview of PCMs and then discusses in detail the possible solutions to address flammability issues associated with phase change materials.

Keywords

Phase Change Materials, Thermal Energy Storage, Flame Retardancy, Encapsulation, Additives

Contents

Materials Research Foundations 190 (2026) 16-49 https://doi.org/10.21741/9781644904077-2

1. Introduction

Fossil fuels are currently the primary source of energy, which unavoidably results in the release of numerous greenhouse gases into the atmosphere, worsening global warming. Furthermore, the non-renewability of fossil fuels and their rapid consumption have produced a concerning picture for sustainability [1]. In light of the aforementioned adverse circumstance, researchers are working to create novel energy conversion and storage strategies in order to solve this problem [2]. Building energy usage, which mostly includes heating and air conditioning, makes for around 32% of all energy consumption worldwide, for this reason thermal energy storage systems have been gained importance for building energy efficiency [3].

Thermal energy storage (TES) is one of the simplest and least expensive strategies to increase energy management efficiency, according to several attempts to do so [4,5]. Phase change- thermal energy storage technology (PCM-TES) has drawn the most attention among the many TES techniques because of its ability to store significantly more energy per unit volume [6,7]. Phase change materials (PCMs) based on latent heat storage have been actively promoted among various thermal storage media because of their high storage density in small thermal fluctuations. They can be considered a great option to increase building energy efficiency and customize the temperature inside buildings. The efficiency of heating and cooling applications is improved by the thermal energy storage at PCMs. One of the most prevalent organic PCMs, fatty acids offer a number of benefits, including consistent melting, chemical stability, high latent heat storage capacity, non-toxicity and, flammability. The flammability of organic PCMs greatly limits the further practical application, for this reason alternative methods to overcome this issue have been improved [8–10]. This chapter provides an overview of PCMs and then discusses in detail the possible solutions to address flammability issues associated with phase change materials for flame retardancy.

2. Phase Change Materials

PCMs are materials that can store high amounts of heat as latent heat when changing phase at a certain temperature that usually from solid to liquid or vice versa and can release this heat back to the environment when necessary (Fig 1.). PCMs absorb or release heat from their surroundings during melting or freezing and thus can keep the ambient temperature constant around the phase change temperature [11]. The latent heat stored by these materials during phase change is very high compared to the sensible heat of the material [12]. For example, freezing water into ice stores 333 kJ/kg of latent heat of melting, while increasing the temperature of the same amount of water by 1 °C requires only ~4.18 kJ/kg of heat [13] Therefore, PCMs provide high heat storage capacity with relatively low temperature differences, allowing heat energy to be stored efficiently and recovered when necessary. Thus, PCMs have applications in wearable technology [13], thermal protection of foods [14], automotive [15], building insulation [16] and solar thermal heat storage [17].

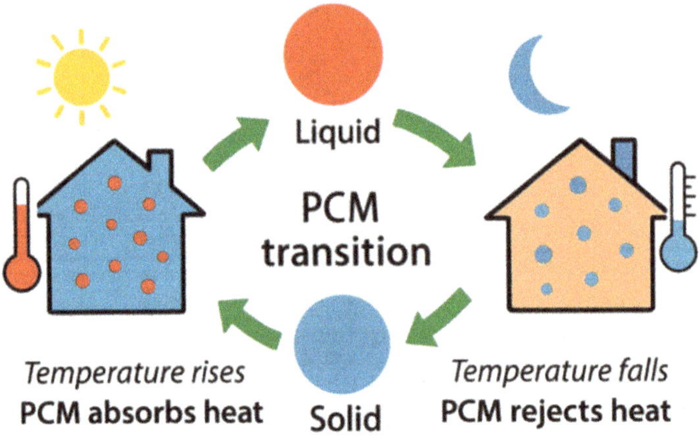

Figure 1. Working mechanism of PCMs

PCMs are divided into various groups according to their chemical structures and phase change properties. They are generally examined in three main categories: organic, inorganic and eutectic PCMs [17]. These classes and their properties are summarized below in Fig 2. :

Materials Research Foundations 190 (2026) 16-49 | https://doi.org/10.21741/9781644904077-2

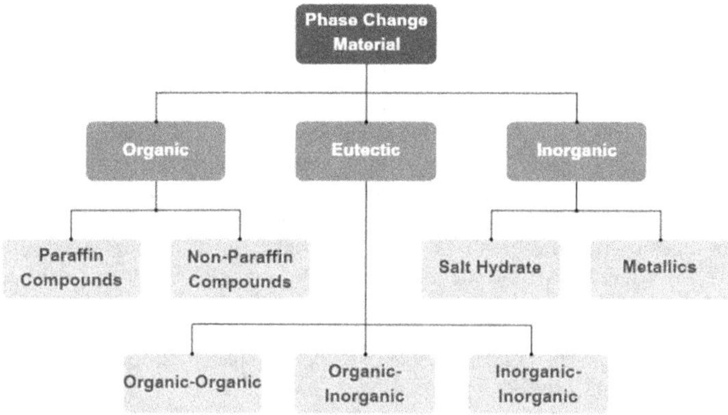

Figure 2. Classification of PCMs

2.1 Organic PCMs

Organic PCMs are carbon-based compounds and are generally substances such as petroleum-derived paraffin waxes or organic fatty acids. Paraffins (alkane waxes) and non-paraffins, i.e. fatty acids, are included in this group. Organic PCMs are materials with low melting points (approximately -5–190 °C [13] and high latent heat values per mass (approximately 150–270 kJ/kg for many paraffins and fatty acids) [18]. Their advantages include their chemical stability and the fact that they do not decompose during phase change; in addition, they are not corrosive to most metal surfaces and are stable in long-term thermal cycles [13]. On the other hand, the biggest disadvantage of organic PCMs is that they are flammable; they can pose a fire risk due to their hydrocarbon structure [18]. Moreover these, their thermal conductivity is low (~0.2 W/m K), which can limit the heat transfer rate [15]. Examples of organic PCMs include paraffin waxes (e.g., octadecane, melting ~28 °C) and fatty acids (e.g., palmitic acid, stearic acid) as given in Table 1.

Table.1 Some of the organic PCMs and their properties

Chemical name	Chemical formulation	Molecular weight, g/mol	Melting point, oC	Latent heat, kcal/g	Specific heat, kcal/g	References
n-Octadecane	$C_{18}H_{38}$	254.49	28.2	0.044	0.00050	[19]
Lauric Acid (dodecanoic)	$C_{12}H_{24}O_2$	200.32	43.8	0.043	0.00043	[20]
Myristic Acid (tetradecanoic)	$C_{14}H_{28}O_2$	228.37	54.3	0.046	0.00038	[20]
PEG600	$H(OCH_2CH_2)_nOH$	600	22 °C	0.0303	0.000595	[21]
Paraffin	C_nH_{2n+2}	341.451	42.0	0.039	0.00048	[22]

2.2 Inorganic PCM

Inorganic PCMs include nonorganic compounds; the most common subgroup in this group is salt hydrates. Inorganic PCMs generally have higher melting points and contain water molecules in their structure. These materials offer much higher thermal conductivity (~0.5 W/m K on average) and their latent heat per mass can be similar to or higher than organics (~250 kJ/kg) [23]. Their important advantages are that they are nonflammable (since they do not contain flammable hydrocarbons) and have a high volumetric heat storage capacity. They are also economical in large-scale applications because they are naturally abundant and relatively inexpensive [24]. However, inorganic PCMs also have disadvantages: Some of them may show supercooling problems or may experience composition separation (salt precipitation, water separation) during phase change, which reduces cycle stability. In addition, some salt hydrates may be corrosive to metals and their volume changes during melting may be greater than organic ones [25]. Examples of inorganic PCMs are sodium acetate trihydrate (melting ~58 °C), calcium chloride hexahydrate ($CaCl_2 \cdot 6H_2O$) (melting ~29 °C) and some low-melting metals/alloys for high-temperature applications as seen from Table.2.

Table.2 Some of the inorganic PCMs and their properties

Chemical name	Chemical formulation	Molecular weight, g/mol	Melting point, °C	Latent heat, kcal/g	Specific heat, kcal/g	References
Sodium Thiosulfate Pentahydrate (STP)	$Na_2S_2O_3 \cdot 5H_2O$	248.19	48–49	0.050	0.00050	[26]
Calcium Chloride Hexahydrate	$CaCl_2 \cdot 6H_2O$	219.08	29.9	0.044	0.00050	[27]
Sodium Acetate Trihydrate	$CH_3COONa \cdot 3H_2O$	136.08	58.0	0.063	0.00052	[28]
Sodium Sulfate Decahydrate	$Na_2SO_4 \cdot 10H_2O$	322.20	32.4 °C	0.060	0.00044	[29]

2.3 Eutectic PCM

Eutectic PCMs are multicomponent mixtures obtained by mixing two or more components in certain proportions and having a fixed melting point like a single compound. Eutectic mixtures can be formed between organic-organic, inorganic-inorganic or organic-inorganic components. When combined in the right proportion, the components of the mixture change phase together at a single melting temperature. The advantage of eutectic PCMs is that they make it possible to obtain a phase change temperature that is exactly compatible with the desired temperature [23]. In this way, they offer sensitive solutions to application-specific temperature ranges. On the other hand, their disadvantages are that they can be high-cost and the ability of each component to provide the same performance in the long term. As an example, please see Table. 3, the organic eutectics capric acid–myristic acid mixture (~18 °C, [30]) or the inorganic eutectics sodium nitrate–potassium nitrate mixture (~300 °C, [31]).

Table.3 Some of the eutectic PCMs and their properties

Chemical name	Chemical formulation	Molecular weight, g/mol	Melting point, °C	Latent heat, kcal/g	Specific heat, kcal/g	References
Stearic Acid + Acetamide (organic/organic)	$C_{18}H_{36}O_2$ / C_2H_5NO	284.48 / 59.07	62.3	0.053	0.00050	[32]
Sodium Nitrate + Potassium Nitrate (inorganic/inorganic)	$NaNO_3$ / KNO_3	84.99 / 101.10	222	0.027	0.00036	[33]
Lauric Acid + Palmitic Acid (organic/organic)	$C_{12}H_{24}O_2$ / $C_{16}H_{32}O_2$	200.32 / 256.43	35.2	0.040	0.00045	[34]
Methyl Palmitate + Methyl Stearate	$C_{17}H_{34}O_2$/ $C_{19}H_{38}O_2$	270.6 / 298.5	23	0.043	-	[35]

The most effective parameters while evaluating the PCMs materials for systems selections were given in details Table.4 that shows the summary of features of the all class of PCMs materials.

Table.4 Summary of features of the all class of PCMs materials

Feature	Organic PCMs: Paraffins & Non-Paraffins	Inorganic PCMs: Salt Hydrates & Metals	Eutectic PCMs: Organic/Inorganic Mixtures
Melting Temperature Range	Generally, it is suitable for use between ~5 – 190°C.	Generally suitable for use between ~15 – 400°C.	It can be adjusted to the desired value. The target temperature is obtained according to the mixture composition.
Latent Heat (Enthalpy of Melting)	150 – 250 kJ/kg (usually medium-high) Example: Paraffin ~180–240 kJ/kg; fatty acids ~150–200 kJ/kg	200 – 350 kJ/kg (salt hydrates, high) Metals: 100–250 kJ/kg (high volumetric capacity with high density)	It may vary depending on the components. It is generally slightly lower than the total enthalpy of the pure components.
Thermal Conductivity	Low: ~0.2 W/m K Additives may be required to increase heat conduction.	For solid salt hydrates, Medium: ~0.4–0.7 W/m K. For metals, Very high: 10–50 W/m K	Medium (depending on composition)
Feature	**Organic PCMs:**	**Inorganic PCMs:**	**Eutectic PCMs:**

Materials Research Foundations 190 (2026) 16-49
https://doi.org/10.21741/9781644904077-2

	Paraffins & Non-Paraffins	Salt Hydrates & Metals	Organic/Inorganic Mixtures
Flammability	Flammable Since they are all carbon based, there is a risk of ignition (especially paraffins).	Non-flammable Salts and metals are non-flammable and do not create a fire load.	Varies depending on composition. If there is organic PCM in the structure, it may act as flammable.
Corrosion Effect	Organics (paraffin, fatty acids) are not corrosive to metal containers.	Many inorganic salts and metals can corrode metal surfaces; special coatings may be required.	Organic-organic eutectics are not corrosive; eutectics with inorganic content carry a corrosion risk due to the salt they contain.
Toxicity	Low Generally non-toxic, many are food/cosmetic derivatives.	Variable Most salt hydrates are non-toxic, but some inorganics (e.g. Li, Pb salts) and metals (Pb, Cd) can be toxic.	Depends on the components Purely organic eutectics are generally nontoxic; in eutectics with inorganic components, toxicity depends on that component.
Cost	Low – Medium Paraffins are inexpensive; fatty acids are moderately costly. They are generally considered cost-effective.	Salts are relatively inexpensive. Metallic PCMs can be very expensive. In general, they can be cheap (salt) or much more expensive (specialty metals) than organics.	Medium – High The use of multiple components and the need for purity increase costs.

2.4 Thermal Properties of PCMs

Thermal properties of PCMs are critical parameters that directly affect their performance in heat storage applications. The phase change temperature of PCMs should be compatible with the operating temperature of the usage area; thus, it can store heat stably by melting and freezing in the desired temperature range [18]. While the latent heat capacity (enthalpy of melting) determines the amount of energy stored per unit mass, density and volumetric storage capacity gain importance especially in systems with space constraints [36]. Specific heat values in solid and liquid phases contribute to the total heat storage performance of PCMs and act as a buffer against temperature fluctuations [24]. In PCMs with low thermal conductivity, it may be necessary to develop composite structures with conductive additives to accelerate heat transfer [37]. In addition, volume changes occurring during phase transition should be considered in system design, and structures that will tolerate expansion should be preferred [38]. All these properties should be evaluated together and the most appropriate PCMs selection for the application should be made.

The main thermal properties that determine the performance of PCMs are summarized with examples in Table 5.

Table.5 Thermal properties that determine the performance of PCM

Feature	Definition	Importance	Example
Phase Change Temperature	Melting or freezing temperature of PCM	Must be compatible with the application temperature range	20–25 °C for indoor use, >200 °C for solar energy
Latent Heat (Enthalpy of Melting)	Heat stored per unit mass of PCM during melting	High value = More energy storage with less material	Paraffin: 150–250 kJ/kg; Salt hydrate: >250 kJ/kg
Density and Volumetric Capacity	The amount of energy stored per unit volume	Important in systems with space limitations	Inorganics generally have higher density
Specific Heat	Energy storage capacity per temperature in solid and/or liquid states	Contributes to sensible heat storage before and after phase change	Creates a buffer effect
Thermal Conductivity	The rate at which PCM conducts heat	Low conductivity = Slow heat transfer; conductivity should be increased	Pure paraffin ~0.2 W/m K; can be increased by adding graphite
Volume Change	Expansion of material during phase change	It can create pressure in closed systems, attention should be paid to system design.	Organic PCMs typically experience 10–15% expansion

In light of the criteria defined above, the most suitable PCM or PCM combination for the application is determined. For example, for the PCM to be placed in electric vehicle battery systems to prevent overheating: It should have a melting point of approximately 35–40 °C [15] , should absorb sudden temperature increases with high latent heat capacity, should be electrically non-conductive, should not contain toxic compounds and, most importantly, should have low flammability so that it will not catch fire at high temperatures [15]. Considering this example, if the PCM used is organic and has a flammability problem despite its thermal advantages, these materials should be modified with flame retardant additives or replaced with inorganic/hybrid alternatives to make them suitable for the application.

2.5 Application Areas

There are many application areas where PCMs benefit from their energy storage properties. The main areas of use and examples are given in Fig.3.

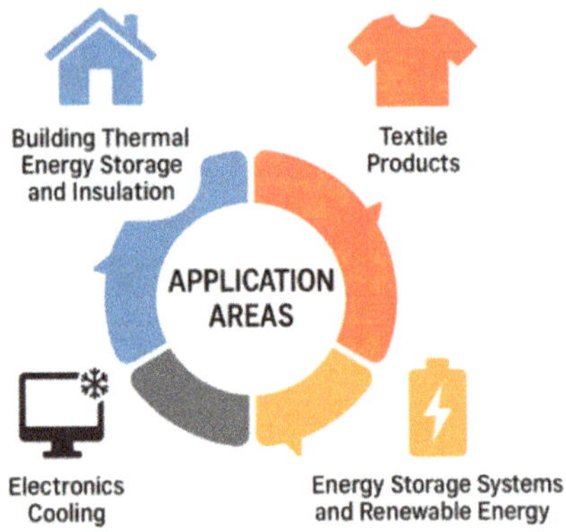

Figure.3 Application area of the PCMs

- **Building Thermal Energy Storage and Insulation:** PCMs are widely researched to increase thermal comfort and reduce energy consumption in buildings. PCMs integrated into walls, ceilings or floors melt and store heat when the environment warms up during the day, preventing the space from overheating; and freeze and return the stored heat when it cools down at night, warming the space. In this way, indoor temperature fluctuations are reduced and HVAC (heating, ventilation, air conditioning) energy loads are reduced. For example, microencapsulated paraffin waxes embedded in plasterboards can turn the walls of a building into heat batteries. PCMs building materials have significant contributions, especially to the concept of zero-energy buildings [39].
- **Textile Products:** In the field of smart textiles, PCMs are used to provide clothing and home textile products with the ability to store heat and release it when necessary. By integrating PCMs into fibers within special microcapsules, for example, a jacket or bed sheet absorbs and stores heat when the ambient or body temperature rises, and regulates thermal comfort by releasing the stored heat when the person is cold . The use of PCMs in sportswear helps to balance body temperature during intense physical activity, while adapting to changes in the external environment in daily wear and keeping the user comfortable. Textiles containing PCMs are being researched in a wide range of areas, from military clothing to astronaut suits [40].
- **Energy Storage Systems and Renewable Energy:** PCMs are used as a critical component of TES systems. In solar power plants, melting and freezing salt mixtures can be used to store heat collected during the day and transfer it to the night. In order to ensure the continuity of intermittent sources such as wind and solar, heat can be stored with the help

of PCMs and later used for electricity generation or heating. In addition, PCMs-based heat batteries prevent energy from going to waste in waste heat recovery projects [41].

- **Cooling in Electronic Devices:** High-performance electronics (e.g. computer processors, power electronics) generate a lot of heat while operating, and instantaneous heat loads need to be balanced. PCMs are positioned adjacent to electronic components, absorbing the heat generated by melting and storing it temporarily, preventing the device from overheating. For example, a thin layer of PCMs placed on the processor of a laptop melts and collects heat during intensive processing; when the processing intensity decreases, the PCMs freeze again and slowly release the heat to the environment. This passive cooling technique can reduce the need for fans or give the system time to react. In addition, PCMs are used in lithium-ion battery packs to reduce the risk of thermal runaway by absorbing heat in case of battery overheating [42].

In addition to those listed above, PCMs are used in the temperature regulation of engine components in the automotive industry, in temperature-controlled packaging in pharmaceutical and food transportation [43] and in the thermal control of special equipment in aerospace [44]. The flexibility of PCM technology creates an application potential in almost every area where heat storage is required, improving energy efficiency and system stability. Each application area may have different priorities in PCM selection. For example, rapid heat transfer (high conductivity) is critical for cooling electronics, while long-term stability and cost may be important in building applications. For this reason, PCMs research continues to develop new application-specific material formulations and composites.

3. Flammability Properties of Phase Change Materials

As mentioned in Section 2, PCMs are divided into three main groups: organic, inorganic, and eutectic compounds. Among these groups, inorganic PCMs (e.g. salt hydrates and metal-based compounds) are quite safe in terms of flammability and generally do not pose a fire risk. However, organic PCMs, especially paraffin waxes and some fatty acids, are highly flammable due to the hydrocarbons in their chemical structures [45]. While these properties make them very efficient in terms of thermal energy storage, they also pose significant risks in terms of fire safety. This chapter discusses the combustion properties of organic PCMs, the fire risks they may cause, and their behavioral parameters during fire; and provides examples of precautions and research developed against these problems.

3.1 Flammability Properties of PCMs

Paraffin (candles) type PCMs are chemically long-chain alkanes and are basically petroleum-derived hydrocarbons. Since it has a petroleum-derived structure, its structure is a flammable fuel. In daily life, paraffins can be burned to provide candlelight. A solid paraffin PCM melts when exposed to direct flame and the vapors from this liquid phase ignite and burn at a certain temperature. The flash point of a typical paraffin candle is around 199 °C, when it reaches this temperature, its vapors can ignite. Its autoignition temperature is around ~245 °C (i.e., it can react with air and start burning without an ignition source at this temperature) [37]. These values show that paraffin will not ignite under normal ambient conditions but will burn easily if the ambient temperature rises to these levels during a fire or if it comes into contact with a flame. It also points out that paraffins, when used as a PCM, can act as a fuel that will rapidly release significant amounts of energy in the event of a fire.

Fatty acids (e.g. stearic, palmitic acid) are also used as organic PCMs and are also flammable; however, their flashover/ignition behavior may be slightly less dangerous than paraffins. In the literature, the latent heat of many fatty acids is in the range of 150–220 kJ/kg and is reported to be more advantageous than paraffin hydrocarbons in terms of fire safety [46]. Nevertheless, a fatty acid PCM can also burn under suitable conditions and create a fire load. Therefore, flammability is a critical problem for all organic PCMs and a design should be made to prevent this feature in the places of use [23].

Inorganic PCMs (salt hydrates etc.) are not flammable; they do not burn when heated, they can only decompose or lose water. Therefore, inorganic PCMs can be an alternative in applications where fire safety is a priority. However, other disadvantages of inorganic PCMs (supercooling, corrosion etc.) and limited options in suitable phase change temperatures mean that they cannot always be used in place of organic PCMs in practice. Therefore, it may be possible to avoid potential problems by choosing composite systems such as eutectic PCMs. As a result, the flammability of these materials is an important problem to be solved, especially in areas where organic PCMs are used, such as buildings, textiles and electronics.

3.2 Fire Risks and Safety of PCM

The evaluation of PCM in terms of fire safety requires the analysis of the effects of these materials on fire behavior. Since organic PCMs are inherently flammable, they add an additional flammable load to the environment in which they are located. For example, a mineral-based wall panel that is normally considered non-flammable may become flammable when paraffin-based PCM is integrated into it. This causes the fire reaction class of the material to decrease. Research shows that most building materials containing PCM remain in low-performance categories such as "D" or "E" in the Euroclass fire classification. Even if flame retardant additives are used, it has been observed that these materials generally do not reach class "C" [47]. These classes indicate that the materials can easily ignite under flame and have poor resistance to the spread of fire. Therefore, the use of PCM is currently limited, especially in areas requiring high fire safety, such as public buildings and escape routes. This situation is one of the important obstacles to the widespread use of PCM technology [47]. Possible fire scenarios in systems containing PCM can occur in different ways. These are;

- **Increased Flame Spread:** PCMs are usually used by dispersing them in a solid matrix (such as plaster, polymer or textile fiber). If the PCM, which melts and becomes liquid during a fire, leaks, the flow of the flammable liquid can cause the fire to spread to other areas. For example, paraffin in the wall can melt and leak through cracks, ignite in the lower layers and cause the fire to expand. This is a critical risk that triggers the rapid spread of fire. In addition, flame propagation can be easier if PCM is present on the surface. Although some studies have shown that well-encapsulated PCMs do not significantly increase flame spread, it is generally accepted that organic PCMs facilitate the growth of fire [47,48].
- **Increase in Heat Release or Fire Intensity:** Flammable PCMs act as an additional fuel source during a fire, increasing the ambient temperature. The heat release rate is an important parameter that indicates how much energy a material contributes to a fire. When materials containing PCM burn, they can cause a high heat release in a short time. For

example, the peak heat release rate of a composite containing an organic PCM can reach values as high as 1000–1200 kW/m^2 [46]. This makes it difficult to control the fire and increases the risk of ignition of other surrounding materials.

- **Prolongation of Ignition and Burning Time:** The ignition behavior of PCMs varies depending on the environment they are in. Pure paraffin can ignite with a certain delay because it needs to melt and vaporize. However, once it catches fire, it tends to burn continuously until the fuel is exhausted [45]. This means that the fire is fed for a long time. When flame retardant additives are added, it has been observed that PCM materials sometimes burn out and re-ignite. For example, adding 20% intumescent additive to paraffin-based PCM delayed ignition and provided self-extinguishing behavior. However, high additive rates such as 40% were required for complete non-flammability [45]. Such high additive rates can reduce the heat storage capacity of the PCM and negatively affect the performance of the material.
- **Production of Smoke and Toxic Gases:** When hydrocarbon-based PCMs, such as paraffin, burn, they usually release carbon dioxide and water vapor. However, if combustion is incomplete, dense smoke and toxic carbon monoxide (CO) can be produced. In enclosed spaces, this can reduce visibility, make evacuation difficult, and can be life-threatening. In microencapsulated PCMs, combustion of the encapsulating material (e.g., polymer) can release additional noxious gases. Some studies have shown that flame retardant additives can also reduce smoke formation. For example, PCM composites with expandable graphite additives produced less smoke and CO during combustion [49]. However, combustion of organic PCMs without additives can lead to significant smoke accumulation, especially in enclosed spaces.

When the above points are evaluated, it becomes clear that multifaceted measures should be taken in terms of fire safety in systems using organic PCM. First, if possible, the design should be made in such a way that the PCM does not come into direct contact with the flame (for example, the PCM layer should be sealed in a fire-resistant panel or covered with a non-flammable material such as plaster). Second, encapsulation techniques can be used to prevent the spread of PCM in the event of fire. Third, ignition can be delayed and heat release can be reduced by adding flame retardant additives to the material composition. The selection of these additives should be made carefully, since not every additive may work with every PCM and may reduce the heat storage capacity [45].

3.3 Combustion Behavior Parameters: Flame Spread, Heat Release, Ignition

Standard fire tests and criteria are used to quantitatively evaluate the flammability performance of organic PCMs. Flame spread rate, heat release rate, ignition temperature/time, and smoke production are the key parameters that characterize the fire response of these materials [46]. These combustion parameters were given in detail as follows:

- **Flame Spread:** Flame spread indicates how fast the fire progresses horizontally/vertically on the surface of the material. In standard tests such as ASTM E84, ISO 5658, the flame spread index is calculated by measuring the distance and speed of the flame front on the material [50] . In PCM-containing composites, it is tested whether the flame spread changes compared to the pure binder material. Theoretically, flammable PCM can fuel the combustion process on the surface and cause the flame to progress faster. However, if the

PCM is in small capsules in a non-flammable matrix, the effect of the PCM on flame spread may be limited as long as the matrix does not burn. Therefore, experimental results vary depending on the design of the system. In general, if the PCM addition is done in an uncontrolled manner, it has the potential to increase flame spread, but this effect can be minimized with appropriate encapsulation and barriers.

- **Heat Release Rate (HRR):** As mentioned above, HRR (Heat Release Rate) is one of the most important parameters that determine the burning intensity of materials in the field of fire science. HRR, measured by cone calorimeter tests, gives the thermal energy released per unit time (usually in kW/m^2) . Since organic PCMs have high heats of combustion, their peak heat release rates are quite high. For example, a shape-stable PCM sample containing paraffin impregnated in polyethylene was measured to have a maximum HRR of ~884 kW/m^2 in a cone calorimeter test; the addition of flame retardant to the same sample reduced this value to ~297 kW/m^2 . Another study reported that the peak HRR of a pure paraffin/HDPE mixture without any additives was 1137 kW/m^2, but this could be reduced to ~392 kW/m^2 with appropriate intumescent additives [46]. These data show that PCM-containing materials can release extremely high heat in their pure state, but can be partially controlled with the right formulation. Nevertheless, compared to standard building materials (e.g. gypsum board, around several hundred kW/m^2), the HRR of PCM-containing materials is still high, which has a negative effect on fire growth.
- **Ignition Temperature and Ignition Time:** Ignition temperature is defined as the temperature at which a material can spontaneously ignite (auto-ignition) or the surface temperature at which it begins to ignite under a certain heat source. We have stated that auto-ignition temperatures for organic PCMs are in the range of 200-300 °C flinnsci.com . Ignition time is the time until the material ignites under a certain flux (e.g. in a cone calorimeter giving a heat flux of 50 kW/m^2). In materials containing PCM, ignition can sometimes occur earlier than in pure polymers because the melted PCM evaporates from the surface and forms the combustible gas mixture. On the other hand, in some cases, the melting and spreading of PCM can also have a cooling effect on the surface and delay ignition somewhat. For example, in one study, it was observed that the ignition delay time increased slightly in 20% intumescent paraffin PCM [46]. However, for the pure form, ignition is generally reported to occur within a short time (a few seconds to a few tens of seconds for small samples). Once ignition occurs, the combustion behavior described above takes over and the material begins to burn.

3.4 Solutions to Flammability Problems of Phase Change Materials

Various material approaches are being developed to reduce the flammability problem of organic PCMs as explained below in Section 3.4.

- **Flame Retardant Additives:** In order to reduce the ignition and burning tendency of organic PCMs, various flame-retardant additives are added to the formulations. Halogenated phosphates, nitrogenous compounds, metal hydroxides and graphite are examples of these additives. While some additives (e.g. brominated compounds) act by suppressing radical reactions in the gas phase, some cool the environment with endothermic reactions (e.g. Al(OH)3 releasing water; some act as a physical barrier by

Materials Research Forum LLC
https://doi.org/10.21741/9781644904077-2

forming carbonized surface layers (e.g. intumescent systems containing ammonium polyphosphate, APP). However, the type and ratio of additives should be selected carefully; since high additive ratios (>30–40%) may negatively affect the thermal performance of PCM [45].

- **Microencapsulation and Coating:** Using PCMs in microencapsulated form within a fireproof shell can also reduce the risk of fire. For example, if paraffin waxes are microencapsulated within a fireproof silica or ceramic shell, the shell acts as a barrier in the event of a fire, making it difficult for the paraffin to evaporate freely and burn. Similarly, direct contact with the fire can be prevented by applying fire-resistant coatings (e.g. fire-retardant paints, gypsum plaster layer) to the surface of PCM-impregnated building elements. These passive measures extend the fire resistance period of the PCM and provide time for the fire brigade to intervene or the extinguishing system to be activated [51].

- **Inorganic/Hybrid Approaches:** Using partially or completely inorganic PCM instead of organic PCM is also a solution. For example, a completely non-flammable salt hydrate PCM can be preferred in an application where the risk of fire is critical. Or, organic PCM is mixed with some inorganic components to increase thermal conductivity and to act as a non-flammable filler during combustion. Research on using filler phases that will limit the combustion of the organic part in hybrid PCM systems is rapidly continuing [52].

3.5 Fire Safety Design

In structures where PCM containing systems will be applied, holistic fire safety strategies should be developed, not only at the material level. Automatic extinguishing systems, smoke detectors, fire zoning solutions and emergency evacuation scenarios are included in this system design. In addition, since laboratory-scale fire tests are not always sufficient, performance evaluation should be carried out by performing large-scale tests and fire scenario simulations [53]. As a result, the flammability problem of organic PCMs limits their use in building and energy storage systems. Therefore, many strategies have been developed such as flame-retardant additives, microencapsulation techniques, hybrid material systems and building-scale fire safety designs. However, the performance of each solution varies according to the type of PCM, application area and usage conditions. Therefore, developing integrated solutions at both material and system levels, ensuring fire safety without compromising thermal performance is one of the most important parameters.

Concerns about flammability have led PCM researchers to focus on this issue in recent years. Many experimental studies and review articles in the literature attempt to measure and improve the fire performance of organic PCMs. McNamee et al. (2023) evaluated the fire performance of PCM-containing building materials through a comprehensive literature review and estimated Euroclass fire classes using small-scale test data. This study revealed that most PCM-containing materials fell within Euroclass D or worse, meaning that their reaction-tilt-fire performance was poor. Even versions with added flame retardants often failed to achieve the desired Class C. This finding scientifically confirms that PCM applications remain limited by current fire regulations [47]. Li et al. (2016) conducted studies using a new flame-retardant system to improve the flame-retardant properties of eutectic mixture paraffin and polypropylene (PP) composites used as form-stable phase change materials (PCM). Supported by various tests such as DSC, TGA and cone calorimeter, this study showed that the composite (PCM6) containing 30% flame retardant provided 126.8 kJ/kg latent heat capacity and reduced the heat release rate (PHRR) of flame-retardant additives from 1189.7 kW/m^2 to 135.9 kW/m^2. It was also emphasized that the dense char layer formed during thermal decomposition increased fire resistance and this new PCM is a potential heat storage material for providing energy efficiency in buildings [54]. In an experimental study conducted by Palacios et al. (2018) (on the fire performance of bulk form PCMs), combustion tests were conducted by adding various flame retardants to pure paraffin and fatty acid eutectic mixtures. The results showed that the addition of magnesium hydroxide or hydro magnesite (rare thermic additives) at very high rates such as 50% significantly shortened the combustion time in fatty acid-based PCMs and provided self-extinguishing behavior. On the other hand, the use of APP even at 40% in paraffin-based PCMs only brought limited improvement and the material could not stop the combustion completely. This study revealed that fatty acids responded better to flame retardants and showed a tendency to self-extinguish, while paraffins were more difficult to control. In the same study, it was reported that PCMs without additives burned continuously when ignited with a single fire source and that it was not possible to extinguish the flame until the fuel was exhausted in the drip test [45].

Zhang et al.'s studies (2010s) measured the fire performance of paraffin/HDPE-based shape-stable PCMs by adding different flame-retardant combinations. In one combination, it was reported that the peak HRR was reduced from ~1130 kW/m^2 to ~430 kW/m^2 by synergizing expandable graphite with ammonium polyphosphate, and the total heat released was also reduced [55]. In another experiment, iron powders were used in combination with an intumescent system, and the HRR was reported to be reduced from ~627 kW/m^2 to ~274 kW/m^2 [46]. These results show that multifunctional additive systems (heat absorbing, char forming, etc.) can significantly reduce the burning severity of PCMs.

These and similar research results show that the flammability problem of PCMs is known and intensive work is being done to solve it. As a result, in order to fully utilize the great potential offered by phase change materials in the field of energy storage, ensuring the fire safety of these materials is of critical importance. Although methods developed to reduce the flammability of organic PCMs (flame retardants, encapsulation, hybrid materials) have achieved some success, there is still a design trade-off for most cases: A balance must be established between heat storage capacity and fire safety. Current research is focused on developing non-flammable or flame-retardant PCM systems without sacrificing thermal performance. When this goal is achieved, PCMs will be much more widely and comfortably used, especially in buildings and other safety-critical environments.

4. Microencapsulation of Phase Change Materials

Microencapsulation, proposed as one strategy to mitigate the inherent flammability of phase change materials (PCMs), traces its early development to the 1950s, when Green and Schleicher produced microencapsulated ink via coacervation of gum arabic and gelatin to create carbonless copy paper. The technique—broadly defined as forming discrete capsules (approximately 1–1000 μm in diameter) by enclosing a solid, liquid, or gaseous core within a synthetic or natural polymeric wall—has since found extensive application across agrochemicals, foods, paints, cosmetics, pharmaceuticals, animal feed, textiles, thermal energy storage, and numerous additional sectors [56].

Although PCMs are widely adopted for latent thermal energy storage because of their favorable heat storage density, their practical deployment is often constrained by leakage during phase transitions. Microencapsulation not only suppresses leakage but can also enhance heat transfer, shield the PCM from environmental or chemical degradation, and increase effective surface area, thereby improving overall performance metrics [57].

A typical microcapsule comprises a core and an enclosing shell, with the shell architecture exerting a decisive influence on functional stability. Dense, continuous, and defect-free shell layers are associated with improved mechanical and thermal robustness [58]. Furthermore, chemical and physical compatibility between the PCM core and the shell material helps accommodate volumetric expansion–contraction cycles, achieve an appropriate wall thickness for the desired diffusion (or release) characteristics, and sustain resistance to shear or other operational stresses under varying service conditions [57].

Microcapsules exhibit diverse morphologies (see Fig. 4), which are strongly governed by the core-to-shell ratio. The mass fraction of encapsulated core typically spans roughly 20–95%, and classification schemes often reflect how the shell envelopes or partitions the core phase.

Mononuclear (Core/Shell) is a single core surrounded by a continuous cladding material. Polynuclear refers to many nuclei surrounded by a continuous shell. Matrix Encapsulation is the structure in which the core material is distributed homogeneously within the shell. Multi-film is a structure in which continuous cores are covered by a multi-layered and continuous shell material[56].

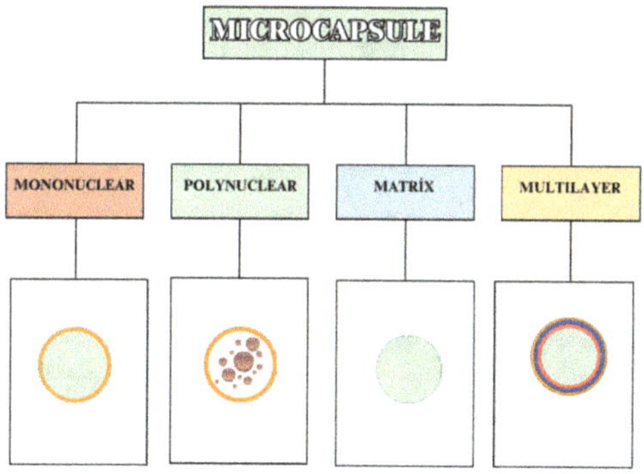

Figure.4 Classification of microcapsule morphology.

Size-based classification of encapsulated phase change systems generally distinguishes macro-scale capsules (> 1000 μm), micro-scale entities (≈ 1 - 1000 μm), and nano-scale capsules (≈ 1 - 1000 nm), with microcapsules historically representing the predominant format in practical applications. Growing recognition of the strong dependence of thermo-physical behavior on particle dimension has accelerated interest in nanoencapsulated PCMs. Owing to their markedly elevated surface-area-to-volume ratio, nanocapsules can promote more rapid heat exchange. At nanometric dimensions, interfacial phenomena—surface chemistry, capillarity, and van der Waals forces—become comparatively more influential and can confer enhanced colloidal stability. Their facile dispersion in liquids also supports formation of persistent, uniform flow fields [57]. Demirbağ and co-workers investigated microencapsulated PCMs incorporating gelatin/sodium alginate shells modified with clay nanoparticles, focusing on flame retardancy and thermal endurance [59]. Comparative analysis demonstrated that the nanocomposite PCM microcapsules exhibited appreciable gains in both thermal stability and resistance to ignition relative to conventional (non–nanofilled) analogues [59]. In a separate effort, Du et al. produced a flame-retardant nano-PCM consisting of an n-octadecane core surrounded by a poly(methyl methacrylate) (PMMA) shell. Computationally assisted thermal assessments revealed that incorporation of these nanoPCMs into gypsum boards improved temperature regulation while markedly lowering heat release and smoke generation, underscoring the promise of flame-retardant nanoencapsulated PCMs for thermal energy storage in controlled-building envelopes and technical textiles. Chen and colleagues embedded an in-situ polymerized, nano-scale encapsulated PCM into ethylene–vinyl acetate (EVA) copolymers, observing enhanced flame retardancy along with retention of thermal performance after 200 heating–cooling cycles. They concluded that such nanoPCM-enhanced composites could enable substantial, large-scale energy savings in building applications [60].

4.1 Encapsulation Techniques

Encapsulation techniques are divided into three groups: physical, chemical and physicochemical methods as given in Fig.5.

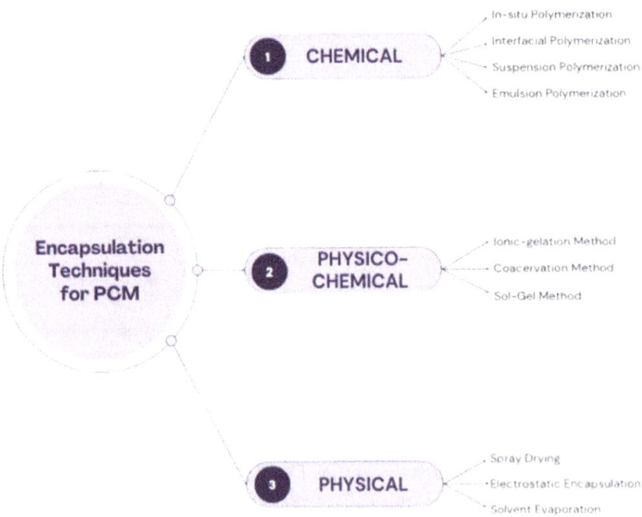

Figure.5 Encapsulation techniques

4.1.1 Physical Methods

Physical micro/nanoencapsulation routes rely on purely mechanical or physicochemical operations rather than de novo chemical reactions, and are generally regarded as comparatively benign in environmental terms for generating particles typically above ~50 µm. The broader family encompasses vibrating (oscillating) nozzle droplet formation, spray drying, centrifugal extrusion, spray cooling (chilling), electrostatic (electrospray / electrostatic precipitation) methods, solvent evaporation processes, air-suspension (fluidized bed) coating, and approaches employing supercritical fluids. Of these, spray drying is the most pervasively implemented platform; specifically for phase change material (PCM) encapsulation, industrial adoption is concentrated on spray drying and electrostatic techniques.

- **Spray Drying:** The process begins with formulation of a stable emulsion (or solution dispersion) in which the PCM (core phase) is finely dispersed within a continuous phase containing the wall-forming substance. Once target droplet size distribution is attained, the feed is atomized into a hot drying chamber, causing rapid solvent (usually water) removal and in situ shell solidification around the dispersed cores. This method combines scalability, relatively low operating cost, and suitability for heat-sensitive actives because of the brief thermal exposure inherent to the rapid moisture flash-off [61].
- **Electrostatic Encapsulation:** This modality employs a high-voltage electric field to overcome interfacial tension and generate charged droplets (or coaxial core–shell droplets) that solidify or gel downstream, commonly yielding capsule diameters on the order of ~100 µm. For successful core–shell formation, the two immiscible liquids must be mutually wettable yet remain phase-separated, and their droplet sizes should be closely matched. The shell fluid is advantageously formulated with a surface tension lower than that of the core phase to promote envelopment. By adjusting solution conductivity, flow rates, and potential difference, either hydrophilic or hydrophobic shells can be produced at micro- or nanometric scales [61,62].
- **Solvent Evaporation:** Also described (in certain contexts) as a liquid–liquid extraction driven encapsulation pathway, this method typically prepares an organic (oil) phase containing a dissolved shell polymer and a volatile, low-boiling solvent, alongside an aqueous phase with an appropriate emulsifier/stabilizer. Upon emulsification, discrete polymer-containing droplets entrap the PCM (or the PCM resides as a dispersed sub-phase within those droplets). Controlled removal of the volatile solvent—often under reduced pressure or gentle heating—induces polymer precipitation and migration toward the core interface, consolidating a coherent capsule wall; the emulsion consequently transitions into a suspension of microcapsules. This strategy is frequently selected when the PCM tolerates moderate thermal inputs and when enhanced barrier (low permeability) properties are desired. Fine-tuning solvent diffusion and evaporation kinetics enables modulation of particle morphology and, in favorable cases, the generation of nanostructured or narrowly distributed capsule sizes. While cost-effective and compositionally versatile (broad range of core–shell chemistries), the approach is most commonly executed at laboratory or pilot scale due to solvent handling and recovery considerations [57,63].

4.1.2 Chemical Methods

Chemical methods are common encapsulation techniques that usually use an organic core and involve chemical reactions. There several alternatives as classified under that based on the formation of the core and shell structure to ensure the capsulation of the PCM as explained below:

- **In Situ Polymerization:** This strategy relies on initiating and propagating polymer (or polycondensate) formation directly at the interface of preformed core droplets. Initially, a dispersion or oil-in-water emulsion of the PCM (organic core phase) is produced under controlled shear. pH adjustment is critical during this emulsification step because it governs both colloidal stability and the size distribution; excessively low pH may promote droplet coalescence or sedimentation. Subsequently, aqueous precondensate (oligomer) streams—often two complementary monomer/oligomer solutions—are metered into the continuously stirred emulsion so that condensation (or addition) proceeds preferentially at the droplet surfaces, progressively yielding an insoluble, cross-linked wall. Sustained agitation helps maintain a narrow size distribution and uniform shell thickness. After achieving the targeted degree of polymer conversion, the system is thermally conditioned (curing stage) and then dried or filtered to isolate the microcapsules [57,63]. Srinivasaraonaik et al. prepared a eutectic stearic acid–capric acid core encapsulated within a melamine–formaldehyde (MF) shell via in situ polymerization, systematically probing pH and impeller speed; optimal encapsulation performance was reported at pH $\approx$ 3.2 and 1500 rpm. The resulting M-PCM exhibited a thermal conductivity of about 0.31 W m^{-1} K^{-1}, a fusion enthalpy near 103.9 kJ kg^{-1}, and durability across roughly 2000 melt–freeze cycles, supporting its applicability to energy-efficient thermal comfort solutions [64]. In a separate advance, Moyun Kang and co-workers incorporated DOPO (9,10-dihydro-9-oxa-10-phosphaphenanthrene-10-oxide) functionalized halloysite nanotubes (c-HNTs) together with copper nanoparticles into melamine–formaldehyde microcapsules using an analogous in situ route. The hybrid additive system significantly elevated thermal stability and improved dynamic temperature regulation, as confirmed by infrared thermography, while imparting flame-retardant functionality suitable for interior thermal management scenarios [65]. Zhe Zhang and colleagues engineered MF-shelled microcapsules embedding paraffin wax (including n-octadecane) with stabilized cellulose nanocrystals (CNC) acting as reinforcing/stabilizing components during interfacial polymerization; the produced microencapsulated PCMs demonstrated self-extinguishing behavior and enhanced fire retardancy relative to unmodified analogues [66].
- **Interfacial Polymerization:** This technique deploys two complementary monomers—one preferentially soluble in an organic (oil) phase and the other in the aqueous phase—brought together within a stabilized emulsion containing an appropriate surfactant. Polymer formation is confined to, and propagated along, the oil–water boundary where the dissolved monomer species meet. Reaction onset and rate are typically governed by thermal input and pH adjustment. Following shell growth to the desired thickness, the dispersion is subjected to separation (filtration or centrifugation) and drying, yielding nano- to microscale capsules; contemporary reports highlight its frequent selection for nanoencapsulation of PCMs [57,63]. Exemplifying the approach, Jian Shi and co-workers developed a rapid, straightforward route to nano-sized PCM entities using paraffin as the core and a poly(methyl methacrylate) (PMMA) shell formed via methyl methacrylate

(MMA) interfacial polymerization. Systematic variation of the MMA/PCM mass ratio and agitation speed produced smooth, dense, compositionally uniform nanocapsules exhibiting an optimal melting enthalpy of 64.93 J g^{-1}, supporting prospective deployment in large-scale thermal energy storage networks [67].

- **Suspension Polymerization:** In suspension polymerization for PCM encapsulation, a water-immiscible monomer (together with the phase change core material) is dispersed as discrete droplets within an aqueous continuum containing protective colloids or dispersants. High-shear agitation first establishes a stable droplet distribution (oil-in-water morphology). Polymerization is then initiated—most commonly via thermal decomposition of an oil-soluble initiator or, less frequently, by pH manipulation—within each suspended organic droplet, which effectively acts as an isolated microreactor. As monomer conversion progresses, diffusional flux toward the droplet interface sustains chain growth until solid or gelled polymer shells (or matrices) entrap high loadings of the PCM. The method is recognized for its comparatively low cost, scalability, and reduced environmental burden, producing encapsulated products spanning approximately 2 to 4000 μm in diameter [57,63]
- **Emulsion Polymerization:** Distinct from suspension polymerization, emulsion polymerization solubilizes (or disperses) a portion of the monomer within surfactant-stabilized micelles in the aqueous phase while a water-soluble initiator generates free radicals directly in the continuous medium. Polymer chains nucleate predominantly inside micelles that have absorbed monomer, enabling the formation of polymeric shells or matrices at nanometric scales suitable for PCM nanoencapsulation [63]. In a representative study, Kazancı et al. produced polystyrene-shelled microcapsules encapsulating paraffin while co-introducing orthophosphoric acid (oPA)—evaluated for the first time as a halogen-free, non-toxic flame retardant—and pentaerythritol (PER) through an emulsion polymerization route. The inclusion of these flame-retardant constituents enhanced both fire resistance characteristics and thermal energy storage performance, indicating suitability for thermoregulatory façade applications in buildings [68].

4.1.3 Physico-Chemical Methods

Within this broader family of encapsulation strategies, purely chemical routes are complemented by physicochemical operations that exploit phase separation, controlled drying or heating steps, and hybrid mechanisms involving condensation, hydrolysis, and ionic or covalent cross-link formation. Collectively, these techniques are adaptable to fabricating both micro- and nano-scale capsules, offering flexibility relative to more conventional single-mode approaches[57].

- **Ionic-gelation Method:** Ionic gelation—applied only sparingly so far for phase change material (PCM) encapsulation in thermal energy storage—typically commences with dissolution or dispersion of the selected PCM in a compatible solvent or continuous medium. A polyelectrolyte solution (e.g., sodium alginate) is then introduced under regulated shear to generate a stable emulsion or droplet distribution. Subsequent transfer (often dropwise addition) of this dispersion into a gelling bath containing multivalent cations induces immediate ionic cross-linking, yielding hydrogel beads or nascent capsule walls around the entrapped core droplets. Post-gelation treatments (washing, controlled thermal conditioning, and partial dehydration) consolidate and mechanically strengthen the formed capsules [57]. Manipulating immersion (residence) time in the ionic solution and increasing the polyelectrolyte concentration generally produces tighter network architectures and can drive a reduction in the final particle size.

- **Coacervation Method:** Coacervation refers to an encapsulation strategy in which two immiscible liquid phases emerge as a consequence of hydrophilic, hydrogen-bonding and electrostatic interactions between polymeric species, and it has been broadly implemented for both micro- and nano-scale encapsulation across food, agricultural, pharmaceutical, and textile applications. While the overarching sequence of steps is analogous, two operational variants—simple (or single) coacervation and complex coacervation—are distinguished by the nature of phase separation driven by the chosen polymer system. In either pathway, the process begins with dissolution of the intended shell-forming polymer (or polymer blend) in a suitable solvent to produce a homogeneous solution. The prospective core (active or PCM) is subsequently introduced and dispersed to generate a stable emulsion or suspension of droplets within this continuous phase. In simple coacervation, one polymer (e.g., ethylcellulose, gelatin) is present and phase separation is induced through the controlled removal or alteration of the solvent environment (such as by adding a non-solvent, adjusting temperature, or modifying pH), leading to polymer-rich coacervate droplets that deposit onto the dispersed core. Complex coacervation, by contrast, relies on the associative complexation of two oppositely charged biopolymers or polyelectrolytes (e.g., gelatin with an anionic partner like alginate); electrostatic neutralization of segments produces a dense coacervate phase capable of forming thinner, more uniform shells, often enabling reduced capsule size relative to simple systems. To trigger or refine phase separation, strategies such as controlled emulsion solvent evaporation or cooling may be employed, after which the nascent coating layer is consolidated. Hardening (or shell fixation) is typically achieved by introducing a cross-linking agent or otherwise promoting intermolecular bonding, yielding an effectively impermeable wall that enhances mechanical integrity and service life. The resulting microcapsules are commonly subjected to sequential washing (e.g., with deionized water, aqueous media, or ethanol) to remove residual reagents, followed by drying—frequently under reduced pressure (vacuum oven)—to obtain the final, free-flowing encapsulated product.
- **Sol-Gel Method:** The sol–gel route is a comparatively straightforward and economical physicochemical encapsulation strategy grounded in the polycondensation-driven transformation of a molecular or colloidal "sol" into a continuous "gel" network. Inorganic oxide formers—most notably silica and titania—are frequently selected as shell matrices due to their favorable thermal conductivity, inherent nonflammability, and structural stability under elevated temperatures. The process is initiated by formulating a clear, metastable colloidal solution comprising suitable precursors, a catalyst (often acid to accelerate hydrolysis and condensation), and an appropriate solvent system (frequently alcohol–water mixtures). Typical silicon-containing precursors include alkoxysilanes such as tetraethoxysilane (TEOS), sodium silicate, or related metal alkoxides; operation at lowered pH values promotes hydrolysis and subsequent polycondensation. After establishing the reactive sol, the target core (e.g., PCM or active agent) is dispersed, generating an emulsion or suspension in which incipient inorganic oligomers nucleate and progressively assemble around the dispersed droplets. Continued condensation yields a three-dimensional, percolating inorganic network that immobilizes and coats the core

phase. Final shell consolidation and attainment of the desired morphological and mechanical characteristics are achieved through controlled drying (to remove solvent and by-products) followed, when required, by thermal treatment or mild sintering to densify the wall material without degrading the encapsulated core [57,63].

4.2 Shell Structure and Importance of PCM Microcapsules

The functional performance of a microencapsulated system is governed not only by intrinsic properties of the core (e.g., molecular or particle size, diffusivity, and volatility) but also by how effectively that core interacts with, and is accommodated by, the surrounding shell. Accordingly, physicochemical compatibility between core and shell constituents is a primary determinant of encapsulation efficiency, stability, and end-use reliability.

Shell architectures can be deliberately engineered to exhibit fully permeable, selectively (semi-) permeable, or essentially impermeable behavior. A permeable wall is tailored to permit controlled liberation of the internal payload; modulation of release kinetics is achieved through judicious adjustment of wall thickness, porosity, and pore size distribution. Semi-permeable shells are fabricated so that they retain the encapsulated core while allowing passage of low–molar mass species (e.g., water, small organics). Such structures underpin uptake–release cycles in which external molecules are sorbed from one environment and subsequently discharged under altered conditions. In contrast, an impermeable (or quasi-impermeable) shell functions principally as a protective barrier, isolating the core from moisture, oxygen, light, or reactive chemicals. In these systems, payload liberation generally requires a triggered disruptive event—melting or softening of the wall, mechanical rupture, photochemical or thermal degradation, desiccation-induced cracking, or dissolution in an appropriate solvent medium [69].

Selection of shell composition is application-specific. For ingestible or food-related delivery formats, generally recognized as safe (GRAS) biopolymers or hydrocolloids such as gum arabic (acacia gum), maltodextrin, or whey proteins are routinely employed; in pharmaceutical or biomedical contexts, synthetic polymers like poly(methyl methacrylate) (PMMA) are common for controlling drug stability and release profiles [70]. Irrespective of sector, an optimal shell for phase change material (PCM) encapsulation should present: (i) sufficient mechanical integrity to suppress cracking and prevent leakage during thermal cycling; (ii) chemical inertness (no adverse reaction with the core or with surrounding matrices); and (iii) adequate thermal conductivity to facilitate efficient heat flux into and out of the PCM during charging and discharging operations. As summarized schematically (Fig.6), widely utilized shell categories encompass conventional synthetic polymers, bio-based macromolecules, inorganic substrates (e.g., silica, clay-derived frameworks), organic–inorganic or polymer–ceramic composites, and metallic or metal–oxide based coatings, each offering distinct trade-offs in permeability control, thermal stability, and mechanical resilience for PCM encapsulation scenarios [71].

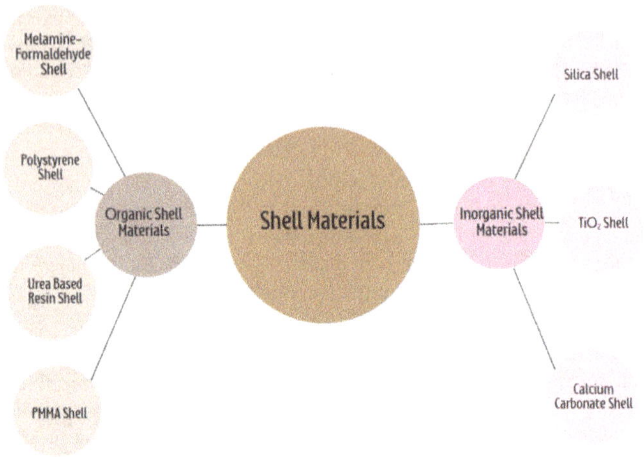

Figure.6 Shell materials classification.

Organic polymeric shells are widely selected for phase change material (PCM) microencapsulation because they can be processed under comparatively mild conditions, are economically attractive, and exhibit good interfacial compatibility with a broad range of latent heat storage cores. Such organic wall materials typically combine ductility with high flexibility, while retaining mechanical robustness and resistance to thermal cycling and deleterious chemical interactions during service. Representative shell-forming polymers include poly(methyl methacrylate) (PMMA), melamine–formaldehyde (MF), polystyrene (PS), polyurea (PU), and urea–formaldehyde resins [72].

- **Melamine-Formaldehyde Shell:** Owing to its densely crosslinked network, MF delivers elevated thermal stability, notable chemical inertness, flame-retardant character, and superior mechanical strength. A conventional fabrication route for MF-walled PCM microcapsules entails (i) preparing an oil-in-water emulsion of the molten or liquid PCM with an appropriate surfactant, (ii) generating a prepolymer condensate from melamine and aqueous formaldehyde, and (iii) inducing in situ polycondensation under acidic conditions (typically pH adjusted with an acid catalyst) at approximately 70–80 °C for on the order of 1–3 h until a coherent shell is formed. The resulting microcapsules are then isolated—commonly by filtration, washing, and drying. Fine control over emulsion droplet size, melamine/formaldehyde molar ratio, reaction pH, temperature profile, and curing duration enables purposeful tuning of shell thickness, morphology, mechanical resilience, encapsulation efficiency, and thermal reliability, making MF-based PCM capsules strong candidates for long-duration thermal energy storage applications.
- **Polystyrene Shell**: Polystyrene remains a frequently employed wall material for PCM encapsulation because it is non-toxic, comparatively inexpensive, chemically and thermally stable within typical operating windows, and amenable to scalable synthetic methodologies. PS shells are commonly produced via suspension polymerization, though

emulsion polymerization routes are also utilized when narrower particle size distributions or specific surface characteristics are desired. Tailoring initiator type, monomer-to-crosslinker ratio, agitation speed, and stabilizer selection allows modulation of capsule integrity and permeability. Incorporation of functional nanoparticles (e.g., magnetic oxides, silver, titanium dioxide, carbon-based nanostructures) into, or onto, the polystyrene shell can elevate thermal conductivity and latent heat exchange rates, while simultaneously imparting auxiliary functionalities such as magnetic responsiveness, antimicrobial activity, photocatalytic capability, or enhanced electrical conductivity [71].

- **Urea Based Resin Shell:** Urea–formaldehyde and related urea-derived resins are employed primarily to entrap organic, macromolecular phase change materials (PCMs) within a protective, conformable wall. These shells simultaneously shield the latent heat storage core from deleterious environmental factors and accommodate volumetric expansion–contraction associated with melting and solidification cycles. Typically fabricated through in situ polymerization routes (e.g., acid-catalyzed condensation in an oil-in-water emulsion), urea-based capsule walls are noted for a favorable balance of toughness, flexibility, and resistance to fatigue under repeated thermal cycling
- **Poly(methyl methacrylate) Shell**: PMMA has gained increasing prominence as a microcapsule wall material owing to its pronounced chemical inertness, biocompatibility, comparatively high Young's modulus, and reliable mechanical performance under operational stresses. PMMA shells can be generated via bulk, solution, suspension, or emulsion polymerization strategies utilizing free-radical or anionic initiation systems; manipulation of process variables (initiator type, dispersant, agitation rate, monomer conversion profile) enables tuning of capsule size distribution and wall morphology. Functional (smart) organic coating architectures are produced by embedding corrosion inhibitors, self-healing agents, or drying oils within PMMA microcapsules; upon environmental stimulus (e.g., pH change, mechanical damage), the encapsulated additive is released, providing active protection in marine, aerospace, and other corrosive service environments [73].

Limitations of purely organic shells and shift toward inorganic walls: While organic polymer shells can endow PCM microcapsules with desirable attributes such as elasticity, ease of processing, and good encapsulation efficiency, they inherently suffer from drawbacks including flammability, comparatively modest chemical resistance to aggressive media, and intrinsically low thermal conductivity, which can retard heat transfer rates during charging/discharging. For applications demanding elevated fire performance, improved barrier characteristics, and higher effective thermal conductivities, inorganic or hybrid inorganic–organic shell materials (e.g., silica, alumina, clay-derived matrices, or ceramic–polymer composites) are therefore preferentially adopted due to their superior mechanical strength, noncombustible nature, and enhanced thermal transport properties.

Oxide-based inorganic shell materials such as silicon dioxide (SiO_2), titanium dioxide (TiO_2), and aluminum oxide (Al_2O_3) exhibit pronounced resistance to corrosive media and maintain structural integrity at elevated temperatures, making them attractive for high-demand thermal environments. When used as encapsulating walls, these inorganic phases can suppress overall heat release and thereby bolster the thermal stability and flame resistance of the enclosed phase change materials (PCMs). A recurring drawback, however, is their intrinsic brittleness; mechanical stresses generated by repeated melt–solidification volume fluctuations may induce fissures and consequent leakage pathways in the capsule shell [78].

- **Silica Shell**: Microcapsules with silica walls are most commonly fabricated via a sol–gel route employing tetraethoxysilane (TEOS) (or analogous alkoxysilane precursors). In a representative workflow, the PCM core (often dispersed as oil droplets) is emulsified in an aqueous continuous phase containing surfactants or stabilizers; subsequently, hydrolysis of the silica precursor and controlled condensation around the dispersed droplets produce an incipient inorganic network that densifies at the interface. After gelation and maturation, the silica-coated capsules are isolated (e.g., through separation, washing, and drying). Silica shells are leveraged to elevate effective thermal conductivity and impart improved flame-retardant performance to the encapsulated system [71].
- **TiO_2 Shell**: Titanium dioxide is incorporated where enhanced thermal transport and multifunctionality are required. Capsule wall formation typically proceeds through hydrolysis–condensation or related polycondensation strategies, often under controlled pH and temperature to direct crystallinity or particle size. Beyond conferring thermal robustness, TiO_2 imparts photocatalytic capability, facilitating the photodegradation of environmentally harmful organic species at the surface. Its comparatively higher raw material and processing costs constitute a limitation. Creation of SiO_2–TiO_2 hybrid shells can synergistically combine mechanical rigidity, thermal conductivity, and functional (e.g., photocatalytic) attributes within a single encapsulation architecture [71].
- **Calcium Carbonate Shell**: Although silica-based shells enhance flame retardance and thermal transport, their fracture susceptibility can still permit leakage under cyclic mechanical or thermal stresses. Calcium carbonate has emerged as a cost-effective alternative offering high chemical stability, excellent mechanical integrity, and superior resistance to rupture compared with several other inorganic matrices. $CaCO_3$ shell formation is commonly achieved via an interfacial precipitation reaction where calcium and carbonate ions nucleate and grow on or around the dispersed PCM droplets. Interest is expanding in $CaCO_3$-walled microcapsules for dual-responsive (e.g., thermally and photo-triggered) functional systems [71].

Studies show that both purely organic and wholly inorganic encapsulation strategies present inherent trade-offs—organic polymer shells are typically combustible and possess low intrinsic thermal conductivity, whereas inorganic shells may be brittle and prone to crack-induced leakage. To reconcile these competing deficiencies, hybrid organic–inorganic shells have been advanced, intentionally integrating complementary material domains to achieve simultaneous mechanical compliance, thermal stability, flame retardancy, barrier function, and controlled release behavior.

Jiyang et al. addressed deleterious microcracking and attendant toxicity concerns in cementitious matrices by engineering a self-healing, flame-retardant, controlled-release microcapsule (SFRM). The capsule wall incorporates hexaphenoxy cyclotriphosphazene together with an epoxy resin component, jointly conferring flame-retardant efficacy and autonomous crack remediation capability through triggered release and polymer network formation [74].

Seeking to mitigate the mechanical deterioration often associated with conventional phosphorus-based flame retardants, Wu and co-workers synthesized a core–shell inorganic–organic hybrid via interfacial chemical modification. The resultant composite demonstrated elevated thermal endurance, improved mechanical performance, and robust flame-retardant behavior relative to unmodified counterparts [75].

Wang et al. produced a structurally tunable hybrid shell by coupling a methyl methacrylate-based polymer phase with a TEOS-derived silica network, yielding capsules with favorable phase transition characteristics and high thermal stability. They highlighted the applicability of these hybrid microcapsules in building energy conservation, acoustic/noise attenuation, battery thermal management, and diverse cooling or heat-regulation platforms [76].

4.3 Effect of Encapsulation on Flame Retardancy

The improvement in fire performance achieved by enclosing a phase change material (PCM) within a protective shell can be interpreted through multiple evaluative criteria: (i) displacement of onset and peak pyrolysis (thermal decomposition) temperatures; (ii) enhancement of interfacial compatibility and dispersion within the host polymer; (iii) alteration of the morphology, apparent physical state, or surface characteristics of the flame retardant (FR); (iv) modulation of aqueous solubility or leaching behavior; and (v) mitigation of smoke and toxic gas evolution. An appropriately engineered encapsulating layer is expected to strengthen these attributes in concert, thereby elevating overall flame resistance and operational safety of the composite system [56].

- **Pyrolysis Temperature:** Forming a micro- or nano-scale shell around a PCM that incorporates or coexists with flame retardant constituents typically raises its effective thermal decomposition threshold by acting as a transient barrier to heat and mass transport. Chlorophosphazene-based species, for instance, are recognized for favorable environmental profiles and robust intrinsic flame-inhibiting efficacy; nevertheless, their relatively modest thermal stability and the deleterious influence they may exert on mechanical integrity upon direct incorporation into a matrix limit practical deployment. Liu Yaqgin and co-workers addressed these deficiencies via microencapsulation of chlorocyclophosphazenes, reporting that the principal decomposition temperature of the encapsulated derivative increased by approximately 200 °C relative to its non-encapsulated counterpart. This significant upward shift evidences that the shell architecture retards volatilization and bond scission events, thereby broadening the feasible processing and service temperature window and expanding application potential [69].
- **Flame Retardant Compatibility with Polymer:** Interfacial compatibility between the flame retardant phase (including PCM–FR assemblies) and the continuous polymer matrix constitutes a decisive factor in achieving durable, mechanically robust, and fire-safe composite materials. Disparities in polarity, surface energy, or chemical functionality frequently produce poor dispersion, agglomeration, or interfacial voids when FR additives are introduced directly, yielding non-uniform distribution and attenuated performance. In a microencapsulated configuration, the shell serves multiple synergistic roles: (i) it isolates the core from adverse interactions with the matrix, (ii) it supplies a tunable surface chemistry that can be engineered (e.g., via functional groups, charge, or roughness) to improve adhesion and wetting, and (iii) it promotes homogeneous dispersion—thereby reducing the risk of phase separation, embrittlement, diminished hardness, or premature failure during processing and service. The most effective flame retardant strategies not only delay ignition and suppress heat release but also preserve or enhance mechanical strength,

thermal cycling resilience, and processability. Where the encapsulation polymer exhibits insufficient affinity for either the FR or the PCM component, defects such as interfacial debonding, microcracking under thermal stress, or exudation may arise. Consequently, rational shell design emphasizing chemical compatibility (e.g., grafted coupling moieties, amphiphilic block segments, or reactive functionalization) and controlled permeability is essential to secure optimal safety margins and operational reliability in thermal energy storage systems employing FR-modified PCMs [78,79].

- **Changing the Appearance and Physical State of the Flame Retardant:** The encapsulation of the PCMs and FRs also in order to make flame retardants easier to mix with polymers, a liquid flame retardant can be made solid by microencapsulation.
- **Effect of Flame Retardant on Water Solubility:** Many flame retardant (FR) additives exhibit pronounced susceptibility to moisture; under prolonged exposure to humid or wet environments they may partially dissolve, migrate through the polymer bulk, and accumulate at (or leach from) the surface, diminishing long-term fire performance and uniformity. Microencapsulation offers an effective mitigation pathway by erecting a diffusion-moderating barrier that physically separates the hygroscopically sensitive FR from ambient water. Illustratively, Guowen Ran and colleagues [80] produced a polylactic acid (PLA) composite incorporating ammonium polyphosphate (APP) that had first been microencapsulated using polysilicone and polyborosilicone (APP as the core within hybrid silicone/borosilicone-derived shells). Their results demonstrated that the encapsulated APP conferred enhanced moisture resistance and concurrently improved the mechanical characteristics of the resulting PLA matrix, in contrast to expectations for the unprotected flame retardant.
- **Release of Toxic Gases:** Toxic gases contained in the flame retardant are reduced or prevented during microencapsulation or precipitation. For example, red phosphorus is a strong flame retardant, but it can react with moisture to release highly toxic phosphine. To prevent this, it is microencapsulated with appropriate materials and protected from moisture.

5. Conclusion

This chapter focus on the flame retardancy solutions for phase changing materials based thermal energy systems. PCMs have wide applications areas such as buildings, vehicles, electronics, textiles, and space applications for energy efficiency applications. Among their benefits, particularly organic PCMs, such fatty acids and paraffin, are very flammable. For providing the fire safety requirements by use of several approach such as flame-retardant additives, microencapsulation techniques, and inorganic or hybrid materials. Microencapsulation technique is not only solving the flammability of PCMs but also inhibiting the leakage and oxidation of PCMs. There are alternative ways for this applications such as chemical, physical, or physico-chemical techniques by use of organic and inorganic shell materials that provide various advantages in terms of thermal stability, mechanical strength, and fire resistance. As a concluding, the providing the fire retardancy properties for safety and keeping the thermal energy storage properties same are important for the applications.

References

[1] W. Aftab, X. Huang, W. Wu, Z. Liang, A. Mahmood, R. Zou, Nanoconfined phase change materials for thermal energy applications, Energy Environ. Sci. 11 (2018) 1392–1424. https://doi.org/10.1039/C7EE03587J.

[2] A. Sternberg, A. Bardow, Power-to-What? – Environmental assessment of energy storage systems, Energy Environ. Sci. 8 (2015) 389–400. https://doi.org/10.1039/C4EE03051F.

[3] D. Ürge-Vorsatz, L.F. Cabeza, S. Serrano, C. Barreneche, K. Petrichenko, Heating and cooling energy trends and drivers in buildings, Renewable and Sustainable Energy Reviews 41 (2015) 85–98. https://doi.org/10.1016/j.rser.2014.08.039.

[4] C. Cárdenas-Ramírez, F. Jaramillo, M. Gómez, Systematic review of encapsulation and shape-stabilization of phase change materials, J. Energy Storage 30 (2020) 101495. https://doi.org/10.1016/j.est.2020.101495.

[5] Y. Li, N. Nord, Q. Xiao, T. Tereshchenko, Building heating applications with phase change material: A comprehensive review, J. Energy Storage 31 (2020) 101634. https://doi.org/10.1016/j.est.2020.101634.

[6] Y. Lin, G. Alva, G. Fang, Review on thermal performances and applications of thermal energy storage systems with inorganic phase change materials, Energy 165 (2018) 685–708. https://doi.org/10.1016/j.energy.2018.09.128.

[7] N. Aslfattahi, A. Zendehboudi, S. Rahman, M.F. Mohd Sabri, S. Mohd Said, A. Arifutzzaman, N.A. Che Sidik, Optimization of Thermal Conductivity of NanoPCM-Based Graphene by Response Surface Methodology, Journal of Advanced Research in Fluid Mechanics and Thermal Sciences 75 (2020) 108–125. https://doi.org/10.37934/arfmts.75.3.108125.

[8] D. Feldman, M.M. Shapiro, D. Banu, C.J. Fuks, Fatty acids and their mixtures as phase-change materials for thermal energy storage, Solar Energy Materials 18 (1989) 201–216. https://doi.org/10.1016/0165-1633(89)90054-3.

[9] N. Han, Z. Li, X. Zhang, W. Yu, X. Chen, D. Wang, J. Li, Synthesis and characterization of cellulose-g-polyoxyethylene (2) hexadecyl ether solid–solid phase change materials, Cellulose 23 (2016) 1663–1674. https://doi.org/10.1007/s10570-016-0901-6.

[10] X. Deng, C. Li, X. Sun, C. Wang, B. Liu, Y. Li, H. Yang, Flame-retardant wood-based composite phase change materials based on PDMS/expanded graphite coating for efficient solar-to-thermal energy storage, Appl. Energy 368 (2024) 123454. https://doi.org/10.1016/j.apenergy.2024.123454.

[11] G. Alva, Y. Lin, G. Fang, An overview of thermal energy storage systems, Energy 144 (2018) 341–378. https://doi.org/10.1016/j.energy.2017.12.037.

[12] S.R.L. da Cunha, J.L.B. de Aguiar, Phase change materials and energy efficiency of buildings: A review of knowledge, J. Energy Storage 27 (2020) 101083. https://doi.org/10.1016/j.est.2019.101083.

[13] N. Sarier, E. Onder, Organic phase change materials and their textile applications: An overview, Thermochim. Acta 540 (2012) 7–60. https://doi.org/10.1016/j.tca.2012.04.013.

[14] E. Oró, A. de Gracia, A. Castell, M.M. Farid, L.F. Cabeza, Review on phase change materials (PCMs) for cold thermal energy storage applications, Appl. Energy 99 (2012) 513–533. https://doi.org/10.1016/j.apenergy.2012.03.058.

[15] J. Jaguemont, N. Omar, P. Van den Bossche, J. Mierlo, Phase-change materials (PCM) for automotive applications: A review, Appl. Therm. Eng. 132 (2018) 308–320. https://doi.org/10.1016/j.applthermaleng.2017.12.097.

[16] S.R.L. da Cunha, J.L.B. de Aguiar, Phase change materials and energy efficiency of buildings: A review of knowledge, J. Energy Storage 27 (2020) 101083. https://doi.org/10.1016/j.est.2019.101083.

[17] Z. Zhou, Z. Zhang, J. Zuo, K. Huang, L. Zhang, Phase change materials for solar thermal energy storage in residential buildings in cold climate, Renewable and Sustainable Energy Reviews 48 (2015) 692–703. https://doi.org/10.1016/j.rser.2015.04.048.

[18] ibrahim Dinçer, H.S. Hamut, N. Javani, Thermal Management of Electric Vehicle Battery Systems, Wiley, 2017. https://doi.org/10.1002/9781118900239.

[19] A. Kurdi, N. Almoatham, M. Mirza, T. Ballweg, B. Alkahlan, Potential Phase Change Materials in Building Wall Construction—A Review, Materials 14 (2021) 5328. https://doi.org/10.3390/ma14185328.

[20] K. Velmurugan, T. Wongwuttanasatian, Preparation and selection of a eutectic phase change material for cooling the PV module under Thailand climatic conditions, E3S Web of Conferences 379 (2023) 03004. https://doi.org/10.1051/e3sconf/202337903004.

[21] E.M. Alawadhi, Thermal analysis of a building brick containing phase change material, Energy Build. 40 (2008) 351–357. https://doi.org/10.1016/j.enbuild.2007.03.001.

[22] F. Liu, S. Gan, Y. Chen, W. Qin, J. Li, Thermal characteristics of Li-ion battery based on phase change material-aluminum plate-fin composite heat dissipation, Energy Sci. Eng. 11 (2023) 62–78. https://doi.org/10.1002/ese3.1357.

[23] L.F. Cabeza, A. Castell, C. Barreneche, A. de Gracia, A.I. Fernández, Materials used as PCM in thermal energy storage in buildings: A review, Renewable and Sustainable Energy Reviews 15 (2011) 1675–1695. https://doi.org/10.1016/j.rser.2010.11.018.

[24] A. Sharma, V.V. Tyagi, C.R. Chen, D. Buddhi, Review on thermal energy storage with phase change materials and applications, Renewable and Sustainable Energy Reviews 13 (2009) 318–345. https://doi.org/10.1016/j.rser.2007.10.005.

[25] M.K. Rathod, Phase Change Materials and Their Applications, in: Phase Change Materials and Their Applications, InTech, 2018. https://doi.org/10.5772/intechopen.75923.

[26] M. Hadjieva, R. Stoykov, T. Filipova, Composite salt-hydrate concrete system for building energy storage, Renew. Energy 19 (2000) 111–115. https://doi.org/10.1016/S0960-1481(99)00024-5.

[27] T. Lee, D.W. Hawes, D. Banu, D. Feldman, Control aspects of latent heat storage and recovery in concrete, Solar Energy Materials and Solar Cells 62 (2000) 217–237. https://doi.org/10.1016/S0927-0248(99)00128-2.

[28] D. Feldman, D. Banu, DSC analysis for the evaluation of an energy storing wallboard, Thermochim. Acta 272 (1996) 243–251. https://doi.org/10.1016/0040-6031(95)02456-5.

[29] T. Wang, S. Wang, R. Luo, C. Zhu, T. Akiyama, Z. Zhang, Microencapsulation of phase change materials with binary cores and calcium carbonate shell for thermal energy storage, Appl. Energy 171 (2016) 113–119. https://doi.org/10.1016/j.apenergy.2016.03.037.

[30] D. Zhou, Y. Zhou, Y. Liu, X. Luo, J. Yuan, Preparation and Performance of Capric-Myristic Acid Binary Eutectic Mixtures for Latent Heat Thermal Energy Storages, J. Nanomater. 2019 (2019) 1–9. https://doi.org/10.1155/2019/2094767.

[31] R. Elarem, T. Alqahtani, S. Mellouli, A. Edacherian, F. Askri, A. Jemni, Numerical analysis of a built-in thermal storage system of metal hydride and nanoparticles enhanced phase change material and nanofluid, Int. J. Energy Res. 45 (2021) 5881–5893. https://doi.org/10.1002/er.6209.

[32] G. Ma, L. Han, J. Sun, Y. Jia, Thermal properties and reliability of eutectic mixture of stearic acid-acetamide as phase change material for latent heat storage, J. Chem. Thermodyn. 106 (2017) 178–186. https://doi.org/10.1016/j.jct.2016.11.022.

[33] Y. Xiong, Z. Wang, P. Xu, C. Hongbing, Y. Wu, Experimental investigation into the thermos-physical properties by dispersing nanoparticles to the nitrates, Energy Procedia 158 (2019) 5551–5556. https://doi.org/10.1016/j.egypro.2019.01.588.

[34] K. TUNCBILEK, A. SARI, S. TARHAN, G. ERGUNES, K. KAYGUSUZ, Lauric and palmitic acids eutectic mixture as latent heat storage material for low temperature heating applications, Energy 30 (2005) 677–692. https://doi.org/10.1016/j.energy.2004.05.017.

[35] A.C. Evers, M.A. Medina, Y. Fang, Evaluation of the thermal performance of frame walls enhanced with paraffin and hydrated salt phase change materials using a dynamic wall simulator, Build. Environ. 45 (2010) 1762–1768. https://doi.org/10.1016/j.buildenv.2010.02.002.

[36] A. Mills, M. Farid, J.R. Selman, S. Al-Hallaj, Thermal conductivity enhancement of phase change materials using a graphite matrix, Appl. Therm. Eng. 26 (2006) 1652–1661. https://doi.org/10.1016/j.applthermaleng.2005.11.022.

[37] S. HIMRAN, A. SUWONO, G.A. MANSOORI, Characterization of Alkanes and Paraffin Waxes for Application as Phase Change Energy Storage Medium, Energy Sources 16 (1994) 117–128. https://doi.org/10.1080/00908319408909065.

[38] H. Fauzi, H.S.C. Metselaar, T.M.I. Mahlia, M. Silakhori, Thermal Reliability of Myristic Acid/Palmitic Acid/Sodium Laurate Eutectic Mixture: A Feasibility Study of Accelerated Aging for Thermal Energy Storage Application, Energy Procedia 61 (2014) 49–54. https://doi.org/10.1016/j.egypro.2014.11.903.

[39] Y. Hu, P.K. Heiselberg, C. Drivsholm, A.S. Søvsø, P.J.C. Vogler-Finck, K. Kronby, Experimental and numerical study of PCM storage integrated with HVAC system for energy flexibility, Energy Build. 255 (2022) 111651. https://doi.org/10.1016/j.enbuild.2021.111651.

[40] Md.T. Hossain, Md.A. Shahid, Md.Y. Ali, S. Saha, M.S.I. Jamal, A. Habib, Fabrications, Classifications, and Environmental Impact of PCM-Incorporated Textiles: Current State and Future Outlook, ACS Omega 8 (2023) 45164–45176. https://doi.org/10.1021/acsomega.3c05702.

[41] R. Chaturvedi, A. Islam, K. Sharma, A review on the applications of PCM in thermal storage of solar energy, Mater. Today Proc. 43 (2021) 293–297. https://doi.org/10.1016/j.matpr.2020.11.665.

[42] S.K. Sahoo, M.K. Das, P. Rath, Application of TCE-PCM based heat sinks for cooling of electronic components: A review, Renewable and Sustainable Energy Reviews 59 (2016) 550–582. https://doi.org/10.1016/j.rser.2015.12.238.

[43] L. Zhao, Q. Yu, M. Li, Y. Wang, G. Li, S. Sun, J. Fan, Y. Liu, A review of the innovative application of phase change materials to cold-chain logistics for agricultural product storage, J. Mol. Liq. 365 (2022) 120088. https://doi.org/10.1016/j.molliq.2022.120088.

[44] S. Wang, X. Hou, J. Yin, Y. Xing, Z. Wang, Comparative Study of the Thermal Enhancement for Spacecraft PCM Thermal Energy Storage Units, Aerospace 9 (2022) 705. https://doi.org/10.3390/aerospace9110705.

[45] A. Palacios, A. De Gracia, L. Haurie, L. Cabeza, A. Fernández, C. Barreneche, Study of the Thermal Properties and the Fire Performance of Flame Retardant-Organic PCM in Bulk Form, Materials 11 (2018) 117. https://doi.org/10.3390/ma11010117.

[46] K. Pielichowska, N. Paprota, K. Pielichowski, Fire Retardant Phase Change Materials—Recent Developments and Future Perspectives, Materials 16 (2023) 4391. https://doi.org/10.3390/ma16124391.

[47] E. Arinaitwe, M. McNamee, M. Försth, Is the fire performance of phase change materials a significant barrier to implementation in building applications?, J. Energy Storage 94 (2024) 112421. https://doi.org/10.1016/j.est.2024.112421.

[48] P. Sun, A. Rodriguez, W. Il Kim, X. Huang, C. Fernandez-Pello, Effect of external and internal heating on the flame spread and phase change of thin polyethylene tubes, International Journal of Thermal Sciences 168 (2021) 107054. https://doi.org/10.1016/j.ijthermalsci.2021.107054.

[49] B. Diaconu, M. Cruceru, L. Anghelescu, Fire Retardance Methods and Materials for Phase Change Materials: Performance, Integration Methods, and Applications—A Literature Review, Fire 6 (2023) 175. https://doi.org/10.3390/fire6050175.

[50] T.-K. Hong, S.-H. Park, Influences of the presence of char layer on flame spreads over wood with different thermal thickness, Journal of Mechanical Science and Technology 37 (2023) 3841–3848. https://doi.org/10.1007/s12206-023-0646-8.

[51] A. Ismail, J. Zhou, A. Aday, I. Davidoff, A. Odukomaiya, J. Wang, Microencapsulation of bio-based phase change materials with silica coated inorganic shell for thermal energy storage, Journal of Building Engineering 67 (2023) 105981. https://doi.org/10.1016/j.jobe.2023.105981.

[52] Q. Huang, C. Li, X. Li, Y. Jin, G. Zhang, J. Deng, Y. Wu, K. Xiong, W. Jiang, Multifunction composite phase change material with inorganic flame retardant and organic form stability for improving battery thermal safety, The Innovation Materials 2 (2024) 100048. https://doi.org/10.59717/j.xinn-mater.2024.100048.

[53] E.K. Asimakopoulou, D.I. Kolaitis, M.A. Founti, Fire safety aspects of PCM-enhanced gypsum plasterboards: An experimental and numerical investigation, Fire Saf. J. 72 (2015) 50–58. https://doi.org/10.1016/j.firesaf.2015.02.004.

[54] L. Li, G. Wang, C. Guo, Influence of intumescent flame retardant on thermal and flame retardancy of eutectic mixed paraffin/polypropylene form-stable phase change materials, Appl. Energy 162 (2016) 428–434. https://doi.org/10.1016/j.apenergy.2015.10.103.

[55] P. Zhang, Y. Hu, L. Song, J. Ni, W. Xing, J. Wang, Effect of expanded graphite on properties of high-density polyethylene/paraffin composite with intumescent flame retardant as a shape-stabilized phase change material, Solar Energy Materials and Solar Cells 94 (2010) 360–365. https://doi.org/10.1016/j.solmat.2009.10.014.

[56] B. Wang, H. Sheng, Y. Shi, W. Hu, N. Hong, W. Zeng, H. Ge, X. Yu, L. Song, Y. Hu, Recent advances for microencapsulation of flame retardant, Polym. Degrad. Stab. 113 (2015) 96–109. https://doi.org/10.1016/j.polymdegradstab.2015.01.008.

[57] K. Ghasemi, S. Tasnim, S. Mahmud, PCM, nano/microencapsulation and slurries: A review of fundamentals, categories, fabrication, numerical models and applications, Sustainable Energy Technologies and Assessments 52 (2022) 102084. https://doi.org/10.1016/j.seta.2022.102084.

[58] T. Wang, X. Liu, J. Luo, R. Liu, G. Sun, Encapsulation of Dyed Alkanes with Polyurethane Microcapsules from Interfacial Polymerization Enables Irreversible Thermochromic Coatings, ACS Appl. Polym. Mater. 6 (2024) 8811–8820. https://doi.org/10.1021/acsapm.4c00889.

[59] S. Demirbağ, S.A. Aksoy, Encapsulation of phase change materials by complex coacervation to improve thermal performances and flame retardant properties of the cotton fabrics, Fibers and Polymers 17 (2016) 408–417. https://doi.org/10.1007/s12221-016-5113-z.

[60] H. Chen, Y. Ma, X. Sheng, Y. Chen, Achieving heat storage coatings from ethylene vinyl acetate copolymers and phase change nano-capsules with excellent flame-retardant and thermal comfort performances, Prog. Org. Coat. 192 (2024) 108478. https://doi.org/10.1016/j.porgcoat.2024.108478.

[61] J. Giro-Paloma, M. Martínez, L.F. Cabeza, A.I. Fernández, Types, methods, techniques, and applications for microencapsulated phase change materials (MPCM): A review, Renewable and Sustainable Energy Reviews 53 (2016) 1059–1075. https://doi.org/10.1016/j.rser.2015.09.040.

[62] A. Jaworek, Electrostatic micro- and nanoencapsulation and electroemulsification: A brief review, J. Microencapsul. 25 (2008) 443–468. https://doi.org/10.1080/02652040802049109.

[63] Y. Huang, A. Stonehouse, C. Abeykoon, Encapsulation methods for phase change materials – A critical review, Int. J. Heat Mass Transf. 200 (2023) 123458. https://doi.org/10.1016/j.ijheatmasstransfer.2022.123458.

[64] S. B, L.P. Singh, I. Tyagi, A. Rawat, S. Sinha, Microencapsulation of a eutectic PCM using in situ polymerization technique for thermal energy storage, Int. J. Energy Res. 44 (2020) 3854–3864. https://doi.org/10.1002/er.5182.

[65] M. Kang, Y. Liu, C. Liang, W. Lin, C. Wang, C. Li, F. Zhang, J. Cheng, Phase change material microcapsules with DOPO/Cu modified halloysite nanotubes for thermal controlling of buildings: Thermophysical properties, flame retardant performance and thermal comfort levels, Int. J. Heat Mass Transf. 207 (2023) 124045. https://doi.org/10.1016/j.ijheatmasstransfer.2023.124045.

[66] Z. Zhang, Z. Zhang, T. Chang, J. Wang, X. Wang, G. Zhou, Phase change material microcapsules with melamine resin shell via cellulose nanocrystal stabilized Pickering emulsion in-situ polymerization, Chemical Engineering Journal 428 (2022) 131164. https://doi.org/10.1016/j.cej.2021.131164.

[67] J. Shi, X. Wu, R. Sun, B. Ban, J. Li, J. Chen, Nano-encapsulated phase change materials prepared by one-step interfacial polymerization for thermal energy storage, Mater. Chem. Phys. 231 (2019) 244–251. https://doi.org/10.1016/j.matchemphys.2019.04.032.

[68] B. Kazanci, K. Cellat, H. Paksoy, Preparation, characterization, and thermal properties of novel fire-resistant microencapsulated phase change materials based on paraffin and a polystyrene shell, RSC Adv. 10 (2020) 24134–24144. https://doi.org/10.1039/D0RA04093B.

[69] S.K. Ghosh, Functional Coatings and Microencapsulation: A General Perspective, in: Functional Coatings, Wiley, 2006: pp. 1–28. https://doi.org/10.1002/3527608478.ch1.

[70] S.H. Sonawane, B.A. Bhanvase, M. Sivakumar, S.B. Potdar, Current overview of encapsulation, in: Encapsulation of Active Molecules and Their Delivery System, Elsevier, 2020: pp. 1–8. https://doi.org/10.1016/B978-0-12-819363-1.00001-6.

[71] T.K. Maiti, P. Dixit, A. Suhag, S. Bhushan, A. Yadav, N. Talapatra, S. Chattopadhyay, Advancements in organic and inorganic shell materials for the preparation of microencapsulated phase change materials for thermal energy storage applications, RSC Sustainability 1 (2023) 665–697. https://doi.org/10.1039/D2SU00116K.

[72] M. Kang, Y. Liu, W. Lin, C. Liang, W. Qu, S. Li, Y. Wang, F. Zhang, J. Cheng, Thermal comfort and flame retardant performance of novel temperature-control coatings with modified phase change material microcapsules, Prog. Org. Coat. 181 (2023) 107579. https://doi.org/10.1016/j.porgcoat.2023.107579.

[73] F. Ahangaran, A.H. Navarchian, F. Picchioni, Material encapsulation in poly(methyl methacrylate) shell: A review, J. Appl. Polym. Sci. 136 (2019). https://doi.org/10.1002/app.48039.

[74] W. Jiang, G. Zhou, J. Duan, D. Liu, Q. Zhang, F. Tian, Synthesis and Characterization of a Multifunctional Sustained-Release Organic–Inorganic Hybrid Microcapsule with Self-Healing and Flame-Retardancy Properties, ACS Appl. Mater. Interfaces 13 (2021) 15668–15679. https://doi.org/10.1021/acsami.1c01540.

[75] S.-G. Wu, C. Liu, Z. Qin, D.-Y. He, Z.-K. Wang, W. Xie, Y.-Z. Wang, L. Chen, Core-shell engineered boehmite-derived organic-inorganic hybrid flame retardant for epoxy resins: synergistically enhanced fire safety and mechanical integrity, Polym. Degrad. Stab. 238 (2025) 111367. https://doi.org/10.1016/j.polymdegradstab.2025.111367.

[76] X. Wang, H. Chen, D. Kuang, X. Huan, Z. Zeng, C. Qi, S. Han, G. Li, Unlocking thermal management capacity: Optimized organic-inorganic hybrid shell phase change microcapsules

with controllable structure and enhanced conductivity, Compos. Sci. Technol. 257 (2024) 110836. https://doi.org/10.1016/j.compscitech.2024.110836.

[77] Y. LIU, G. ZHAO, Study on the Properties of Microcapsulated Chlorocyclophosphazene Polypropylene Composites, Chin. J. Chem. Eng. 15 (2007) 429–432. https://doi.org/10.1016/S1004-9541(07)60103-7.

[78] S. Liang, N.M. Neisius, S. Gaan, Recent developments in flame retardant polymeric coatings, Prog. Org. Coat. 76 (2013) 1642–1665. https://doi.org/10.1016/j.porgcoat.2013.07.014.

[79] Y. Hou, Z. Xu, F. Chu, Z. Gui, L. Song, Y. Hu, W. Hu, A review on metal-organic hybrids as flame retardants for enhancing fire safety of polymer composites, Compos. B Eng. 221 (2021) 109014. https://doi.org/10.1016/j.compositesb.2021.109014.

[80] G. Ran, X. Liu, J. Guo, J. Sun, H. Li, X. Gu, S. Zhang, Improving the flame retardancy and water resistance of polylactic acid by introducing polyborosiloxane microencapsulated ammonium polyphosphate, Compos. B Eng. 173 (2019) 106772. https://doi.org/10.1016/j.compositesb.2019.04.033.

Materials Research Foundations 190 (2026) 50-76 https://doi.org/10.21741/9781644904077-3

Chapter 3

Biomass-Based Flame Retardants: An Overview

Meenu Mariam Jacob[1], Muthamilselvi Ponnuchamy[1*]

[1]Department of Chemical Engineering, SRM Institute of Science and Technology, Kattankulathur, Tamil Nadu, India - 603202

muthamip@srmist.edu.in

Abstract

Increased need for sustainable, non-toxic, and high-performance flame retardants has prompted intense research into biomass-based alternatives to replace conventional halogenated systems. Obtained from renewable resources like lignin, cellulose, chitosan, and phytic acid, biomass-based flame retardants provide several benefits such as environmental compatibility, carbon neutrality, and in many cases, inherent flame-retardant activity. These materials act through diverse mechanisms such as char formation, thermal insulation, endothermic decomposition, and radical scavenging. Improvements in chemical modification and nano structuring processes have made possible the design of these bio-derived compounds for better thermal stability, polymer matrix compatibility, and higher fire resistance. This chapter offers a detailed introduction to the underlying principles of flame retardancy, physicochemical characteristics of prominent biomass feedstocks, and the latest advances in their chemical functionalization for flame-retardant purposes. The incorporation of biomass-derived retardants into thermoplastics, thermosets, composites, and textiles is addressed in conjunction with performance metrics of evaluation and mechanistic understanding. Additionally, material processing challenges, efficiency of flame retardants, and regulatory issues are critically examined, and green flame retardant design directions of the future are envisioned. The chapter highlights the potential of biomass-derived systems to help create safer, more sustainable products across a variety of industries.

Keywords

Biomass, Flame Retardants, Sustainable Polymers, Fire Safety, Eco-Friendly

Contents

1. Introduction

The growing use of polymeric materials in contemporary society ranging from construction to electronics, transport, and textiles, and packaging has exacerbated fire safety concerns. Synthetic polymers, especially petrochemical-based polymers, are themselves combustible because of their high contents of hydrocarbons and low thermal stability. They not only spread fire but also emit heat, smoke, and harmful gases like carbon monoxide, hydrogen cyanide, and volatile organics [1]. In order to counteract these risks, flame retardant additives have become key players in polymer mixtures, limiting ignition, lowering flammability, retarding flame spread, and suppressing heat release. Nonetheless, numerous typical flame retardants, especially halogenated products, are linked to severe environmental and health hazards [2]. Their combustion and breakdown products tend to be persistent, bio accumulative, and toxic. Additionally, various flame retardants like polybrominated diphenyl ethers (PBDEs), hexabromocyclododecane (HBCD), and some phosphorus-based products have been prohibited or restricted under international regulatory regimes like the European Union's REACH regulation and the Stockholm Convention on Persistent Organic Pollutants (POPs) [3–5].

Under such circumstances, the shift towards biomass-derived flame retardants has come forth as a green and scientifically viable option. These materials are based on renewable, non-toxic, and frequently biodegradable renewable resources like lignin, cellulose, hemicellulose, chitosan, starch, and phytic acid [6]. Their inherent characteristics such as high char yield, heteroatom content such as phosphorus and nitrogen, and endothermic degradability naturally position them for flame retardant uses. In addition, these biomass fractions can be chemically modified to become more heat stable, more effective flame inhibitors, and compatible with a wide range of polymer matrices [7]. This renders them not only effective substitutes for synthetic retardants but also multifunctional agents that add to mechanical reinforcement, barrier, and environmental durability of the host polymer system [8].

Lignin, for example, is a highly aromatic macromolecule rich in phenolic hydroxyl and methoxyl groups and thus a good carbon source for char formation. Thermal decomposition of lignin produces a hard, insulating char coating that acts as a physical shield, shielding the polymer beneath it from heat and oxygen [9]. In a similar vein, phytic acid, a naturally occurring phosphorus-containing compound isolated from seeds, can serve as an acid source and a blowing agent within intumescent systems, favouring the growth of expanded, foamed char morphologies that resist flame propagation [10]. Chitosan, a nitrogen polysaccharide derived from crustacean shells, has potential to make contributions in the form of gas-phase radical trapping and condensed-phase charring, particularly when functionalized with phosphorus-containing groups. The synergistic effect of combined nitrogen and phosphorus in such systems improves flame retardant performance via both physical and chemical means [11].

Functionalization and incorporation of biomass-derived materials into flame retardant systems have also been facilitated by recent developments in green chemistry, material science, and nanotechnology. Chemical modification processes like phosphorylation, sulfonation, and graft copolymerization have been applied to increase flame retardance along with interfacial adhesion with synthetic and bio-based polymers. Further, nanostructured hybrids like lignin-clay nanocomposites, cellulose-nanotube blends, and chitosan–graphene oxide conjugates have been utilized to achieve better dispersion, control of thermal conductivity, and flame suppression efficacy at reduced loading levels. These materials not only strengthen the flame retardant mechanism but also decrease the total additive level needed, limiting negative impact on mechanical performance and processability of the host polymer [12].

Apart from technical performance, the attraction of biomass-based flame retardants is that they support sustainability objectives. In contrast to conventional retardants derived from finite fossil fuel sources and causing environmental toxicity, biomass-based materials are within the natural carbon cycle. Several of these are sourced from agricultural waste, forestry residues, or industrial waste streams, providing value-added routes for biomass valorisation [13–16]. Their low toxicity and biodegradability ensure they are safer for end-of-life disposal through incineration, composting, or landfilling. In addition, their utilization supports circular economy models, where waste streams are re-circulated into production cycles, and environmental footprints are minimized throughout the material lifecycle [17–20].Although these benefits exist, biomass-based flame retardants are not without drawbacks. Drawbacks such as low thermal stability at processing temperatures, hydrophilicity, low compatibility with hydrophobic polymers, and inconsistency in raw material composition need to be overcome by thoughtful formulation and material design. Recent research has started to break down these barriers by using compatibilizers, surface treatments, nano structuring, and multi-component synergistic systems. As the field develops, the acceptance of biomass-based flame retardants into mainstream products will be based on their capacity to comply with tight fire safety requirements, mechanical performance specifications, and economic competitiveness.

Table 1. *Comparison of Halogenated vs. Biomass based flame retardants*

Property	Halogenated FRs	Biomass Based FRs
Source	Petrochemical	Renewable
Mechanism	Gas-phase radical quenching	Char formation, intumescence
Toxicity	High (e.g. dioxins)	Low
Environmental Impact	Persistent, bio accumulative	Biodegradable, circular
Cost	Mature market	Moderate, improving

This chapter offers an overview of biomass-based flame retardants, starting with a review of the underlying principles of flame retardancy and combustion chemistry. It then reviews the sources, chemical structure, and modification options for major biomass-derived materials. Applications across various classes of polymers thermoplastics, thermosets, foams, and textiles are critically examined, with a focus on performance measurement metrics and mechanisms of flame retardancy. The chapter is rounded off with environmental and regulatory aspects, existing challenges, and future directions for development of sustainable, bio-based fire protection technologies. Through this review, the chapter seeks to shed light on the scientific underpinnings

Materials Research Foundations 190 (2026) 50-76 https://doi.org/10.21741/9781644904077-3

and application possibilities of biomass-derived flame retardants as key building blocks of future-generation fire-safe and eco-friendly materials.

2. Principles of Flame Retardancy

Comprehension of the fundamentals of flame retardancy requires an initial understanding of the combustion process and the thermochemical mechanisms of the ignition and sustenance of fire in polymeric materials. Polymer combustion is a self-propagating process initiated by thermal decomposition. When subjected to an external heat source, polymers pyrolyze, a complicated series of endothermic reactions that result in bond scission, depolymerization, and the creation of volatile and combustible fragments. These volatiles migrate from the condensed phase to the gas phase, where they combine with atmospheric oxygen and burn, producing heat, light, and smoke. The heat produced continues to drive the degradation of the material below it, creating a positive feedback mechanism that sustains the process of burning [21]. Hence, the disruption of any part of this chain pyrolysis, volatile formation, combustion, or heat feedback is the essence of obtaining flame retardancy.

Polymer combustion occurs in a standard sequence of four steps: heating, decomposition, ignition, and flame spread. During the heating process, the polymer takes up heat energy, and as its temperature reaches near-decomposition, macromolecular chain random scission starts [22]. This decomposition, or pyrolysis, leads to the formation of flammable gases, tars, and solid residues (char). With adequate accumulation of volatile gases, ignition is triggered when the gas-phase mixture attains the lower flammability limit (LFL) and is subjected to an ignition source. This is succeeded by flame spread, where the volatiles are still providing exothermic energy for continued degradation [23]. The strategies of flame retardants seek to break this cycle of heat and oxidation either in the condensed phase, where the solid is present, or the gas phase, where combustion is taking place.

Flame retardants work by a number of mechanisms, commonly classified as physical, chemical, or thermal modes of action, and these can act in the gas phase, condensed phase, or both. Physical mechanisms usually include heat absorption or dilution of fuel gases. Endothermic thermal decomposition of some flame retarding additives like aluminium hydroxide or magnesium hydroxide withdraws a lot of heat from the polymer matrix and hence decreases the temperature to below the ignition point [24]. At the same time, emission of non-burning gases like water vapour or carbon dioxide dilutes the level of burning volatiles and oxygen in the combustion area . Although effective, such mechanisms tend to be associated with high additive loadings that could be detrimental to mechanical properties or processability of the polymer [25].

Materials Research Foundations 190 (2026) 50-76 https://doi.org/10.21741/9781644904077-3

Table 2. *Classification of flame retardants mechanisms*

Mechanism	Mode of action	Phase	Example additives	Characteristics
Physical	Endothermic decomposition and dilution of flammable gases	Condensed phase	$Al(OH)_3$, $Mg(OH)_2$	Absorbs heat, releases water vapor/CO_2; requires high loading
Chemical (gas phase)	Radical scavenging and flame inhibition	Gas phase	Halogenated FRs, organophosphorus compounds	Interrupts H/OH chain reactions; reduces flame propagation
Chemical (condensed phase)	Catalytic char formation and barrier creation	Condensed phase	Phosphorous and nitrogen based FRs, biomass derived FRs	Forms stable carbonaceous layer; reduced volatile release
Intumescent system	Expansion and insulation via acid char, and blowing agent	Both	Ammonium polyphosphate + pentaerythritol + melamine	Produces foamed, protective char layer that insulated polymer

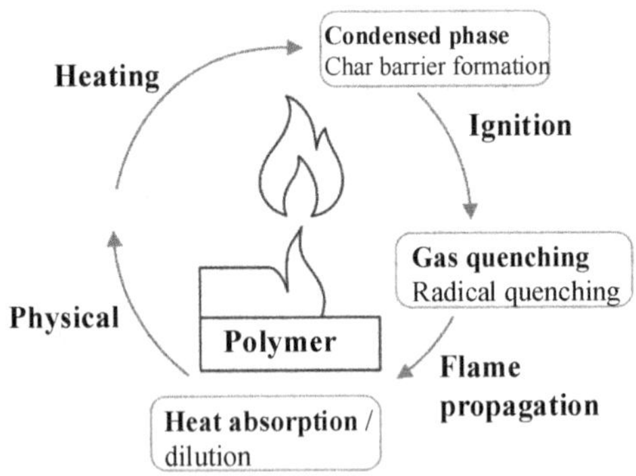

Figure 1. *Combustion cycle and flame retardant intervention points*

Chemical flame retardancy is produced through inhibition of the radical chain reactions that propagate combustion. Flame spreading in the gas phase is fuelled by free radicals H•, OH•, and O•, which are produced from the volatiles of hydrocarbons. Halogenated flame retardants, for instance, suppress combustion by emitting halogen radicals that react with the active hydrogen or hydroxyl radicals and extinguish the flame. As a result of environmental issues, however,

Materials Research Foundations 190 (2026) 50-76
https://doi.org/10.21741/9781644904077-3

halogenated systems are being substituted with halogen-free systems more and more, most of which work in the condensed phase [26].

Condensed-phase flame retardation is based on the enhancement of char formation on the polymer surface. The creation of a thermally stable carbonaceous layer also provides a physical barrier that protects the underlying material from heat and oxygen and decreases volatile degradation products' release. Intumescent flame retardants are a well-explored type of condensed-phase system that works by a synergistic effect among three ingredients: an acid donor (e.g., phosphoric acid), a char producer (e.g., polyols or carbohydrates), and a blowing agent (e.g., melamine or ammonia-generating compounds). When heated, the acid triggers dehydration reactions on the carbon source, resulting in the production of a carbonous, viscous char. At the same time, the blowing agent gas decomposes and emits gases that make the char expand to create an insulating foam-like blanket that acts to protect the material from heat and oxygen efficiently [27].

Biomass-derived flame retardants naturally predispose themselves towards condensed-phase action because of their chemistry and thermal degradation characteristics. Lignin, being a multifunctional polyphenolic architecture, forms thermally induced cross-links and aromatic condensation upon pyrolysis, resulting in a hard and protective char. Phytic acid, with its high phosphorus content, is both a source of acid and a charring agent, facilitating dehydration of nearby biomass components and increasing the structural strength of the carbon layer formed. Chitosan and starch derivatives, which are high in hydroxyl and amino group contents, also facilitate carbonization and intumescence when used with other constituents. In addition, the occurrence of heteroatoms like nitrogen, phosphorus, and sulphur in the derivatives of modified biomass improves radical scavenging in the gas phase as well as catalytic charring in the condensed phase [28].

Fire performance of flame retardants is tested through different fire test procedures, each measuring different features of fire behaviour. Limiting Oxygen Index (LOI), represented by the lowest concentration of oxygen that supports combustion, is a common measure. Materials with an LOI in excess of 26% tend to be self-extinguishing. The UL-94 tests for vertical and horizontal burning present qualitative ratings related to flame spread and dripping tendencies. Cone calorimetry, a more quantitative technique, involves measurement of heat release rate (HRR), total heat release (THR), smoke generation, and time to ignition under conditions of controlled irradiation, yielding a detailed description of fire performance. Thermal gravimetric analysis (TGA) and differential scanning calorimetry (DSC) are used to determine thermal degradation kinetics, activation energy, and residue formation, which are important to determine the flame retardant mechanism at the molecular level [29].

Overall, flame retardancy is an interdisciplinary science combining heat transfer, combustion chemistry, materials science, and polymer physics. Successful flame retardants have to interfere with the combustion cycle by heat absorption, radical quenching, or protective barrier formation. Biomass-based flame retardants, which possess high carbon content, diverse functional groups, and natural occurrence of phosphorus- and nitrogen-containing moieties, provide attractive mechanisms in both condensed and gas phases. An intimate understanding of these basic principles is crucial for the rational design and optimization of bio-based systems capable of fulfilling the rigorous fire safety requirements in contemporary materials engineering.

Materials Research Forum LLC
https://doi.org/10.21741/9781644904077-3

3. Biomass Sources for Flame Retardants

The utilization of biomass as a flame retardant precursor is founded on the abundance, renewability, and chemical functionality of various naturally derived macromolecules. These biopolymers, obtained from plant, marine, or microbial origins, possess inherent structural features such as high aromaticity, nitrogen or phosphorus content, and dense hydroxyl group networks that make them suitable for application as flame retardant agents. The most prominent categories of biomass for such applications include lignin, cellulose and hemicellulose, chitosan, and phytic acid [30]. Each of these compounds exhibits distinct decomposition behaviour and flame-retardant action and can be employed as such or chemically modified to enhance their compatibility with polymer matrices and overall efficacy.

Table 3. *Comparison of major biomass sources used in flame retardant development*

Biomass Source	Elements	Key features	Decomposition / change formation	Mechanism
Lignin	C, H, O	Aromatic phenolic macromolecule	Dehydrogenation ➔aromatic char	Condensed phase char formation, radical trapping
Cellulose	C, H, O	Linear β-(1-4)-linked D-glucose polymer	Dehydrogenation ➔levoglucosan + char	Intumescence, dehydration catalysis
Hemicellulose	C, H, O	Branched polysaccharide with mixed sugars	Early decomposition ➔volatiles + char	Acid catalysed char
Chitosan	C, H, O, N	β-(1-4)- linked D-glucosamine	Crosslinking and N_2 release	Gas phased radical inhibition + condensed phase char
Phytic acid	C, H, O, P	Inositol ring with six phosphate esters	Dehydration and foamed char formation	Acid source +has phase P-radical quenching
Starch/ Tannins/ Proteins	C, H, O, N (variable)	Hydroxyl-rich / aromatic / peptide-based	Dehydration and aromatic crosslinking	Carbon source + synergistic flame retardancy

Lignin, the second most abundant natural polymer after cellulose, is an amorphous, highly branched, and heterogeneous phenolic biopolymer found in the secondary cell walls of vascular plants. Structurally, lignin is composed of methoxylated phenylpropanoid units derived from coniferyl, sinapyl, and p-coumaryl alcohols [31]. Its highly aromatic structure imparts excellent

Materials Research Foundations 190 (2026) 50-76 https://doi.org/10.21741/9781644904077-3

thermal stability and char-forming ability, making lignin a valuable carbon source in condensed-phase flame retardant systems. During thermal decomposition, lignin undergoes dehydrogenation, crosslinking, and aromatization, leading to the formation of a thermally stable and insulating char layer. This char not only acts as a physical barrier to oxygen and heat but also minimizes the evolution of combustible volatiles [32]. Moreover, lignin can be chemically modified through phosphorylation, sulfonation, or incorporation of nitrogen-containing groups to improve its flame retardant efficiency and to introduce gas-phase flame inhibition mechanisms. Phosphorylated lignin derivatives, for instance, have demonstrated enhanced intumescence behaviour and increased char yield in polypropylene and epoxy composites [33]. Due to its availability as an industrial by-product from paper and biorefinery operations, lignin also offers a cost-effective and sustainable route for large-scale flame retardant production.

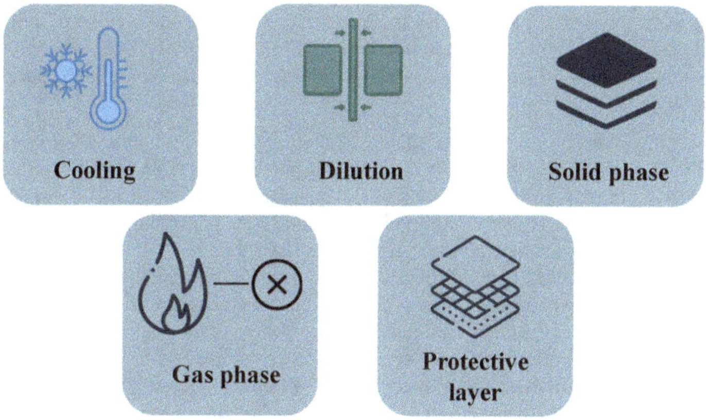

Figure 2. *Mechanisms of Action in Flame Retardancy*

Cellulose and hemicellulose, both polysaccharides derived from plant biomass, are widely investigated as char-forming agents due to their high carbon and hydroxyl content. Cellulose, a linear polymer of β-(1→4)-linked D-glucose units, exhibits high crystallinity and thermal stability, with a degradation onset typically above 300°C. Upon heating, cellulose undergoes dehydration and depolymerization, leading to the formation of levoglucosan and other volatile compounds as well as residual char [34]. Hemicellulose, in contrast, is amorphous and branched, composed of various sugar monomers including xylose, arabinose, and mannose. It exhibits a lower thermal degradation threshold and produces more char and volatiles at lower temperatures compared to cellulose. The inclusion of both cellulose and hemicellulose in flame retardant systems facilitates the formation of an intumescent char layer when combined with acid sources and blowing agents [35]. Moreover, chemical modification of cellulose via phosphorylation or grafting with nitrogenous functionalities has led to improved flame retardant activity, often allowing such materials to function as both the carbon source and acid donor in intumescent systems. Microcrystalline cellulose, cellulose nanofibers, and cellulose nanocrystals are also being explored for their ability to act as nano-reinforcing agents that contribute mechanical strength and thermal shielding properties in flame-retardant polymer composites [34].

Chitosan, derived from the deacetylation of chitin obtained from crustacean shells, is a nitrogen-containing polysaccharide consisting of β-(1→4)-linked D-glucosamine and N-acetyl-D-glucosamine units. It is unique among biomass-based flame retardants due to its intrinsic nitrogen content, which allows for dual-phase flame inhibition [36]. In the condensed phase, chitosan undergoes dehydration and crosslinking to form a cohesive char structure. In the gas phase, the release of nitrogen-containing volatiles interferes with the flame's radical chain reactions, reducing the concentration of H• and OH• radicals and thereby suppressing combustion. Chitosan also possesses functional groups primary amines and hydroxyls that enable facile chemical modification [37]. Phosphorylation of chitosan enhances its acid-releasing ability and char yield, making it a multifunctional intumescent component. The use of chitosan in polylactic acid (PLA), polyamide, and polyurethane matrices has demonstrated improved flame retardancy, as evidenced by reduced peak heat release rate (PHRR), total heat release (THR), and increased limiting oxygen index (LOI) [37]. Furthermore, the biocompatibility and antimicrobial properties of chitosan offer added value in applications such as flame-retardant coatings for medical textiles, packaging, and interior furnishings.

Phytic acid, or myo-inositol hexakisphosphate, is another promising biomass-derived compound that has gained considerable attention for its high phosphorus content and intumescent action. It is primarily sourced from plant seeds and cereals and constitutes a major form of phosphorus storage in plants. Structurally, phytic acid contains six phosphate ester groups attached to an inositol ring, providing both high acidity and thermal reactivity. Upon heating, phytic acid decomposes to release phosphoric acid and polyphosphoric acid species, which catalyse the dehydration and charring of adjacent polymers or biomass components. This action results in the formation of a dense, expanded char that insulates the material and prevents flame propagation [38]. In addition to its role as an acid donor, phytic acid contributes to gas-phase radical quenching by releasing phosphorus-containing volatiles that scavenge reactive flame intermediates. Phytic acid has been used directly as a flame retardant or as a reactive intermediate in the synthesis of bio-based flame retardant resins, including phosphorylated epoxies, polyesters, and polyurethanes [39]. When incorporated into composite systems, it has shown excellent synergistic effects with other bio fillers, enhancing both thermal stability and mechanical integrity.

Other emerging biomass-derived sources include starch, tannins, proteins such as casein and keratin, and agricultural residues like peanut shells, bagasse, and rice husks. These materials, while less studied, offer valuable structural motifs for carbonization and flame inhibition. Starch, a polysaccharide composed of amylose and amylopectin, undergoes thermal degradation to yield char and levoglucosan and is often used in combination with phosphorus compounds in intumescent formulations [39]. Tannins, polyphenolic compounds found in tree bark and plant exudates, contribute aromatic content and free radical scavenging capacity, aiding in condensed- and gas-phase flame inhibition [39]. Proteins, due to their peptide bonds and side-chain functionalities, can be functionalized with phosphorus- or nitrogen-containing groups to enhance thermal decomposition pathways favourable for char formation [40]. Agricultural residues, depending on their lignocellulosic composition, can be processed into biochar or used as reinforcing fillers in polymer matrices, contributing both structural and flame retardant benefits [41].

Materials Research Forum LLC
https://doi.org/10.21741/9781644904077-3

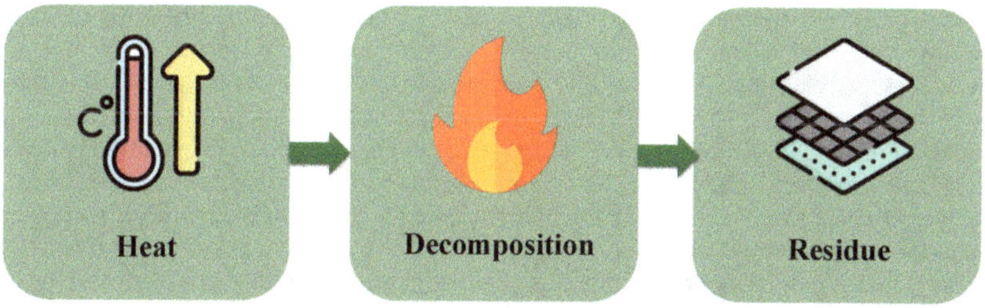

Figure 3. *Thermal degradation of polymers*

The diversity of biomass sources provides a rich platform for developing sustainable flame retardant systems tailored to specific material applications. Through careful selection, modification, and integration, these naturally occurring compounds can be engineered to deliver performance metrics comparable to or in some cases exceeding those of traditional synthetic additives. The continued exploration of these bio-based materials, supported by advances in characterization, functionalization, and processing technologies, is essential for realizing the full potential of biomass as a cornerstone of green flame retardant innovation.

4. Synthesis and Functionalization of Biomass-Based Flame Retardants

The straightforward integration of the original biomass into flame retardants frequently results in limited performance because of lack of thermal stability, hydrophilicity, and insufficient compatibility with polymer matrices. Therefore, the chemical functionalization and modification of the bio-derived molecules are essential for changing them into high-performance flame retardants. These changes are aimed at introducing or reinforcing critical functionalities like phosphorus or nitrogen content, char-forming ability, thermal degradation behaviour, and dispersibility in hydrophobic media [42]. Functionalization methods are categorized as phosphorylation, nitrogen incorporation, sulfonation, esterification, cross-linking, and the formation of hybrid and nanocomposite systems. By using these methods, it becomes feasible to control the pathways of degradation of the precursors in the biomass so as to encourage condensed-phase carbonization, inhibition of gas-phase radicals, or intumescence [41].

Table 4. *Summary of functionalization methods and their outcomes*

Modification method	Biomass	Introduced Functionality	Effect of Flame retardancy
Phosphorylation	Lignin, cellulose	P-O-C groups	Higher char yield
Nitrogen incorporation	Chitosan	Amine groups	Gas-phase radical trapping
Sulfonation	Lignin	$-SO_3H$ groups	Enhanced dehydration catalysis
Esterification	Starch	Hydrophobicity, crosslinking	Improved dispersion

One of the most common modification methods used in flame retardant synthesis from biomass is phosphorylation, mainly because phosphorus has a key role to play in radical quenching and condensed-phase charring. This process usually involves the introduction of phosphate or phosphonate groups on to the backbone of the biomass through esterification or condensation reactions. Phosphorylation, for instance, of lignin or cellulose may be achieved through reaction with phosphoric acid, phosphorus oxychloride, or cyclic phosphates like 2-chloro-1,3,2-dioxaphospholane. The reaction goes through nucleophilic attack of the hydroxyl groups on the phosphorus centres and the formation of phosphate esters covalently linked to the biomass [43]. Phosphorylated lignin has more thermal degradation patterns, resulting in higher char yield and higher ignition resistance. Additionally, thermally generated acidity during the decomposition of phosphate groups acts as a catalyst for dehydration processes, enhancing the creation of thermally stable aromatic carbon skeletons. Phytic acid, which is naturally phosphorylated, may be chemically attached to polysaccharides or employed as a reactive monomer in the production of phosphorus-rich polyesters and polyurethanes [44]. This functionalization allows these materials to act proactively in intumescent fire-retarder systems, where they are both acid sources and carbon-producing agents.

Nitrogen functionalization is another viable method to increase the fire-retarder effectiveness of biomass. Nitrogen groups introduce gas-phase activity by releasing inert gases or radical-scavenging gases like ammonia, nitrogen, or nitric oxides upon heat decomposition. Chitosan, having primary amines along its polysaccharide backbone, may be crosslinked with urea, melamine, or cyanuric acid derivatives to enhance nitrogen content and form thermally labile bonds that endothermically decompose [45]. Further, amino alkylation of hydroxyl-abundant biomolecules like cellulose or starch through reaction with amines or amino acids gives nitrogen-abundant frameworks that are responsible for char formation as well as flame inhibition. With phosphorus groups, nitrogen functionalities have been found to generate synergistic effects, increasing flame retardancy through condensed and gas-phase processes [46]. This has been seen in cases where lignin is functionalized with ammonium polyphosphate and urea, leading to enhanced thermal stability and peak heat release rate suppression in polyolefin composites.

Sulfonation, although less prevalent than phosphorylation, offers a further mechanism for increasing the flame retarding capability of biomass by incorporating sulphur-containing groups that are capable of inducing charring and inhibiting combustion. Sulfonated lignin, obtained by reaction with sulfuric acid or chlorosulfonic acid, has increased solubility, acidity, and catalytic dehydration activity [47]. When heated, sulfonated materials can emit sulphur oxides that function as scavengers for oxygen and prevent the flame. While sulphur-based flame retardants have gone out of Favor in certain applications because of toxicity issues, when utilized in low levels or blended with nitrogen and phosphorus functionalities, they can still provide performance advantages [48].

Esterification and etherification reactions are also utilized to alter the physicochemical properties of biomass for flame retardant applications. Such reactions are generally carried out by reacting hydroxyl-bearing biopolymers like starch or cellulose with anhydrides, acid chlorides, or epoxides to introduce hydrophobicity, improve thermal resistance, or facilitate cross-linking [49]. For instance, maleic anhydride grafting on lignin or cellulose is reported to enhance the tendency towards carbonization and dispersion in polymer matrices. Epoxide-functionalized phytic acid can also be reacted with chitosan to produce crosslinked flame-retardant networks with improved water resistance and char formation ability [49]. These chemical modifications also open the door to the design of reactive flame retardant systems that covalently bond into the host polymer matrix, limiting additive migration and enhancing mechanical integrity.

In addition to molecular-level functionalization, much effort has focused on the use of bio-based hybrid and nanocomposite flame retardants. These include composite materials that integrate biomass-derived elements with inorganic nanostructures like layered double hydroxides (LDHs), montmorillonite clays, graphene oxide, and metal–organic frameworks (MOFs). The addition of such nanomaterials can increase barrier performance, thermal conductivity, and reinforcement in structure. For example, lignin–clay nanohybrids have been found to synergistically enhance char strength and decrease the total heat release of polypropylene composites. Likewise, phosphorus group-functionalized cellulose nanocrystals (CNCs) have been found to exhibit enhanced dispersion in polymer matrices and enhanced flame retardant performance at low loadings. The nanoscale size of such additives enhances interfacial area and facilitates the creation of tortuous pathways that impede oxygen and volatile degradation product diffusion.

Reactive extrusion is being adopted with greater frequency as a processing technique for incorporating functionalized biomass into polymeric systems via melt processing. This solventless process facilitates in situ chemical reactions between modified biomass and thermoplastics to allow grafting or copolymerization [50]. Reactive extrusion has been employed to incorporate phosphorylated starch into polylactic acid, forming bio-based composites with improved flame resistance, lowered flammability, and preserved mechanical properties. Other low-temperature or solventless synthesis paths like ball milling, microwave-assisted synthesis, and supercritical fluid processing are also being explored to satisfy industrial scalability and environmental safety requirements [51].

It should be noted that functionalization should be performed with an optimal balance of reactivity, thermal stability, and environmental compatibility. Over-functionalization can impair the structural integrity of the biomass, lower biodegradability, or deposit toxic residues. Thus, synthesis of biomass-derived flame retardants should adhere to the principles of green chemistry such as the utilization of non-toxic reagents, minimal usage of solvents, and gentle reaction conditions. Life-cycle analysis and environmental impact assessments are necessary to confirm the sustainability assertions of these materials.

The synthesis and functionalization of biomass-derived flame retardants involve a wide range of chemical and processing strategies for achieving optimal fire resistance with environmental stewardship. Through the incorporation of functional groups like phosphate, amine, sulfonate, or ester linkages and their integration into hybrid nanocomposites, it is possible to design multifunctional flame retardant systems with customized performance features. Such progress is essential for overcoming the built-in deficiencies of unmodified biomass and for allowing them to be used in demanding industrial and safety-critical situations.

5. Applications in Polymeric Systems

The addition of biomass-derived flame retardants to polymeric matrices has grown as a renewable and efficient way to promote fire resistance while avoiding environmental toxicity. The use of these flame retardants is rapidly being utilized in a variety of polymer matrices such as thermoplastics, thermosets, elastomers, and textile fibres. The choice and performance of a flame retardant system are significantly influenced by the thermal properties of the host polymer, its

Materials Research Forum LLC
https://doi.org/10.21741/9781644904077-3

decomposition pattern, and the additive-polymer processing compatibility. Biomass-derived materials, when properly modified or blended with synergistic agents, can efficiently decrease flammability, inhibit smoke and toxic gas emission, and preserve or even enhance mechanical integrity. This section discusses the use of such systems in prominent polymer families, highlighting the interaction between bio-based flame retardants and the polymer matrix, the mechanisms of flame suppression, and related performance parameters.

In thermoplastics like polypropylene (PP), polyethylene (PE), polystyrene (PS), and poly(lactic acid) (PLA), biomass-based flame retardants are used to combat the inherent flammability of these materials. Thermoplastics melt and flow when heated, which can aid dripping and flame spread. Incorporation of biomass-derived flame retardants, especially those with high char-forming capacity and intumescent properties, is essential to controlling this behaviour. Lignin, for instance, has been extensively applied in polypropylene composites as a source of carbonaceous char. When it is adapted with phosphorus-containing additives, lignin not only improves the development of an ordered, bloated char matrix upon burning but also assists with gas-phase radical quenching. In those formulations, lignin is normally blended with APP and pentaerythritol or other polyols in order to design a completely bio-based intumescent system. These systems show dramatic decreases in peak heat release rate (PHRR), total heat release (THR), and time to ignition (TTI), with acceptable mechanical properties like tensile strength and elongation at break. PLA, a biodegradable thermoplastic, is especially suitable for bio-based flame retardant systems because of its chemical similarity with most biomass-derived additives. Phytic acid and phosphorylated starch have been added to PLA matrices by melt compounding and in situ reactive extrusion, resulting in composites with enhanced limiting oxygen index (LOI), UL-94 V0 rating, and smoke reduction [51].

In thermosetting resins like epoxy, unsaturated polyester, and phenolic resins, the flame retardancy challenge is compounded by their high crosslink density and exothermic curing behaviour, which can enhance combustion. Biomass-derived flame retardants for such systems need to be reactive or compatible with the resin curing chemistry in order to avoid phase separation and ensure homogeneity. Lignin has been effectively grafted with reactive epoxy groups or blended into bio-based epoxy formulations to act as both a carbon source and a structural component. Phytic acid and its derivatives have been used as reactive flame retardant hardeners, integrating with the epoxy matrix through phosphate ester bonds. Such covalent integrations reduce additive migration and assist in thermal stability in the long term. In epoxy systems, these bio-based formulations have shown noteworthy char residue enhancements, decrease in average heat release rates, and improved resistance against dripping and ignition. In phenolic resins, already fire-resistant by nature, incorporation of modified lignin or tannin derivatives has added to char stability and minimized volatile emission upon combustion.

Flexible and rigid polyurethane foams are two of the most combustible polymeric materials owing to their large surface area, open-cell structure (in flexible foams), and low thermal conductivity. Such systems usually need both gas-phase and condensed-phase mechanisms of flame retardancy. Flame retardants derived from biomass have also been used to meet this double requirement, such as chitosan, lignin, and phosphorylated saccharides. For example, chitosan nanoparticles have been used in water-blown polyurethane foams, where they serve as a char former and an endothermic decomposer. Upon phosphorus or nitrogen-rich group functionalization, these materials enhance the development of an insulating char layer and the emission of non-flammable gases that dilute the flame atmosphere. Synergism between chitosan and expandable graphite or intumescent agents has been found to decrease flammability and enhance smoke suppression. In insulation applications of rigid foams, the incorporation of phosphorylated lignin or phytic acid flame retardants enhances thermal stability and decreases peak heat release without having a significant influence on thermal conductivity or compressive strength.

In fibre applications like textiles, nonwovens, and paper, flame retardancy should be obtained with minimal effect on fabric hand, colour, flexibility, and breathability. Bio-based flame retardant treatments and coatings provide a green alternative for conferring flame-resistant durability. Phytic acid-based flame retardants have found extensive application in textile finishing as part of sol–gel formulations or through crosslinking with citric acid, urea, or cellulose derivatives. The treatments deposit stable phosphate networks on the fibre surface, which catalyse char formation and minimize after flame and afterglow durations in routine textile flammability tests. Chitosan and its derivatives have also been employed as coating agents owing to their film-forming capacity, biocompatibility, and reactivity with natural and synthetic fibres. In cotton fabrics, chitosan-phytic acid complexes have shown lasting flame retardant properties even after repeated cycles of washing. In synthetic fibres like nylon or polyester, which are more stable to aqueous treatments, flame retardant treatment usually includes the use of bio-based polyurethane binders or reactive extrusion of phosphorus-containing additives into the fibre itself during spinning.

Hybrid materials and composites provide another significant field of application of biomass-based flame retardants. In wood–plastic composites (WPCs), plastics filled with agricultural residues, and bio-based structural composites, the addition of lignocellulosic fillers naturally enhances char yield, but other flame retardant treatments are still necessary to satisfy fire safety requirements. In such systems, mixtures of phosphorylated lignin, phytic acid, and nano-reinforcements like modified montmorillonite or layered double hydroxides have been used to attain synergistic flame quenching. These hybrids enhance the thermal barrier characteristics and slow down the heat and mass transfer at combustion. In addition, functionalized biomass application in epoxy glass fibre composites and carbon-fibre reinforced plastics has demonstrated that flame retardant functionality can be incorporated without fatally weakening structural performance or thermal properties.

The effectiveness of these compounds is usually measured by standardized tests for flame retardancy like Limiting Oxygen Index (LOI), UL-94 vertical and horizontal burning tests, cone calorimetry, thermogravimetric analysis (TGA), and scanning electron microscopy (SEM) of char morphology. In the majority of instances, biomass-derived flame retardants make a significant contribution to char yield and decrease heat release rates, although issues exist in realizing reproducible performance over a wide variety of polymer systems. Compatibility, processing conditions, additive migration, and moisture sensitivity are vital factors that have to be handled through proper formulation and surface engineering.

In summary, biomass-derived flame retardants have shown high flexibility and efficiency in a broad spectrum of polymeric applications. Their potential to function through multiple mechanisms—such as intumescence, radical trapping, and char residue formation—renders them equally applicable for bulk and surface treatments. Future development based on chemical modification, nano-hybrid structures, and fabrication techniques is predicted to further increase their utility in high-performance materials, making the dream of sustainable, fire-resistant polymers an industrial reality.

6. Performance Evaluation and Mechanistic Studies

Design and optimization of flame retardants from biomass necessitate stringent performance testing to confirm their effectiveness and decipher the mechanisms of flame inhibition. A thorough evaluation of flame retardants involves both macroscopic fire tests under regulated fire conditions and microscopic examinations that uncover structural and chemical changes through thermal degradation. The effectiveness of a flame retardant system is usually measured by a number of complementary techniques, such as Limiting Oxygen Index (LOI), UL-94 vertical and horizontal burning tests, cone calorimetry, thermogravimetric analysis (TGA), Fourier-transform infrared spectroscopy (FTIR), and scanning electron microscopy (SEM). Each technique offers important information on flammability behaviour, thermal stability, char structure, gas-phase interactions, and condensed-phase transformations.

The Limiting Oxygen Index (LOI) test is usually the first screening test for determining flammability. It determines the minimum oxygen concentration in some oxygen–nitrogen mixture that is needed to support combustion of a vertically oriented polymer test sample. A material with an LOI greater than 26% is typically considered self-extinguishing. Biomass-derived flame retardants tend to enhance the LOI of polymer composites through the encouragement of condensed-phase char formation, which serves as an insulating barrier against heat. For instance, adding phosphorylated lignin or chitosan to polypropylene or PLA matrices has regularly yielded LOI values above 28%, reflecting higher levels of resistance against continued burning. The test offers a simple comparison between neat polymers and flame-retardant formulations but does not reflect dripping behaviour or flame propagation dynamics under actual fire conditions.

As a complement to LOI measurements, the UL-94 flammability test assesses burning behaviour of materials under vertical and horizontal orientations. The test classifies materials in terms of after-flame time, dripping, and capacity to extinguish following ignition. Biomass-derived formulations containing intumescent systems or crosslinked phosphate networks often meet V-0 ratings, the highest in the UL-94 vertical burn test. Cotton fabrics treated with phytic acid–chitosan coatings, for example, show fast self-extinction and char development, which restrict dripping and after flame. Such qualitative ratings, although helpful for regulatory purposes and product marking, need to be supplemented by more quantitative information to guide material design and process development.

Cone calorimetry is accepted as the most quantitative and all-encompassing technique for assessing flame retardant performance under simulated fire-like conditions. The method exposes a sample of material to a controlled external heat flux and measures parameters like heat release rate (HRR), peak heat release rate (PHRR), total heat release (THR), time to ignition (TTI), smoke production rate (SPR), and mass loss rate. Biomass-based flame retardant systems work by lowering the PHRR and THR values through improved char layer formation and reduced volatile combustible production. For instance, PLA composites with phosphorylated starch or lignin have shown PHRR reductions of more than 40% relative to the neat polymer. Char layer formation slows the diffusion of oxygen and protects the original polymer from heat, hence decreasing the rate of pyrolysis and burning. Ignition time can be retarded owing to the flame retarding effect of the char layer, although this is more related to composite thermal diffusivity. Smoke generation, a serious risk in fires, is also minimized in most biomass-based systems, especially when blended with charring agents and synergists stabilizing thermal degradation and inhibiting aromatic volatile evolution.

Materials Research Foundations 190 (2026) 50-76 https://doi.org/10.21741/9781644904077-3

Thermal gravimetric analysis (TGA) is essential to the knowledge of the degradation profile and thermal stability of biomass-based flame retardants. TGA determines the weight loss of a material as a function of temperature under inert or oxidative conditions. Flame retardants derived from biomass have multistage degradation patterns because of the presence of separate organic phases, e.g., polysaccharides, phenolics, and functionalized moieties. For instance, phosphorylated cellulose degrades initially through dehydration and char production, followed by decomposition of phosphate ester functionalities. The presence of flame retardants tends to raise the onset degradation temperature and the residual char at elevated temperatures, both of which are desirable signs of condensed-phase flame retardancy. Derivative thermogravimetric (DTG) curves provide kinetic information on the decomposition process, which can be utilized to model activation energies and reaction pathways through Kissinger or Flynn–Wall–Ozawa methods.

Spectroscopic methods, particularly FTIR and Raman spectroscopy, offer molecular-scale information on the chemical processes that are operative under pyrolysis and combustion. FTIR examination of char residues typically indicates the presence of aromatic C=C bonds, P=O and P–O–C structures, and carbonyl functions, which attest to successful production of thermally stable phosphate and carbonaceous forms. The decrease in aliphatic C–H and hydroxyl groups indicates that dehydration and aromatization reactions have taken place during combustion, catalysed by phosphorus or acid groups introduced by biomass modification. In situ FTIR with TGA (TG-FTIR) can track the evolution of volatile products, showing the emission of ammonia, water, CO_2, and phosphorus-containing volatiles in nitrogen- and phosphorus-modified biomass flame retardants. Raman spectroscopy further defines the carbonaceous char structure by probing the D and G bands that refer to disordered and graphitic carbon domains, respectively. An increased intensity ratio of G/D bands indicates an increased graphitization and thermal stability of the char, which in turn relates to improved flame retardance.

Microscopic analyses like SEM and transmission electron microscopy (TEM) are required for the assessment of char residue morphology and the flame retardant additive dispersion. Surface morphology of intumescent char and its barrier behaviour is ascertained from SEM imaging. The dense, cohesive, and continuous nature of the char layer is favoured for hindering heat and mass transfer during combustion. SEM and TEM in nanocomposite materials can also verify the homogeneous dispersion of nanoparticles derived from biomass (e.g., cellulose nanocrystals, nanoscale modified lignin) in the polymer matrix. Agglomeration, low interfacial adhesion, or phase segregation can compromise flame retardant effectiveness by inducing defects that weaken thermal insulation.

Mechanistic investigations have revealed that biomass-derived flame retardants can act through various mechanisms, frequently exhibiting both condensed-phase and gas-phase activity. In the dense phase, the hydroxyl-rich biopolymers like starch, cellulose, and chitosan are dehydrated with the formation of carbon and a thermally insulating char layer due to dehydration. The presence of phosphorus-containing groups adds acidic catalytic sites for increased enhancement of the process, with further decrease in the evolution of volatile fuel gases. Gas phase nitrogenous functionalities decompose to release ammonia or inert gases, which dilute the flame environment and extinguish free radical reactions. Further, phosphorus volatiles can react with H• and OH• radicals, limiting the availability of these essential intermediates necessary to provide the sustaining action for combustion. The combination of phosphorus and nitrogen, especially, has been found to result in better flame retardant performances by activating thermal shielding and chemical inhibition mechanisms at the same time.

Over the past decade or so, there has been growing use of state-of-the-art characterization techniques including pyrolysis–gas chromatography–mass spectrometry (Py-GC–MS), X-ray photoelectron spectroscopy (XPS), and nuclear magnetic resonance (NMR) spectroscopy to enlarge mechanistic insights. Py-GC–MS detects volatile pyrolysis degradation products that help trace out the pyrolysis route. XPS assists in determining the quantitative elemental composition and oxidation states of phosphorus and nitrogen in residues of char, providing information regarding stability and chemical nature of the flame retardant coating. Solid-state NMR describes the molecular structure of crosslinked biomass-derived networks and variations under thermal processing.

Together, these characterization and evaluation methods offer a multiscale understanding of biomass-derived flame retardant systems. They enable the quantification of performance, the detection of failure mechanisms, and the iterative development of materials with enhanced flame retardant characteristics. With the advancing technology moving toward commercial readiness, standardized test protocols and mechanistic modelling will become necessary to provide reproducibility, predictability, and regulatory acceptability of these green flame retardant technologies.

7. Environmental and Regulatory Considerations

The shift from traditional halogenated flame retardants to biomass alternatives is primarily driven by growing realization of the environmental, health, and safety risks associated with legacy chemicals. Traditionally, flame retardants like polybrominated diphenyl ethers (PBDEs), tetrabromobisphenol A (TBBPA), and hexabromocyclododecane (HBCD) were widely applied because of their effectiveness and broad polymer compatibility. But with time, there has been a large body of evidence that indicates many of these chemicals are persistent in the environment, bio accumulative in wildlife and humans, and potentially toxic to endocrine, neurological, and reproductive systems. Consequently, a number of these compounds have been banned or limited under global treaties like the Stockholm Convention on Persistent Organic Pollutants (POPs), the European Union's REACH (Registration, Evaluation, Authorisation and Restriction of Chemicals) regulation, and the United States' Toxic Substances Control Act (TSCA).

Conversely, biomass-based flame retardants are frequently derived from naturally sourced materials like lignin, cellulose, starch, phytic acid, and chitosan, which are non-toxic, biodegradable, and renewably sourced. These inherent characteristics comparatively diminish the environmental and human health concerns that usually accompany synthetic flame retardants. In contrast to the halogenated compounds that, upon combustion, produce toxic products like dioxins and furans, some biomass-based flame retardants pyrolyze to give innocuous chemicals, e.g., carbon dioxide, water, and phosphate esters. More importantly, also the strong char-forming capacity of bio-based systems can in addition to act as smoke suppression and decrease release of incomplete burning products, lessening toxic release during fire episodes.

However, environmental friendliness cannot be anticipated on the pure basis of naturality. Biomass-derived flame retardants are often chemically modified to improve flame retardant effectiveness, polymer matrix compatibility, and thermal stability. Such modifications usually include reagents like phosphorus oxychloride, chlorosulfonic acid, epoxides, and acid anhydrides, which are toxic or hazardous. Organic solvents, energy-consuming processing, and auxiliary stabilizers or crosslinkers can further complicate the environmental profile. Thus, a complete

 Materials Research Forum LLC
https://doi.org/10.21741/9781644904077-3

environmental evaluation needs to consider not just the source of the raw material but also the whole life cycle, encompassing extraction, functionalization, inclusion in polymers, in-use practices, and end-of-life treatment.

Life cycle assessment (LCA) is an important method for comparing quantitatively the environmental effects of biomass-based and conventional flame retardants. LCA entails cradle-to-grave analysis of environmental parameters like greenhouse gas emissions, water usage, energy requirements, acidifying potential, and ecotoxicity. Early LCA research has shown that biomass flame retardants tend to have lower global warming potential (GWP) and fossil fuel demand than their halogenated and some phosphorus-based synthetic counterparts. For example, phosphorylated lignin and starch-based intumescent systems lowered total GWP by over 40% when applied to polypropylene composites, primarily as a result of the renewable carbon content and lowered dependence on fossil origin chemicals. Nevertheless, the environmental benefit is extremely processing efficiency and purification step- and waste generation during functionalization-dependent. Green chemistry innovations like solvent-free processing, enzymatic phosphorylation, or microwave-assisted synthesis are the key to reducing such environmental costs.

Table 5. *Life cycle comparison of flame retardant systems*

Aspect	Halogenated	Phosphorous based synthetic	Biomass based
Carbon source	Fossil	Fossil	Renewable
Toxic emissions	High	Moderate	Low
Biodegradability	Poor	Moderate	High
Regulatory acceptance	Declining	Conditional	Strongly favoured

Biodegradability is also a feature often linked to biomass-based flame retardants. In theory, natural polymers like lignin, starch, and chitosan undergo microbial degradation in soil or composting conditions. Yet, chemical adjustments—especially additions of aromatic phosphorus or halogen functionality—are known to hinder biodegradability or yield release of transformation products whose ecological toxicity is poor. Hence, biodegradability testing employing standardized protocols, like OECD 301 series methodology or ASTM D5988, must be done to authenticate environment degradability claims. In addition, toxicity tests with aquatic animals (e.g., Daphnia magna, zebrafish embryos, or microalgae) and soil organisms are needed to determine the possible ecological effect of leachates from flame retardants, particularly when they are applied outdoors or in wet environments.

National and international safety and materials regulations control regulatory acceptance and market uptake of biomass-based flame retardants. Within the European Union, flame retardant chemicals and formulations are required to meet the REACH directives, which involve manufacturers and importers proving safety through extensive toxicological and environmental information. The Biocidal Products Regulation (BPR) can also come into play in the event of antimicrobial properties of the flame retardant, such as those found in some chitosan-based systems. In the US, flame retardants are controlled by the Environmental Protection Agency (EPA) under TSCA and the Safer Choice initiative, which promotes the use of low-hazard, sustainable substitutes. Furthermore, industry-specific fire safety regulations—like UL-94, NFPA 701, ASTM E84, and EN 13501-1 need to be met in order for materials to qualify for application in electronics, textiles, construction, or transportation industries. Flame retardants from biomass sources need to satisfy not only these fire performance standards, but also prove non-toxic, stable, and robust through usual aging and weathering tests.

It should be noted that the regulatory landscape of flame retardants is active and more precautionary. The demand for non-toxic, circular, and climate-neutral materials has accelerated policy actions like the European Green Deal, Circular Economy Action Plan, and several state-level halogenated flame retardant bans in the United States. These policy frameworks are having a greater impact on industrial R&D priorities and public procurement habits. Hence, to obtain general acceptance among biomass-based flame retardants, it is imperative to present exhaustive safety dossiers, toxicological information, and environmental performance metrics. Open reporting of formulation ingredients, migration tendencies, and long-term stability will be essential to generate consumer and regulatory confidence.

Overall, biomass-derived flame retardants show considerable promise as eco-friendly alternatives to traditional systems, especially given tightening international regulation and growing demands for sustainability. Nonetheless, the environmental and regulatory situation calls for an integral consideration of the material beyond their renewable origin. Functionalization pathways, processing technologies, degradation by-products, and in-use emissions all need to be taken into account when considering their actual environmental impact. Adoption of life cycle thinking, meeting safety requirements, and conformity to circular economy guidelines will be paramount to making sure that biomass flame retardants are not just functional but also accountable and future-proof ingredients of new fire-safe materials.

8. Challenges and Future Prospects

In spite of the encouraging developments in biomass-based flame retardants, a myriad of technical, economic, and systemic issues remain to limit their commercial uptake into large-scale polymer systems. Even though the bio-sustainability lure is compelling, deployment in practical use necessitates performance equivalence to or overness compared to conventional systems, in addition to being economically viable, compatible with process, and complying with changing fire protection standards. These complex demands place heavy constraints on material synthesis methods, material design, and formulation strategies and call for an interdisciplinary approach to innovation.

A particularly basic problem is the thermal instability and process sensitivity of many biomass-derived substances. Natural polymers like lignin, starch, and chitosan tend to have restricted thermal stability in standard polymer process conditions, such as during melt compounding, extrusion, or injection moulding, where temperature levels can range up to 200–250 °C. Unmodified biomass components would degrade, depolymerize, or release volatile compounds under such conditions, resulting in discoloration, gas evolution, and non-uniform dispersion. Such thermal incompatibility not only restricts the variety of host polymers into which they can be introduced but also impacts the reproducibility and long-term stability of the resultant composites. To meet this, there is an urgent requirement to create thermally stable derivatives via selective cross-linking, chemical grafting, or surface modification methods that maintain flame retardancy while securing molecular integrity at high-temperature processing.

Severely interrelated with this is the concern of incompatibility with hydrophobic polymer matrices, like polyolefins, polystyrene, and some engineering plastics. Biomass-derived flame retardants, which are inherently hydrophilic owing to the presence of excess hydroxyl, amino, or carboxyl groups, tend to exhibit poor interfacial adhesion and dispersion in non-polar polymers. Such incompatibility can cause phase separation, agglomeration, and void formation, all of which degrade mechanical properties and homogeneity of flame retardant distribution. Although compatibilizers such as maleic anhydride-grafted polymers or surfactant-mediated approaches have been used, they tend to add complexity of processing and are not necessarily able to completely eliminate interfacial tension. Future research needs to be directed toward the design of tunable surface-modified biopolymers that exhibit controlled miscibility for a wider range of polymer chemistries without needing to use additional coupling agents.

Table 6. *Key challenges and future strategies*

Challenge	Cause	Proposed strategy	Expected outcome
Thermal instability	Low decomposition temperature	Chemical crosslinking, nanohybrids	Improved processability
Poor compatibility	Hydrophilicity	Surface modifications	Uniform dispersion
High additive loading	Low intrinsic FR efficiency	Synergistic blends	Reduced loading requirement
Standardization	Feedstock variability	ISO protocols, ML modelling	Consistent data

The other challenge is the balance between flame retardancy and material properties. High levels of loading of bio-based additives—commonly needed to provide adequate char yield or intumescence—can negatively impact the mechanical strength, elasticity, transparency, or impact resistance of the host polymer. In contrast to many synthetic flame retardants that are effective at comparatively low loadings (10–15 wt%), biomass-based systems can need 20–40 wt% to provide similar performance, particularly in single-component systems. Such high loadings create processing viscosity, flow behaviour, and dimensional stability problems. In addition, char-forming catalysts also have the potential to plasticize the matrix or lower molecular mobility, and consequently influence tensile and flexural performance. Such constraints drive the need to design synergistic systems, whereby biomass-derived phases are blended with nano-reinforcements, inorganic catalysts, or low-load reactive additives in order to meet the necessary flame retardancy at lower overall loading. Custom hybrid systems combining gas-phase and condensed-phase activity, flame retardancy, and structural reinforcement are a very promising avenue.

Standardization and performance benchmarking are another major impediment to market readiness. Flame retardant performance is usually measured through a range of techniques—LOI, UL-94, cone calorimetry, TGA, and others. In the absence of harmonized testing protocols for bio-based systems, comparison between studies becomes problematic. In addition, differences in biomass feedstock composition based on regional, seasonal, or process variations add variability that can influence batch-to-batch consistency and reproducibility of the results. The variability is particularly significant for industrial lignin whose structural heterogeneity and molecular weight

 Materials Research Forum LLC
https://doi.org/10.21741/9781644904077-3

distribution are a function of the pulping process (kraft, sulphite, organosolv, etc.). The establishment of standardized extraction, purification, and characterization protocols for biomass flame retardants will be critical to allowing industrial scalability and regulatory approval. In addition, predictive modelling frameworks that correlate structural features with flame retardant efficacy—e.g., machine learning algorithms based on large experimental data sets—potentially could enable the rapid design of next-generation bio-retardants.

Cost-effectiveness and supply chain scalability continue to be ongoing issues. Although biomass-derived flame retardants are commonly referred to as cost-effective because they are renewable-based and plentiful, the true picture is that functionalization, purification, and formulation processes can make production costly. Compounds like phosphorus oxychloride, solvents, and cross-linkers are not just pricey but also pose safety and environmental handling issues. Moreover, most synthesis pathways are batch-oriented and not easily amenable for large-scale continuous processing. Scaling up low-cost, green synthetic methodologies—like solvent-free phosphorylation, reactive extrusion, supercritical fluid processing, or microwave-assisted functionalization—will play a crucial role in reducing the cost of production and bringing biomass flame retardants into line with industrial throughput demands. Utilizing agricultural by-products, forestry residue, and food process residues as raw material sources following strategic disposition can further contribute to sustainability while lowering raw input expenditures.

Scientifically, a better mechanistic insight into flame retardancy in biomass systems is critical for innovation with purpose. Although condensed-phase processes like char production and thermal insulation are well characterized, gas-phase processes in bio-based systems are less well understood and also less fully exploited. Volatile species produced by thermal breakdown, their interaction with fire radicals, and their role in flame quenching are active research topics. Sophisticated in situ methods like pyrolysis–GC–MS, TG–FTIR, and synchrotron-based spectroscopies can illuminate such dynamic processes. Additionally, the use of stimuli-responsive or "smart" flame retardants, which respond or reorganize their structure to heat, mechanical stress, or fire stimuli, has not yet been fully investigated in the field of biomass-derived materials. These systems might provide adaptive protection and self-healing capabilities, new avenues for functional fire-safe materials.

Lastly, regulatory integration and public perception will mold the future roadmap for biomass-based flame retardants. As regulations across the world trend towards prohibiting or limiting conventional halogenated systems, demand is increasing for alternatives that are safer to use. Yet, to qualify as alternative products, biomass-based flame retardants not only have to match or surpass fire safety requirements, but also prove long-term stability, low toxicity, minimum additive migration, and compatibility with circular economy strategies. The capacity for recycling or biodegradation of flame retardants, recovery of valuable components like phosphorus, and reduction of toxic emissions during burning or waste disposal will be essential to total lifecycle sustainability. Public acceptance is also becoming dependent on open-labelling, health data reporting, and verifiable environmental advantages, particularly in critical applications like consumer electronics, children's products, and building materials.

In summary, although considerable progress has been made in the production of biomass-based flame retardants, the sector is confronted by intricate and interdependent challenges which demand a collaborative and interdisciplinary response. Convergence of materials science, polymer engineering, surface chemistry, environmental science, and regulatory policy will be critical to

Materials Research Forum LLC
https://doi.org/10.21741/9781644904077-3

address existing limitations and realize the complete potential of bio-based flame retardant technologies. In the future, the next generation of biomass-based flame retardants will most probably be multi-functional, molecularly precise, and integrated at a system level, making it possible to design safer, smarter, and greener fire-resistant products for a wide range of applications.

Conclusion

The advent of biomass-derived flame retardants represents a watershed moment in the evolution of eco-friendly fire-resistant materials. In answer to mounting regulatory pressure, environmental concerns, and consumer calls for more environmentally friendly alternatives to traditional halogenated systems, biomass-derived compounds present a powerful combination of sustainability, functionality, and tunability. Borrowed from lignin, cellulose, starch, chitosan, phytic acid, and other renewable natural biopolymers, these compounds confer both condensed-phase and gas-phase flame retardancy through modes of action such as char promotion, catalytic dehydration, radical quenching, and intumescence. Their chemical amenability per se—expressed by numerous hydroxyl, amino, and phosphate groups—allows for numerous chemical modifications to customize their performance in diverse polymer matrices.

Over the past decade, research into biomass-based flame retardants has evolved from a conceptual exercise in green chemistry to a materials design discipline grounded in structure–property relationships and application-oriented engineering. Advances in functionalization techniques have enabled the precise tuning of thermal stability, hydrophobicity, and interfacial adhesion, which are critical for compatibility with thermoplastics, thermosets, elastomers, and fibre-based materials. The incorporation of nitrogen and phosphorus moieties into biomass frameworks improved their flame retardant efficacy, allowing them to achieve or surpass industry standards like UL-94 V-0 ratings and high limiting oxygen index (LOI) requirements. These advances have been supplemented by the integration of biomass flame retardants into hybrid formulations with inorganic nanofillers or reactive polymers, broadening their functionality to encompass not only fire resistance but also mechanical reinforcement, barrier improvement, and lower smoke toxicity.

In spite of this advancement, there are still challenges that moderate the shift from laboratory-scale innovation to industrial application. These are thermal degradation of raw biomass during polymer processing, poor dispersion in hydrophobic matrices, and high additive loadings necessary to attain regulatory compliance. In addition, whereas bio-derived materials are usually assumed to be environmentally friendly, chemical modification routes, processing solvents, and end-of-life emissions need to be critically evaluated through life cycle analysis and ecotoxicological testing to confirm their sustainability credentials. Standardization of raw material supply, performance assessment, and regulatory compliance is still patchy, requiring mutually coordinated action from the academic, industrial, and policy-making communities.

However, future opportunities for biomass-based flame retardants are significant. Green functionalization technologies, including solvent-free phosphorylation, enzyme-based modification, and microwave-assisted processing, provide avenues towards cheaper, scalable, and greener production. The use of high-throughput characterization and computational screening technologies—especially machine learning and predictive modelling-based tools—can speed up material discovery and optimization. Additionally, the merging of flame retardancy with the principles of the circular economy has opened up new possibilities for combining bio-based flame retardants into closed-loop systems, where products are designed for disassembly, recycling, or biodegradation. Here, the recycling of key elements like phosphorus from bio-flame-retardant wastes could be central to future materials circularity approaches.

Moving forward, it is envisioned that next-generation flame retardants will be multi-functional, modular in design, and environmentally friendly right from the start. Materials from biomass are well suited to be the platform for these systems owing to their renewability, intrinsic reactivity, and chemical diversity. With advances in the field, future research will increasingly not only target enhancing fire performance but also on allowing integration with additive manufacturing, responsive materials, and digital sensing platforms. Through this, flame retardant systems will shift from being passive protective coatings to active ingredients of smart, sustainable, and high-performance materials architecture.

In summary, biomass-derived flame retardants are not merely a safer substitute for halogenated compounds—they reflect a paradigm shift in the design, conception, and application of flame retardancy in the 21st century. By combining green chemistry, polymer science, and fire engineering, these systems can potentially characterize the next generation of fire-safe materials that fulfil both functional performance requirements and planetary boundaries. The ongoing commitment to scientific rigor, interdisciplinary synthesis, and life cycle thinking will be necessary to achieve this vision globally.

Reference

[1] S. Lambert, M. Wagner, Environmental performance of bio-based and biodegradable plastics: The road ahead, Chem. Soc. Rev. 46 (2017) 6855-6871. https://doi.org/10.1039/C7CS00149E

[2] M.A. Dubé, S. Salehpour, Applying the principles of green chemistry to polymer production technology, Macromol. React. Eng. 8 (2014) 7-28. https://doi.org/10.1002/mren.201300103

[3] C.A. Wilkie, A.B. Morgan, Fire Retardancy of Polymeric Materials, CRC Press, Boca Raton, 2024. https://doi.org/10.1201/9781003380689

[4] S. Levchik, Phosphorus-Based Flame Retardants, in: Non-Halogenated Flame Retard. Handb. 2 Nd Ed., Wiley, 2021: pp. 23-99. https://doi.org/10.1002/9781119752240.ch2

[5] J. de Boer, S. Harrad, M. Sharkey, The european regulatory strategy for flame retardants - the right direction but still a risk of getting lost, Chemosphere 347 (2024) 140638. https://doi.org/10.1016/j.chemosphere.2023.140638

[6] Y. Liang, H. Jian, C. Deng, J. Xu, Y. Liu, H. Park, M. Wen, Y. Sun, Research and application of biomass-based wood flame retardants: A review, Polymers (Basel). 15 (2023) 950. https://doi.org/10.3390/polym15040950

[7] Y. Xiao, X. Mu, S. Chen, G. Jiang, Z. Xu, C. Ma, L. Song, Y. Hu, Biomass-derived polyphosphazene toward simultaneously enhancing the flame retardancy and mechanical properties of epoxy resins, Chemosphere 311 (2023) 137058. https://doi.org/10.1016/j.chemosphere.2022.137058

[8] X. Zheng, Y. Tao, L. Jin, Z. Zheng, J. Sun, Y. Sun, J. Du, X. Zheng, S. Wu, Y. Pang, Z. Shen, P. Du, H. Chen, Flame retardant mechanism-inspired fabrication of all-biomass-derived

porous carbons for zinc-ion hybrid supercapacitors, Ind. Crops Prod. 218 (2024) 119034. https://doi.org/10.1016/j.indcrop.2024.119034

[9] M. Li, P. M.N., J. Song, Effect of synthesized lignin-based flame retardant liquid on the flame retardancy and mechanical properties of cotton textiles, Ind. Crops Prod. 212 (2024) 118283. https://doi.org/10.1016/j.indcrop.2024.118283

[10] L. Li, Z. Chen, J. Lu, M. Wei, Y. Huang, P. Jiang, Combustion behavior and thermal degradation properties of wood impregnated with intumescent biomass flame retardants: phytic acid, hydrolyzed collagen, and glycerol, ACS Omega 6 (2021) 3921-3930. https://doi.org/10.1021/acsomega.0c05778

[11] X.-L. Li, X.-H. Shi, M.-J. Chen, Q.-Y. Liu, Y.-M. Li, Z. Li, Y.-H. Huang, D.-Y. Wang, Biomass-based coating from chitosan for cotton fabric with excellent flame retardancy and improved durability, Cellulose 29 (2022) 5289-5303. https://doi.org/10.1007/s10570-022-04566-x

[12] Y. Xu, K. Zhang, J. Wang, Y. Wang, Biopolymer-based flame retardants and flame-retardant materials, Adv. Mater. 37 (2025) 2414880. https://doi.org/10.1002/adma.202414880

[13] M.M. Jacob, M. Ponnuchamy, A. Roshin, A. Kapoor, Adsorptive removal of oxytetracycline hydrochloride using bagasse-based biochar powder and beads, Chemosphere 363 (2024) 143016. https://doi.org/10.1016/j.chemosphere.2024.143016

[14] S.W. Sherpa, M. Ponnuchamy, A. Kapoor, M.M. Jacob, P. Sivaraman, Facile removal of sulfamethoxazole antibiotic from contaminated water using bagasse-derived pyrolytic biocarbon: Parametric assessment, mechanistic insights and scale-up analysis, J. Water Process Eng. 60 (2024) 105110. https://doi.org/10.1016/j.jwpe.2024.105110

[15] M.M. Jacob, M. Ponnuchamy, A. Kapoor, P. Sivaraman, Bagasse-based biochar for the adsorptive removal of chlorpyrifos from contaminated water, J. Environ. Chem. Eng. 8 (2020) 103904. https://doi.org/10.1016/j.jece.2020.103904

[16] M. Ponnuchamy, A. Kapoor, P. Senthil Kumar, D.-V.N. Vo, A. Balakrishnan, M. Mariam Jacob, P. Sivaraman, Sustainable adsorbents for the removal of pesticides from water: a review, Environ. Chem. Lett. 19 (2021) 2425-2463. https://doi.org/10.1007/s10311-021-01183-1

[17] M.M. Jacob, M. Ponnuchamy, A. Kapoor, D.B. Pal, P. Sivaraman, Biochar-Based Mixed Matrix Membranes: A Novel Approach to Water Treatment, in: Clean Energy Prod. Technol., 2024: pp. 229-247. https://doi.org/10.1007/978-981-97-0847-5_11

[18] M.M. Jacob, M. Ponnuchamy, A. Kapoor, D.B. Pal, P. Sivaraman, Biochar Innovations for Adsorption of Water Contaminants in Water Treatment, in: Clean Energy Prod. Technol., 2024: pp. 183-201. https://doi.org/10.1007/978-981-97-0847-5_9

[19] M.M. Jacob, M. Ponnuchamy, A. Kapoor, P. Sivaraman, Adsorptive membrane separation for eco-friendly decontamination of chlorpyrifos via biochar-impregnated cellulose acetate mixed matrix membrane, Environ. Sci. Pollut. Res. 31 (2024) 56314-56331. https://doi.org/10.1007/s11356-024-34912-4

[20] M.M. Jacob, M. Ponnuchamy, A. Kapoor, P. Sivaraman, Achieving up to 95% removal efficiency of chlorpyrifos pesticide using sugarcane bagasse-based biochar alginate beads in a

continuous fixed-bed adsorption column, Environ. Res. 269 (2025) 120902. https://doi.org/10.1016/j.envres.2025.120902

[21] M. Abdur Rashid, W. Liu, Y. Wei, Q. Jiang, Review of reversible dynamic bonds containing intrinsically flame retardant biomass thermosets, Eur. Polym. J. 173 (2022) 111263. https://doi.org/10.1016/j.eurpolymj.2022.111263

[22] S.S. Ray, L.T. Temane, J.T. Orasugh, Principle of Fire/Flame-Retarding Polymer Materials, in: 2024: pp. 37-41. https://doi.org/10.1007/978-3-031-51924-6_4

[23] S. Xie, Z. Liu, A. Feng, X. Hao, R. Ou, L. Sun, T. Liu, Q. Wang, Flame retardant modification of poplar wood based on sustainable impregnation solution with high biomass content, Ind. Crops Prod. 215 (2024) 118616. https://doi.org/10.1016/j.indcrop.2024.118616

[24] M. Lewin, E.D. Weil, Mechanisms and modes of action in flame retardancy of polymers, in: Fire Retard. Mater., Elsevier, 2001: pp. 31-68. https://doi.org/10.1533/9781855737464.31

[25] F. Laoutid, L. Bonnaud, M. Alexandre, J.-M. Lopez-Cuesta, P. Dubois, New prospects in flame retardant polymer materials: From fundamentals to nanocomposites, Mater. Sci. Eng. R Reports 63 (2009) 100-125. https://doi.org/10.1016/j.mser.2008.09.002

[26] D.. van Krevelen, Some basic aspects of flame resistance of polymeric materials, Polymer (Guildf). 16 (1975) 615-620. https://doi.org/10.1016/0032-3861(75)90157-3

[27] Y. Xue, J. Feng, Z. Ma, L. Liu, Y. Zhang, J. Dai, Z. Xu, S. Bourbigot, H. Wang, P. Song, Advances and challenges in eco-benign fire-retardant polylactide, Mater. Today Phys. 21 (2021) 100568. https://doi.org/10.1016/j.mtphys.2021.100568

[28] L. Liu, Y. Xu, Y. Pan, M. Xu, Y. Di, B. Li, Facile synthesis of an efficient phosphonamide flame retardant for simultaneous enhancement of fire safety and crystallization rate of poly (lactic acid), Chem. Eng. J. 421 (2021) 127761. https://doi.org/10.1016/j.cej.2020.127761

[29] X. Hu, J. Sun, X. Li, L. Qian, J. Li, Effect of phosphorus-nitrogen compound on flame retardancy and mechanical properties of polylactic acid, J. Appl. Polym. Sci. 138 (2021). https://doi.org/10.1002/app.49829

[30] M.N.A.M. Taib, P. Antov, V. Savov, W. Fatriasari, E.W. Madyaratri, R. Wirawan, L.M. Osvaldová, L.S. Hua, M.A.A. Ghani, S.S.A.O. Al Edrus, L.W. Chen, D. Trache, M.H. Hussin, Current progress of biopolymer-based flame retardant, Polym. Degrad. Stab. 205 (2022) 110153. https://doi.org/10.1016/j.polymdegradstab.2022.110153

[31] H. Yang, B. Yu, X. Xu, S. Bourbigot, H. Wang, P. Song, Lignin-derived bio-based flame retardants toward high-performance sustainable polymeric materials, Green Chem. 22 (2020) 2129-2161. https://doi.org/10.1039/D0GC00449A

[32] N.N. Solihat, A.F. Hidayat, M.N.A.M. Taib, M.H. Hussin, S.H. Lee, M.A.A. Ghani, S.S.O. Al Edrus, H. Vahabi, W. Fatriasari, Recent developments in flame-retardant lignin-based biocomposite: Manufacturing, and characterization, J. Polym. Environ. 30 (2022) 4517-4537. https://doi.org/10.1007/s10924-022-02494-2

[33] H. Meng, M. Wen, J. Shi, Y. Liang, H. Jian, Biomass flame retardants with application of cellulose, lignin, and hemicellulose from wood resources and their flame retardant

technologies in related materials：A review, Sustain. Chem. Pharm. 42 (2024) 101816. https://doi.org/10.1016/j.scp.2024.101816

[34] A.L. Mohamed, A.G. Hassabo, Flame Retardant of Cellulosic Materials and Their Composites, in: 2015: pp. 247-314. https://doi.org/10.1007/978-3-319-03467-6_10

[35] M.N. Azman Mohammad Taib, T.S. Hamidon, Z.N. Garba, D. Trache, H. Uyama, M.H. Hussin, Recent progress in cellulose-based composites towards flame retardancy applications, Polymer (Guildf). 244 (2022) 124677. https://doi.org/10.1016/j.polymer.2022.124677

[36] W. Tang, Z. Qin, X. Li, T. Liu, X. Cao, H. Yang, Z. Cao, X. Jin, P. Yuan, Eco-friendly flame retardant comprising chitosan-based effectively enhances the flame retardancy of wood with low weight percent gain, Int. J. Biol. Macromol. 282 (2024) 136868. https://doi.org/10.1016/j.ijbiomac.2024.136868

[37] J. Wang, J. Chen, W. Wang, J. Lu, Y. Li, T. Bai, J. Chen, Z. Zhu, D. Wang, Preparation of boron-containing chitosan derivative and its application as intumescent flame retardant for epoxy resin, Cellulose 30 (2023) 4663-4681. https://doi.org/10.1007/s10570-023-05174-z

[38] G.-Z. Yin, X.-M. Yang, J.L.D. Palencia, J. Hobson, A.M. López, D.-Y. Wang, Phytic acid as a biomass flame retardant for polyrotaxane based phase change materials, J. Energy Storage 56 (2022) 105853. https://doi.org/10.1016/j.est.2022.105853

[39] T.C. Mokhena, E.R. Sadiku, S.S. Ray, M.J. Mochane, K.P. Matabola, M. Motloung, Flame retardancy efficacy of phytic acid: An overview, J. Appl. Polym. Sci. 139 (2022). https://doi.org/10.1002/app.52495

[40] S.H. Jeong, C.H. Park, H. Song, J.H. Heo, J.H. Lee, Biomolecules as green flame retardants: Recent progress, challenges, and opportunities, J. Clean. Prod. 368 (2022) 133241. https://doi.org/10.1016/j.jclepro.2022.133241

[41] V.A. Yiga, M. Lubwama, S. Pagel, J. Benz, P.W. Olupot, C. Bonten, Flame retardancy and thermal stability of agricultural residue fiber-reinforced polylactic acid: A Review, Polym. Compos. 42 (2021) 15-44. https://doi.org/10.1002/pc.25835

[42] X. Wang, W. Xing, X. Feng, B. Yu, L. Song, Y. Hu, Functionalization of graphene with grafted polyphosphamide for flame retardant epoxy composites: synthesis, flammability and mechanism, Polym. Chem. 5 (2014) 1145-1154. https://doi.org/10.1039/C3PY00963G

[43] B.G. Fiss, L. Hatherly, R.S. Stein, T. Friščić, A. Moores, Mechanochemical phosphorylation of polymers and synthesis of flame-retardant cellulose nanocrystals, ACS Sustain. Chem. Eng. 7 (2019) 7951-7959. https://doi.org/10.1021/acssuschemeng.9b00764

[44] Y. Zhang, Y. Ren, X. Liu, T. Huo, Y. Qin, Preparation of durable flame retardant PAN fabrics based on amidoximation and phosphorylation, Appl. Surf. Sci. 428 (2018) 395-403. https://doi.org/10.1016/j.apsusc.2017.09.155

[45] W. Xing, W. Yang, W. Yang, Q. Hu, J. Si, H. Lu, B. Yang, L. Song, Y. Hu, R.K.K. Yuen, Functionalized carbon nanotubes with phosphorus- and nitrogen-containing agents: Effective reinforcer for thermal, mechanical, and flame-retardant properties of polystyrene nanocomposites, ACS Appl. Mater. Interfaces 8 (2016) 26266-26274. https://doi.org/10.1021/acsami.6b06864

[46] F. Fang, P. Song, S. Ran, Z. Guo, H. Wang, Z. Fang, A facile way to prepare phosphorus-nitrogen-functionalized graphene oxide for enhancing the flame retardancy of epoxy resin, Compos. Commun. 10 (2018) 97-102. https://doi.org/10.1016/j.coco.2018.08.001

[47] R. Yang, J. Zhou, X. Meng, H. Lu, A. Dong, L. Qi, X. Yang, C. Mei, C. Xia, J. Li, Preparation of straw/PLA composites with excellent flame retardancy based on deep eutectic solvent sulfonation, Int. J. Biol. Macromol. 279 (2024) 135238. https://doi.org/10.1016/j.ijbiomac.2024.135238

[48] X. Ye, C. Zhan, B. Wang, T. Sai, C. Zhang, J. Li, Z. Guo, S. Huo, Engineering sulfonated carbon black for flame-retardant and smoke-suppressive polycarbonate with well-preserved mechanical performances, Polym. Degrad. Stab. 230 (2024) 111012. https://doi.org/10.1016/j.polymdegradstab.2024.111012

[49] Y. Kim, C. Jung, Reaction mechanisms applied to starch modification for biodegradable plastics: Etherification and esterification, Int. J. Polym. Sci. 2022 (2022) 1-10. https://doi.org/10.1155/2022/2941406

[50] H. Vahabi, F. Laoutid, K. Formela, M.R. Saeb, P. Dubois, Flame-retardant polymer materials developed by reactive extrusion: present status and future perspectives, Polym. Rev. 62 (2022) 919-949. https://doi.org/10.1080/15583724.2022.2052897

[51] S. Bourbigot, G. Fontaine, A. Gallos, S. Bellayer, Reactive extrusion of PLA and of PLA/carbon nanotubes nanocomposite: processing, characterization and flame retardancy, Polym. Adv. Technol. 22 (2011) 30-37. https://doi.org/10.1002/pat.1715

Flame Retardant Materials Materials Research Forum LLC
Materials Research Foundations 190 (2026) 77-117 https://doi.org/10.21741/9781644904077-4

Chapter 4

Carbon-Based Nanomaterials in Flame Retardancy

Adila Adila[1], Muntasa Muntasa[1], Muhammad Zeeshan Ali janjua[1], Raif Azeem[1], Saria Fatima[1], Muhammad Ravish khan[1], Madiha Habib[1], Sana Rubab Mir[2], Lutf ur Rahman[1], Asma Rehman[1]*

[1]National Institute for Biotechnology & Genetic Engineering College, Pakistan Institute of Engineering & Applied Sciences (NIBGE-C, PIEAS), Faisalabad 38000, Pakistan

[2]Department of Chemistry and Chemical Engineering, SBASSE LUMS, Lahore

asmanano@gmail.com

Abstract

Carbon nanomaterials, such as graphene, carbon nanotubes, pristine, fullerenes carbon black and nano-diamonds, are emerging as advanced flame retardants for polymers and composites. They provide superior flame retardancy, enhanced mechanical and thermal properties, and multifunctionality at low loadings compared to conventional additives. Through mechanisms such as char formation and radical scavenging, carbon nanomaterials effectively reduce flammability and smoke production. Their synergy with traditional flame retardants further improves performance, offering a promising, halogen-free, and sustainable alternative for fire safety applications.

Keywords

Carbon Nanotubes, Graphene, Fullerene, Flame Retardancy, Nano-Diamonds, Carbon Black, Char Formation

List of Abbreviations

CNT	Carbon Nanotube
MWCNT	Multi-Walled Carbon Nanotube
SWCNT	Single-Walled Carbon Nanotube
GO	Graphene Oxide
CB	Carbon Black
LOI	Limiting Oxygen Index
THR	Total Heat Release
PHRR	Peak Heat Release Rate
TPU	Thermoplastic Polyurethane
PP	Polypropylene
PLA	Polylactic Acid
PMMA	Poly(methyl methacrylate)

Materials Research Foundations 190 (2026) 77-117 https://doi.org/10.21741/9781644904077-4

Contents

Materials Research Foundations 190 (2026) 77-117 https://doi.org/10.21741/9781644904077-4

1. Overview of Flame Retardant Polymers

1.1 Polymer Flammability and Fire Safety Challenges

Polymers play a vital role in modern engineering due to their lightweight, non-corrosion nature, and adjustable properties [1,2]. Nonetheless, Polymers tend to burn easily due to their chemical structure, organic and combustible nature, which leads to serious fire safety issues as described in Fig. 1. Many fires have been triggered by flammable polymeric materials, resulting in over 40,000 deaths globally and making this an important issue for public safety experts to overcome [3]. In addition, polymers burn quickly and cause a lot of smoke, which can be very dangerous for people and their surroundings. That's why, when ignited, polymers like polyethylene (PE) and polypropylene (PP) quickly spread heat and help fire to advance faster. Polyethylene pipes burn up in just 3 minutes and produce straight "fire highways" inside buildings. The use of synthetic materials (e.g., polyurethane foam) in modern houses lowers the time it takes for a room to burst into flames from 30 minutes to only 3–5 minutes. In aerospace and military areas, heat causes fiber-reinforced polymer (FRP) composites to melt or delaminate, bringing fire farther into structures and leading to huge fires. Polymer burning produces carcinogenic PM2.5 dust as well as the gases hydrogen cyanide (HCN) and carbon monoxide (CO) from polyurethane foams and nylon, which can cause people to lose consciousness, become dizzy, and suffer heart failure [4]. Volatile Organic Compounds such as benzene and styrene can make eyes, skin, and the respiratory system uncomfortable and may lead to asthma attacks [5]. Cancer, problems with the endocrine system, and damage to the nervous system may be caused by dioxins and polycyclic aromatic hydrocarbons (PAHs) from PVC and polystyrene [6]. These risks can be overcome by using flame retardants, which are used to slow or prevent combustion so that both bulk and composite polymer materials are less likely to catch fire [7].

Materials Research Foundations 190 (2026) 77-117
https://doi.org/10.21741/9781644904077-4

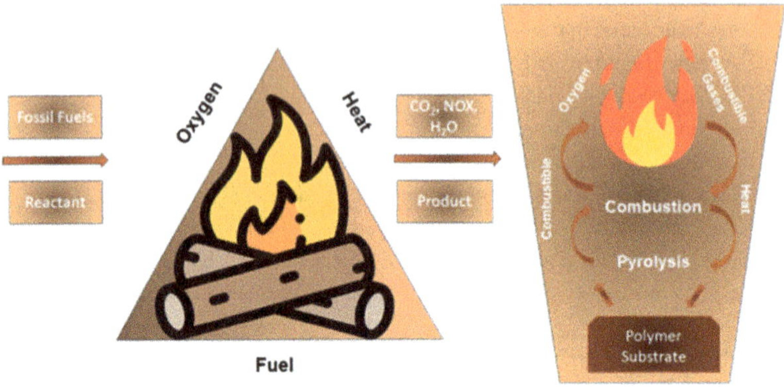

Figure 1: Schematic for Polymer Combustion Process and Pyrolysis, illustrates the thermal degradation and combustion process, such as absorption of heat, scission of chains, release of volatile, ignition and combustion

1.2 Conventional Flame Retardants

Conventional flame retardants are chemical additives incorporated into materials to reduce their flammability and slow the spread of fire. They are widely used in consumer products, electronics, furniture, textiles, and building materials, and are typically classified based on their chemical composition and mechanism of action [8]. Brominated Flame Retardants (BFRs) include polybrominated diphenyl ethers (PBDEs), hexabromocyclododecane (HBCD), and tetrabromobisphenol A (TBBPA). These compounds release bromine radicals upon heating, which interfere with combustion chain reactions by capturing free radicals necessary for fire propagation. The same is for chlorinated Flame Retardants, such as chlorinated paraffins and other organochlorine compounds, which function similarly to BFRs, releasing chlorine radicals that inhibit combustion [9]. Phosphorus-based flame-retardants, such as inorganic phosphorus compounds (ammonium polyphosphate (APP) and red phosphorus), act by forming a protective char layer on the material surface during combustion, which isolates the substrate from heat and oxygen. Triphenyl phosphate (TPP) and resorcinol bis(diphenyl phosphate) (RDP) retard fires through condensed phase (char formation) and the gas phase (radical scavenging) mechanism [10]. Melamine, melamine urate, and melamine salts are included in nitrogen-based flame retardant materials, which release inert gases (such as nitrogen oxides) upon decomposition, diluting oxygen and flammable gases in the combustion zone. The primary mechanism of these classes is through gas-phase action by reducing the concentration of oxygen and flammable gases available for combustion [11]. Intumescent flame retardants, a combination of acid, carbon, and gas sources, include ammonium polyphosphate (acid), pentaerythritol (carbon), and melamine (gas). Upon heating, these systems expand to form a thick, insulating char that protects the underlying material from heat and flame [12].

Materials Research Foundations 190 (2026) 77-117 https://doi.org/10.21741/9781644904077-4

1.3 Environmental and Health Concerns

Conventional flame retardants, especially halogenated and certain organophosphates, are linked to significant environmental and health concerns. These issues arise from their persistence, bioaccumulation, mobility, and toxicity.

1.3.1 Environmental Concerns

Flame retardants, particularly brominated and chlorinated compounds, do not break down easily in the environment. PBDEs and NBFRs are highly persistent in the environment and accumulate in living organisms, particularly in the food chain. For instance, PBDEs are found in wildlife, house dust, and even human breast milk [13]. Flame retardants readily leach from products and can be transported globally, even to remote areas. This widespread distribution complicates efforts to control their environmental impact [14]. Excessive use of some retardants, such as the ammonium phosphate-based aerial wildfire retardants (e.g., Phos-Chek), can cause nutrient pollution in water bodies, leading to algal blooms and oxygen depletion, which harms aquatic life [15]. Moreover, certain additives, such as sodium ferrocyanide, can release toxic cyanide under sunlight, posing a severe risk to aquatic life [16]. Products containing hazardous flame retardants are often exported to developing countries as e-waste, where primitive recycling methods like open burning release toxic compounds into the environment and expose local populations to health risks [17].

1.3.2 Health Concerns

Several flame retardants, such as tetrabromobisphenol A (TBBPA) and tris(1,3-dichloro-2-propyl)phosphate (TDCPP), are classified or suspected as carcinogens. TBBPA has been classified by the International Agency for Research on Cancer (IARC) as “probably carcinogenic to humans”. High exposure to PBDEs is linked to a nearly 300% increased risk of cancer mortality, with evidence pointing to oxidative stress and DNA damage as underlying mechanisms [18]. Brominated flame retardants (BFRs) such as polybrominated diphenyl ethers (PBDEs) and novel brominated flame retardants (NBFRs) interfere with hormone systems, particularly thyroid function. Studies show that PBDEs disrupt adrenal hormones like aldosterone, potentially leading to cardiovascular issues [13,19]. Exposure to BFRs is linked to neurobehavioral impairments, including reduced memory and motor coordination, and increased anxiety. Animal studies demonstrate specific effects on the hippocampus, a brain region that is critical for memory [20].

Organophosphate flame retardants (OPFRs) and BFRs are associated with reproductive toxicity, including reduced fertility and adverse effects on fetal and child development [21]. Both BFRs and OPFRs can alter immune system function, increasing susceptibility to infections and autoimmune diseases [21]. Brominated flame retardants can accumulate in the respiratory tract, inhibit cell vitality, promote apoptosis, induce DNA damage, and foster inflammation and oxidative stress, leading to decreased pulmonary function. Elevated serum concentrations of PBDEs are associated with impaired pulmonary function, including reduced lung ventilation and increased inflammation [21].

1.4 Nanotechnology-based Flame Retardants:

Nanotechnology-based flame retardants possess several clear advantages over conventional flame retardants, primarily due to their unique properties and mechanisms enabled by nanoscale

Materials Research Foundations 190 (2026) 77-117 https://doi.org/10.21741/9781644904077-4

materials such as clays, carbon nanotubes (CNTs), graphene, and other engineered nanoparticles. In recent years, nanocarbon materials have attracted the interest of researchers due to their excellent properties. Nanocarbon-based flame-retardant polymer composites have enhanced thermal stability and mechanical properties compared with traditional flame-retardant composites [22]. They also synergize with traditional flame retardants, such as phosphorus or nitrogen compounds, to further reduce flammability and improve overall performance. For example, carbon nanotube-based intumescent systems have demonstrated three-stage flame retardant action, from ignition to self-extinguishment, by combining char formation, gas dilution, and radical scavenging [23].

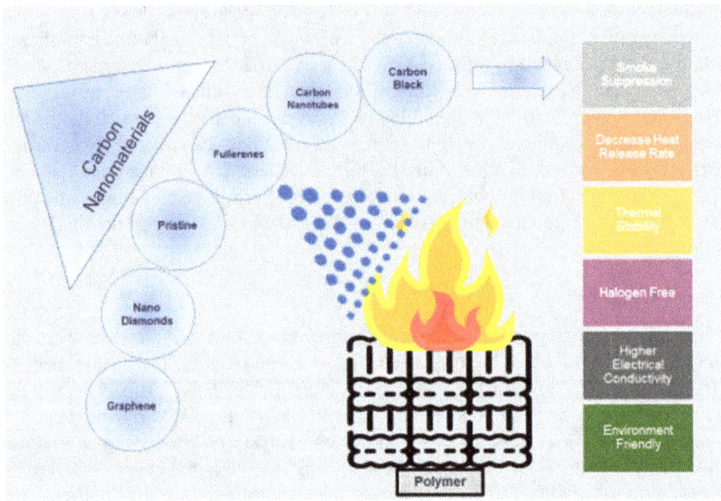

Figure 2: Carbon Nanomaterials, their Composite and Characteristics as Potential Flame Retardant.

1.4.1 Key Advantages of Carbon Nanomaterials as Flame Retardants

Nanomaterials provide effective flame retardancy at much lower concentrations (often <5% by weight) compared to conventional additives, which may require 20–60% loading. This is due to their high surface area and improved dispersion within polymer matrices, which enhances their interaction with the material and fire suppression mechanisms [24]. Combining nanoparticles with conventional flame retardants (e.g., phosphorus-based or nitrogen-based compounds), as shown in Fig. 2, results in synergistic effects, further improving performance at even lower additive levels [24,25]. Conventional flame retardants often degrade mechanical properties due to high loadings. In contrast, Carbon nanomaterials can enhance or at least maintain the mechanical strength, stiffness, and toughness of the host material [24]. Nanocomposites can offer additional benefits by providing multifunctionality such as improved thermal stability, electrical conductivity, and barrier properties, making them suitable for advanced applications in electronics, automotive, and aerospace [25]. Carbon-based nanoparticles promote the formation of a robust, thermally

Materials Research Foundations 190 (2026) 77-117 https://doi.org/10.21741/9781644904077-4

insulating char layer during combustion. This char acts as a physical barrier, shielding the underlying material from heat and oxygen and preventing dripping and flame spread [25]. It can also significantly reduce the peak heat release rate (PHRR) and total smoke production, which are critical for fire safety in enclosed spaces. Carbon nanomaterial-based flame retardants are halogen-free, avoiding the release of toxic, persistent, and bioaccumulative compounds associated with conventional halogenated retardants. These are environmentally benign, simplifying disposal and reducing environmental impact [25]. The small size and high surface area of nanoparticles enable uniform dispersion within polymers, improving compatibility and processability [24]. Moreover, nanotechnology allows for innovative application methods such as layer-by-layer (LbL) assembly, which can create multifunctional, ultrathin flame-retardant coatings on complex substrates [26]. This chapter provides an overview of the main categories, workings, and performance traits of carbon-based nanomaterials as efficient flame-retardant polymer system additives.

1.5 General Mechanisms of Flame Retardancy

In the last decade, a lot of progress in flame retardancy has made it an impressive and progressive technique; advancements have made it eco-friendly and crucial nowadays [27]. The basic mechanism of flame retardancy was discovered in 1947, which includes the thermal degradation of the polymer in order to check the effect of additives on it. Moreover, the removal of oxygen from the surface of the polymer by coating of the retardant material is done through the production of fuel fuel-poor pyrolytic pathway. Many other processes were used to stop the production of combustible gases through internal barriers. All these processes led to the discovery of the two basic mechanisms of flame retardancy, one is the condensed phase and the other is the gas phase mechanism, as shown in Fig. 3. In order to achieve flame retardancy, factors that are responsible for the fire are crucial. Heat, oxygen, and fuel are the main causes of fire, and flame retardancy needs to remove any one of it [28]. Combustion occurs when a material can burn, get heated, and reach its ignition temperature. Flame-retardant material absorbs that heat in that particular time period and leads to flame retardancy because it stops the spread of fire and reduces smoke production. The endothermic effect is responsible for this process, which is the first step in flame retardancy. The second step is covering effect; the retardant material has the property to form a covering layer, which prevents the escape of gases from the surface and stops fire. In the process of combustion, the free radical is necessary, and when fire retardant reacts with these free radicals the flame density is reduced and as a result, burning rate is decreases. The last is suffocation effect when the retardant material burns, it produces non-combustible gases which decrease the volume of combustible one's, which results in suffocation thus decreasing the rate of burning.

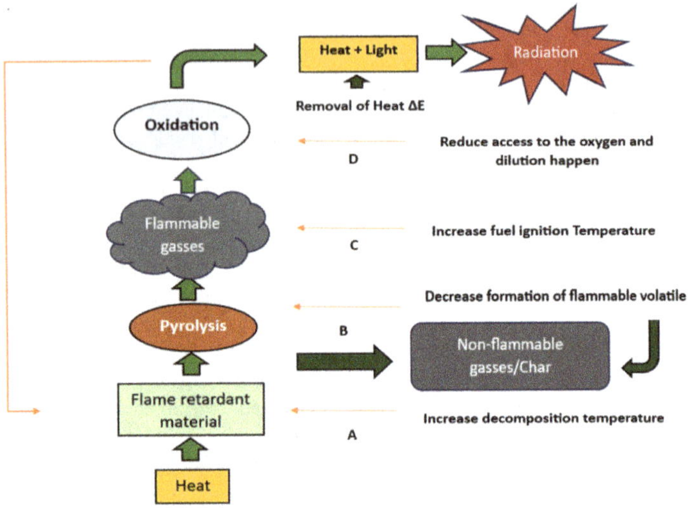

Figure 3: Key Difference between the Gas Phase and Condense Phase Mechanism of Flame Retardancy. Condense Phase; A: flammable retardant material increases the decomposition temperature, B: flame retardant makes non-flammable gases and char instead of flammable gases, Gas Phase; C: Flammable retardants increase fuel ignition temperature, slowing or stopping combustion, D: Non-flammable gases dilute the oxygen or reactive gases in the flame, slow down oxidation reaction.

1.5.1 The Gas Phase Mechanism

The flame-retardant activity of the substance in the gas phase mechanism is interfered by the reaction of the flame retardant with the flame. The heat is turned off from the material; this activity depends on the flame and the amount of inhibitor. Main mechanism involves the dilution of inert gases, quenching of free radicals, heat removal, oxygen isolation, and blowing out effects. The combustion of polymer materials involves a chain reaction, which is driven by free radicle. These radicles are captured by the substances which are released by gas phase flame retardants, due to which the chain reaction is interrupted, and the burning process is stopped, this is called the quenching effect. Dilution effect reduced the oxygen and other combustion gases, making the fire retardancy process easy. Mostly metal hydroxide and carbonate additives act in such a way that they dilute the oxygen, which is a main component of the burning process and other combustible gases. The non-combustible gases like ammonia, nitrogen, and water vapors produced by the burning of flame-retardants dilute the flammable gases and lower the oxygen level, which reduces the process of burning. Flame-retardant materials have the ability to show the endothermic process by heat removal. They absorb the heat or produce water vapors, which cool down the burning process. Some polymer retardant materials (magnesium/aluminum hydroxides, chlorinated paraffin, and antimony oxides) can melt the polymers, and the molten polymers can reduce heat. Meanwhile, oxygen isolation by some compounds (antimony trioxide, brominated flame

Materials Research Foundations 190 (2026) 77-117 https://doi.org/10.21741/9781644904077-4

retardants, and phosphorous) created a barrier of non-combustible gases that cover the material and reduce the oxygen entry near to the combustible gases and thus eliminate the process of burning. Lastly, Blowing-out effect produced by which flame retardants get decomposed and produced bubbles. These burst at high velocity and release the non-combustible gases, which reduces the process of burning.

1.5.2 The Condense Phase Mechanism

Condense phase mechanism of flame retardancy is also called pyrolysis zone. This mechanism plays an important role in commercial flame-retardant polymers. In many halogens free flame-retardants, protective layer formation is an important and main process. Many phosphorous-based flame retardants are successfully used nowadays and the reason is increased charr formation like red phosphorous in glass fiber reinforced polyamine 66. Halogen based flame retardancy is also known as the most typical example of flame inhibition and metal hydrates for heat sink by using condense phase mechanism [29]. Phenomena such as intumescence, charring and protective layer formation all depends on chemical reactions [30]. The condense phase refers to the formation of different carbon layers on the surface of solid layers which prevent the production of combustible gases and thus reduce the burning process. The fiber material used as a flame-retardant gets depolymerized and thus reduces the melting temperature which become the cause of difference between the ignition temperature and melting temperature. The material used in flame retardancy get soften shrink, and become molten liquid drops; thus the heat of the burning process is taken away, and the whole burning process is slowed down [31].

2. Classifications and Types of Carbon-based Nanomaterials

Carbon-based nanomaterials can be categorized based on their dimensionality and structural characteristics, encompassing carbon nanotubes, pristine carbon nanotubes, surface functionalized carbon nanotubes, graphene and graphene oxide, fullerenes, carbon black, and nano-diamond structures. These have been explored below in terms of their ability to enhance the flame-retardant properties of polymer composites. Each category of material has its own structural properties and processes, which determine its usefulness in providing thermal stability and fire resistance.

2.1 Carbon Nanotubes (CNTs)

Fire resistance is greatly improved in polymers by the addition of high performance carbon nanotubes additives and their sub fractions, single walled carbon nanotubes and multi-walled carbon nanotubes. Carbon nanotubes are suitable multi-functional nanomaterials that deliver high thermal conductivity and mechanical strength, increase electrical and structural quality and make the polymer composites flame resistant [32]. Because CNTs have a tubular nano form, they can serve as thermal insulators, building materials, and char stabilizers, so their use in advanced flame-retardant systems is extremely important [33].

2.1.1 Mechanism of Carbon Nanotubes in Flame Retardancy

At the burning and destructive stages of polymers, carbon nanotubes (CNTs) perform different functions related to stopping or slowing down the process. A protective char layer is widely recognized as one of the best flame-retardant features. Heating effects in CNTs result in the formation of an insulating layer of carbon on the surface of polymers because of their high thermal

Materials Research Foundations 190 (2026) 77-117 https://doi.org/10.21741/9781644904077-4

conductivity and great stability. Physically, the fire retardant protects the polymer by reducing its exposure to heat and also covering up flammable gases, as well as oxygen [34].

Carbon nanotubes (CNTs) act to interrupt or minimize the process of polymers burning and being damaged. It is well recognized that the formation of a protective char layer by CNTs greatly helps with fire retardance. This is shown in Fig. 4, where it is clear that CNTs block flammable gases, trap oxygen, and protect the polymer by creating a strong heat shield.

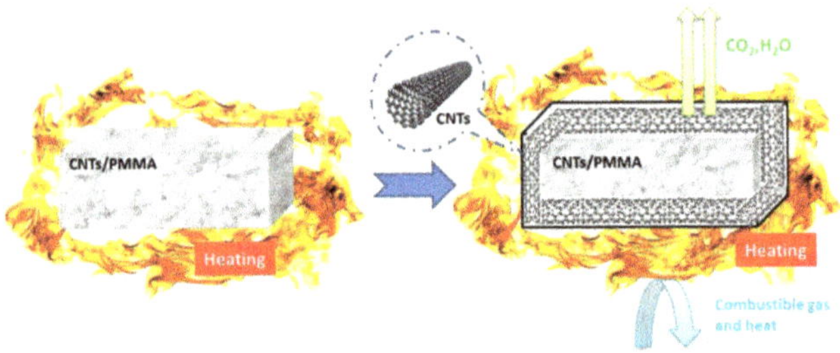

Figure 4: Flame Retardant Mechanism of CNTs in Polymers (This Diagram Illustrates How Carbon Nanotubes (CNTs) Enhance Flame Retardancy in Polymers. It Highlights the Formation of a Protective Char Layer, the Barrier Effect against Heat and Oxygen, and the Suppression of Volatile Gas Release.

The second important reaction is suppression of pathway degradation. In the process of thermal decomposition, CNTs tackle formed polymeric radicals to prevent their destabilization and decelerate chain-breaking processes. The radical-trapping mechanism affects the decomposition rate processes and promotes the conversion of fuel molecules into condensed-paste residues instead of volatile decomposition products [35]. The large aspect ratio and the nanosized diameters of CNTs make the mass transportation course convoluted. The labyrinth effect creates non-burning performances due to the prevention of combustible gas release and rate-of-release control to delay ignition of the combustible gas, hence minimizing flame spread [36].

The chaotic penetration of carbon nanotubes in the body of the polymer reinforces the adhesive qualities of the protective char covering. The regular flame retardants provide poor porous residues while burning. CNTs produce a reinforced char that further grows stronger and is not mechanically susceptible. The structural stability is one of the key elements that provide the sustainability of effective flame protection for a long period under fire conditions [37].

Based on the electrical properties of carbon nanotubes, flame retardancy can be realized utilizing the electro-thermal methods. The systems that are responsible for the Joule heating result in early carbonization, and preferential degradation of the region that is rich in CNTs to develop a protective char layer before fire spreads [23].

CNTs exhibit enhanced protective characteristics against fire, along with other known flame-retardant additives, using additive effects. The addition of CNTs as an additive in intumescent flame retardants (IFRs) enhances their expansion and thermality of intumescent char in the system. CNTs enable a better dispersion of flame-retardant phases in the layered silicate systems, resulting in a prudent barrier against fires. Hybrid fuel retardant systems relying on the use of CNTs show

Materials Research Foundations 190 (2026) 77-117 https://doi.org/10.21741/9781644904077-4

a considerable decrease in the heat release rate, total smoke production, and toxic gas emission, which makes these systems the ones best suited for important safety implementations [33].

Flame retardant activity of CNTs depends on many complex factors such as the characteristics of polymer matrices, CNT type, and dispersion state, as well as along with synergic additives. The synergistic combination of char formation with radical stabilization and mass transport barrier, and synergistic enhancement makes CNTs one of the choicest nanomaterial additives for the polymers of the coming generations' flame retardance development [38]. This implies that the mechanism of action of CNTs is mainly a condensed-phase process.

2.1.2 Properties of Carbon Nanotubes as Flame Retardant Additives

Why does merging thermal, mechanical, and electrical properties explain the stabilizing properties of carbon nanotubes against fire? Because their decomposition point is above 600 °C, carbon nanotubes have a higher decomposition temperature than polymers. Integrating them with polymer matrices brings about increased resistance to heat by allowing the polymer to resist burning for a longer duration. There is a need for radical changes to the properties of polymers that burn in some settings [39]. By using CNTs, a polymer material gains greater strength, rigidity, and resistance to impacts. Because both shape and matrix interfaces play a role, minimal CNT concentrations (often less than 5 wt% %) can make a structure much more stable under heat stress. With the help of CNTs, polymer composites increase their range of important multifunctional abilities because of their ability to conduct electricity. It is possible to have fire protection, electrostatic discharge (ESD) defense, and electromagnetic interference (EMI) shielding, all brought by CNT-based conductive networks in resins.

2.1.3 Applications of Carbon Nanotubes in Flame Retardant Polymers

Fire-retardant properties are enhanced in several thermosetting and thermoplastic matrices when CNT is used as an additive. When CNTs are mixed with Polylactic Acid (PLA), the material displays both a raised Limiting Oxygen Index and a lower maximum heat release rate. Therefore, materials with an improved resistance to ignition and slower spreading of flames are suitable for packaging products, electronic devices, and textiles.

The mechanical and dimensional stability of polyamide 66 (PA66) increases when introduced with CNTs, though this only leads to a decrease in fire resistance. In the automotive and electrical industries, both the flame retardant and conductive abilities of Fiber-Reinforced Polymer (FRP) help make their products more valuable. Adding nanotubes to the material improves how flexibly and elastically Thermoplastic Polyurethane (TPU) responds. These composites show excellent resistance to fire and have conductive properties which makes them suitable for insulating wires and building barriers to protect electronics and safety equipment [40].

When polymers are added to carbon nanotubes, the materials formed display stability under extreme conditions, including high temperatures, oxygen and UV-Ray or weather exposure. Due to the excellent properties of CNTs, these businesses are able to ensure fire safety while using their products in multiple commercial areas.

Different experiments reveal that CNTs improve the way materials resist catching fire. A combination of 3 wt. % MWCNTs with PLA caused a 30% decrease in Peak Heat Release Rate (PHRR) and a rise in the quantity of char. The combination of CNTs in intumescent coatings greatly raised the performance against fire.

Materials Research Forum LLC
https://doi.org/10.21741/9781644904077-4

The use of CNT with TPU composites in cable insulation is now known to confer advantages. The addition of just 2% CNTs led the composite to pass the UL-94 V-0 flame test and gain improvement in tensile strength and reduced the smoke produced during burning. Thanks to these properties, it is possible to maintain fire safety and structural integrity because they make it possible to carry out environmental uses that need both mechanical strength and resistance against electricity.

Various polymer studies have confirmed that CNTs raise all aspects of flame retardancy while also making the material more heat tolerant and less likely to produce smoke. It is obvious from the research that CNTs can be used effectively as flame retardant nanofillers in today's polymer applications [23].

2.2 Pristine Carbon Nanotubes

Using unchanged "as-made" carbon nanotubes (CNTs) for fire protection in polymers is gaining interest. The intact carbon nanotubes, having unexposed and unaltered surfaces, keep their natural structure and key properties such as strong electrical conductivity, good mechanical strength and satisfactory thermal performance. CNTs as-produced have characteristics that make them suitable for fire protection without influencing the original substance. Since they naturally stick together and hate water, they do not move very well which reduces their performance. By including CNTs in polymers, their flame-retarding effects can still be noticed despite their several drawbacks [41]. A transmission electron microscopic image of tubular structure of pristine CNTs are shown in Fig. 5.

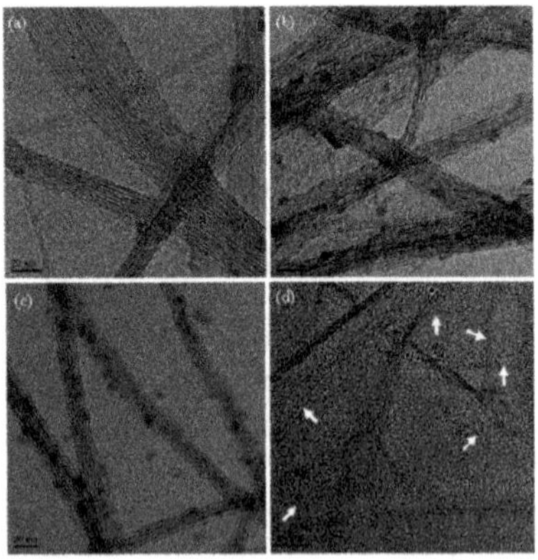

Figure 5: A High-resolution TEM Image Displaying the Tubular Structure of Pristine CNTs. This Visual Aids in Understanding Their Nanoscale, Morphology and Potential for Agglomeration [42].

2.2.1 Mechanism of Flame Retardancy of Pristine Carbon Nanotubes

The predominant action of pristine CNTs is also determined by the condensed phase because it is stable and non-volatile. The flame-retardant qualities of a single CNT are due to its stability and

Materials Research Foundations 190 (2026) 77-117
https://doi.org/10.21741/9781644904077-4

resistance to heat during combustion. According to research, adding pure CNTs to polymers leads to improved ability to resist burning. The carbon-based residue formed initially supports the development of a stable protective char layer through the process of thermal breakdown. The heat barrier property of char allows it to repel heat and oxygen and hold the by-products in the reaction mass. Pristine CNTs produce a solid and resilient boundary which helps them prevent char growth better than most clean polymers, especially because of their long, connected structure [23].

There are two routes by which defects in CNTs can be created by thermal action, altering the ways the molecules are destroyed. The structure and thermal stability of the polymer do not change early due to the presence of such CNTs. Interference in heat diffusion due to the network-forming ability of char reduces the rate of the polymer chain diffusion needed for pyrolysis. The undamaged CNTs can interact with the products of decomposition, changing them, so smoke and hazardous gas are less likely to be produced during thermal decomposition [42]. CNTs in their pure form have an inert surface because they do not undergo chemical reactions, unlike functionalized CNTs during polymer and additive blending. Because unadulterated CNTs are not reactive, they are suitable for applications that need reliable and chemically stable platforms. Compared to others, chemically pure carbon nanotubes remain stable during polymerization and continue to serve over a longer period in applications that require ongoing use. The properties of pure CNTs greatly benefit polymer-based flame-retardant materials. Bare CNTs have very stable thermal features due to their ability to resist both oxidation and damage by heat above 600 °C in an inert atmosphere. Ensuring the proper addition of these fillers to polymers increases the temperature at which polymers degrade and gives them better ability to resist fires.

Research in engineering shows that adding pristine CNTs can improve the tensile strength, increase the elastic modulus, and enhance impact resistance in a host polymer. With strong van der Waals forces, Young's modulus ensures that oxides remain reliable under stressful loads. Even tiny amounts of antioxidants in composite materials help keep polymers strong and safe from damage caused by heat [43]. Because the functionalization does not occur on the surface of pristine CNTs, their high conductivity is not impacted. Pure carbon nanotubes in plastic form offer effective dissipation of static charge and electromagnetic shielding by connecting different areas with continuous conductivity. The flame-resistant qualities and electrical performance of nylon help improve many types of electronic devices, cables, and antistatic packaging.

2.2.2 Applications in Polymer Matrices

Pristine CNTs have been added into several polymer resins, for example, PLA, TPU, and PA66. Adding pristine type CNT to PLA results in better heat protection and first char formation of the resulting biodegradable plastic. The addition of clean CNTs improves the PLA materials' resistance to heat and makes them suitable for higher temperatures. The dual effects of Pure CNTs make it suitable for automobile companies and the electric industry to use PA66 for engineering plastics. The strong attraction of top CNTs and polymer chain blocks chain movement in the polymer, thereby lowering the flammability of the material [44].

Systems containing pristine CNTs function as versatile additives, defending against flames, scratches, and heat. These composites demonstrate key safety features such as lower heat release and more time before ignition when exposed to fire in the wire sheathing and protective fabrics, as well as shoe applications. When we have higher humidity, higher temperatures, and oxidative environments, composite polymers still stay stable. Because CNTs are both sturdy and independent from chemicals, they maintain their thermal stability and are useful whether used outside or inside a building. Pristine CNTs application in different areas is summarized in Fig. 6.

Figure 6: Application of Flame Retardant in Polymer,showing different polymer products that use flame-retardant systems, including coating, fabrics, electronic enclosures, and construction products. The number highlights the popularity of flame-retardant technology in the industrial application in enhancing fire safety.

Once clean CNTs are added to polymer composites, they are effective in protecting them from fire, as proved through investigations. Composites containing 2 wt. % pristine MWCNTs had a higher limiting oxygen index and lower peak heat release rate compared to those without them. Due to the CNT network, the material performed better because it gained stronger char-forming and thermal barrier effects. When synthesizing these materials, the TPU composites were bonded with originally pure CNTs without the need for chemical modifications. Adding low volumes of CNT led to a stronger result in V-0 testing for this material during vertical burning tests. Tensile strength and electrical conductivity both showed that adding pure CNTs could help the composite material.

While integrating pristine CNTs has succeeded, achieving true dispersions is the main obstacle we encountered. If it were not for sophisticated techniques that use high shearing of molten polymers and include polymerization with agents that help keep the CNTs separate, pristine CNTs would often link together. Flame-retarding improvement is better seen in CNT arrangements than in known fire retardants when the efficiency and function were compared [45].

2.3 Surface Functionalized Carbon Nanotubes:

Functionalized carbon nanotubes (fCNTs) are made of CNTs with organic groups attached to their surface and tips, making them easier to mix with polymers and bond at their interface. As a result

Materials Research Foundations 190 (2026) 77-117
https://doi.org/10.21741/9781644904077-4

of this modification, CNTs are less likely to clump and are easier to handle, with improved mechanical strength and processability. The process of functionalization makes it possible to attach biomolecules to CNTs, offering more usability in environmental areas.

2.3.1 Mechanisms Functionalized Carbon Nanotubes as a Flame Retardant

These carbon nanotubes increase the polymer's flame retardancy by various mechanisms, which are affected through specific functionalization methods useful for the carbon nanotubes. They make a protective network inside the matrix of polymer, behaving as a physical hurdle. This network minimizes the diffusion of combustible gases and heat as well. It reduces oxygen permeation as well and increases the layer of char when combustion occurs. For example, Ammonium polyphosphate, joining with multi-walled carbon nanotubes, which is amino functionalized, forms as nanocoating on ramie fabric. Flame retardancy and thermal stability increase with the help of this network [46]. Graphitized char layer, which is stable, behaves like a thermal insulator. It prevents the entry of heat into the underlying material. It minimizes flammable volatiles. For instance, joining of PDSPB poly (diaminodiphenyl methane spirocyclic pentaerythritol bisphosphonate) with CNTs makes core-shell network which increases flame retardancy and formation of char [47]. Decomposition of CNTs releases flame retardant materials, which trap free radicals and reduce flame propagation. This network also stops the process of combustion. For example, CNTs having phosphorous-containing groups emit phosphorous-containing compounds, which stop the production of free radicals, reducing flame propagation. Heat release rates are reduced and char formation increases by using metal hydroxides along with the graphene oxide in the polymers. Surface-modified CNTs are shown to promote the formation of a barrier at the condensed phase and possibly to some degree aid in radical entrapment in the gas phase. The flame retardancy mechanism of these carbon nanotubes is explained in Fig. 7.

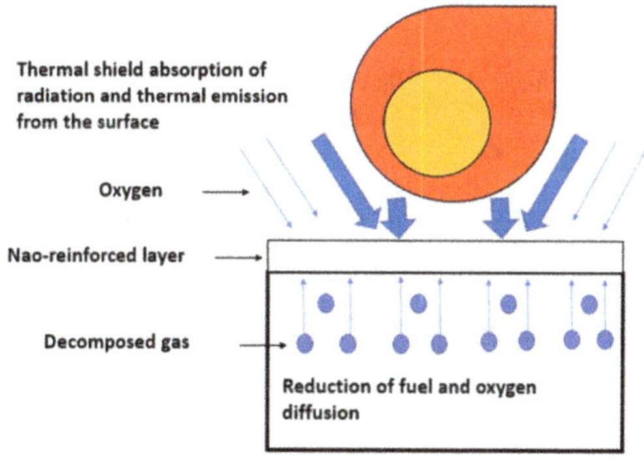

Figure 7: Flame-retardancy mechanism of Carbon nanotubes, carbon nanotubes (CNTs) form an entangled network of thermally stable nanotube structures which enhance the strength of the char layer, restrict heat and mass transport, and release heat during combustion. This condensed-phase mechanism is effective in reducing polymer degradation and flame propagation.

Materials Research Foundations 190 (2026) 77-117 https://doi.org/10.21741/9781644904077-4

2.3.2 Properties of Surface Functionalized Carbon Nanotubes as Flame Retardant Polymers

They increase the activity of flame-retardant polymers by increasing electrical conductivity, mechanical stability, and thermal stability. These improvements in turn increase interfacial interactions and the dispersion inside the matrix of the polymer. A combination of Polystyrene C_{60} with CNTs enhances the temperature up to 68 °C and the highest weight loss temperature up to 87 °C. This enhancement is because of the CNTs' radical trapping ability and effective dispersion [48]. Phosphorous-containing gases in CNTs increase char yield and thermal stability as well. Char layer becomes stable due to CNTs' strong hydrogen bonding, which provides excellent cross-linking when decomposition happens [49]. They enhance residual mass up to high temperatures in the styrene-butadiene rubber and acrylic rubber as well. This increases thermal resistance [50].

Carbon nanotubes in epoxy resins, along with graphite sheets, showed a 33.5% enhancement in tensile strength. On the other side, 44.8% enhancement in impact strength at a filler loading of 3.9wt.% [51]. Young's modulus increases up to 50% when we functionalize epoxy resins with CNTs. It is because compressibility reduces, and interfacial bonding becomes strong [52]. Up to 21% tensile strength increases and up to 11% Young's modulus increases when we functionalize epoxy resins with CNTs [53]. Polymers' electrical conductivity increases with the help of CNTs. This functionalization is important for applications needing electromagnetic interference shielding characteristics and antistatic characteristics. 1.7×10^{-3} S/cm electrical conductivity is achieved by functionalize epoxy resins with CNTs and sheets of graphite of 11.0wt.%, showing the matrix conductive networks formation [51]. Up to 92% longitudinal direction and up to 59% electrical conductivity increase when we use CNTs in a composite of CFRP. These improvements increase the applications of composites in aerospace structures by protecting against lightning strikes [53].

2.3.3 Applications of Surface Functionalized Carbon Nanotubes in Flame Retardant Polymers:

The flame resistance of polylactic acid (PLA) increases by functionalizing it with the CNTs. In PLA, LOI increases up to 255.3% from 20.0% and UL-94 V-0 rating is achieved due to the incorporation of carboxymethyl chitosan with hexachlorocyclotriphosphazene with CNTs of 1 wt.%. The thermal stability of Polyamide 66 increases by functionalizing it with the CNTs of 0.5wt. %. High-temperature applications become possible due to this modification [22]. Elongation break and tensile strength of Thermoplastic Polyurethane increase due to CNTs functionalization. They showed increased flame retardancy and minimal flammability temperature, and enhanced decomposition temperature. The thermal stability of the composite increased at high temperatures. Interactions inside the matrix remain stable in humid conditions. Flexibility improvement is due to the mechanical stress [54].

According to recent research, char layer forms and delays the ignition when amino-functionalized CNTs are infused with Polylactic Acid composites. This composite improves the dimensional stability of the substrate. Heat rate drops up to 35% by introducing surface-modified CNTs by 0.3 wt. % in case of Polyamide 66. Mechanical stability and flexibility are improved by using Thermoplastic Polyurethane. As a result, char layer forms which stops fire and there is a noticeable decrease in smoke production was also observed [55].

Materials Research Foundations 190 (2026) 77-117 https://doi.org/10.21741/9781644904077-4

2.4 Graphene and Graphene Oxide:

2.4.1 Graphene and Graphene Oxide Mechanisms of Flame Retardancy:

Graphene and materials that are based on graphene, such as graphene oxide, are primarily controlled by the condensed-phase mechanism as they exhibit excellent barrier and char-forming characteristics. Oxygen diffusion limits, heat transfer slows down and combustible gases release minimizes when polymers having GO act as a physical hurdle [56]. Flammable volatiles release reduces, heat do not affect material present internally, and insulated the polymer thermally when polymers having GO form char layers inside the polymers [22]. Free radicals are trapped by graphene oxide during the combustion process by producing phosphorous-containing compounds. Heat release rates are reduced and char formation increases by using metal hydroxides along with the graphene oxide in the polymers.

2.4.2 Properties of Graphene and Graphene Oxide as Flame Retardant Polymers

Graphene contains a honeycomb structure, which is hexagonal and two-dimensional. It is made through carbon monoatoms sp^2 hybridization. Excellent mechanical properties, high thermal stability, and large aspect ratio of graphene grab the attention of Scientists to use it in different polymers to increase their strength, durability and flexibility. Zigzag path effect produces in the polymer due to the graphene incorporation. Pyrolysis products diffusion path changes significantly due to this effect. Mass loss rate reduction becomes non-significant. Polymer host properties increase due to graphene nanosheets effective dispersion. Furthermore, graphene increases the formation of homogenous and dense char layers of condensed phase during the polymer matrix combustion. These structured char layers stop the decompositions products thermally present in the internal from entering the area of flaming. It also decreases the oxygen and heat entrance into the polymer.

Adding graphene oxide or graphene into polymers could increase their properties of flame retardant. Graphene layers have strong π-π stacking and high van der Waals interactions between the layers of atoms as well. Aggregation of graphene layers become easy in the matrix of polymer due to these strong forces. This causes poor dispersion and compatibility. Flame retardancy decreases by using graphene simply. Synthetic graphene-based polymers contain less LOI values and UL94 ratings. It is because that it only performs its function by becoming a physical barrier. Polymers combustion do not easily prevent by it. Conventional flame-retardant combination with graphene is an effective strategy to increase the flame-retardant properties of the polymers having graphene. Various synergistic flame retardants and surface treatments are applied for the reinforcement of graphene-based systems flame retardance performances.

Graphene derivatives modified with metal oxides are also utilized as flame retardants in various polymers. Graphene-zinc stannate is made with the help of hydrothermal methods. It is useful in toxic gases reduction produced during polymers burning. Graphene oxide surfaces decorate with nanosheets of $Ni(OH)_2$ with the help of strong affinity of NH_2 groups to Ni^{2+}. This decoration is helpful in the reduction of total smoke production, THR and PHRR. Polymer nanocomposites having graphene has high flame retardancy due to the effective strategy of surface modification techniques such as joining of flame retardants like B, N, P and Si with the graphene [57].

Materials Research Foundations 190 (2026) 77-117
https://doi.org/10.21741/9781644904077-4

2.4.3 Applications of Graphene and Graphene Oxide in Flame Retardant Polymers

Polylactic acid (PLA) is not flame resistant and it can be modified with the use of graphene and graphene oxide to make it flame resistant and less flammable. Limiting oxygen index value of 36.0 and V0 rating of UL-94 achieves by 3wt% loading of the combination of polyethyleneimine-modified graphene oxide with the polyphosphonate which is bio-based. This strategy increases the PLA's tensile toughness, including an elongation at 13.1% break and 39.1 MPa tensile strength.

Ternary hybrid nanoflake having ZIF-8, PPA and GO at loading of 2wt% decreases the PHRR up to 39.5% and enhances 27% LOI. Reinforcing effects and good dispersion also cause improvement in the toughness and tensile strength. Branched polyethyleneimine and 9,10-dihydro-9-oxa-10-phosphaphenanthrene-10-oxide modification with GO decreases specific extinction area up to 92%. It also decreases total smoke release up to 79% [58].

Phytic acid and lignin addition with GO in Polyamide 66 fabrics increase flame retardancy. It increases 27% LOI and decreases 25% PHRR. This treatment does not affect fabric properties, including exhibited wash durability, tensile strength, handle and color. After five washing cycles, 17% PHRR reduces [59].

Graphene oxide having Thermoplastic Polyurethane increases smoke repression and flame retardancy. TPU having graphene 1 wt.%, melamine polyphosphate, pentaerythritol and ammonium phosphate PCOP (peak carbon monoxide production) by 91.3%, PSPR (peak smoke production rate) by 92.1% and PHRR by 85.3%. 29% LOI and UL-94 V-0 also achieves. Flame retardant systems having graphene oxide exhibit strong thermal stability for the protective char layers formation, increasing the composites flame retardancy. TPU composites and PLA toughness and tensile strength increase due to the graphene oxide incorporation. Wash durability in PA66 maintains after various wash cycles [22]

The mechanical properties of polyurethane foams and flame retardancy are enhanced by the incorporation of phytic acid-functionalized graphene oxide into the intumescent flame retardants. PA-GO having phytic acid with graphene oxide is the new modification. It is then joined into RPUF and IFRs, including expandable graphite and ammonium polyphosphate. PA-GO of 0.3 wt.% enhances 26.5% from 25.7% LOI UL-94 rating also improves. It declines THR and PHRR 22.2% from 28.5%. It enhances char amount up to 24.6 wt.% from 22.2 wt.%, making a dense and continuous char layer having minimal holes, increasing the barrier effect when combustion happens [60].

Graphene oxide addition with the phosphaphenanthrene makes a new flame retardant. Researchers evaluate its effectiveness by increasing the epoxy resins composites flame retardancy. The loading amount is 4%. Composites get UL-94 rating and exhibiting outstanding flame-retardant properties. It provides physical shield and chemical char formation. It also reduces decomposed volatiles rates and combustion heat rate and release of oxygen rate as well. Overall, it increases mechanical properties [61].

2.5 Fullerenes

Following diamond and graphite, fullerene is the third most important form of carbon. Several members of the fullerene family are useful, but C60 is the most significant one [62]. Paul J. Krusic and his team initially reported on C60's high-capacity radical capture capability in 1991. Since

Materials Research Foundations 190 (2026) 77-117 https://doi.org/10.21741/9781644904077-4

then, research has progressively started on how C60 affects polymeric materials' thermal and flame-retardant qualities [63].

2.5.1 Mechanisms of Flame Retardancy

Fullerenes make polymers flame-retardant mainly by trapping free radicals. Combustion is an intricate process that involves several chain reactions. Free radicals, which are very reactive intermediates, are the main force behind these processes [64]. Fullerenes, especially C60, have a network of 30 conjugated double bonds and a distinctive spherical shape. Because of their structure, they can function as "radical sponges". Multiple free radicals can be efficiently removed from the combustion process by reacting with and being trapped by a single fullerene molecule. Acting as radical sponges as shown, fullerenes can retard the combustion by termination of a chain reaction followed by modification of degradation pathways.

Fullerenes absorb free radicals, which stops the chain processes that keep burning of the flame. This lowers the quantity of heat emitted and the speed of combustion [65]. This radical-trapping activity is associated with a gas-phase flame-retardant process. Fullerenes can modify the thermal breakdown routes in some polymers. For example, they can encourage non-radical breakdown pathways, which are frequently less exothermic and generate less flammable volatiles, and inhibit degradation caused by free radicals [66]. Mechanism of flame retardancy for fullerenes are described in Fig.8 below.

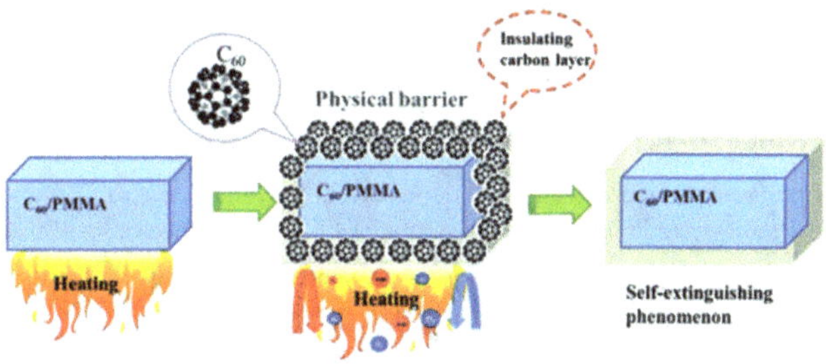

Figure 8: Mechanism of Flame Retardance for Fullerenes [68], when it burns, fullerene molecules are act as radical scavengers, which neutralize the active species in the flame zone. The mechanism breaks the chain reactions, reduces the temperature of a flame, and suppresses the burning intensity, in general.

2.5.2 Properties of Fullerenes as Flame Retardant Polymers

The fullerenes C60 and others have demonstrated potential for enhancing polymers' ability to withstand flames. These establish the polymer composites' mechanical characteristics, electrical conductivity, and thermal stability.

Thermal Stability

Fullerenes can increase the thermal stability of polymers in several ways. However, for them to be effective, they can neutralize free radicals. These radicals of the chain reaction, or combustion, were successfully trapped by the fullerenes, which stopped, or slowed down or stopped, the chain

Materials Research Foundations 190 (2026) 77-117 https://doi.org/10.21741/9781644904077-4

reaction. Fullerenes, on the other hand, can also promote the development of a char layer on the surface of the burning polymer. This char pile restricts the movement of heat, which is a component of combustion, and oxygen, which is also a component of combustion, by creating a barrier between the heat and the underlying substance. Fullerenes have been shown to increase thermal stability by raising the early breakdown temperature and lowering the peak heat release rate when added to polymers like polymethyl methacrylate (PMMA) [67].

Mechanical Properties

In order to enhance the mechanical characteristics of the underlying material, fullerenes are added as reinforcing agents to structural materials and polymers. These attributes include durability against impacts, elastic modulus, and elasticity of reinforcement. However, the amount of improvement that can be obtained is dependent on the homogeneous distribution of the fullerenes in the polymer. Applications where the polymer is under stress or receiving an impact, such as protective coatings or structural elements, can benefit from the enhanced mechanical properties [22].

Electrical Conductivity

By themselves, they have little conductivity, but they can assist in making certain polymer composites electrically more conductive. Fullerenes can create a conductive network inside the polymer matrix if they are present in high enough concentration. The percolation threshold is the name given to this concentration. Applications for improved electrical conductivity include energy storage devices, sensors, and antistatic materials. For example, the usage of polymers modified with fullerene in supercapacitors has been investigated [68].

2.5.3 Applications of Fullerenes in Flame Retardant Polymers

Fullerenes' extraordinary properties and unique spherical form have garnered plenty of interest in their potential as flame-retardant additives in a range of polymer matrices.

Use in Polylactic Acid (PLA)

As a sustainable substitute for conventional plastics made from petroleum, PLA is a biobased and biodegradable polymer that has grown in popularity. Its intrinsic flammability, however, restricts its use in several fields, like building and electronics. Researchers have looked into several methods, including adding fullerenes, to improve PLA's flame resistance. According to the study, utilizing fullerenes might significantly increase PLA polymers' resistance to flame. The limiting oxygen index (LOI) of the polymer that is the lowest amount of oxygen necessary for combustion, is greatly raised by modifications to the fullerenes because they improve the dispersibility into the PLA matrix. Additionally, the system's peak release rate of heat (PHRR), a crucial fire intensity measure, is reduced when fullerene is added. The fullerenes work together to scavenge radicals and create char when the PLA's fire resistance is increased [69].

Use in Polyamide 66 (PA66)

PA66, a thermoplastic with superior mechanical and chemical resistance, is a high-performance engineering thermoplastic. It is widely used in textiles, electronics, and automobile applications. In some cases, however, it is flammable, which poses a serious fire danger. Researchers have studied the use of fullerenes to improve the incombustibility of PA66. Upon burning, the fullerene molecules may cause the polymer chains to crosslink, and the so-formed carbonaceous char

Materials Research Foundations 190 (2026) 77-117 https://doi.org/10.21741/9781644904077-4

becomes an oxygen and a heat barrier. Fullerenes can also scavenge free radicals, which is the critical process in combustion, and can be used as flame retardants. Together, these methods significantly improve the flame retardant ability of PA66 [70].

Use in Thermoplastic Polyurethane (TPU)

TPU is an elastomer with outstanding impact strength, abrasion resistance, and flexibility. It is used in electronic gadgets, automobile parts, footwear, etc. However, due to its flammability, TPU cannot be used in fire-sensitive situations. Studies show that when fullerenes are created, TPU's mechanical properties can be improved, the flammability of the material can be decreased, and the thermal stability can be increased. Fullerenes, however, may increase TPU's flame resistance in a number of methods. Because fullerenes are radical scavengers and trap unstable molecules created in the burning process, they can initially slow down burning. Second, fullerenes have the ability to encourage the development of a char layer, which serves as an obstacle to the passage of heat as well as oxygen. Another benefit is that fullerenes can enhance the TPU's physical characteristics, such as its elasticity and tensile resistance [71].

It is suggested by a study that fullerene C60 in polypropylene (PP) nanocomposites leads to a higher flame-retardant property, reflected by lower THR and PHRR. The PP matrix's temperature resilience was enhanced by fullerene nanoparticles, which also made it possible for the defensive char coating to grow on the surface, protecting the material from heat and oxygen [72]. A follow-up investigation discovered that the Styrene-Butadiene-Styrene (SBS) block copolymer system was enhanced in the 85% region of vertical combustion grade by the use of C60 in conjunction with traditional flame retardants such as aluminum hydroxide (ATH). Fullerenes can effectively be utilized in conjunction with additional flame proofing agents to further improve performance, according to the Multiplicative outcome [73].

2.6 Carbon Black

The material known as carbon black (CB) is made up of small particles, the majority of which are elemental carbon. It results from thermal breakdown or incomplete combustion of hydrocarbon feedstock's such as oil, natural gas, or coal tar. These processes then produce primary particles that aggregate to form aggregates, which may further agglomerate into larger structures. The main characteristics of carbon black important for its uses are particle size, structure (the degree of aggregation), and surface chemistry. These properties can be adjusted during the manufacture of everything from pigments in inks and coatings to reinforcing fillers in tires and rubbers to achieve certain performance attributes in a diversity of applications.

2.6.1 Mechanisms of Flame Retardancy

More recently, carbon black is being used as functional additives in polymers specially for flame retardancy [74]. Several processes as to how in polymers, carbon black works as a flame retardant are outlined as follows:

Development of a Layer of Protective Char

When exposed to elevated temperatures and flame, carbon black can help a polymer form a stable char coating on its exterior [23]. The quantity of oxygen that may reach the combustion zone is physically limited by this char coating, which also shields the material underneath it from further heat by eliminating oxygen from that area. Research indicates that carbon black can activate a

Materials Research Foundations 190 (2026) 77-117 https://doi.org/10.21741/9781644904077-4

polymer matrix carbonization process and synthesize a more robust and cohesive char structure [75].

Suppression of Smoke

It is thought that the increased char formation brought on by carbon black reduces the emission of volatile organic chemicals, which make up the majority of smoke [76]. Additionally, some research suggests that carbon black and other flame retardants work in concert to reduce the formation of smoke [77]. The principal mechanism by which carbon black undergoes catalytic char formation is condensed-phase, but with some dilution effects in the gas-phase.

Free Radical Barrier

By capturing free radicals produced during the thermal breakdown of polymers, carbon black nanoparticles can improve flame retardancy [78]. The mechanism of carbon black based flame retardancy is explained in Fig.9.

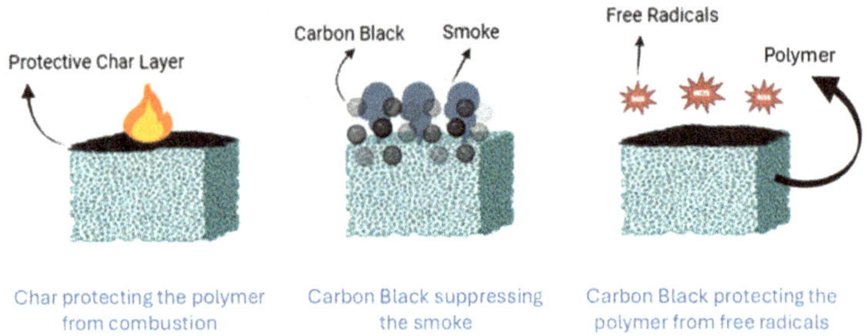

Figure 9: Mechanism of Flame Retardancy by Carbon Black, the condensed-phase action of carbon black, catalyst of char formation and a stabilizer of the protective layer made of carbonaceous material. The figure shows also partial effects of gas-phase, where carbon black particles dilute oxygen concentration and decrease the development of smoke.

2.6.2 Properties of Carbon Black as Flame Retardant Polymers

Several unique properties that affect polymer behavior during combustion contribute to carbon black's effectiveness. These properties are as follows:

Thermal Stability

Because of its strong thermal stability, carbon black can tolerate high temperatures without decomposing much. The CB particles in a polymer stay stable and help to build a protective char coating when it is exposed to fire. This char coating slows down the breakdown process by separating the deeper material from the temperature source. As a result, CB's exceptional thermal stability helps to postpone ignition and reduce the burning rate of the polymer [79].

Materials Research Foundations 190 (2026) 77-117 https://doi.org/10.21741/9781644904077-4

Mechanical Properties

CB can occasionally improve stiffness and mechanical strength. By serving as reinforcing fillers, the fine CB particles can increase the polymer's resistance to deformation. High levels of CB, however, might occasionally result in a reduction in toughness and flexibility. When designing flame-retardant polymer composites, striking a compromise between mechanical performance and flame retardancy is crucial [80].

Electrical Conductivity

CB can greatly improve the electrical conductivity of a composite when added to a polymer matrix. Although unrelated to flame retardancy, this characteristic may be useful in some situations where electrical conductivity and flame resistance are both necessary. For instance, CB can assist in dissipating static charge and lowering the risk of fire in cables and electronic packages [81].

2.6.3 Applications of Carbon Black in Flame Retardant Polymers

Carbon black is utilized in many diverse sectors, including technological devices, automotive, as well as construction, since it may be helpful in enhancing fire safety in various polymer matrices.

Thermoplastic Polyurethane (TPU)

The flame resistance and the reduction of smoke emissions of TPU composites are commonly achieved by using the CB. CB is likely to work cooperatively with another flame retardant, such as APP, in TPU to promote the char formation and decrease smoke emission during the combustion process. This is critical in any application where it is required, like cable jacketing and automobile parts etc. [82].

Polyamide 66 (PA66)

Furthermore, CB may greatly enhance PA66's endurance to temperature and resistance to fire. A char-forming agent, it forms a protective coating that prevents the material from burning further. PA66 is used in the production of connectors and any other electrical components for its mechanical strength and thermal resistance, which is very useful [83].

Polylactic Acid (PLA)

PLA is combustible by nature, despite being a bio-based polymer with expanding uses. CB could be added to PLA polymers to increase their durability against flames. Studies show that CB can successfully lower PLA's flammability and raise its limiting oxygen index (LOI), enabling it to be used in more demanding applications [84]. Depending on the environment around them, CB-filled flame-retardant polymers may operate differently.

High-Temperature Environments

The great thermal stability of CB is essential in hot conditions or areas that might catch fire. It lessens the chance of a fire spreading and helps preserve the integrity of the polymer composite. For example, CB-filled polymers offer an extra degree of protection in automotive and aerospace applications where materials are subjected to high temperatures [85].

Humid Environments

Certain polymers' ability to withstand flames might be impacted by humidity. On the other hand, CB performs consistently in humid environments. During the compressed phase, it mostly promotes the formation of char, an event which is not much influenced by moisture [86].

Chemical Exposure

The stability of the flame-retardant ingredient is essential in settings where polymers are subjected to chemicals. Carbon black (CB) chemical inertness and resistance to several chemicals ensure long-term flame-retardant effectiveness. This is crucial in industrial environments where corrosive materials may come into contact with polymers [87].

Several studies show that carbon black works well as a flame retardant in polymers. For example, scientists have looked into how CB might improve polypropylene (PP) composites' ability to withstand flames. Researchers found that combining CB with carbon fibers significantly reduced the rate of heat escape and improved thermal stability. This enhancement is due to the creation of a robust 3-D web framework that effectively prevents combustion [88]. Additionally, studies on high-impact polystyrene (HIPS) have demonstrated that flame-retarded HIPS composites may greatly increase their thermal and thermo-oxidative durability by using a modest quantity of CB. Interestingly, CB successfully stops leaking while burning, which is an important aspect of fire safety [89]. Additionally, an example research examined the effects of CB on the fire resistance of a number of ethylene vinyl acetate (EVA) blends including magnesium hydroxide (MH) and huntite hydromagnesite (HH). The study says CB improves these mixes' flame retardancy, giving wire and cable applications a better fire retardant characteristic [90]. These case studies highlight how versatile and successful CB is in improving the flame retardancy of different polymer systems.

2.7 Nano-Diamonds in Flame Retardant Polymers

Nanodiamonds (NDS) represent diamond particles consisting of particles where one dimension extends less than 100 nanometers. Nano-diamonds manifest distinctive characteristics because they have exceptional hardness, thermal conductivity, and a high surface area that sets them apart from bulk diamond crystals. The production of Nano-diamonds arises from detonation procedures or carbon-rich materials high-pressure, high-temperature heat treatment. The technical potential of nano-diamonds has increased in recent times because of their exceptional mechanical and chemical traits and thermal behavior that makes them useful for electronics and medical uses, and material science requirements. Their diameter sizes, starting from 4–5 nm extending to 20 nm, make nano-diamonds suitable for enhancing the properties of various polymers and composites [91].

Various industries use polymers extensively because they provide economic advantages, lightweight qualities, and flexible features. These materials naturally combust and thus create major fire risks across multiple industries, including electronic devices, textiles, automotive components and construction products. Trusted safety-sensitive functions depend on flame retardancy as an essential characteristic for polymers [92]. Flame retardants comprising halogenated compounds, phosphates, and metal hydroxides have been used for long periods. The materials have unwanted drawbacks, including toxicity effects, environmental issues, and reduced mechanical characteristics. The market requires better flame retardants which both safeguard the matrix performance and maintain environmental safety standards. Flame retardant capabilities of

Materials Research Foundations 190 (2026) 77-117 https://doi.org/10.21741/9781644904077-4

polymers receive substantial improvement through nano-diamond incorporation as a result of their fundamental properties [93].

2.7.1 Mechanisms of Flame Retardancy

The flame retardancy mechanisms apply physical and chemical means to block burning processes. The mechanisms responsible for fire resistance in polymers are supported by nano-diamonds (NDS) because they bolster the thermal properties and suppress fire development.

2.7.1.1 Physical Mechanisms

Formation of Protective Char Layer

Heat exposure enables nano-diamonds to create a protective carbon char layer that forms on the surface of polymers. A protective barrier layer forms on top of the underlying material, protected by the combustion zone, because of nano-diamonds. This slows down flame propagation. Under high temperatures, the persistent protective layer develops from nano-diamonds because these compounds retain their stability. A thick and durable char layer formation acts as a flame-delaying system, protecting the bulk polymer from thermal transfer by minimizing heat propagation. Studies confirmed that modification of polystyrene (PS) and polyamide 66 (PA66) with nano-diamonds produced substantial char development in thermal conditions, leading to enhanced flame protection because of decreased material heat influx [94]. Mostly, nano-diamond products are inert in the gaseous state and they function at the condensed state by increasing the thermal stability and char development.

Intumescence and Barrier Formation

Nano-diamonds enable polymers to show intumescent behavior during heating processes that form protective foam insulation. The enlarged char layer is a thermal protector and physical defender that hinders oxygen access to burning polymers. The uniformity and stability of the intumescent layer improved through nano-diamonds that enhance material retention and prevention of thermal degradation and physical cracking.

2.7.1.2 Chemical Mechanisms

Radical Scavenging

The burning process accelerates because combustion produces high-energy free radicals, including hydroxyl (OH•), alkyl (R•) and oxygen (O•) types. The radical scavenging property of nano-diamonds traps reactive species, thus stopping them from continuing the combustion reaction. The flame retardancy benefits from nano-diamonds stem from their ability to eliminate free radicals by interaction, weakening the combustion reaction intensity [95].

Heat Release Reduction

Adding nano-diamonds to polymer composites makes them more effective at heat transfer because it speeds up the heat dissipation during flame conditions. Heat distribution within localized regions remain minimal because of which ignition temperature and combustion rate. Nano-diamonds provide superior thermal conductivity that protects the crystalline structure of polymers during extended hot conditions.

Materials Research Foundations 190 (2026) 77-117
https://doi.org/10.21741/9781644904077-4

Interference with Combustion Gases

When polymers fuel combustion reactions, they generate the toxic substances carbon monoxide (CO) and carbon dioxide (CO2) and volatile organic compounds (VOCS). Scientific studies indicate that nano-diamonds act as chemicals, which permit them to capture combustion gases and lower dangerous fume and smoke production in fires. The material safety in fire-prone conditions becomes better because of this fire suppression achievement. During combustion tests, polyamide-based composite research showed nano-diamonds reduced flame-retardant and smoke suppression properties by reducing toxic gas emissions, particularly CO and CO2 [96].

2.7.2 Properties of Nano-Diamonds as Flame Retardant Polymers

Nano-diamonds offer unique properties and strong enhancement capabilities for polymers in flame retardancy applications. The flame-retardant polymer composites receive significant value from nano-diamonds because of their multiple properties, including thermal stability, mechanical reinforcement, and electrical conductivity.

Thermal Stability

Nano-diamonds retain thermal stability because the strong sp^3 carbon-carbon bond framework resists decomposition at high temperatures. Adding nano-diamonds to polymers improves thermal stability through elevated decomposition temperature, enhancing fire resistance. PA66 polyamide receives a 40°C boost in its degradation temperature when nano-diamonds are added to the material structure. The improved thermal resistance begins later because nano-diamonds extend the material's resistance to burning during high-heat conditions [97]. When polylactic acid (PLA) composites incorporate nano-diamonds, they become more resistant to thermal decomposition as the thermal decomposition onset progresses by more than 100°C. Numerous studies have confirmed that nano-diamonds enhance PLA matrix thermal stability because of their remarkable heat dissipation properties.

Mechanical Properties

Incorporating nano-diamonds in polymer matrices enables reinforcement by improving mechanical qualities, including strength, toughness and elasticity. The extensive surface area of nano-diamonds enables strong interface bonds with the polymer molecules, producing a tougher composite structure. When nano-diamonds are added to their structure, polypropylene (PP) composites become stronger against impact forces. The composite's fracture toughness surpassed neat PP values by 30% yet maintained equivalent flame retardancy levels.

Electrical Conductivity

Nano-diamonds offer exceptional electrical features that enable their use in application environments needing flame-retardant materials, which are also electrically conductive. The electrical conductivity improvement capability of nano-diamonds in polymer composites creates opportunities for electronics applications and the automotive and aerospace industries. Incorporating nano-diamonds into epoxy-based composites improves electrical conductivity, making the material appropriate for environments needing fire resistance and electrical conductivity capabilities. Nano-diamond-infused epoxy composites serve electronic components well due to their enhanced conductivity and superior flame retardancy characteristics in heating environments and potential fire threat situations. Incorporating nano-diamonds into conductive polymers allows for simultaneous flame retardancy maintenance and required electrical functionality. Such capabilities make them excellent for circuit boards, automotive electronics, and sensors because they fulfil fire safety requirements without compromising electrical functioning [98]. A brief description of nano-diamonds based flame retardants with their application, characteristics, properties and mechanism is shown in Fig. 10.

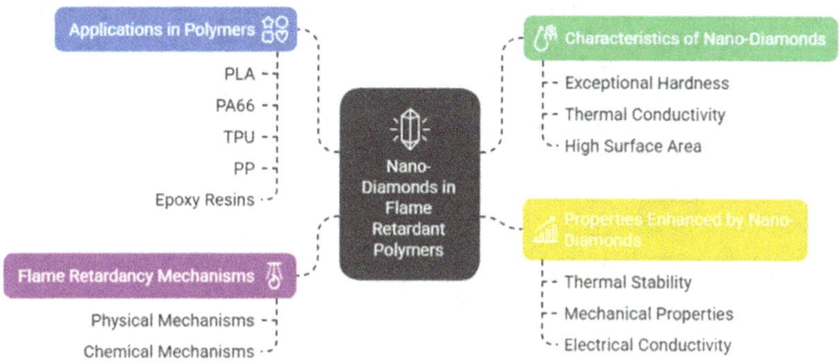

Figure 10: Flow Diagram for Nano-Diamonds in Flame Retardancy.

2.7.3 Applications of Nano-Diamonds in Flame Retardant Polymers

Researchers have used nano-diamonds to boost the flame-retardant properties, thermal stability, mechanical performance, and overall properties of different polymers. The electronics, automotive, aerospace, and construction industries strongly benefit from using nano-diamonds because fire safety remains their top priority. Several important applications of nano-diamonds exist when they are used with different polymer matrices.

2.7.3.1 Use in Various Polymer Matrices

Polylactic Acid (PLA)

PLA emerges as a sustainable, biodegradable material that can be sourced from corn starch and sugarcane. The flame-retardant properties of PLA fail to reach satisfactory levels for particular uses despite its eco-friendly characteristics. Biodegradability remains intact when researchers combine nano-diamonds into polylactic acid to enhance its fire-resistant performance [99].

The combination of PLA polymeric materials containing nano-diamonds shows diminished peak heat release rate (PHRR) and reduced smoke output according to both TGA and cone calorimeter results, thus making this eco-friendly material ready for sustainable packaging and electronic product applications.

Polyamide 66 (PA66)

PA66 is a high-performance engineering thermoplastic suitable for automotive applications, electrical use, and industrial needs because it possesses both mechanical strength and excellent chemical resistance. The natural combustibility of this material restricts its application in particular

Materials Research Foundations 190 (2026) 77-117 https://doi.org/10.21741/9781644904077-4

high-risk scenarios. Incorporating Nano-diamonds enables PA66 to become flame retardant, yet maintains all its mechanical characteristics [100]. Adding nano-diamonds to PA66 composites produces advanced thermal degradation resistance because the material retains elevated decomposition temperatures while generating better char stability during fire conditions.

Thermoplastic Polyurethane (TPU)

The deformable material called Thermoplastic polyurethane (TPU) exhibits strength, adaptability, and resistance to friction damage. TPU material gains enhanced flame resistance and natural flexibility when nano-diamonds are added to its composition. TPU composites reinforced with nano-diamonds attain superior tensile strength and better elongation at break characteristics, qualifying them for functional uses that need flexible fire-resistant materials, including protective coatings, textiles, and automotive components. The char forming process strengthens because nano-diamonds minimize flame growth while decreasing heat release in fire conditions [100].

Polypropylene (PP)

Polypropylene (PP) is a leading thermoplastic polymer that provides low production expenses and superlative resistance against chemical substances. The property of poor flame retardancy constitutes a barrier to using this material in situations where safety is essential. Nano-diamonds succeed in enhancing PP flame retardancy capabilities while leaving mechanical properties along processing characteristics unaffected. Adding nano-diamonds to PP creates materials with improved thermal stability because the degradation temperature changes and heat release rate decreases during combustion. Integrating nano-diamonds with polypropylene results in complete resistance to fire while reducing smoke volume, enabling its application in building materials, automotive parts, and electrical components [101].

Epoxy Resin Composites

The widespread use of epoxy resins fails to meet fire safety standards, restricting their applications where flammability safety is critical. When epoxy resin receives nano-diamond additions, its thermal stability and mechanical qualities experience better outcomes. Incorporating nano-diamonds into epoxy composites results in better flame retardant characteristics because these composites form more effective char and dissipate heat better, reducing heat release rates and smoke output [102]. Nano-diamonds embedded into epoxy composites give the material superior impact tolerance and breakage resilience. This enables suitable application in aerospace and automotive areas, together with electronics, where material reliability and fire protection matter most [103].

2.7.4 Representative Studies and Findings of Nano-diamonds as a Flame Retardants

Numerous industries study the implementation of nano-diamonds within flame-retardant materials that constitute polymers. The following case studies show countless practical examples and benefits of using nano-diamonds in flame-retardant polymer composites for sustainable packaging.

This research aimed to develop improved flame retarding abilities for Polylactic Acid (PLA) as a biodegradable material so it could become suitable for sustainable consumer packages and packaging applications [104].

Materials Research Foundations 190 (2026) 77-117 https://doi.org/10.21741/9781644904077-4

An investigation included admixing nano-diamonds with PLA through a melt blending process. The researchers assessed the flame retardance of PLA modified with nano-diamonds through tests performed with cone calorimeters and UL 94 vertical burning methods.

Nano-diamonds resulted in a 30% decrease of peak heat release rate (PHRR) and smoke production rate (SPR) compared to pure PLA. The thermal decomposition temperature of PLA increased by more than 100 degrees Celsius through this modification. Research outcomes showed that nanodiamonds provided improved fire protection to PLA materials without compromising their decomposition properties [96].

The use of nano-diamonds represents an optimal strategy to enhance PLA fire resistance qualities, making it well-suited for sustainable applications that require fire safety.

When applied to automotive engineering components, a research study investigates Flame Retardancy in Polyamide 66 (PA66).PA66 received nano-diamond reinforcement through an industrial extrusion technique. Vertical burning tests and thermogravimetric analysis (TGA) and impact testing method assessed the flame retarding capabilities of the produced PA66 composites.

PA66 plastic reinforced with nano-diamond achieved better thermal resistance because its decomposition point shifted upwards by 40°C from pure PA66. The material achieved better tensile strength performance and impact resistance characteristics due to its exposure to harsh conditions. The tested specimen experienced a marked decrease in both the spread of flame rate and the smoke production values. PA66 material became fit for automobile interior components and electrical housings after nano-diamond reinforcement improved its combination of flame retardancy and mechanical strength [105].

In order to give a better picture of the presented studies, comparative performances of different carbon based nanocomposites in enhancing flame retardancy are as summarized in Table 1. This table shows important materials, polymer matrices, and the resulting increase in thermal stability and limiting oxygen index (LOI). This type of a comparative analysis highlights how diverse carbon-based nanomaterials are, as well as highlighting their different mechanisms and effectiveness in the promotion of flame-retardant behavior.

Materials Research Foundations 190 (2026) 77-117 https://doi.org/10.21741/9781644904077-4

Table 1: Comparison of the flame-retardant properties of several carbon-based nanocomposites.

Sr. No.	Polymer System	Carbon Nanofiller (Loading)	Synergistic Additive (Loading)	Key Performance Improvement	Primary Flame-Retardant Mechanism	Reference
1	**Epoxy Resin**	Multi-Walled Carbon Nanotubes (0.5-1 wt%)	Phosphorus-based compound (e.g., DOPO)	Achieves top self-extinguishing rating (UL-94 V-0). Oxygen required for burning increases significantly (LOI > 30%).	Synergistic; Phosphorus quenches flames in the gas phase, while CNTs form a protective, heat-resistant char layer.	[106]
2	**Polypropylene (PP)**	Graphene Nanoplatelets (1-3 wt%)	Intumescent Flame Retardant (IFR)	Peak heat release rate is drastically reduced (by over 70%). Creates a strong, swollen char that insulates the polymer.	Synergistic; Graphene reinforces the char, making it a more effective and stable physical barrier against heat and flames.	[107]
3	**Polylactic Acid (PLA)**	Graphene Oxide (GO) (2-5 wt%)	Metal Hydroxides (e.g., ATH) or Clays	Becomes self-extinguishing (UL-94 V-0). The temperature at which it decomposes increases, enhancing stability.	GO creates a "tortuous path" that slows down fuel and heat, while also forming a protective char layer.	[108]
4	**Polyurethane Foam**	Carbon Nanotubes (1-2 wt%)	-	Resists melting and dripping when exposed to fire. Peak heat release rate is reduced significantly.	CNTs form a network that stabilizes the foam structure, promotes char formation, and acts as a heat shield.	[109]
5	**Synthetic Fabrics (e.g., Polyester)**	Carbon Nanotubes or Graphene (Coating)	-	The fabric stops burning immediately after the flame is removed and shows minimal damage.	The carbon coating acts as a physical barrier, shielding the fabric from the flame and dissipating heat.	[110]
6	**Polyurethane Foam**	Carbon Nanotubes (1-2 wt%)	-	Resists melting and dripping when exposed to fire. Peak heat release rate is reduced significantly.	CNTs form a network that stabilizes the foam structure, promotes char formation, and acts as a heat shield.	[111]

3. Challenges and Limitations

Carbon-based nanomaterials (such as graphene, carbon nanotubes, and carbon black) have demonstrated remarkable promise as flame retardants in polymers and composites, offering superior performance, multifunctionality, and environmental benefits compared to conventional

Materials Research Foundations 190 (2026) 77-117 https://doi.org/10.21741/9781644904077-4

additives. However, their widespread adoption and practical implementation face several critical challenges and limitations that must be carefully addressed.

Dispersion and Compatibility Issues

Achieving a uniform distribution of carbon nanomaterials within polymers is a major problem. The strong van der Waals and π-π stacking interactions in carbon nanotubes (CNTs) and graphene often result in aggregation, making it challenging for them to stick to other materials and reducing the uniformity of their flame retardation. Agglomeration weakens the reinforcement and it also lowers the quality of char and the fire barrier. Functionalizing the surface with polymers or surface groups usually improves compatibility and dispersion, but the complex chemistry needed may add to both the costs and extra steps in the process.

Impact on Optical and Mechanical Properties

Carbon nanomaterials can make a material stronger and more resistant to heat, but their addition might negatively change other features of the material. As an example, CNTs addition in transparent polymers such as poly(methyl methacrylate) (PMMA) can greatly decrease transparency which may be unwanted for its use as coatings or films. While tensile and impact strength are usually preserved or increased, it is still challenging to find a balance between flame retardancy, transparency and how easily these materials can be processed.

Processing and Scalability Challenges

It is challenging and expensive to add functionalization and bring together carbon nanomaterials and polymers on an industrial scale. The process of large-scale, defect-free nanomaterial production is still very costly and uses a lot of energy. Additionally, obtaining uniform results and performance in common manufacturing processes (such as extrusion and injection molding) is not easy and may need additional tools or chemicals, resulting in increased costs and more complexity.

Environmental and Health Concerns

Carbon nanomaterials are believed to have fewer negative effects on health and the environment than most other flame retardants, yet concerns about their influence at the nanoscale are not fully resolved. Handling, manufacturing or disposing of nanomaterials may put workers at risk of inhaling them and we do not yet know the long-term effects these materials have on the environment and health. As regulations and standard risk assessment procedures are not yet complete for nanomaterials, there is confusion among manufacturers and users of such materials.

Limited Synergy without Functionalization or Hybridization

Although carbon nanomaterials work well at low concentrations, their best flame retardant results are achieved when combined with phosphorus or nitrogen compounds. Still, to ensure efficient synergy, hybrid systems should be well designed and may need challenging processing or alterations. For illustration, CNTs modified with tannic acid and coated on ammonium polyphosphate (APP) have improved flame resistance and lessened loss of strength, but such systems must be carefully managed to maintain the required structures and interactions.

Cost and Market Penetration

It is still quite costly to use carbon nanomaterials like graphene and high-purity CNTs for flame retardancy which keeps them from being used more widely by industry. Although the cost of phosphorus-based flame retardants has dropped, it is still much higher than for standard flame retardants, making them suitable only for high-value or specialized goods.

Durability and Long-Term Performance

Durability and stability of Carbon nanomaterial-based flame retardants have not been proven under normal conditions like sun exposure, moisture and wear. Nanomaterials may gradually degrade or leach which could weaken their flame retardant properties and bring environmental risks.

Recyclability and End-of-Life Management

When polymers are mixed with carbon nanomaterials, recycling them may become more difficult. Although it has been suggested that recycling nanocomposites is possible without much loss of properties, having nanomaterials in waste may make it harder for standard recycling technologies and requires special methods.

4. Future Perspective

Carbon nanomaterials, including graphene, carbon nanotubes, pristine, fullerenes, nano-diamonds and carbon black, have been found to offer better flame retardancy, improved strength and multiple useful properties at low loadings than traditional halogenated and phosphorus-based substances. Future research is likely to concentrate on fixing existing challenges, like making sure nanoparticles are evenly distributed in polymer matrices and finding ways to reduce the cost of high-quality nanomaterials. Such advances may result in better compatibility, stronger performance and the possibility of combining these new formulations with standard retardants. To succeed in the market, companies will require advanced manufacturing techniques and the development of standard testing methods to guarantee safety, performance and environmental friendliness. Concerns about sustainability and the environment will remain challenging. Although carbon nanomaterials are less likely to be toxic than conventional retardants, they should go through life-cycle and toxicity testing to identify potential risks from release of nanoparticles during each step. Combining carbon nanomaterials with phosphorus or nitrogen additives in intumescent systems should result in more powerful, hardy and environmentally friendly flame retardants. For carbon nanomaterial-based flame retardants to be used, more research, industry work and strong regulations are necessary to develop better and more environmentally friendly fire safety products.

Conclusion

Carbon nanomaterials could become the new generation of flame retardants because they offer multiple benefits, are environmentally friendly and perform better than traditional options at lower concentrations. Nevertheless, it is important to find solutions for dispersion, processing, cost, risks to the environment and health and durability before their use can spread widely. Investigating functionalization, hybrid combinations and practical manufacturing is needed to address these issues and take advantage of carbon nanomaterials for fire safety.

References

[1] V. Mittal, Advances in synthesis and properties of engineering polymers, J. Compos. Mater. 2015, 1-13. https://doi.org/10.1016/B978-1-78242-308-9.00001-X

[2] C.R. Lima, N.F. de Souza, F. Camargo, Study of wear and corrosion performance of thermal sprayed engineering polymers, Surf. Coat. Technol. 220 (2013) 140-143. https://doi.org/10.1016/j.surfcoat.2012.05.051

[3] B. Yu, B. Yuan, B. Tawiah, W. Wu, Y. Hou, Editorial: fire risk and prevention technologies of polymeric and composite materials, Front. Mater. 10 (2023) 168. https://doi.org/10.3389/fmats.2023.1129598

[4] M. Borucka, K. Mizera, J. Przybysz, P. Kozikowski, A. Gajek, Analysis of flammability and smoke emission of plastic materials used in construction and transport, Mater. 16 (2023) 2444. https://doi.org/10.3390/ma16062444

[5] N. Tagiyeva, A. Sheikh, Domestic exposure to volatile organic compounds in relation to asthma and allergy in children and adults, Expert Rev. Clin. Immunol. 10 (2014) 1611-1639. https://doi.org/10.1586/1744666X.2014.972943

[6] A. Valavanidis, N. Iliopoulos, G. Gotsis, K. Fiotakis, Persistent free radicals, heavy metals and PAHs generated in particulate soot emissions and residue ash from controlled combustion of common types of plastic, J. Hazard. Mater. 156 (2008) 277-284. https://doi.org/10.1016/j.jhazmat.2007.12.019

[7] R.N. Walters, R.E. Lyon, Flammability of polymer composites, Off. Aviat. Res. Dev. Fed. Aviat. Admin. 2008.

[8] B.W. Liu, H.B. Zhao, Y.Z. Wang, Advanced flame-retardant methods for polymeric materials, Adv. Mater. 34 (2022) 2107905. https://doi.org/10.1002/adma.202107905

[9] M. Venier, A. Salamova, R.A. Hites, Halogenated flame retardants in the Great Lakes environment, Acc. Chem. Res. 48 (2015) 1853-1861. https://doi.org/10.1021/acs.accounts.5b00180

[10] S. Levchik, Phosphorus-based flame retardants, Non-Halogenated Flame Retard. Handb. (2021) 23-99. https://doi.org/10.1002/9781119752240.ch2

[11] A.B. Morgan, M. Klatt, Nitrogen-based flame retardants, Non-Halogenated Flame Retard. Handb. (2021) 236-270. https://doi.org/10.1002/9781119752240.ch5

[12] K.-S. Lim, S.-T. Bee, L.T. Sin, T.-T. Tee, C. Ratnam, D. Hui, A. Rahmat, A review of application of ammonium polyphosphate as intumescent flame retardant in thermoplastic composites, Compos. Part B Eng. 84 (2016) 155-174. https://doi.org/10.1016/j.compositesb.2015.08.066

[13] P. Singh, S. Pandit, M. Sinha, D. Yadav, R. Parthasarathi, Computational risk assessment of persistence, bioaccumulation, and toxicity of novel flame-retardant chemicals, J. Phys. Chem. A 127 (2023) 10747-10757. https://doi.org/10.1021/acs.jpca.3c04160

[14] S. Yasin, N. Behary, M. Curti, G. Rovero, Global consumption of flame retardants and related environmental concerns: a study on possible mechanical recycling of flame-retardant textiles, Fibers 4 (2016) 16. https://doi.org/10.3390/fib4020016

[15] D.Q. Brito, C. Henke-Oliveira, E.C. Oliveira-Filho, Acute toxicity of commercial wildfire retardants to two daphniid species (Ceriodaphnia dubia and Daphnia magna), Toxics 12 (2024) 548. https://doi.org/10.3390/toxics12080548

[16] D.M. Backer, S.E. Jensen, G.R. McPherson, Impacts of fire-suppression activities on natural communities, Conserv. Biol. 18 (2004) 937-946. https://doi.org/10.1111/j.1523-1739.2004.494_1.x

[17] E. Awere, P.A. Obeng, A. Bonoli, P.A. Obeng, E-waste recycling and public exposure to organic compounds in developing countries: a review of recycling practices and toxicity levels in Ghana, Environ. Technol. Rev. 9 (2020) 1-19. https://doi.org/10.1080/21622515.2020.1714749

[18] P.J. Chen, P.C. Lai, Y.C. Lu, B.L. Pan, W.T. Huang, C.T. Kung, J.C. Chiang, F.J. Cheng, L.J. Wang, S.H. Li, Association between exposure to organophosphate flame retardants and epidermal growth factor receptor expression in lung cancer patients, Thorac. Cancer 15 (2024) 1805-1814. https://doi.org/10.1111/1759-7714.15411

[19] J. Feiteiro, M. Mariana, E. Cairrão, Health toxicity effects of brominated flame retardants: from environmental to human exposure, Environ. Pollut. 285 (2021) 117475. https://doi.org/10.1016/j.envpol.2021.117475

[20] N.E. Kramer, C.E. Fillmore, E.G. Slane, L.M. Barnett, J.J. Wagner, B.S. Cummings, Insights into brominated flame retardant neurotoxicity: mechanisms of hippocampal neural cell death and brain region-specific transcriptomic shifts in mice, Toxicol. Sci. 201 (2024) 282-299. https://doi.org/10.1093/toxsci/kfae090

[21] H. Mao, T. Lin, S. Huang, Z. Xie, S. Jin, X. Shen, Y. Jin, Y. Ding, The impact of brominated flame retardants (BFRs) on pulmonary function in US adults: a cross-sectional study based on NHANES (2007-2012), Sci. Rep. 14 (2024) 6486. https://doi.org/10.1038/s41598-024-57302-9

[22] Y. Yang, J.L. Díaz Palencia, N. Wang, Y. Jiang, D.-Y. Wang, Nanocarbon-based flame-retardant polymer nanocomposites, Molecules 26 (2021) 4670. https://doi.org/10.3390/molecules26154670

[23] Q. Qu, J. Xu, H. Wang, Y. Yu, Q. Dong, X. Zhang, Y. He, Carbon-nanotube-based intumescent flame retardants achieve high-efficiency flame retardancy and simultaneously avoid mechanical property loss, Polymers 15 (2023) 1406. https://doi.org/10.3390/polym15061406

[24] F. Teles, G. Martins, F. Antunes, Fire retardancy in nanocomposites by using nanomaterial additives, J. Anal. Appl. Pyrolysis 163 (2022) 105466. https://doi.org/10.1016/j.jaap.2022.105466

[25] W. Soutter, Nanomaterials for environmentally friendly flame retardants, 2012.

[26] K.M. Holder, R.J. Smith, J.C. Grunlan, A review of flame-retardant nanocoatings prepared using layer-by-layer assembly of polyelectrolytes, J. Mater. Sci. 52 (2017) 12923-12959. https://doi.org/10.1007/s10853-017-1390-1

[27] L. Chen, Y.Z. Wang, A review on flame-retardant technology in China. Part I: development of flame retardants, Polym. Adv. Technol. 21 (2010) 1-26. https://doi.org/10.1002/pat.1550

[28] N.G. Silva, N.C. Zanini, A.G. de Souza, R.F. Barbosa, D.S. Rosa, D.R. Mulinari, Halogen-based flame retardants in polyurethanes, Mater. Chem. Flame-Retard. Polyurethanes Vol. 1:

A Fundam. Approach, ACS Publ. (2021) 141-171. https://doi.org/10.1021/bk-2021-1399.ch007

[29] S.V. Levchik, Introduction to flame retardancy and polymer flammability, Flame Retard. Polym. Nanocompos. (2007) 1-29. https://doi.org/10.1002/9780470109038.ch1

[30] J. Mena, S. Vera, J.F. Correal, M. Lopez, Assessment of fire reaction and fire resistance of Guadua angustifolia Kunth bamboo, Constr. Build. Mater. 27 (2012) 60-65. https://doi.org/10.1016/j.conbuildmat.2011.08.028

[31] M. Lewin, E.D. Weil, Mechanisms and modes of action in flame retardancy of polymers, Fire Retard. Mater. (2001) 31-68. https://doi.org/10.1533/9781855737464.31

[32] G.H. Yeoh, I.M. De Cachinho Cordeiro, W. Wang, C. Wang, A.C.Y. Yuen, T.B.Y. Chen, J.B. Vargas, G. Mao, U. Garbe, H.T. Chua, Carbon-based flame retardants for polymers: a bottom-up review, Adv. Mater. 36 (2024) 2403835. https://doi.org/10.1002/adma.202403835

[33] Y. Zhang, C. Shi, X. Qian, J. Jing, L. Jin, DOPO/silicon/CNT nanohybrid flame retardants: toward improving the fire safety of epoxy resins, Polymers 14 (2022) 565. https://doi.org/10.3390/polym14030565

[34] F. Chen, Y. Huang, R. Li, S. Zhang, Q. Jiang, Y. Luo, B. Wang, W. Zhang, X. Wu, F. Wang, Super-durable and fire-retardant structural coloration of carbon nanotubes, Sci. Adv. 8 (2022) 5882. https://doi.org/10.1126/sciadv.abn5882

[35] P. Cui, A.-J. Wang, Synthesis of CNTs/CuO and its catalytic performance on the thermal decomposition of ammonium perchlorate, J. Saudi Chem. Soc. 20 (2016) 343-348. https://doi.org/10.1016/j.jscs.2014.09.010

[36] P. Yang, H. Liu, L. Meng, Y. Bai, J. He, Constructing independent CNT conductive networks to boost combustion performance of metastable intermixed composites, Polym. Compos. 45 (2024) 4994-5008. https://doi.org/10.1002/pc.28104

[37] S. Lee, H.M. Kim, D.G. Seong, D. Lee, Synergistic improvement of flame-retardant properties of expandable graphite and multi-walled carbon-nanotube-reinforced intumescent polyketone nanocomposites, Carbon 143 (2019) 650-659. https://doi.org/10.1016/j.carbon.2018.11.050

[38] X. Ji, D. Chen, Q. Wang, J. Shen, S. Guo, Synergistic effect of flame retardants and carbon nanotubes on flame-retarding and electromagnetic-shielding properties of thermoplastic polyurethane, Compos. Sci. Technol. 163 (2018) 49-55. https://doi.org/10.1016/j.compscitech.2018.05.007

[39] W. Zhan, L. Ni, Z. Gu, F. Cui, J. Jiang, L. Chen, The influences of graphene and carbon nanotubes on properties of waterborne intumescent fire-resistive coating, Powder Technol. 385 (2021) 572-579. https://doi.org/10.1016/j.powtec.2021.03.018

[40] M.R. Mousavi, A.C. Tehran, K. Shelesh-Nezhad, Study on morphology, mechanical, thermal and viscoelastic properties of PA6/TPU/CNT nanocomposites, Plast. Rubber Compos. 49 (2020) 400-413. https://doi.org/10.1080/14658011.2020.1784588

[41] Y. Zhang, C. Ma, M. Yang, D. Yuan, K. Tang, C. Li, B. Mao, Can additive carbon nanotubes reduce the PMMA fire risk?, Fire Saf. J. 136 (2023) 103757. https://doi.org/10.1016/j.firesaf.2023.103757

[42] J. Spièce, K. Lulla, P. de Crombrugghe de Picquendaele, L. Divay, O. Bezencenet, B. Hackens, P. Gehring, A.J. Robson, C. Evangeli, O.V. Kolosov, Nanoscale heat transport of vertically aligned carbon nanotube bundles for thermal management applications, ACS Appl. Mater. Interfaces 16 (2024) 45671-45677. https://doi.org/10.1021/acsami.4c07913

[43] H. Du, X. Chen, H. Ding, X. Yin, Y. Su, G.J. Weng, The effect of adding graphene oxide into CNT/polymer systems on CNT dispersion and mechanical properties of hybrid nanocomposites, Compos. Commun. 53 (2025) 102196. https://doi.org/10.1016/j.coco.2024.102196

[44] A. Blokhin, T. Dyachkova, A. Maksimkin, R. Stolyarov, A. Suhorukov, I. Burmistrov, A. Kharitonov, Polymer composites based on epoxy resin with added carbon nanotubes, Fuller. Nanotub. Carbon Nanostruct. 28 (2020) 45-49. https://doi.org/10.1080/1536383X.2019.1671364

[45] Y. Hou, C. Liao, S. Qiu, Z. Xu, X. Mu, Z. Gui, L. Song, Y. Hu, W. Hu, Preparation of soybean-root-like CNTs/bimetallic oxide hybrids to enhance fire safety and mechanical performance of thermoplastic polyurethane, Chem. Eng. J. 428 (2022) 132338. https://doi.org/10.1016/j.cej.2021.132338

[46] T. Zhang, H. Yan, M. Peng, L. Wang, H. Ding, Z. Fang, Construction of flame-retardant nanocoating on ramie fabric via layer-by-layer assembly of carbon nanotubes and ammonium polyphosphate, Nanoscale 5 (2013) 3013-3021. https://doi.org/10.1039/c3nr34020a

[47] H. Pan, W. Ma, Z. Zhang, Y. Liu, F. Lu, B. Yu, X. Zhang, Co-effect flame retardation of Co_3O_4-loaded titania nanotubes and α-zirconium phosphate in an epoxy matrix, ACS Omega 5 (2020) 28475-28482. https://doi.org/10.1021/acsomega.0c02584

[48] P. Song, L. Zhao, Z. Cao, Z. Fang, Polypropylene nanocomposites based on C_{60}-decorated carbon nanotubes: thermal properties, flammability, and mechanical properties, J. Mater. Chem. 21 (2011) 7782-7788. https://doi.org/10.1039/c1jm10395d

[49] S.-E. Zhu, L.-L. Wang, H. Chen, W. Yang, A.C.-Y. Yuen, T.B.-Y. Chen, C. Luo, W.-M. Bi, E.-Z. Hu, J. Zhang, Comparative studies on thermal, mechanical, and flame-retardant properties of PBT nanocomposites via different oxidation-state phosphorus-containing agents modified amino-CNTs, Nanomaterials 8 (2018) 70. https://doi.org/10.3390/nano8020070

[50] P. Rybiński, R. Anyszka, M. Imiela, M. Siciński, T. Gozdek, Effect of modified graphene and carbon nanotubes on the thermal properties and flammability of elastomeric materials, J. Therm. Anal. Calorim. 127 (2017) 2383-2396. https://doi.org/10.1007/s10973-016-5841-8

[51] S. Zhang, Y. Jiang, Y. Sun, J. Sun, B. Xu, H. Li, X. Gu, Preparation of flame-retardant and conductive epoxy resin composites by incorporating functionalized multi-walled carbon nanotubes and graphite sheets, Polym. Adv. Technol. 32 (2021) 2093-2101. https://doi.org/10.1002/pat.5239

[52] K.S. Khare, F. Khabaz, R. Khare, Effect of carbon-nanotube functionalization on mechanical and thermal properties of cross-linked epoxy-carbon-nanotube nanocomposites:

Materials Research Forum LLC
https://doi.org/10.21741/9781644904077-4

role of strengthening the interfacial interactions, ACS Appl. Mater. Interfaces 6 (2014) 6098-6110. https://doi.org/10.1021/am405317x

[53] Y. Akcin, S. Karakaya, O. Soykasap, Electrical, thermal and mechanical properties of CNT-treated prepreg CFRP composites, Mater. Sci. Appl. 7 (2016) 465. https://doi.org/10.4236/msa.2016.79041

[54] A. Kausar, I. Rafique, B. Muhammad, Significance of carbon nanotube in flame-retardant polymer/CNT composite: a review, Polym.-Plast. Technol. Eng. 56 (2017) 470-487. https://doi.org/10.1080/03602559.2016.1233267

[55] V. Sharma, S. Agarwal, A. Mathur, S. Singhal, S. Wadhwa, Advancements in nanomaterial-based flame retardants for polymers: a comprehensive overview, J. Ind. Eng. Chem. 133 (2024) 38-52. https://doi.org/10.1016/j.jiec.2023.12.010

[56] S. Lu, W. Hong, X. Chen, Nanoreinforcements of two-dimensional nanomaterials for flame-retardant polymeric composites: an overview, Adv. Polym. Technol. 2019 (2019) 4273253. https://doi.org/10.1155/2019/4273253

[57] S. Huo, Y. Guo, Q. Yang, H. Wang, P. Song, Two-dimensional nanomaterials for flame-retardant polymer composites: a mini review, Adv. Nanocompos. 2024. https://doi.org/10.1016/j.adna.2024.07.001

[58] P. Yang, H. Wu, F. Yang, J. Yang, R. Wang, Z. Zhu, A novel self-assembled graphene-based flame retardant: synthesis and flame-retardant performance in PLA, Polymers 13 (2021) 4216. https://doi.org/10.3390/polym13234216

[59] Y. Pan, N. Hong, J. Zhan, B. Wang, L. Song, Y. Hu, Effect of graphene on the fire and mechanical performances of glass-fiber-reinforced polyamide 6 composites containing aluminum hypophosphite, Polym.-Plast. Technol. Eng. 53 (2014) 1467-1475. https://doi.org/10.1080/03602559.2014.909483

[60] X. Zhou, F. Jiang, Z. Hu, F. Wu, M. Gao, Z. Chai, Y. Wang, X. Gu, Y. Wang, Study on the flame retardancy of rigid polyurethane foam with phytic-acid-functionalized graphene oxide, Molecules 28 (2023) 6267. https://doi.org/10.3390/molecules28176267

[61] H. Kim, D.W. Kim, V. Vasagar, H. Ha, S. Nazarenko, C.J. Ellison, Polydopamine-graphene oxide flame-retardant nanocoatings applied via an aqueous liquid crystalline scaffold, Adv. Funct. Mater. 28 (2018) 1803172. https://doi.org/10.1002/adfm.201803172

[62] H.W. Kroto, J.R. Heath, S.C. O'Brien, R.F. Curl, R.E. Smalley, C_{60}: Buckminsterfullerene, Nature 318 (1985) 162-163. https://doi.org/10.1038/318162a0

[63] P. Krusic, E. Wasserman, P. Keizer, J. Morton, K. Preston, Radical reactions of C_{60}, Science 254 (1991) 1183-1185. https://doi.org/10.1126/science.254.5035.1183

[64] J. Jia, Y. Wu, D. Zhao, F. Wang, D. Wang, Q. Yang, Y. Xing, Study on combustion reaction products and free radicals of aliphatic hydrocarbon functional groups in coal based on ML potential MD, Sci. Rep. 15 (2025) 10144. https://doi.org/10.1038/s41598-025-88563-7

[65] B. Troitskii, L. Troitskaya, A. Yakhnov, M. Lopatin, M. Novikova, Retardation of thermal degradation of PMMA and PVC by C_{60}, Eur. Polym. J. 33 (1997) 1587-1590. https://doi.org/10.1016/S0014-3057(97)00048-7

Materials Research Forum LLC
https://doi.org/10.21741/9781644904077-4

[66] V.V. Zuev, F. Bertini, G. Audisio, Fullerene C_{60} as stabiliser for acrylic polymers, Polym. Degrad. Stab. 90 (2005) 28-33. https://doi.org/10.1016/j.polymdegradstab.2005.02.011

[67] Y. Pan, Z. Guo, S. Ran, Z. Fang, Influence of fullerenes on the thermal and flame-retardant properties of polymeric materials, J. Appl. Polym. Sci. 137 (2019) 47538. https://doi.org/10.1002/app.47538

[68] A. Kausar, Fullerene reinforced polymeric nanocomposites for energy storage-status and prognoses, Front. Mater. 9 (2022) 874169. https://doi.org/10.3389/fmats.2022.874169

[69] D. Zhang, M. Pei, K. Wei, F. Tan, C. Gao, D. Bao, S. Qin, Flame-retardant properties and mechanism of polylactic acid-conjugated flame-retardant composites, Front. Chem. 10 (2022) 894112. https://doi.org/10.3389/fchem.2022.894112

[70] W. Lv, J. Lv, C. Zhu, Y. Zhang, Y. Cheng, L. Zeng, L. Wang, C. Liao, Thermal stabilities and flame retardancy of polyamide 66 prepared by in situ loading of amino-functionalized polyphosphazene microspheres, Polymers (Basel) 15 (2022) 218. https://doi.org/10.3390/polym15010218

[71] Y. Kanbur, U. Tayfun, Development of multifunctional polyurethane elastomer composites containing fullerene: mechanical, damping, thermal, and flammability behaviors, J. Elastomers Plast. 51 (2018) 009524431879661. https://doi.org/10.1177/0095244318796616

[72] X. Wang, E.N. Kalali, J.-T. Wan, D.-Y. Wang, Carbon-family materials for flame-retardant polymeric materials, Prog. Polym. Sci. 69 (2017) 22-46. https://doi.org/10.1016/j.progpolymsci.2017.02.001

[73] X. Zhou, S. Ran, H. Hu, Z. Fang, Improving flame-retardant efficiency by incorporation of fullerene in styrene-butadiene-styrene block copolymer/aluminum hydroxide composites, J. Therm. Anal. Calorim. 125 (2016) 199-204. https://doi.org/10.1007/s10973-016-5354-5

[74] C.M. Long, M.A. Nascarella, P.A. Valberg, Carbon black vs. black carbon and other airborne materials containing elemental carbon: physical and chemical distinctions, Environ. Pollut. 181 (2013) 271-286. https://doi.org/10.1016/j.envpol.2013.06.009

[75] Y. Hu, S. Wang, Z. Ling, Y. Zhuang, Z. Chen, W. Fan, Preparation and combustion properties of flame-retardant nylon 6/montmorillonite nanocomposite, Macromol. Mater. Eng. 288 (2003) 272-276. https://doi.org/10.1002/mame.200390017

[76] H. Yang, Y. Guan, L. Ye, S. Wang, S. Li, X. Wen, X. Chen, E. Mijowska, T. Tang, Synergistic effect of nanoscale carbon black and ammonium polyphosphate on improving thermal stability and flame retardancy of polypropylene: a reactive network for strengthening carbon layer, Compos. Part B Eng. 174 (2019) 107038. https://doi.org/10.1016/j.compositesb.2019.107038

[77] M. Zhu, Z. Yan, S. Haibo, W. Bibo, Y. Hu, Effect of carbon-black-microencapsulated ammonium polyphosphate on the flame retardancy and mechanical properties of polyurethane composites, Polym.-Plast. Technol. Mater. 59 (2020) 83-94. https://doi.org/10.1080/25740881.2019.1625384

[78] P. Nymark, K.A. Jensen, S. Suhonen, Y. Kembouche, M. Vippola, J. Kleinjans, J. Catalán, H. Norppa, J. van Delft, J.J. Briedé, Free radical scavenging and formation by multi-walled

Materials Research Forum LLC
https://doi.org/10.21741/9781644904077-4

carbon nanotubes in cell-free conditions and in human bronchial epithelial cells, Part. Fibre Toxicol. 11 (2014) 4. https://doi.org/10.1186/1743-8977-11-4

[79] J. Liu, Y. Zhang, Y. Guo, C. Lu, B. Pan, S. Peng, J. Ma, Q. Niu, Effect of carbon black on the thermal degradation and flammability properties of flame-retarded high-impact polystyrene/magnesium hydroxide/microencapsulated red phosphorus composite, Polym. Compos. 39 (2016) e23998. https://doi.org/10.1002/pc.23998

[80] J. Wang, Q. Li, C. Wu, H. Xu, Thermal conductivity and mechanical properties of carbon-black-filled silicone rubber, Polym. Polym. Compos. 22 (2014) 393-400. https://doi.org/10.1177/096739111402200405

[81] P.-C. Ma, M.-Y. Liu, H. Zhang, S.-Q. Wang, R. Wang, K. Wang, Y.-K. Wong, B. Tang, S. Hong, K.-W. Paik, J.-K. Kim, Enhanced electrical conductivity of nanocomposites containing hybrid fillers of carbon nanotubes and carbon black, ACS Appl. Mater. Interfaces 1 (2009) 1090-1096. https://doi.org/10.1021/am9000503

[82] L. Li, X. Zhao, C. Ma, X. Chen, S. Li, C. Jiao, Smoke suppression properties of carbon black on flame-retardant thermoplastic polyurethane based on ammonium polyphosphate, J. Therm. Anal. Calorim. 126 (2016) 5815. https://doi.org/10.1007/s10973-016-5815-x

[83] A. Ciampaglia, R. Ciardiello, F. Cesano, G. Belingardi, V. Brunella, Multifunctional material design for strain sensing: carbon black effect on mechanical and electrical properties of polyamides, Compos. Struct. 304 (2023) 116373. https://doi.org/10.1016/j.compstruct.2022.116373

[84] N. Wang, X. Zhang, X. Ma, J. Fang, Influence of carbon black on the properties of plasticized poly(lactic acid) composites, Polym. Degrad. Stab. 93 (2008) 1044-1052. https://doi.org/10.1016/j.polymdegradstab.2008.03.023

[85] H. Yang, J. Gong, X. Wen, J. Xue, Q. Chen, Z. Jiang, N. Tian, T. Tang, Effect of carbon black on improving thermal stability, flame retardancy and electrical conductivity of polypropylene/carbon fiber composites, Compos. Sci. Technol. 113 (2015) 31-37. https://doi.org/10.1016/j.compscitech.2015.03.013

[86] G. Shrestha, S. Traina, C. Swanston, Black carbon's properties and role in the environment: a comprehensive review, Sustainability 2 (2010) 294. https://doi.org/10.3390/su2010294

[87] M.S. Murad, A.K. Hamzat, E. Asmatulu, R. Asmatulu, Flame-retardant fiber composites: synergistic effects of additives on mechanical, thermal, chemical, and structural properties, Adv. Compos. Hybrid Mater. 8 (2024) 31. https://doi.org/10.1007/s42114-024-01111-1

[88] H. Yang, X. Wen, J. Xue, Z. Jiang, N. Tian, T. Tang, Effect of carbon black on improving thermal stability, flame retardancy and electrical conductivity of polypropylene/carbon fiber composites, Compos. Sci. Technol. 113 (2015) 31-37. https://doi.org/10.1016/j.compscitech.2015.03.013

[89] E. Jakab, M.A. Uddin, T. Bhaskar, Y. Sakata, Thermal decomposition of flame-retarded high-impact polystyrene, J. Anal. Appl. Pyrolysis 68 (2003) 83-99. https://doi.org/10.1016/S0165-2370(03)00075-5

[90] E.M. Alosime, A.A. Basfar, A systematic investigation on the effect of carbon nanotubes and carbon black on the mechanical and flame-retardancy properties of polyolefin blends, Polymers (Basel) 16 (2024) 417. https://doi.org/10.3390/polym16030417

[91] A.M. Schrand, S.A.C. Hens, O.A. Shenderova, Nanodiamond particles: properties and perspectives for bioapplications, Crit. Rev. Solid State Mater. Sci. 34 (2009) 18-74. https://doi.org/10.1080/10408430902831987

[92] J. Lopez-Cuesta, Flame-retardant polymer nanocomposites, in: Adv. Polym. Nanocompos., Elsevier, 2012, pp. 540-566. https://doi.org/10.1533/9780857096241.3.540

[93] A. Kausar, Polymer/Nanodiamond Nanocomposites: Fundamentals, Properties and Applications, Elsevier, 2024. https://doi.org/10.1016/B978-0-443-15908-4.00005-6

[94] S. Chauhan, N. Jain, U. Nagaich, Nanodiamonds with powerful ability for drug delivery and biomedical applications: recent updates on in vivo study and patents, J. Pharm. Anal. 10 (2020) 1-12. https://doi.org/10.1016/j.jpha.2019.09.003

[95] Ü. Tayfun, M. Doğan, Flame-retardant properties of fullerene and nanodiamond-based polymer nanocomposites, in: Flame Retard. Nanocompos., Elsevier, 2024, pp. 263-286. https://doi.org/10.1016/B978-0-443-15421-8.00004-5

[96] J. Zhang, Z. Li, L. Zhang, J.G. Molleja, D.-Y. Wang, Bimetallic metal-organic frameworks and graphene oxide nanohybrids for enhanced fire-retardant epoxy composites: a novel carbonization mechanism, Carbon 153 (2019) 407-416. https://doi.org/10.1016/j.carbon.2019.07.003

[97] W. Ahmed, S. Gul, M. Awais, Z.U. Hassan, S. Jabeen, M. Farooq, A review: novel nanohybrids of epoxy/polyamide with carbon nanotube/nanodiamond, Polym.-Plast. Technol. Mater. 60 (2021) 579-600. https://doi.org/10.1080/25740881.2020.1819314

[98] B. Tawiah, S. Ullah, I.M.D.C. Cordeiro, A.C. Yuen, Y. Ming, M.Z. Rahman, D. Chen, W. Cai, Z. Guangping, B. Amirbek, Nanodiamond-decorated phosphor-silicone thermally conductive PA6 composites with excellent flame retardancy, nanomechanical and thermo-pyroelectric response, Chem. Eng. J. 2025 162249. https://doi.org/10.1016/j.cej.2025.162249

[99] Q. Li, Y.-T. Pan, Zero-dimensional nanomaterials for flame-retardant epoxy thermosets and composites, in: Non-Halogenated Flame-Retard. Technol. Epoxy Thermosets Compos., Elsevier, 2024, pp. 271-289. https://doi.org/10.1016/B978-0-443-16046-2.00013-4

[100] F. Pan, M. Khan, L. Tiehu, E. Javed, A. Hussain, A. Zada, D. Alei, Z. Wahab, Effect of nanodiamond particles on the structure, mechanical, and thermal properties of polymer-embedded ND/PMMA composites, J. Polym. Eng. 42 (2022) 795-807. https://doi.org/10.1515/polyeng-2022-0012

[101] F. Namini, M. Nouri, J. Dorranian, Effect of nanodiamond on morphology, crystallinity and thermal behavior of electrospun nanofibers of polyamide-6, Polymer (Korea) 38 (2014) 773-780.

[102] P. Li, G. Wei, H. Zhang, L. Guo, R. Han, X. Zhu, W. Yang, A novel intumescent flame-retardant thermoplastic polyurethane based on triazine-phosphorus and its nanocomposites with carbon nanotube, Compos. Sci. Technol. 188 (2020) 107977.

[103] X. Bai, F. Ma, J. Li, The advances of carbon nanomaterials in flame-retardant polymer materials, Adv. Compos. Hybrid Mater. 3 (2020) 442-459.

[104] S. Wang, S. Chen, J. Zhang, F. Dong, C. Liu, S. Dai, The development of carbon nanomaterials in flame-retardant polymer materials, Polymers (Basel) 15 (2023) 1703.

[105] A. Kausar, Advances in carbon nanomaterial-based flame-retardant polymer composites, J. Thermoplast. Compos. Mater. 37 (2024) 3050-3075. https://doi.org/10.1177/08927057241233568

[106] S. Amirabadi, Developing nano-fibrillated flame-retardant polymer composites and foams, University of Toronto (Canada), 2024.

[107] G. Huang, W. Chen, T. Wu, H. Guo, C. Fu, Y. Xue, K. Wang, P. Song, Multifunctional graphene-based nano-additives toward high-performance polymer nanocomposites with enhanced mechanical, thermal, flame retardancy and smoke suppressive properties, Chem. Eng. J. 410 (2021) 127590. https://doi.org/10.1016/j.cej.2020.127590

[108] X. Hu, S. Han, Y. Zhang, T. Sai, P. Song, X. Wen, T. Tang, L. Chen, Y.-Z. Wang, Green synthesis of inorganic fire retardants, J. Inorg. Mater. (2023). https://doi.org/10.1039/BK9781839169793-00218

[109] Y. Kim, S. Lee, H. Yoon, Fire-safe polymer composites: flame-retardant effect of nanofillers, Polymers 13(4) (2021) 540. https://doi.org/10.3390/polym13040540

[110] Q. Ge, J. Chu, W. Cao, F. Yi, Z. Ran, Z. Jin, B. Mao, Z. Li, K.S. Novoselov, Graphene-based textiles for thermal management and flame retardancy, Adv. Funct. Mater. 32(42) (2022) 2205934. https://doi.org/10.1002/adfm.202205934

[111] C.I. Idumah, Recent advancements in fire retardant mechanisms of carbon nanotubes, graphene, and fullerene polymeric nanoarchitectures, J. Anal. Appl. Pyrolysis 174 (2023) 106113. https://doi.org/10.1016/j.jaap.2023.106113

Materials Research Foundations 190 (2026) 118-152 https://doi.org/10.21741/9781644904077-5

Chapter 5

Carbon-Based Nanomaterials for Flame Retardant Polymers

Md. Enamul Kabir, Most. Israt Jahan, Muhammed Shah Miran*

Department of Chemistry, University of Dhaka, Dhaka-1000, Bangladesh

shahmiran@du.ac.bd

Abstract

Polymers are frequently employed materials; however, they are exceedingly combustible, which presents significant fire hazards. Many conventional flame retardants (FRs), particularly halogenated ones, are harmful to the environment and reduce mechanical qualities. A multipurpose alternative, carbon nanotubes, graphene, Fullerene, and carbon black improve flame resistance, thermal stability, mechanical strength, and conductivity at low loadings. This chapter covers carbon reinforced polymer composites' flame retardancy mechanisms, synthesis methods, and synergistic effects with typical FRs. Addressing present issues and future research objectives, it emphasizes the potential of carbon nanomaterials to produce flame-retardant polymers that are safer, more performant, and less harmful to the environment.

Keywords

Carbon-based materials, Limiting Oxygen Index, Peak heat release rate, Flame retardant polymers, Carbon Nanotubes

Contents

Materials Research Forum LLC
https://doi.org/10.21741/9781644904077-5

1. Introduction

Polymers find extensive applications in construction, building materials, consumer goods, and transportation. A significant downside, though, is that the hydrocarbons used in polymers production make them very combustible. Fig. 1 shows how polymers burn compared to their cost per pound. It highlights that common plastics can keep burning on their own, which poses a fire risk. These plastics make up about 95% of all polymer uses. Importantly, most deaths in building fires happen due to inhaling smoke and toxic gases [1]. To increase fire resistance, certain composite materials include a lot of fillers, which reduces mechanical strength of the polymer [2].

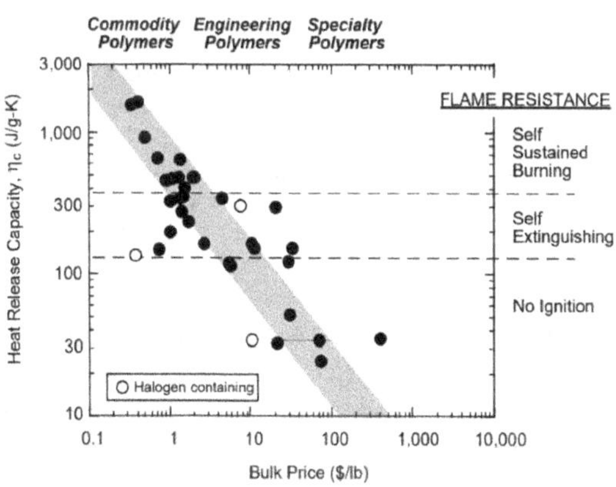

Figure 1: The combustion characteristics, heat release, and cost of commercially available polymers [1].

Besides flammability, pure polymers usually have low mechanical strength and poor thermal and electrical conductivity [3]. Conventional flame retardants (FRs) can only improve flame resistance and do not provide multifunctional benefits. They often need to be added in large amounts to be effective, which raises costs and harms mechanical properties. Also, halogenated FRs are widely used because they work well, pose health and safety risks and have been restricted by the European Union. The use of these halogen-containing compounds is being eliminated globally as a result of growing health and environmental concerns. Instead, non-halogenated alternatives like aluminum trihydroxide and phosphorus-based FRs are increasingly used. Phosphorus FRs are growing fast and are expected to continue expanding, offering good cost and performance benefits as a safer alternative. Although traditional FRs help with fire resistance, they still face challenges with toxicity and environmental impact. This creates an urgent need for new FRs that work efficiently at low loadings, are safer, and less harmful to the environment. To achieve this, researchers are exploring innovative solutions. One promising approach is adding nanofillers to polymers to create multifunctional nanocomposites. This can boost mechanical strength, thermal and electrical conductivity, and flame resistance. Thanks to nanomaterials, polymer nanocomposites can achieve these improvements with relatively low filler content (usually 3–5 wt.%), preserving the lightweight nature of polymers. Layered silicates were the first successful nanofillers used in polymer nanocomposites and were commercialized by Toyota in 1991 [4]. The first FR polymer nanocomposite, nylon/silicate layers, was reported in 1996 [5]. When exposed to flame for 30 seconds, this nanocomposite stopped burning immediately after the flame was removed, unlike pure nylon, which kept burning. Although formal testing was limited back then, layered silicates showed promise, leading to research on other nanofillers like carbon-based materials [1,6] and, more recently, MXene sheets [7]. Nonetheless, carbon-based fillers exhibit outstanding performance even at extremely low loadings, according to Wang et al. They are very suitable for use in multifunctional polymer nanocomposites due to their exceptional strength, barrier efficiency, conductivity, and significantly decreased flammability [8]. To produce sophisticated carbon materials from banana plant pseudo-stems on a large scale, M.M. Rahman et al. developed an easy, cheap, and eco-friendly process. These materials showed improved electrical conductivity. Building on earlier efforts to enhance the flame retardancy of polymers by means of carbon nanotubes, graphene, carbon black, and carbon nanofibers, this study gathers the most recent results on the utilization of carbon nanomaterials as a reinforcing agent in multifunctional polymer nanocomposites. The rapid advancement of nanotechnology and the integration of carbon nanoparticles into polymer manufacturing are considered in this paper. The report evaluates the current level of FR technology, with a focus on the most up-to-date findings regarding multifunctional FR polymer nanocomposites made of graphene, carbon black, and carbon nanotubes (CNTs).

2. Characterization Techniques for Flame-Retardant Performance

Standardized techniques frequently employed to evaluate flammability encompass the UL94 test, limiting oxygen index (LOI) test, and cone calorimetry. These evaluations provide essential parameters for classifying the flammability of polymers and for assessing the effectiveness of FRs incorporated into polymer materials.

Materials Research Foundations 190 (2026) 118-152 https://doi.org/10.21741/9781644904077-5

2.1 Limiting Oxygen Index

An oxygen/nitrogen mixture's LOI is the lowest oxygen concentration that can support three minutes of sample combustion or burn a 5-centimeter piece. This is estimated using Eq. 1.

$$\mathrm{LOI} = \frac{O_2}{O_2+N_2} \times 100\% \quad (1)$$

According to ISO 4589, the test involves vertically mounting a sample (80 mm × 10 mm × 4 mm) in a glass column and igniting it from the top. Because it requires very tiny sample volumes, LOI testing is an important and economical way to screen flame-retardant polymer candidates and control quality. It fails to replicate actual fire conditions, though, because of its artificially raised oxygen level and minimal heat input.

2.2 UL-94 Vertical Burning Test

To determine whether electronic devices or appliance plastic parts are combustible, Underwriters Laboratories created the UL-94 test series. The vertical burning (UL-94 V) test is the gold standard for testing the combustibility and flame spread of a polymer. Ten seconds are spent with a blue flame that is 20 mm in diameter and 50 W strong, held beneath the base of the specimen. After recording the amount of time that it requires for the flame to go out (t_1), a second ignition is performed for ten additional seconds, and the flame-out time (t_k) and afterglow time (t_t) are also recorded. Testing five samples and classifying the findings as V-0, V-1, or V-2 according to criteria is done in accordance with International Electrotechnical Commission (IEC) 60695-11-10 [8].

2.3 Cone Calorimetry

An extensively utilized bench-scale technique for assessing the fire characteristics of polymers is cone calorimetry, which is standardized under ISO 5660. A conical heater emits radiant heat onto a sample, which is then lit with an electric spark. The heat release rate (HRR) is determined by gas flow and the amount of oxygen present. A critical measure of a material's combustibility is its peak HRR. Further metrics include mass loss, total smoke released (TSR), adequate heat of combustion (EHC), time to ignition (TTI), time to flame out (TOF), and total heat released (THR). This test provides detailed insights into how fires behave in hypothetical real-world scenarios.

2.4 X-ray Photoelectron Spectroscopy

The chemical composition of char can be quantitatively and qualitatively analyzed using X-ray photoelectron spectroscopy (XPS). Spectra showing changes in binding energy allow it to distinguish between elements in different chemical situations, including C-O, C=O, C-P, and C=C. To infer reactions involving condensed phases, XPS is an invaluable tool. The presence of phosphorus acids with P-NH bonds in the char, as revealed by XPS analysis, suggests that these acids play a role in the rapid development of char in ethylene-vinyl acetate (EVA) when melamine phosphate and a cyclic phosphate are utilized. Also, it can determine the proportions of aliphatic/aromatic hydrocarbons, ethers, hydroxyls, and carbonyls. The oxidized to total carbon ratio (C_{ox}/C_a) provides insights into thermal oxidative resistance [9]. Elemental migration during

Materials Research Foundations 190 (2026) 118-152
https://doi.org/10.21741/9781644904077-5

decomposition, such as silicon movement, can also be studied to evaluate the formation of protective ceramic-like layers.

2.5 Raman Spectroscopy

Raman spectroscopy is extensively used to characterize carbon-based chars. Typical spectra display two primary peaks: the G band (~1590 cm^{-1}), associated with crystalline graphite vibrations, and the D band (~1360 cm^{-1}), linked to disordered or amorphous carbon. The intensity ratio of the G to D band (I_G/I_D) indicates the level of graphitization, with higher values reflecting more ordered carbon structures. Graphitized char is essential for improving thermal stability and controlling heat and mass transfer [10].

3. Role of Carbon-Based Nanomaterials in Flame Retardancy

The mechanical strength of the polymer matrix can be compromised due to the high loading levels required to achieve fire safety regulations for traditional flame-retardant additives. One promising solution to this problem is the emergence of carbon-based nanoparticles as flame-retardant alternatives, particularly in the field of nanotechnology. For quite some time, several polymeric systems have used expandable graphite (EG) as a fire retardant component. Graphene and CNTs are two examples of nanostructured carbon materials that have recently become renowned for their synergistic flame-retardant properties and remarkable thermal stability. Graphene, for instance, exhibits excellent thermal stability and acts as an effective physical barrier, contributing to its inherent flame-retardant capabilities. Nevertheless, pristine graphene alone is generally insufficient to bring polymer matrices to standard flame-retardancy levels. Therefore, it is usually employed as a synergist in combination with traditional flame-retardant systems. Experimental studies have demonstrated the effectiveness of graphene in enhancing the performance of flame retardants. For instance, the addition of graphene and molybdenum disulfide (MoS_2) to an EVA matrix made it possible to lower the amount of aluminum hydroxide used, while still meeting the UL-94 V-0 flammability standard. In comparison, EVA composites with only aluminum hydroxide and MoS_2 attained just a V-2 rating, indicating that graphene facilitates MoS_2 dispersion and enhances flame retardancy. Furthermore, the performance of graphene in intumescent FR systems is influenced by its concentration. The addition of a small amount of graphene to a polypropylene/IFR composite greatly reduced the peak heat release rate (PHRR) compared to the composite lacking graphene. Excessive graphene loading, however, elevates melt viscosity, hindering the development of an extended carbon layer and reducing the entire flame retardant efficacy. Inorganic nanomaterials such as montmorillonite (MMT) have also been explored for their synergistic flame retardant potential. Although MMT has good thermal stability, it tends to agglomerate in polymers, reducing its flame-retardant and mechanical performance. A study showed that combining graphene and MMT in polyimide aerogels significantly increased flame retardancy, with LOI values surpassing 50, demonstrating the synergy of the hybrid system. Graphene Oxide (GO) and MMT hybrids offer improved dispersion and compatibility in the matrix [11]. Natural fiber-reinforced polypropylene, despite its mechanical benefits, is highly flammable. Traditional additives like ammonium polyphosphate (APP) and magnesium hydroxide (MH) require high loadings that may impair properties. Exfoliated graphene nanosheets improve the efficiency of these additives through barrier effects and enhanced carbonization. CNTs also serve as effective synergists. Thermally oxidized CNTs in epoxy resins show better flame resistance than untreated ones, with similar benefits observed in electromagnetic shielding materials using

Materials Research Foundations 190 (2026) 118-152 https://doi.org/10.21741/9781644904077-5

polystyrene and thermoplastic polyurethane (TPU) matrices. In TPU systems, the combination of CNTs and IFRs enhanced both flame resistance and conductivity by facilitating CNT dispersion [12]. CNTs have also improved the flame retardancy of nylon fabrics and fibers, reinforcing the carbon layer structure during combustion. However, excessive CNT loading can hinder performance due to increased melt viscosity affecting the foaming action of APP. An intumescent flame retardant with low nitrogen/phosphorus, grafted onto fullerene (C_{60}), enhanced the flame resistance of polypropylene through radical dissipation and char formation. Even small additions of this fullerene-based material significantly enhanced stability. Similarly, aluminum hydroxide combined with fullerene achieved top flame-retardant classification in SBS composites, even when the aluminum hydroxide content was reduced. DOPO, a phosphorus-based flame retardant, suffers from migration issues in polymers. To address this, it was chemically grafted onto hydroxyl-functionalized carbon black, forming a stable hybrid. When added to polylactic acid (PLA), this hybrid improved both flame retardancy and electrical conductivity [11].

4. Synthesis and Incorporating Carbon Nanomaterials into Polymers

4.1 Fullerene

Fullerene is a spherical carbon allotrope with excellent thermal stability and mechanical strength. Its incorporation into polymer matrices, particularly polypropylene (PP), represents a notable advancement in flame-retardant materials. Although research on fullerene-based FR polymers is still in its early stages, recent studies have shown promising results. For example, fullerene films have been used as insulating layers to dissipate heat and prevent material ignition efficiently. Polypropylene/fullerene nanocomposites prepared by melt compounding have demonstrated improved flame-retardant properties with increasing fullerene content. As noted in one study, the addition of just 2 wt.% fullerene reduced the peak HRR by 46%. This enhancement is attributed to two primary mechanisms: (1) fullerene functions as a "radical sponge," capturing free radicals during combustion, and (2) it stabilizes PP radicals during thermal degradation, facilitating the formation of a cross-linked gelled-ball structure. This structure increases gel content and viscosity, reinforcing the resistance of the polymers to thermal breakdown. In addition to improved flame resistance, fullerene also enhances the thermal stability of PP in oxidative environments. At a 2 wt.% loading, the onset and maximum decomposition temperatures of PP were increased by 20 °C and 62 °C, respectively, due to fullerene's radical-scavenging properties. Researchers have also explored functionalized and hybrid fullerenes for advanced applications. A dendrimer-like intumescent fire retardant altered with fullerene markedly enhanced the thermal-oxidative resilience and flame retardancy of polypropylene nanocomposites. Likewise, fullerene-decorated carbon nanotubes (C_{60}-d-CNTs) integrated into polypropylene matrices resulted in a significant 71% decrease in PHRR with merely 1 wt.% loading. As indicated in Table 1, this outstanding flame-retardant effect results from the synergistic interaction between fullerene's ability to trap radicals and the barrier properties of the CNT network [13].

Table 1: The flammability of polymer nanocomposites containing fullerene [13].

Type of filler	Loading (wt.%)	Improvement in T_{ign} (s)	Reduction in Peak HRR (%)
Fullerene	0.5	13	33
	1.0	15	41
	2.0	14	14
C_{60}-d-PDBPP	0.5	4	46
	1.0	6	52
	2.0	8	59
C_{60}-d-CNTs	1.0	-	71

4.2 Carbon Nanotubes

CNTs have emerged as an up-and-coming alternative to traditional flame retardants (Fig. 2). Incorporating small quantities of CNTs into various polymer matrices such as polyethylene (PE), PP, PLA, polystyrene, EVA, polymethyl methacrylate (PMMA), polyamide-6, polycarbonate (PC), polyether ether ketone, polybenzoxazine, and epoxy resins has been shown to mitigate fire risks significantly [14]. An example of a composite material with excellent flame-retardant properties is multi-walled carbon nanotube (MWNT)/PP that was made by melt blending without the use of surface treatment or other additives. There was about a 73% reduction in peak HRR when compared with pure PP at a loading level of only 1 vol.% [15]. Similarly, research on PMMA has shown that flammability is significantly affected by CNT type and concentration. Gasification tests indicated that PMMA containing 0.5 wt.% single-walled carbon nanotubes (SWNTs) had the lowest mass loss rate among various nanocomposite formulations, followed closely by MWNT-filled PMMA.

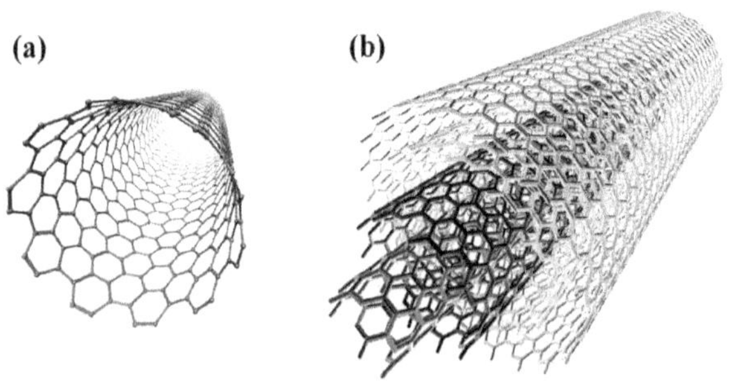

Figure 2: Illustrative diagram of (a) a single-walled carbon nanotube (SWCNT) and (b) a multi- walled carbon nanotube (MWCNT) [13].

Materials Research Foundations 190 (2026) 118-152 https://doi.org/10.21741/9781644904077-5

Other fillers, such as carbon nanofibers (CNFs) and carbon black particles, did not show significant improvements compared to unmodified PMMA. These results suggest that the effectiveness of CNTs stems from their ability to form a protective, network-like char layer during combustion. Multifunctional SWNT/cellulose composite papers produced via papermaking techniques also demonstrated good electrical conductivity and reduced heat release rates. Solvent-free methods were used to prepare MWNT/polybenzoxazine nanocomposites, where even at low CNT concentrations, significant improvements in rheological, electrical, and flame-retardant behavior were observed. Recently, CNT bucky papers, which are formed through filtration of CNT suspensions, have been applied as surface flame-retardant barriers for polymeric materials. Incorporating SWNT/MWNT bucky papers onto polyimide/carbon fiber composites via compression molding proved more effective than direct CNT blending. MWNT bucky paper/epoxy/carbon composites significantly reduce peak HRR by over 60% and smoke production by 50%, due to the dense nanotube network and small pores that limit heat and gas flow during burning. Functionalizing CNTs enhances their dispersion in polymers and strengthens their interaction with the matrix, improving composite performance. Three key strategies to develop flame-retardant CNT/polymer nanocomposites include: (1) surface modification with coupling agents, (2) covalent bonding with flame-retardant molecules, and (3) hybridization with inorganic particles. Functionalized CNTs have been successfully integrated into various polymers, with epoxy composites containing 9 wt.% functionalized CNTs showing improved flame resistance. Similarly, polyvinylphosphonic acid/MWNT/cotton composites exhibited a notable decrease in burning rate compared to untreated cotton. Another practical approach entails the covalent attachment of organic fire retardants to CNTs, which has demonstrated an improvement in flame-retardant efficacy across multiple polymer systems, including PP, acrylonitrile butadiene styrene (ABS), EVA, PC, epoxy, and cyanate ester. To enable this bonding, CNTs must first undergo surface treatment, for example, hydroxylation using potassium hydroxide and ethanol, or acid oxidation to introduce carboxyl groups. These functional groups allow flame retardants to be chemically grafted onto the CNTs. An example includes the conversion of carboxyl-functionalized MWNTs into acid chlorides using $SOCl_2$, followed by reaction with amino-terminated flame retardants. Functionalization of CNTs not only makes them more dispersed, but it also makes their interface with the polymer stronger, leading to composites that are more flame resistant and have a higher Young's modulus than untreated CNTs. Inorganic particle hybridization with CNTs is a third method that is gaining popularity. This method includes phosphazene, layered double hydroxides (LDH), zirconia, silica, and aluminum nitride, among others. These hybrid CNT systems have been successfully used to fabricate flame-retardant polymer composites with superior performance in terms of fire resistance, thermal stability, and dielectric behavior, surpassing the effectiveness of pristine CNTs alone. Recently, CNTs have been combined with other additives to develop highly effective flame-retardant surface coatings for polymers using a layer-by-layer (LbL) assembly technique. Because pristine CNTs tend to aggregate due to their high aspect ratio, they are typically modified through processes such as amination or carboxylation to introduce ionic charges that enable their stable dispersion in aqueous media for electrostatic LbL assembly. As illustrated in Fig. 3, this technique involves the sequential deposition of multiple thin layers on a polymer surface. The resulting CNT-based network substantially enhances fire resistance; for instance, when applied to polyurethane (PU) foam, it reduces the peak HRR by 35% and helps prevent pool fires by forming a robust, insulating barrier.

 Materials Research Forum LLC
https://doi.org/10.21741/9781644904077-5

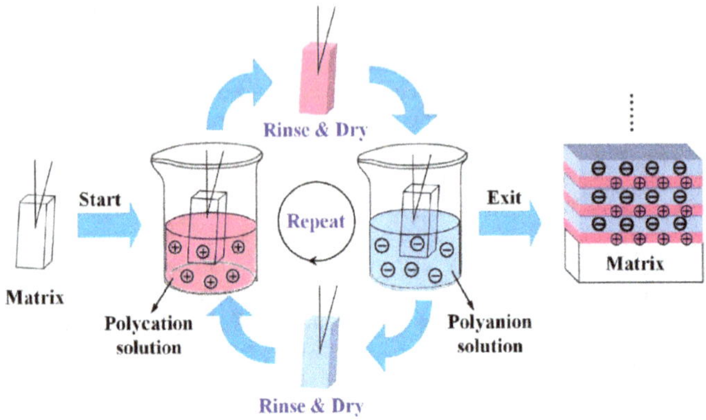

Figure 3: Diagram showing the layer-by-layer assembly process for CNT/polymer multilayer coatings, where the substrate is alternately immersed in positively charged (cationic) and negatively charged (anionic) solutions, with a rinsing step performed after each immersion to remove loosely bound materials [13].

4.3 Graphene

Graphene has been employed directly to fabricate flame-retardant polymer composites by disintegrate compounding and solution blending due to its inherently superior flame-retardant properties. Melt extrusion has allowed us to produce flame-retardant polypropylene composites based on graphene, for example. With a high HRR decrease of 76%, graphene appears to be more dispersed than conventional nano- and micro-scale carbon fillers while concurrently improving firmness, conductivity of electricity, and flame resistance [13]. In solution-blended graphene/waterborne polyurethane nanocomposites, significant enhancements in flame retardancy and smoke suppression were achieved, including a 25% decrease in total smoke release at just one weight percent of graphene, along with lowered Peak HRR and THR. An innovative method for functionalizing graphene with flame retardants is sol-gel chemistry. In this process, flame retardants and silane coupling agents react to producing siloxane-based flame retardants, which then link to graphene or GO. Hydrolysis and condensation processes are used to graft those siloxane flame retardants atop silane-modified graphene or GO. GO compounds functionalized with silicon and phosphorus have been developed using this approach; these materials have enhanced the flame retardancy of polymers like polyurethane, PMMA, PVA, and epoxy resin.

4.4 Graphite

One common method for creating EG, an intercalation product of graphite, is to combine concentrated sulfuric acid with strong oxidizing chemicals and treat flake graphite. This treatment oxidizes the graphite and inserts sulfate anions between its layers. EG retains many advantageous properties of natural graphite, including low cost, abundant availability, electrical conductivity, and the ability to enhance the mechanical strength of polymers. One of its standout characteristics is excellent flame retardancy, and its commercial availability facilitates easy incorporation into polymer matrices. Microencapsulation and the sol-gel technique are the two main approaches to increasing compatibility with polymers of EG. A silane coupling agent is used to functionalize EG and a pre-polymer or a monomer in the sol-gel technique. This is followed by hydrolysis and condensation reactions that establish covalent bonds between EG and the polymer, resulting in better integration. Composites such as EG/epoxy and EG/PMMA made using this approach have demonstrated strong flame-retardant properties. Another strategy to enhance the interface between EG and polymers is microencapsulation, where EG particles are coated with a polymer shell. For example, polyurethane-coated EG has been incorporated into EVA composites, as depicted in Fig. 4. These composites achieved a UL-94 V-0 flame-retardant rating even after soaking in water at 70°C for 168 hours. These materials outperformed composites generated from untreated EG in terms of both mechanical and dynamical thermal properties, due to the enhanced adherence that existed between the encapsulating EG and the EVA matrix. In addition, rigid polyurethane foams have been found to effectively contain flame retardants when EG is microencapsulated with melamine-formaldehyde resin or methyl methacrylate-acrylic acid copolymers [13].

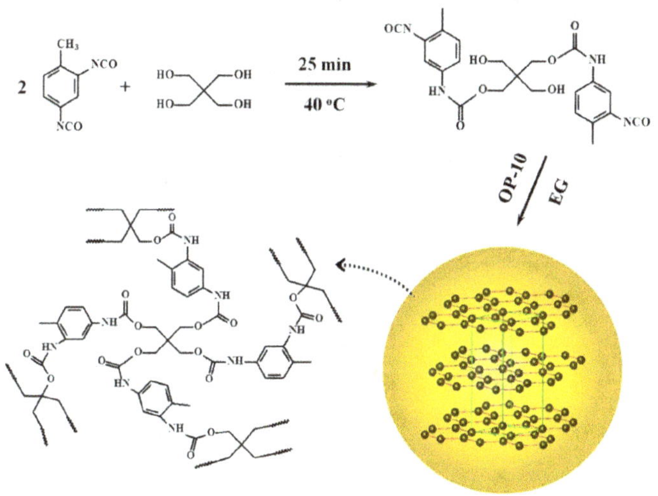

Figure 4: Illustration showing the synthesis process of EG microcapsules coated with polyurethane; OP-10 refers to polyoxyethylene octylphenol ether used in the formulation [13].

5. Mechanisms of Flame Retardancy in Carbon Nanomaterial-Polymer Composites

Combustion is a complex interplay of physical and chemical processes, where multiple mechanisms operate simultaneously, making it difficult to isolate a single dominant pathway. Flame retardancy is generally attributed to mechanisms occurring in both the gas and condensed phase, as illustrated in Fig. 5, which outlines four key regions: the flame zone, molten polymer, char layer, and the underlying intact polymer. Polymer volatile breakdown products react with oxygen to form radicals and release heat in the gas phase, which is also known as the flame zone.

Materials Research Foundations 190 (2026) 118-152 https://doi.org/10.21741/9781644904077-5

The char layer controls the flow of mass and heat from the gas to the condensed state. Subsequently, volatile compounds are released into the flame from the molten polymer layer (the zone of thermal breakdown) [16]. The underlying polymer remains largely unaffected but lies in contact with the decomposing layer.

5.1 Condensed-Phase Mechanism

Flame retardants (FRs) that act in the condensed phase primarily enhance flame resistance by promoting the formation of a stable, insulating char layer on the material surface. This char layer plays several critical roles during combustion: it reduces the amount of flammable material available for burning, releases water vapor that dilutes combustible gases and serves as a thermal barrier due to its low thermal conductivity. The char layer reflects heat radiation, thereby slowing down polymer degradation, and restricts the flow of volatile compounds and heat to the flame zone, effectively suppressing the overall combustion process. An essential element is the incorporation of flame-retardant materials that promote char formation and enhance the carbon layer. The reinforced carbon layer functions as an efficient barrier, obstructing heat and oxygen from accessing the underpinning polymer and hindering the dispersion of the active free radicals, thus markedly improving flame retardancy. Phosphorus-containing FRs are notable for forming pyro- or polyphosphoric species upon decomposition, which catalyzes the development of a robust, carbonaceous char barrier. Similar behavior is observed in systems incorporating boric acid, borates, or silicon-based additives. The quality of the char, not just the yield, is essential; it depends on the graphitic content, morphological features, and resistance to thermal oxidation. Carbon-based materials are effective "carbonifics," improving char density, integrity, and thermal shielding performance [16].

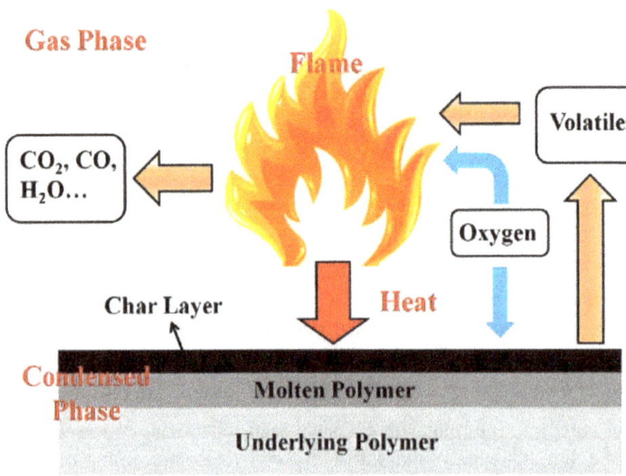

Figure 5: Schematic diagram of the flame-retardant process in action. [16].

Materials Research Foundations 190 (2026) 118-152 https://doi.org/10.21741/9781644904077-5

5.2 Gas-Phase Mechanism

FRs operating in the gas phase interfere with the combustion of volatile pyrolysis products by disrupting radical-driven reactions. Combustion sustains itself through chain-branching processes such as:

(1) $H^{\bullet} + O_2 \rightarrow OH^{\bullet} + O^{\bullet}$

(2) $O^{\bullet} + H_2 \rightarrow OH^{\bullet} + H^{\bullet}$

Halogenated FRs (e.g., chlorine- or bromine-based) are known for releasing $Cl^{\bullet}$ or $Br^{\bullet}$ radicals that capture highly reactive $H^{\bullet}$ and $OH^{\bullet}$, terminating the radical chain reaction. This suppression of exothermic reactions reduces flame temperature, slows pyrolysis, and limits the release of flammable gases. Certain flame retardants operate by addressing the concentration of active free radicals produced during polymer combustion. These radicals are crucial for maintaining the chain reaction. Specific additives function by scavenging or sequestering these radicals, so interrupting the chain reaction and ceasing flame growth. Others release inert gases such as NH_3 upon decomposition, which helps dilute oxygen and free radicals in the flame zone, delaying flame development and reducing flammability [17].

Xin Wang et al. proposed the flame-retardant mechanism of a composite material composed of polyphosphamide-grafted graphene and epoxy resin (PPA-g-GNS/EP). The enhanced fire-resistant properties of this composite are illustrated in Fig. 6. Firstly, phosphate species released during the thermal decomposition of polyphosphamide catalyze the degradation of the macromolecular network within the thermoset matrix, generating various fragment ions. Under high-temperature conditions, some of these fragments undergo recombination. For example, benzene and aniline can react to form carbazole, which may further react with benzene to yield several polycyclic aromatic hydrocarbons (PAHs). These PAHs have been identified by mass spectrometry at m/z 404 and 493. Secondly, the reactive PAH species are adsorbed onto the graphene surface, where they undergo further polymerization. Graphene acts as a template for the growth of micro-char structures. The accumulation of numerous micro-char particles results in the formation of a persistent and dense char coating. This protective barrier efficiently restricts the transmission of mass and heat between the gaseous phase and the underlying polymer. The multi-stage procedure is essential for the superior flame-retardant capabilities exhibited by the PPA-g-GNS/EP composites [17].

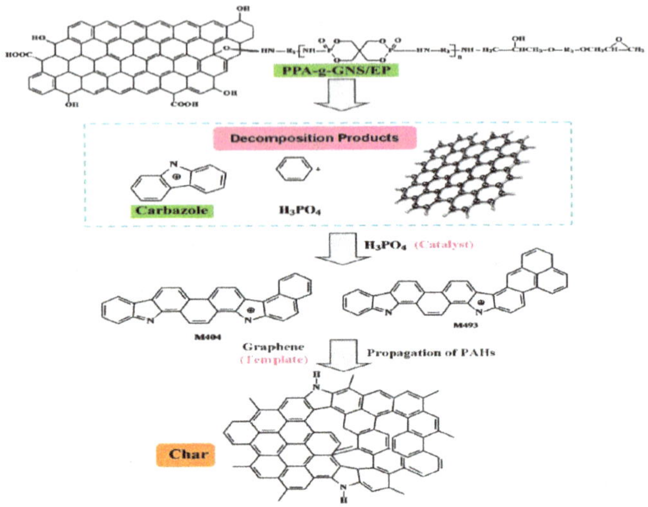

Figure 6: The proposed mechanism for char formation in PPA-g-GNS/EP composites [17].

Qian Leng and colleagues created a flame-retardant nanocomposite composed of alpha-zirconium phosphate, cerium phosphate, and GO (ZCG) using a hydrothermal synthesis method. Structural studies showed that hexagonal flaky α-ZrP was anchored onto the surface of CPO rods. This ZCG nanocomposite was then integrated with APP to develop an intumescent flame-retardant system for polyvinyl alcohol (PVA). The proposed flame-retardant mechanism is illustrated in Fig. 7. During combustion, APP decomposes and releases polyphosphoric acid, which forms a protective layer on the substrate, promoting dehydration and carbonization of the surface and enhancing bonding with the flame retardant. In the ZCG system, α-ZrP, supported on thermally stable CPO, acts as a solid acid catalyst that accelerates matrix carbonization and promotes char layer formation. α-ZrP serves as a bridging agent, connecting GO and CPO via hydrogen bonding to form a synergistic "brick-and-cement" structure. This configuration increases the tortuosity of oxygen and gas diffusion pathways, significantly enhancing the barrier properties of the PVA film. Thermal decomposition of APP releases non-combustible gases such as NH_3 and N_2, which dilute atmospheric oxygen and assist in expanding the carbon layer, thereby impeding combustion [3]. Qian Leng and colleagues created a flame-retardant nanocomposite composed of alpha-zirconium phosphate, cerium phosphate, and GO (ZCG) using a hydrothermal synthesis method. Structural studies showed that hexagonal flaky α-ZrP was anchored onto the surface of CPO rods. This ZCG nanocomposite was then integrated with APP to develop an intumescent flame-retardant system for polyvinyl alcohol (PVA) [19].

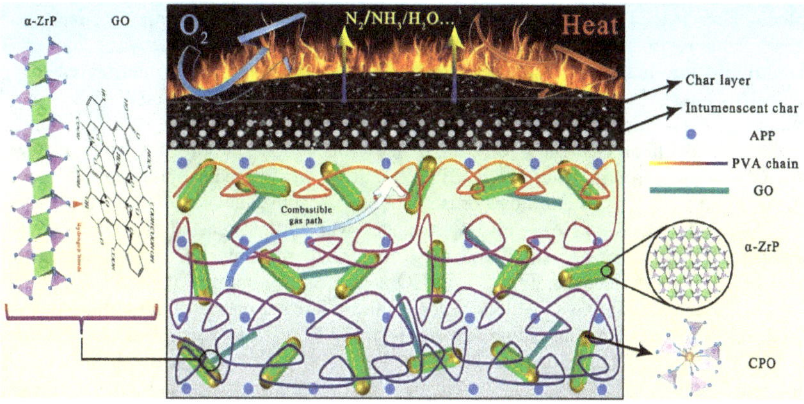

Figure 7: Schematic representation of the potential flame retardant mechanism of ZCG/APP in PVA [18].

6. Graphene in Flame Retardant Polymers

Graphene nanosheets often tend to restack within polymer matrices due to strong π–π interactions, which leads to poor dispersion. To address this issue, surfactant modification is commonly employed to enhance the compatibility of graphene with polymers. More recently, in efforts to achieve both uniform dispersion and improved flame retardancy, researchers have explored surface modification or decoration of graphene with conventional FR agents [20]. For instance, reduced GO nanosheets were functionalized in a single-step process using $POCl_3$ and 4,4′-diaminodiphenylmethane, introducing phosphorus and nitrogen elements. The modified nanosheets significantly decreased the PHRR and overall heat release when added to epoxy systems. However, it's important to note that further verification is needed, as comparisons with unmodified reduced GO at the same concentration were not included. In a separate study, Fang and Wang's research groups chemically modified GO to boost synergistic effects. They developed a straightforward and eco-friendly approach using phytic acid and piperazine, both rich in phosphorus and nitrogen, through a water-based self-assembly technique. This procedure involved stirring a watery piperazine solution overnight as GO was gradually added. It was then mixed with a water-based phytic acid solution and stirred again. After washing, the final compound, which is phytic acid@piperazine@GO (PPGO), was obtained. The incorporation of a modest quantity of PPGO into epoxy resin resulted in a significant decrease in both PHRR and carbon monoxide emissions, surpassing the effects reported with unmodified GO. Visual evidence from cone calorimetry testing supported these findings: the epoxy containing PPGO maintained a continuous char layer, whereas the neat epoxy and the sample with unmodified GO showed incomplete layers.

Scanning electron microscopy (SEM) images further confirmed that the PPGO-modified composite developed a more cohesive char structure. Specifically, the char layer in the PPGO sample appeared dense and non-porous, in contrast to the porous and vented structures seen in the other two samples. In the gas phase, phosphorus-containing FRs can inhibit radical chain reactions, but their primary activity is in the condensed phase, where they encourage the burning of an

Materials Research Foundations 190 (2026) 118-152 https://doi.org/10.21741/9781644904077-5

insulating char layer. By chemically modifying GO and uniformly distributing the phosphorus–nitrogen flame retardant within the polymer, a more consistent char barrier was achieved. This enhanced structure helped to insulate the material, reducing heat transfer and the emission of combustion gases. In a related approach to developing multifunctional flame-retardant polyurethane nanocomposites, GO was chemically grafted with surfactants containing nitrogen, phosphorus, and silicon elements. To begin, amino groups found in polyethylenimine interacted with GO's epoxide groups. After that, DOPO-PEI@GO was formed by bonding the remaining amino groups to a phosphorus-containing molecule (9,10-dihydro-9-oxa-10-phosphaphenanthrene-10-oxide). This intermediate was further reacted with an isocyanate-functionalized silane compound, producing the final multifunctional GO (mfGO). As demonstrated visually, incorporating this mfGO significantly improved the flame resistance of polyurethane, enabling it to pass vertical flame testing where untreated or unmodified composites failed [21]. There was a notable improvement in tensile strength and a considerable decrease in PHRR, overall heat discharge, and smoke production at a specific mfGO loading. In comparison, incorporating unmodified GO into polyurethane resulted in a minor reduction in heat release and only minimal decreases in total heat and smoke output. First, the complex maze-like char structure that the GO created slowed down the release of flammable gases. Second, the non-combustible gases released during the disintegration of the customized GO diluted the volatile substances and oxygen in the combustion area. Third, the stable oxide layers that were formed during burning by the phosphorus and silicon features attached to the graphene reinforced the protective char layer. These three factors mainly contribute to enhanced flame-retardant performance (Fig. 8). In addition to improved flame resistance, the modified polyurethane also exhibited multifunctional properties. A diselenide-based polyurethane matrix was synthesized using a special chain extender, imparting the material with both self-healing and shape memory behavior. The resulting nanocomposites demonstrated high shape memory performance and substantial self-repair capability. This study clearly highlights how chemical modification of GO can lead to multifunctional polymer nanocomposites with excellent flame retardancy. In another example, Baek J. et al. [23] produced graphene phosphonic acid (GPA) by ball-milling graphite with red phosphorus for 12 hours, followed by exposure to atmospheric moisture. The resulting GPA was highly water-dispersible, confirmed by zeta potential measurements. When applied as a coating on Hanji paper, GPA demonstrated superior flame resistance compared to conventional GO, as shown in Fig. 9.

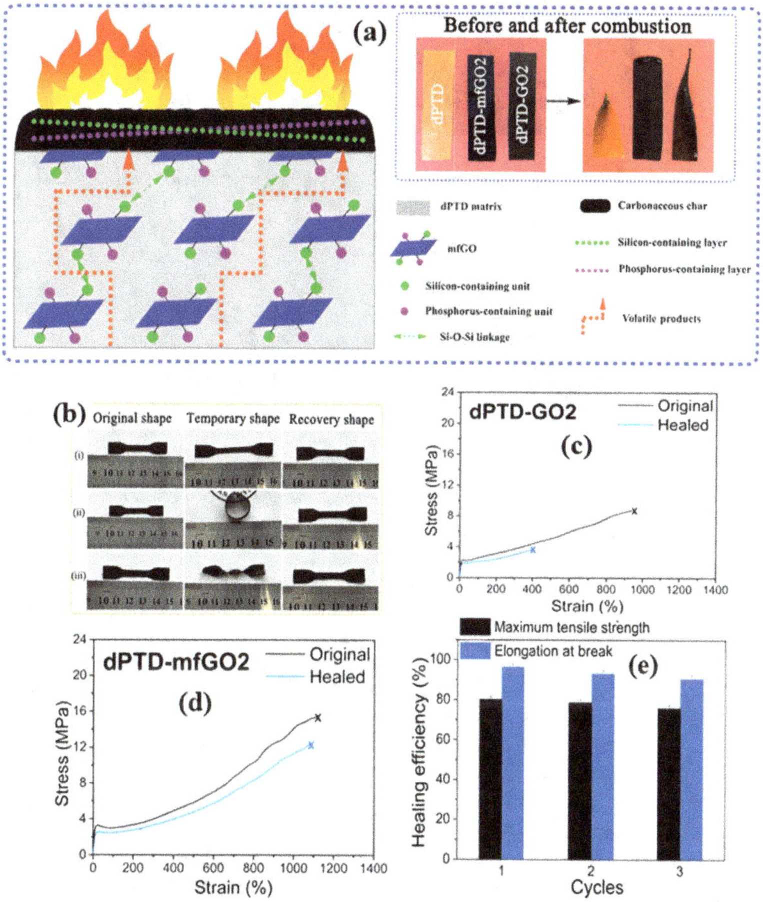

Figure 8: (a) Flame-retardant behavior of nanocomposites based on dPTD, along with their shape memory and self-healing capabilities (images b–g). In this context, dPTD refers to a polyurethane matrix containing diselenide groups, mfGO₂ indicates the incorporation of 2 wt.% multifunctionalized GO, and GO₂ refers to 2 wt.% unmodified GO [22].

Materials Research Foundations 190 (2026) 118-152
https://doi.org/10.21741/9781644904077-5

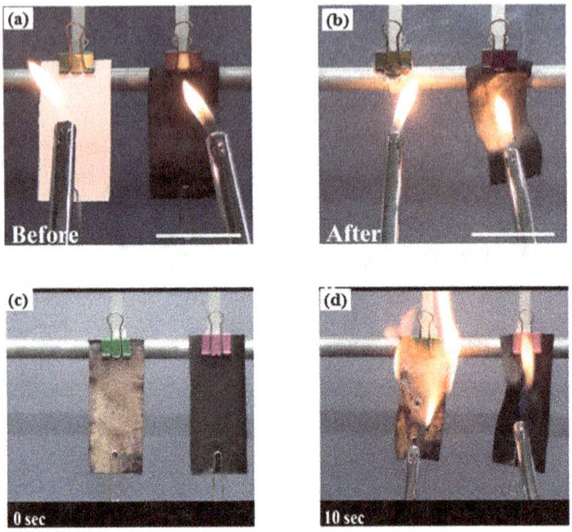

Figure 9: Differences in flame resistance between white, untreated Hanji paper and black, graphene-and phosphonic acid-coated Hanji paper: (a) at the start (0 seconds) and (b) after 20 seconds of flame exposure. Comparison of flame-retardant behavior between two Hanji papers coated with GO (left) and graphene–phosphonic acid composite (right): (c) prior to ignition and (d) following flame exposure [23].

Another study found that by using decreased GO, the flame-retardant characteristics of ABS might be enhanced without the use of halogen-based additives. Initially, graphite oxide underwent bromination with tetrabromide to yield brominated GO (Br-GO) through covalent bonding. A subsequent step involved the deposition of a molybdenum-and antimony-containing molecule, Sb-Mo/Br-GO, onto the Br-GO surface by non-covalent interactions. Fig. 10 shows that even tiny amounts of these nanocomposites added to ABS significantly improved performance, especially in terms of tensile strength and stiffness, while also significantly lowering the peak rate of heat discharge and smoke emissions [24].

The level of interaction across GO and halogens is unclear; nevertheless, since composites including solely Br-Sb-Mo or GO have not been assessed. Further synergy was observed when reduced GO was combined with nickel ammonium phosphate monohydrate (ANP) in PVC. The ANP-modified rGO, prepared via a solvothermal method, reduced Peak HRR by 70% at 13 wt.% loading, which is much higher than the 29.7% and 24.9% reductions achieved with ANP or rGO alone. This hybrid improved the mechanical strength of PVC by 6.7%, whereas ANP by itself weakened the matrix. Smoke production was also significantly reduced: 59% with ANP-rGO, compared to 54% and 51% with ANP and rGO alone [25]. Achieving optimal synergy is essential for maximizing flame retardancy without sacrificing mechanical integrity. A critical component in minimizing building fire deaths is the collaborative suppression of smoke during polymer burning, which can be achieved by integrating graphene or CNTs with metal oxides or salts. As an example, whereas rGO alone raised smoke by 52% and Ni + Ce oxides individually by 3.2%

Materials Research Forum LLC
https://doi.org/10.21741/9781644904077-5

[26], combining both with PP reduced overall smoke release by 36%. Similarly, total smoke was decreased by 41% in epoxy resin mixed with Ce-MnO_2-modified graphene, as compared to 13.3% in epoxy with unaltered graphene and 27.7% in Ce-MnO_2 alone [27]. Combining conventional FRs with LDHs, which are ionic layered materials defined by interchangeable anions between their layers, has further increased their flame retardancy. It has been demonstrated that graphene treated with LDHs through non-covalent interactions can reduce smoke emission, increase TTI, and lower peak HRR and THR. These modifications have synergistic effects. Graphene nanosheets (GNS) were coated with Ni-Fe LDH using a self-assembly process, for instance. The flame-retardant efficacy of epoxy was greatly improved by this hybrid of Ni-Fe LDH and GNS, surpassing that of either component alone [28].

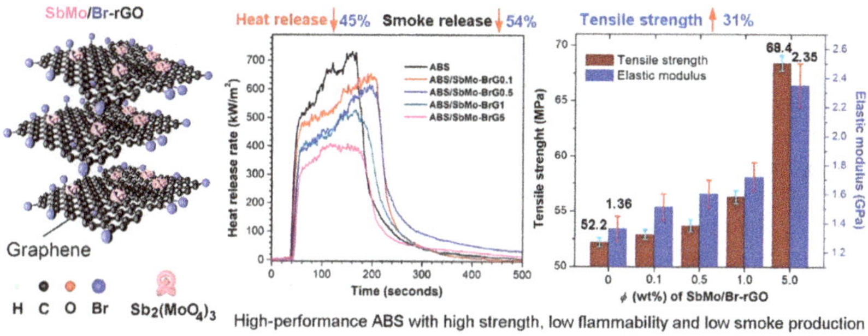

Figure 10: Flame resistance and mechanical characteristics of ABS/Sb–Mo/Br-GO composite materials [24].

7. Carbon Nanotubes in Flame-Retarded Polymers

Both multi-walled and single-walled CNTs show great promise as flame-retardant additives for polymer systems. Epoxy resins, polymethyl methacrylate (PMMA), epoxy, polypropylene, polylactide, and polystyrene are just a few of the many polymers that have these nanoparticles added to them. They help form a stable char layer, like graphene, during the condensing phase, where their flame-retardant activity is most pronounced. The polymer below is protected from thermal breakdown by this char layer. Also, CNTs function as physical barriers by introducing complex, tortuous pathways that hinder the permeation of volatile gases. Their inherently high aspect ratio interrupts direct gas-phase transport routes, thereby restricting the supply of combustible gases. In many cases, this process improves the LOI and decreases the peak HRR by encouraging the formation of a thick, solid char layer. Typically, SWCNTs outperform MWCNTs in comparison. This advantage is attributed to their higher surface area per unit volume, which decreases the spacing between nanotube fillers and increases the interfacial contact area with the surrounding polymer matrix. The development of a char layer that is both less dense and efficient is aided by this improved interaction [29]. Since the flame-retardant mechanism of CNTs closely resembles that of graphene, an in-depth discussion is not provided further. Despite being considered weak, van der Waals forces are highly significant for CNTs owing to their large surface area-to-mass ratio. Consequently, CNTs tend to aggregate into rope-like bundles or nested

Materials Research Foundations 190 (2026) 118-152 https://doi.org/10.21741/9781644904077-5

structures in many solvents [30,31]. The dispersion of CNTs significantly influences the mechanical properties and flame-retardant effectiveness of the composites. If the nanotubes are not evenly distributed, there will be uneven char production, which can allow gases and heat to penetrate. Kashiwagi and co-workers [32] conducted a comparative analysis of PMMA samples containing well-dispersed and poorly dispersed SWCNTs, observing that uniform dispersion led to notable decreases in both Peak HRR and mass loss rate (Fig. 11a–d).

Chemical modifications like covalent functionalization or π-π interactions can significantly enhance CNT dispersion in polymers. However, such changes may damage the CNT structure by shortening the tubes or unzipping them, which reduces their mechanical load-bearing ability and hinders the formation of a continuous, dense network. Since such a network is critical for forming strong char layers, a balance between surface functionalization and structural preservation must be maintained when pristine MWCNTs and high-density PE-coated MWCNTs are incorporated into EVA. The coated versions significantly enhanced the mechanical strength and flame-retardant behavior. As an example, the inclusion of virgin MWCNTs significantly raised the Young's modulus of EVA, while the coated ones showed an even more dramatic rise. The char residue in EVA/MWCNT nanocomposites (Fig. 11e) was dense and cohesive when coated MWCNTs were used, in contrast to the brittle and cracked char formed with pristine MWCNTs (Fig. 11f). This compact char structure effectively reduced smoke production rates. These enhancements were credited to the improved dispersion and stronger interfacial bonding provided by the coating, as observed in SEM images (Fig. 11g & h) [34]. Yang et al. [35] reported similar findings in PP composites, where functionalized MWCNTs (f-CNTs) led to the formation of uniform and complete char layers (Fig. 11i), while unmodified CNTs produced incomplete and fractured char structures (Fig. 11j). Another important factor affecting the flammability of the nanocomposite is the component ratio of carbon nanotubes. For instance, according to [33], MWCNTs with larger aspect ratios significantly reduced the maximum mass loss rate among polystyrene and generated a strong char sheet (Fig. 11k). In contrast, MWCNTs with a lesser aspect ratio produced a fragmented char structure and produced less noticeable benefits (Fig. 11l). The concentration of CNTs also plays a key role in enhancing both flame-retardant and mechanical performance of polymers. Typically, increasing CNT content improves flame resistance. Owing to their thermal stability, CNTs do not contribute to fuel during combustion. Instead, their volume replaces flammable polymer portions and supports the development of continuous and voluminous char layers. As depicted in Fig. 11m, a minimal CNT content does not significantly reduce heat release in PMMA, but moderate increases substantially lower the peak HRR. However, further additions beyond an optimal level offer no extra advantage due to agglomeration and poor dispersion. CNTs are also effective in reducing melt-dripping in polymers. Figures 12 (a) and 12 (b) illustrate that PLA/MWCNT nanocomposites exhibit prolonged burn durations and resist dripping, unlike pure PLA, which melts and drips quickly when burned. This improvement is linked to the increase in melt viscosity caused by the CNT network in PLA. Nonetheless, combustion still occurs in both cases [36]. Fig. 12c demonstrates that incorporating CNTs raised the LOI of elastomeric foam, granting it self-extinguishing capabilities [37].

Materials Research Forum LLC
https://doi.org/10.21741/9781644904077-5

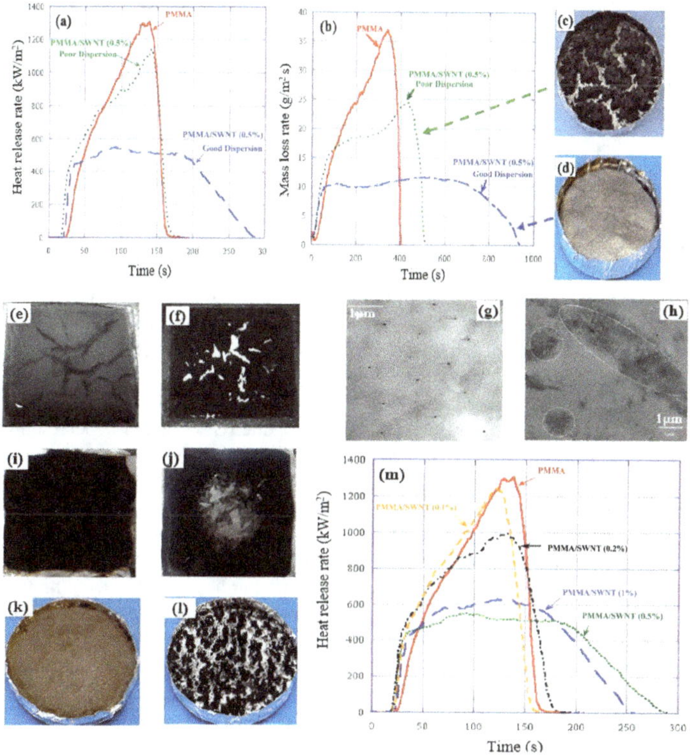

Figure 11: Nanocomposites comprising polymers and carbon nanotubes can achieve flame resistance by modifying several parameters, including: (1) the dispersion quality of CNTs, which affects (a) the heat release rate (HRR), (b) the mass loss rate in PMMA/CNT systems, and (c, d) the morphology of the resultant char residues; (2) the characteristics of char residues observed in (e) EVA composites with PE-coated MWCNTs and (f) those containing unmodified MWCNTs, alongside SEM images illustrating the dispersion states of (g) PE-coated and (h) pristine MWCNTs within the EVA matrix, as well as char residues from (i) polypropylene/functionalized-CNT and (j) polypropylene/pristine CNT composites; (k) the aspect ratio of CNTs, demonstrated within char remains establishes of polystyrene/MWCNT composite materials with (k) a high aspect ratio of 150 and (l) a lower aspect ratio of 49; and (m) the CNT concentration in PMMA and its relationship with HRR (m) [33].

CNTs and graphene are both carbon-based, but their structural variances cause them to interact with host polymers in different ways, affecting their mechanical behavior, flame retardancy, and electrical and thermal conductivity, among other physical qualities. As an example, when it comes to flame retardancy in PP, thermally reduced GO (TRGO) outperforms MWCNTs. When 5 wt.% MWCNTs were added to PP, Peak HRR was reduced by 57.5%, and TRGO was reduced by 73.5%.

Materials Research Foundations 190 (2026) 118-152 https://doi.org/10.21741/9781644904077-5

Additionally, HRR-time testing revealed that the PP/TRGO nanocomposite burned for approximately 100 seconds longer. While TRGO did not affect the UL94 test, MWCNTs dropped the PP grade from HB to "not rated" [38]. These distinctions arise because CNTs create looser structures through line-contacts, in contrast to graphene, whose plate-like structure provides a larger interface area with the polymer, leading to more compact composites. When it comes to reducing the conduction of heat and mass from heated surfaces, the structure of graphene is superior. Dittrich B. et al. [38] confirmed this, finding that Peak HRR lowering was more strongly associated with the BET total area of carbon allotropes. A low CNT concentration significantly improves both the mechanical characteristics and the flame retardancy. Enhanced mechanical performance is exhibited by the majority of polymer/CNT nanocomposites. For example, PP with 2 wt.% MWCNTs exhibits an enhanced storage modulus and GO decreases from 80 °C to 120 °C [39]. The high aspect ratio and rigidity of CNTs limit the unraveling of polymer chains, making them more resistant to deformation. The transfer of stress between the matrixes to the CNTs, which is essential for these advantages, is greatly aided by strong interfacial bonding. Bhattacharya [40] compiled results from several studies that looked at how carbon nanotubes affected the modulus of elasticity and tensile strength of various polymers. Adding CNTs to insulating polymers makes them more conductive, thanks to their high electrical conductivity. There is a percolation threshold below which composites made of polymer and carbon nanotubes do not conduct electricity. Once this point is passed, even small amounts of carbon nanotubes can significantly increase conductivity by creating a dense network of conductivity. One example is the percolation achieved by aligned epoxy/CNT nanocomposites at a weight percentage of only 0.0025 wt.% [41]. With more CNTs, the space between the fillers gets smaller, which means more electrons can move about. The conductivity characteristic of these composites has been the subject of multiple reviews. Finally, CNTs are functional fillers in polymers because they are flame-resistant, reinforce the material, and facilitate electrical conduction. The kind, proportion, alignment, and composition of the polymer determine its total effectiveness. These materials significantly improve flame retardancy, thermal, and mechanical strength, and electrical conductivity, and have similar properties to graphene.

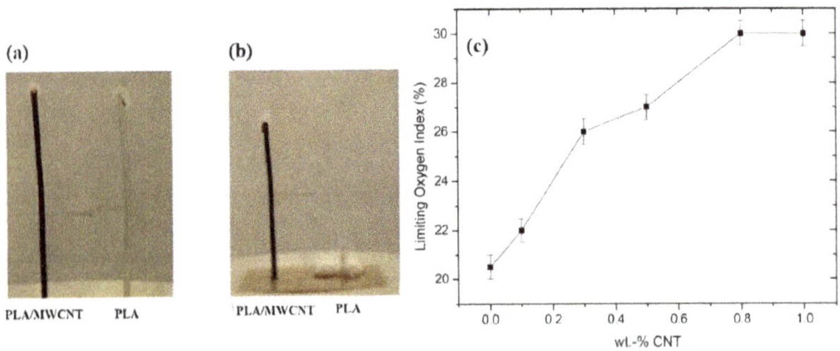

Figure 12: Flame retardant performance of pure PLA and its nanocomposites with MWCNTs: (a) immediately upon ignition, (b) 15 seconds post-ignition, and (c) LOI values observed in elastomer-based composites containing MWCNTs. [37]

Materials Research Foundations 190 (2026) 118-152 https://doi.org/10.21741/9781644904077-5

8. Activated Carbon and Carbon Black in Flame-Retarded Polymers

One of the first carbon compounds, carbon black (CB), is made by thermal cracking or partial burning of heavy petroleum byproducts. Its many uses include adding strength to elastomers, coloring, and enhancing the conductivity of a wide range of polymers. The tire industry and other items made of elastomers, like belts and hoses, are the primary consumers of CB. Researchers studying polymer nanocomposites for enhanced conductivity of electricity, thermal resilience, and flame resistance continue to center their attention on CB because of its wide availability, low cost, and well-established production processes. Polymer composites with CB were created by Tang et al. [42]. These composites showed improved electrical conductivity, thermal durability, flame retardancy, and mechanical strength. Maximum HRR and THR were significantly reduced, and LOI was increased when CB was added to PP. During the thermo-oxidative breakdown of PP, CB caught peroxy radicals and formed a gel-like conjugated network, which was thought to be responsible for these increases. The UL94 vertical flame test still failed PP/CB composites despite these upgrades, primarily due to the melt pouring process. To overcome this, CB was mixed with traditional flame retardants such as carbonates, phosphorus-based compounds, intumescent agents, and metal hydroxides to fulfill essential flame retardancy requirements, especially UL94 and LOI standards. Unlike CB-only formulations, which were unrated, PP composites containing CB and an intumescent flame retardant, along with a compatibilizer specifically maleic anhydride-grafted polyolefin elastomer, attained a V-0 classification in the UL94 vertical test [43]. The CB/PP system exhibited an electrical percolation threshold [46]. Before being integrated into PLA, CB underwent surface treatment involving hydrogen peroxide hydroxylation and chemical-based grafting with a flame retardant containing phosphorus, as detailed in an additional study [44]. This method of modification produced a composite (Fig. 13b) that outperformed neat PLA in terms of mechanical characteristics, electrical conductivity, and flame resistance, as shown in Fig. 13a.

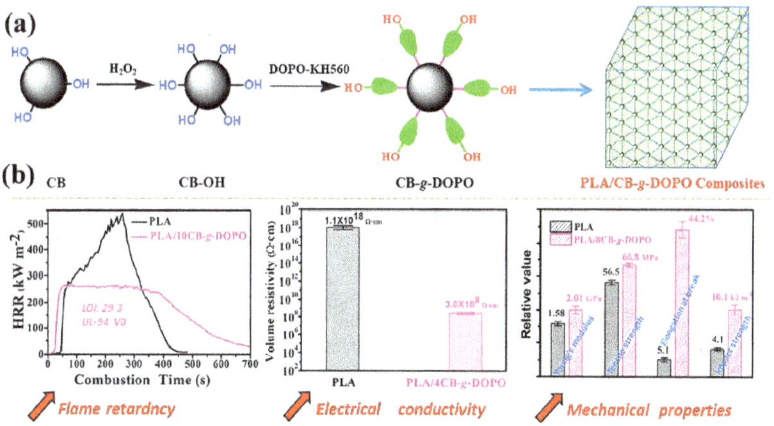

Figure 13: (a) Illustration of the modification process of CB using DOPO and the subsequent fabrication of its PLA-based composites, and (b) the resulting composites' flame retardant performance, electrical conductivity, and mechanical characteristics [44].

Using CB in polymer matrices alongside traditional FRs has often shown synergistic benefits. A mix of high-impact polystyrene (HIPS), microencapsulated red phosphorus (MRP), and a small quantity of CB enhanced the UL94 vertical combustion categorization from a mid-level to the highest grade, as seen in the cited figure.

When used with conventional FRs in polymer systems, CB has shown significant synergistic effects. The UL94 flame rating was raised from V-1 to V-0 when a minor quantity of CB was added to a HIPS blend that already contained MH and MRP. Using scanning electron microscopy and thermo-oxidative degradation studies, Liu J. et al. studied the HIPS/MH/MRP composites. Their findings demonstrated that CB improved heat stability by decreasing the degradation rate. By way of illustration, during thirty minutes of thermal processing at 350 °C, the composite lacking CB lost a significant amount of weight, whereas the one with CB retained a substantial portion. The scanning electron micrographs (SEMs) revealed numerous huge pores in the CB-containing sample, but only a handful of tiny vents in the CB-free sample. Based on the results, it appears that CB helps create a denser and more fire-resistant structure by crosslinking the polymer matrix [45]. In addition, CB is an efficient method for decreasing TSR from TPU composites. The TPU/ammonium polyphosphate (13.6 wt.%) mixture with 2 wt.% CB reduced TSR by 72.6%, but ammonium polyphosphate alone reduced TSR by just 43.6% [46]. The enhanced performance was associated with a decrease in smoke generation rate and an increase in melting viscosity at high temperatures caused by the CB.

9. Fullerenes and Their Role in Polymer Flame Retardancy

Fullerenes are a class of zero-dimensional (0D) carbon nanomaterials that were initially developed during the 1990s. Their exceptional geometric structure, marked by an abundance of π-electrons on the outer surface, confers a wide range of remarkable properties. The configuration of fullerenes can vary based on the number of carbon atoms in each cluster, including C_{60}, C_{70}, C_{78}, C_{82}, C_{84}, C_{90}, and C_{96}. These fullerenes exhibit notable characteristics such as unique optical behaviors, improved lubrication efficiency, catalytic activity, superconductivity, biocompatibility, and exceptional resistance to oxidation, among other outstanding performance attributes. The inherent structure of fullerenes, with their double bonds, enables them to function as radical scavengers due to their capacity to trap free radicals, which contributes to superior thermal and thermo-oxidative stability when incorporated into polymer substrates within nanocomposites. This section will focus on novel methodologies and significant progress in enhancing the fireproofing of polymer composites utilizing fullerenes, as prior research by Pan et al. [47] and Tayfun et al. [48] has already summarized the thermal stability and flame-retardant applications of fullerenes. In earlier studies, pristine fullerenes were commonly added directly into polymer matrices through melt compounding as single additives to improve thermal stability. Fang et al. [49] applied 2 wt.% of C_{60} to PP and HDPE/EVA blends, resulting in 47% and 45% reductions in peak HRR, respectively. This substantial decline in HRR was attributed to the mechanism of free radical trapping. To enhance the flame-retardant performance of fullerenes, functionalization with various elements and compounds has been widely explored. Due to its 30 carbon-carbon double bonds, C_{60} has a high affinity for free radicals, allowing it to form in situ gel-like networks through interactions with macromolecules. Song et al. [51] chemically grafted C60 with nitrogen- and phosphorus-containing PDBPP (Fig. 14a) and incorporated the modified C_{60}-PDBPP into a PP matrix via melt blending. This modification significantly improved the composite's flame retardancy compared to unmodified C_{60}/PP systems. Guo et al. [52] prepared C_{60}-DOPO via a basic reflux process with

DOPO (Fig. 14b), which, when added to PP, led to notable reductions in maximum heat release and smoke production. Beyond thermoplastics, Wang et al. [53] synthesized C_{60}-anchored PEI-RGO hybrids through a two-step sonication approach and incorporated them into an epoxy (EP) matrix via mold curing. Even at low loadings, this hybrid greatly enhanced flame resistance. Song et al. [54] explored synergistic effects by modifying CNTs with hydroxyl, amine, and C_{60} groups, and blending them with PP. As a result of gas-phase inhibition and condensed-phase char production, the consequent C_{60}-CNT/PP composites showed a significant improvement in flame retardancy. Zhou et al. [55] combined C_{60} with ATH and SBS in different mass ratios, producing composites that passed V-0 vertical burning tests and showed enhanced char residue. The improved performance was attributed to both ATH's dehydration effects and the radical-trapping ability of fullerene. Similarly, Pan et al. [47] altered C_{60} with iron and integrated it with APP/PER to create an intumescent flame retardant (IFR) for HDPE. Tsai et al. [56] developed layered inorganic materials by incorporating C_{60} into a system with benzalkonium chloride and N-methyl pyrrolidine, which increased MMT interlayer spacing and improved the flame retardancy of novolac epoxy resin. While promising, these C_{60}-based systems often show slightly lower efficiency compared to in situ polymerized nanocomposites.

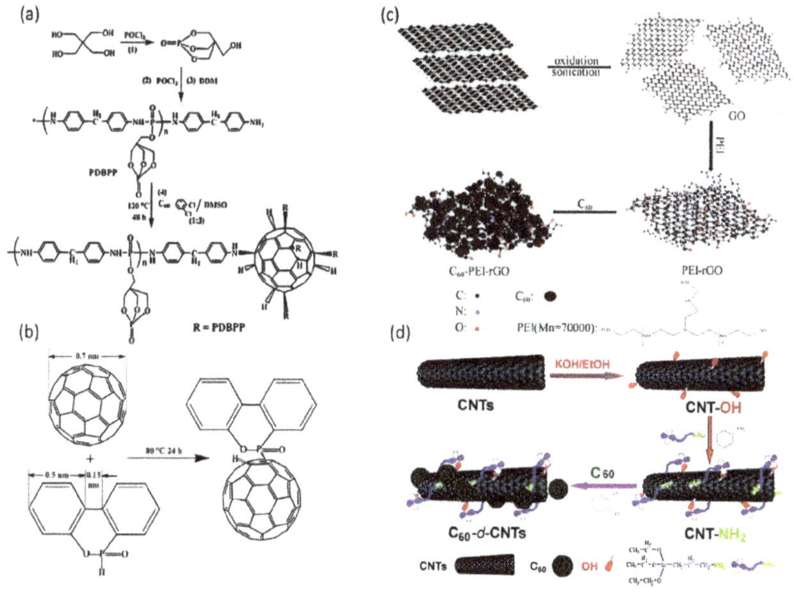

Figure 14: Synthesis methods and molecular structures of functionalized fullerene-based composites: a) C_{60}-PDBPP; b) C_{60}-DOPO; c) C_{60}-PEI-RGO ; d) C_{60}-CNT; [50,51,52,65]

Materials Research Foundations 190 (2026) 118-152 https://doi.org/10.21741/9781644904077-5

10. Environmental and Toxicological Aspects of Carbon Nanomaterial Flame Retardants

FRs are engineered to reduce the likelihood of fires igniting when exposed to minor heat sources, such as cigarettes, candles, or electrical malfunctions. They are extensively utilized in the construction, electronics, automotive, and transportation industries, in addition to applications in wires, cables, textiles, and others. Nevertheless, escalating environmental concerns, toxicity, and threats to human health are driving initiatives to restrict or eradicate the use of these compounds. Currently, most FRs are halogen-based, but formulations based on inorganic acids, nitrogen, phosphorus, and intumescent compounds remain significant in flame retardant research. Nevertheless, rising awareness of environmental and health issues is driving interest in safer, more sustainable alternatives [57]. One such emerging alternative is EG, which is gaining attention due to its environmental and economic benefits. Despite its promise, studies have not thoroughly evaluated the potential health and ecological risks associated with EG. According to existing literature, EG has not been classified under European regulatory frameworks, and most available data is limited to the Chinese market [58]. Graphite, EG's raw material, is also unclassified under EC regulations, though occupational exposure, particularly in the graphite industry, has been linked to conditions like dust-induced bronchitis. While sulfuric acid used in EG production is corrosive in concentrated form, current findings do not show any significant adverse environmental or health impacts from using EG in polyolefin-based FRs. EG is considered a leading material among environmentally friendly alternatives. Furthermore, composites like EG combined with manganese ferrite have shown promise in applications such as oil spill cleanup, effectively removing heavy oils from water. The stability of graphite under ambient conditions and the unique expandability of EG make it especially valuable for designing flame-retardant systems. When used as a flame retardant, the ability of materials to expand and their high thermal stability are critical performance attributes [59]. Cristina Alonso et al. conducted an in vitro toxicological assessment of EG as a flame retardant in polyester (PET) textiles. A control PET fabric and a PET fabric treated with EG were examined before any skin contact tests. Fabric pieces were placed on glass slides, and specific areas were imaged. Optical images from the Raman microscope revealed large, well-distributed graphite domains, approximately 10 μm in size across the surface of the treated fabric (Fig. 15-a) [60]. Subsequently, Raman mapping of an 80 μm × 80 μm area within the graphite regions identified three components: graphite, an organic substance, and small carbonate particles (likely calcium carbonate). Fig. 15-b shows the Raman spectrum and mapping results, with graphite appearing predominantly (in red), and spherical calcium carbonate particles visualized in green. The graphite spectrum featured three characteristic peaks-D, G, and 2D bands isolated using Lorentzian fitting (Fig. 15-c) [60]. The D-band (around 1360 cm^{-1}) signified the presence of structural defects, while the G-band (around 1580 cm^{-1}) corresponded to an ordered graphite structure, and the 2D-band (around 2700 cm^{-1}) was also detected. Following this, a percutaneous absorption study was carried out. Three samples of the treated fabric were applied to the skin using Franz diffusion cells, ensuring the untreated side was in contact with the skin. For comparison, a control PET sample without EG was also included. After a 24-hour exposure period, the fabric was removed, and the outermost skin layer (stratum corneum) was sequentially stripped. The deeper layers, comprising the inner stratum corneum, viable epidermis, and dermis, were then examined using confocal Raman microscopy, in accordance with established in vitro skin permeation guidelines. Raman spectra typical of untreated skin were also recorded (see Fig. 15-d) [60].

Materials Research Foundations 190 (2026) 118-152 https://doi.org/10.21741/9781644904077-5

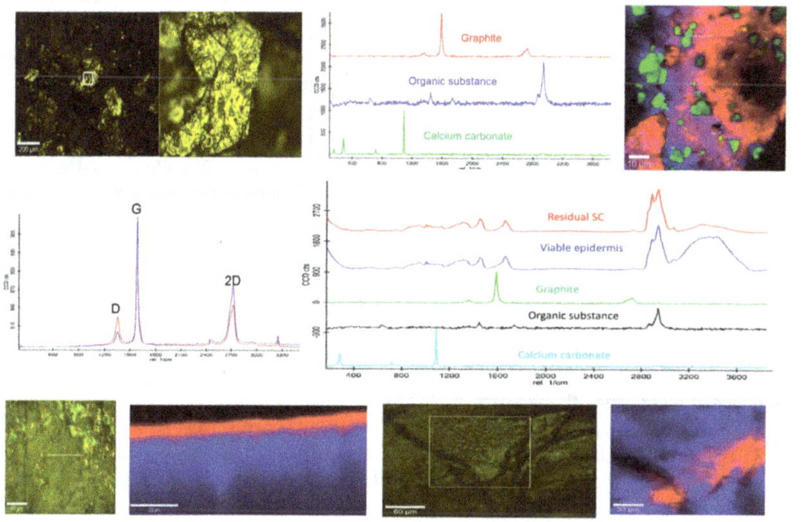

Figure 15: (a) Graphite coatings on PET fabric, consisting of graphite scales, were observed using optical microscopy, (b) The Raman spectrum (left) and the microscopy mapping image (right) of the fabric surface treated with EG reveal the presence of graphite (highlighted in red), an organic substance (shown in blue), and calcium carbonate (depicted in green), (c) The Raman spectrum of graphite displays three characteristic peaks corresponding to the D, G, and 2D bands, (d) Raman spectra of the control samples of both dried and hydrated skin, along with the spectra of potential compounds that may result from exposure to the EG-treated textile, (e) A microscopy image and Raman mapping captured at a depth of 30 μm beneath the skin surface, following 24 hours of exposure to the EG50 textile, display residual stratum corneum (SC) in red and the viable epidermis and dermis in blue, (f) A microscopy image and XY Raman mapping taken 20 μm beneath the skin surface, after 24 hours of contact with the EG50 textile, reveal residual stratum corneum (SC) highlighted in red and the viable epidermis shown in blue [60].

Post-stripping spectra showed residual stratum corneum (red), while deeper epidermal layers (20 μm depth) showed stronger water signals between 3300-3600 cm^{-1} (blue spectrum), indicating more hydrated tissue. For skin samples exposed to EG-treated fabric, Raman mapping was performed over a 100 μm length and 30 μm depth using a ×100 objective, with a 5 mW laser and 0.5 s exposure time (Fig. 15-e) [60]. As expected, drier skin near the surface showed residual stratum corneum signals (red), while deeper layers (30 μm) exhibited higher water content (blue). Additional XY mapping (150 × 150 μm) at ~20 μm depth (Fig. 15-f) revealed primarily viable epidermis (blue) with minimal residual stratum corneum (red). No graphite (green signal) was detected in any of the skin samples, nor was any graphite visible on the skin surface via optical microscopy. This confirmed that no EG was absorbed through the skin. Thus, after 24 hours of contact, EG from the treated PET fabric did not penetrate the skin. Exposure assessment based on the detection limit confirmed zero absorption. The margin of safety (MoS) exceeded 1032, well

Materials Research Foundations 190 (2026) 118-152 https://doi.org/10.21741/9781644904077-5

above the benchmark of 100, demonstrating that graphite-treated polyester fabric poses no toxicological risk. These findings support the safe use of EG as a sustainable and effective flame retardant for upholstery applications [60]. While CNTs have shown promise as FRs and are being considered as alternatives to halogenated FRs [61], there remains limited understanding of their behavior during fires. Additional research is needed to evaluate how CNTs perform under various combustion conditions, particularly since they may contribute to smoke production, promote fire growth, or elevate airborne nanoparticle concentrations, posing respiratory health risks [62]. CNTs, owing to their large surface area and capacity to adsorb organic substances and metals particularly when oxidized or hydroxylated may interact with byproducts of combustion and potentially serve as carriers within the respiratory system (Nowack & Bucheli, 2007). In a study by Kashiwagi et al., the flame-retardant behavior of multi-walled carbon nanotubes (MWCNTs) in PP composites was investigated. Combustion residues were examined using SEM and energy-dispersive X-ray (EDX) spectroscopy. In the PP/MWCNT composite containing a small volume of MWCNTs, diffuse black and red residues were identified, which were linked to partially oxidized CNTs and iron oxide. Fuzzy, black residues made of carbon nanotube structures that were locally oriented were more abundant at a 2% loading. These results make one wonder whether CNTs could become aerosolized after burning or after mechanical disturbance. Researchers Nyden et al. (2010) looked examined the effects of oxidized nanofibers of carbon and polyurethane foam (PUF) on emission of particles and char formation in well-ventilated environments. In contrast to CNF-infused PUF, which created an interconnected network of CNFs that decreased flammability and dripping, pure PUF produced very little char when burned. The CNF-PUF composite, when burned, produced solely round soot particles devoid of any protruding features. Nevertheless, the presence of lengthy shapes in the char indicates that the CNFs were thermally destroyed when burned. Within one minute of being warmed to 650°C in atmosphere, these structures vanished from the char. A large number of submicron-size particles were emitted when the char was mechanically agitated about 10,000 particles/cm^3 which is much greater than background levels. These particles were believed to be CNF-based, as the char was largely composed of them. However, the study did not clarify the aggregation state of the released particles and used CNFs rather than CNTs, limiting direct comparisons [62]. Understanding the toxicity, biodistribution, fate, and exposure of graphene-based materials (GBMs) is crucial not only for evaluating their environmental risks but also for assessing their overall sustainability. Life cycle assessment (LCA) provides a systematic framework for assessing sustainability by combining models of manufacturing processes and supply chains with the environmental interactions of graphene-based materials (GBMs). Recent LCA research has focused on evaluating the environmental effects linked to the production of materials such as graphene, graphite nanoplatelets, and reduced GO, employing different synthesis methods including chemical reduction, ultrasonication, thermal exfoliation with ball milling, chemical vapor deposition, and epitaxial growth. For example, research under the MISTRA Environmental Nano safety project compared graphene made via ultrasonication and chemical reduction, noting differences in energy and water usage, as well as in human and ecological toxicity. Although the findings vary due to differences in materials, methods, and production scales, all studies consistently point to high energy use and the involvement of harmful chemicals (such as diethyl ether and methane) as major contributors to environmental impacts like global warming, water pollution, and toxicity. Despite these insights, important gaps in the sustainability evaluation of GBMs remain unresolved [63].

Materials Research Foundations 190 (2026) 118-152 | https://doi.org/10.21741/9781644904077-5

11. Challenges and Future Directions

Recently, carbon-based nano compounds have garnered significant interest in improving flame retardancy in polymeric matrices. Compared to conventional FRs, these nanomaterials offer substantial improvements not only in flame resistance but also in mechanical strength, thermal stability, and electrical conductivity, resulting in high-performance multifunctional flame-retardant nanocomposites. The unique morphologies of various carbon nanostructures are closely linked to their exceptional properties. These materials can be incorporated into a wide range of polymer matrices or fabricated into single-component systems with inherent functional characteristics. For instance, CNTs have been employed as a structural matrix to produce carbon-based paper, suggesting innovative directions for the development of advanced nanocomposites. Despite their promising properties, pristine carbon nanomaterials generally serve as synergistic agents rather than standalone FRs. Furthermore, achieving uniform dispersion within polymer matrices remains a significant challenge. Inadequate dispersion can negatively impact the mechanical performance of the composites. To mitigate these constraints, methods for surface modification are utilized with two primary aims: (i) to reduce interparticle interactions and enhance dispersion within the underlying matrix; and (ii) to include flame-retardant groups with functional properties into the nanomaterial exterior [64]. Although many nano-carbon materials exhibit superior functional characteristics, their large-scale application is hindered by the lack of economically viable production methods. Continued efforts in developing cost-effective synthesis techniques are essential for broader implementation. Pristine, defect-free forms such as CNTs and graphene typically require costly and low-yield synthesis methods. In contrast, top-down fabrication techniques, though producing materials with structural imperfections like vacancies, interstitials, or lattice distortions, offer higher yields and cost-effectiveness. Interestingly, these defects can serve as active sites for functionalization, facilitating better interfacial compatibility with polymer matrices. The surface integration of carbon-based nanomaterials is essential for the progression of flame-resistant technologies and is anticipated to continue as a prominent research domain. While the integration of carbon nanomaterials with traditional FRs has gained traction, several research gaps remain. Priority areas include elucidating the synergistic mechanisms in both condensed and gas phases, exploring covalent and π–π interactions to tailor molecular compatibility, and establishing a comprehensive understanding of the structure–property relationships within polymer matrices. Furthermore, there is a pressing need to develop scalable, environmentally benign synthesis strategies, such as incorporating phosphorus- and nitrogen-rich surfactants in conjunction with carbon nanomaterials. Another pressing issue is the recyclability of carbon-based nanocomposites, which has received limited attention to date. In the context of increasing environmental concerns and energy efficiency demands, improving material recyclability, particularly for thermoset-based nanocomposites, is vital. A viable technique entails reusing surplus thermoset carbon nanocomposites into highly valuable carbon-based products. Emerging nanomaterials like MXene and boron nitride are also being investigated as multifunctional flame-retardant fillers [1]. These materials aim to impart not only flame resistance but also additional features such as electrical conductivity, self-healing, and multi-signal sensitivity, thereby aligning with the growing demand for advanced multifunctional polymer composites. Despite the superior performance of some nanomaterials compared to traditional FRs, their formal certification and market implementation remain limited. This is due primarily to concerns over cost, as well as occupational health and safety implications. Current flame retardancy standards are primarily designed for conventional materials and do not yet fully account for the unique characteristics of nanomaterials. Therefore, comprehensive interdisciplinary collaboration is essential to identify potential health risks, establish standardized manufacturing protocols, and develop regulatory frameworks to safely and effectively integrate nanomaterials into commercial flame-retardant systems.

Conclusions

The emergence of various carbon allotropes in recent decades has significantly advanced the design of carbon-based FRs for polymers. These materials not only offer remarkable flame resistance and low toxicity emissions, but their structural diversity—particularly in nanoforms-enables broad compatibility with numerous polymers and flame-retardant additives, often resulting in synergistic effects and multifunctional benefits. This comprehensive review aims to provide a molecular-level understanding of how carbon-based FRs function in polymers, using an integrative approach that combines detailed analytical techniques with computational materials science. Carbon-based FRs have been shown to enhance polymer flame resistance by improving thermal stability, thermal conductivity, and mechanical strength through preserved structural integrity. Notably, materials with exfoliated layered structures, such as GO, demonstrate superior performance by forming effective barrier layers, outperforming other morphologies like spherical particles, tubes, or platelets. The effectiveness of these retardants primarily arises from two mechanisms occurring in both the "condensed phase" and the "gas phase": (1) The formation of a thick char coat on the polymer exterior safeguards the underlying material and restricts the emission of combustible degradation byproducts; and (2) the capture of reactive radicals, thereby disrupting the combustion chain reaction. Historically, research on carbon-based FRs has followed a top-down strategy, synthesizing and integrating pure carbon materials such as graphite, graphene, CNTs, and fullerenes into polymer systems, then experimentally evaluating their fire-resistant properties. Despite these efforts, the detailed chemical and physical behavior of flame-retardant polymers during thermal degradation, especially the transition from condensed to gas phase, remains poorly understood. Addressing this gap, this review adopts a bottom-up perspective, utilizing molecular dynamics simulations to investigate flame-retardant mechanisms at the atomic level. This approach reveals the decomposition pathways and the generation of flammable volatiles, providing valuable insights for optimizing the design and synthesis of next-generation carbon-based FRs. Current flame retardancy standards and regulations were developed with traditional FRs in mind. While some nanomaterials have demonstrated superior performance compared to these conventional FRs, they have not yet been certified for commercial use. This is probably because of the worries about OHS and the significant expenses associated with it. The only way to overcome these obstacles and adequately evaluate the hazards that nanoparticles and nanocomposites may pose to human health and safety is through extensive multidisciplinary cooperation. Researchers in academia and professionals in the business world must work together to develop standards for the manufacture, processing, use, and assessment of the performance of nanomaterials as a new class of FRs.

References

[1] S. Araby, B. Philips, Q. Meng, J. Ma, T. Laoui, C.H. Wang, Recent advances in carbon-based nanomaterials for flame retardant polymers and composites, Compos. B: Eng. 212 (2021) 108675. https://doi.org/10.1016/j.compositesb.2021.108675

[2] J. Giebułtowicz, M. Rużycka, P. Wroczyński, D.A. Purser, A.A. Stec, Analysis of fire deaths in Poland and influence of smoke toxicity, Forensic Sci. Int. 277 (2017) 77-87. https://doi.org/10.1016/j.forsciint.2017.05.018

[3] G. Beyer, Nanocomposites: a new class of flame retardants for polymers. Plastics, Additives and Compounding 4 (2002) 22-28. https://doi.org/10.1016/S1464-391X(02)80151-9

[4] T. Kurauchi, A. Okada, T. Nomura, T. Nishio, S. Saegusa, R. Deguchi, R. Degushi, Nylon 6-clay hybrid-synthesis, properties and application to automotive timing belt cover. SAE Transactions (1991) 571-577. https://doi.org/10.4271/910584

[5] E.P. Giannelis, Polymer layered silicate nanocomposites. Adv. Mater. 8 (1996) 29-35. https://doi.org/10.1002/adma.19960080104

[6] S. Araby, J. Li, G. Shi, Z. Ma, J. Ma, Graphene for flame-retarding elastomeric composite foams having strong interface. Compos. A: Appl. Sci. Manuf. 101 (2017) 254-264. https://doi.org/10.1016/j.compositesa.2017.06.022

[7] H. Huang, D. Dong, W. Li, X. Zhang, L. Zhang, Y. Chen, X. Lu, Synergistic effect of MXene on the flame retardancy and thermal degradation of intumescent flame retardant biodegradable poly (lactic acid) composites. Chin. J. Chem. Eng. 28 (2020) 1981-1993. https://doi.org/10.1016/j.cjche.2020.04.014

[8] X. Wang, E.N. Kalali, J.T. Wan, D.Y. Wang, Carbon-family materials for flame-retardant polymeric materials. Prog. Polym. Sci. 69 (2017) 22-46. https://doi.org/10.1016/j.progpolymsci.2017.02.001

[9] S. Bourbigot, M. Le Bras, R. Delobel, L. Gengembre, XPS study of an intumescent coating: II. Application to the ammonium polyphosphate/pentaerythritol/ethylenic terpolymer fire retardant system with and without synergistic agent. Appl. Surf. Sci. 120(1997) 15-29.

[10] X. Wang, H. Yang, L. Song, Y. Hu, W. Xing, H. Lu, Morphology, mechanical and thermal properties of graphene-reinforced poly (butylene succinate) nanocomposites. Compos. Sci. Technol. 72 (2011) 1-6. https://doi.org/10.1016/j.compscitech.2011.05.007

[11] Y. Yang, J.L. Díaz Palencia, N. Wang, Y. Jiang, D.Y. Wang, Nanocarbon-based flame retardant polymer nanocomposites. Molecules 26 (2021) 4670. https://doi.org/10.3390/molecules26154670

[12] H. Parsimehr, M. Enayati, A.E. Langroudi, Recent Developments in Green Flame Retardants Based on Carbon Nanotubes. Materials and Chemistry of Flame-Retardant Polyurethanes Volume 2: Green Flame Retardants, 2021, p. 47-63. https://doi.org/10.1021/bk-2021-1400.ch004

[13] X. Wang, E.N. Kalali, J.T. Wan, D.Y. Wang, Carbon-family materials for flame-retardant polymeric materials. Prog. Polym. Sci. 69 (2017) 22-46. https://doi.org/10.1016/j.progpolymsci.2017.02.001

[14] A.N. Muranov, V.R. Lysenko, M.A. Kocharov, Rheological behavior features of feedstocks with a two-component wax-polyolefin binder compared to analogs based on polyoxymethylene. J. Compos. Sci. 8 (2024) 199. https://doi.org/10.3390/jcs8060199

[15] M. Ciesielski, B. Burk, C. Heinzmann, M. Döring, Fire-retardant high-performance epoxy-based materials. In Novel Fire Retardant Polymers and Composite Materials, Woodhead Publishing, 2017, pp. 3-51. https://doi.org/10.1016/B978-0-08-100136-3.00002-9

[16] Y. Yang, J.L. Díaz Palencia, N. Wang, Y. Jiang, D.Y. Wang, Nanocarbon-based flame retardant polymer nanocomposites. Molecules 26 (2021) 4670. https://doi.org/10.3390/molecules26154670

[17] X. Wang, W. Xing, X. Feng, B. Yu, L. Song, Y. Hu, Functionalization of graphene with grafted polyphosphamide for flame-retardant epoxy composites: synthesis, flammability, and mechanism. Polym. Chem. 5 (2014) 1145-1154. https://doi.org/10.1039/C3PY00963G

[18] Q. Leng, J. Li, Y. Wang, Structural analysis of α-zirconium phosphate/cerium phosphate/graphene oxide nanocomposites with flame-retardant properties in polyvinyl alcohol. New J. Chem. 44 (2020) 4568-4577. https://doi.org/10.1039/C9NJ06253J

[19] W. Luo, Y. Li, H. Zou, M. Liang, Study of different-sized sulfur-free expandable graphite on morphology and properties of water-blown semi-rigid polyurethane foams. RSC Adv. 4 (2014) 37302-37310. https://doi.org/10.1039/C4RA05559D

[20] B. Yu, Y. Shi, B. Yuan, S. Qiu, W. Xing, W. Hu, Y. Hu, Enhanced thermal and flame retardant properties of flame-retardant-wrapped graphene/epoxy resin nanocomposites. J. Mater. Chem. 3 (2015) 8034-8044. https://doi.org/10.1039/C4TA06613H

[21] F. Fang, S. Ran, Z. Fang, P. Song, H. Wang, Improved flame resistance and thermo-mechanical properties of epoxy resin nanocomposites from functionalized graphene oxide via self-assembly in water. Compos. B: Eng. 165 (2019). 406-416. https://doi.org/10.1016/j.compositesb.2019.01.086

[22] W. Du, Y. Jin, S. Lai, L. Shi, Y. Shen, H. Yang, Multifunctional light-responsive graphene-based polyurethane composites with shape memory, self-healing, and flame retardancy properties. Compos. A: Appl. Sci. Manuf. 128 (2020) 105686. https://doi.org/10.1016/j.compositesa.2019.105686

[23] M.J. Kim, I.Y. Jeon, J.M. Seo, L. Dai, J.B. Baek, Graphene phosphonic acid as an efficient flame retardant. ACS nano. 8 (3) 2820-2825. https://doi.org/10.1021/nn4066395

[24] G. Huang, S. Huo, X. Xu, W. Chen, Y. Jin, R. Li, H. Wang, Realizing simultaneous improvements in mechanical strength, flame retardancy and smoke suppression of ABS nanocomposites from multifunctional graphene. Compos. B: Eng. 177 (2019) 107377. https://doi.org/10.1016/j.compositesb.2019.107377

[25] J. Li, W. Meng, W. Wu, Z. Zhou, W. Zhang, P. Song, H. Qu, Nickel ammonium phosphate and reduced graphene oxide two-dimensional hybrid material for improving the fire safety and mechanical properties of poly (vinyl chloride). Polym. Int. 69 (2020) 1227-1236. https://doi.org/10.1002/pi.6066

[26] N. Hong, Y. Pan, J. Zhan, B. Wang, K. Zhou, L. Song, Y. Hu, Fabrication of graphene/Ni-Ce mixed oxide with excellent performance for reducing fire hazard of polypropylene. RSC Adv. 3 (2013) 16440-16448. https://doi.org/10.1039/c3ra42095g

[27] S.D. Jiang, Z.M. Bai, G. Tang, L. Song, A.A. Stec, T.R. Hull, Y. Hu, Fabrication of Ce-doped MnO2 decorated graphene sheets for fire safety applications of epoxy composites:

flame retardancy, smoke suppression and mechanism. J. Mater. Chem. 2 (2014) 17341-17351. https://doi.org/10.1039/C4TA02882A

[28] X. Wang, S. Zhou, W. Xing, B. Yu, X. Feng, L. Song, Y. Hu, Self-assembly of Ni-Fe layered double hydroxide/graphene hybrids for reducing fire hazard in epoxy composites. J. Mater. Chem. 1 (2013) 4383-4390. https://doi.org/10.1039/c3ta00035d

[29] T. Kashiwagi, F. Du, J.F. Douglas, K.I. Winey, R.H. Harris Jr, J.R. Shields, Nanoparticle networks reduce the flammability of polymer nanocomposites. Nat. mater. 4 (2005) 928-933. https://doi.org/10.1038/nmat1502

[30] J.F. Colomer, L. Henrard, G. Van Tendeloo, A. Lucas, P. Lambin, Study of the packing of double-walled carbon nanotubes into bundles by transmission electron microscopy and electron diffraction. J. Mater. Chem. 14 (2004) 603-606. https://doi.org/10.1039/b311551h

[31] I.V. Obronov, V.I. Kleshch, E.A. Smolnikova, D.A. Bandurin, A.N. Obraztsov, Field emission properties of single-walled carbon nanotube films. J. nanoelectron. Optoelectron. 8 (2013) 71-74. https://doi.org/10.1166/jno.2013.1438

[32] T. Kashiwagi, F. Du, K.I. Winey, K.M. Groth, J.R. Shields, S.P. Bellayer, J.F. Douglas, Flammability properties of polymer nanocomposites with single-walled carbon nanotubes: effects of nanotube dispersion and concentration. Polymer. 46 (2005)471-481. https://doi.org/10.1016/j.polymer.2004.10.087

[33] S. Peeterbroeck, F. Laoutid, J.M. Taulemesse, F. Monteverde, J.M. Lopez-Cuesta, J.B. Nagy, P. Dubois, Mechanical properties and flame-retardant behavior of ethylene vinyl acetate/high-density polyethylene coated carbon nanotube nanocomposites. Adv. Funct. Mater. 17 (2007) 2787-2791. https://doi.org/10.1002/adfm.200600936

[34] H. Yang, L. Ye, J. Gong, M. Li, Z. Jiang, X. Wen, T. Tang, Simultaneously improving the mechanical properties and flame retardancy of polypropylene using functionalized carbon nanotubes by covalently wrapping flame retardants, followed by linking polypropylene. Mater. Chem. Front. 1 (2017) 716-726. https://doi.org/10.1039/C6QM00172F

[35] B.H. Cipiriano, T. Kashiwagi, S.R. Raghavan, Y. Yang, E.A. Grulke, K. Yamamoto, J. F. Douglas, Effects of aspect ratio of MWNT on the flammability properties of polymer nanocomposites. Polymer. 48 (2007) 6086-6096. https://doi.org/10.1016/j.polymer.2007.07.070

[36] S. Bourbigot, G. Fontaine, A. Gallos, S. Bellayer, Reactive extrusion of PLA and of PLA/carbon nanotubes nanocomposite: processing, characterization and flame retardancy. Polym. Adv. Technol. 22 (2011) 30-37. https://doi.org/10.1002/pat.1715

[37] R. Verdejo, F. Barroso-Bujans, M.A. Rodriguez-Perez, J.A. de Saja, M. Arroyo, M.A. Lopez-Manchado, Carbon nanotubes provide self-extinguishing grade to silicone-based foams. J. of Mater. Chem. 18 (2008) 3933-3939. https://doi.org/10.1039/b805943h

[38] B. Dittrich, K.A. Wartig, D. Hofmann, R. Mülhaupt, B. Schartel, Flame retardancy through carbon nanomaterials: Carbon black, multiwall nanotubes, expanded graphite, multi-layer graphene and graphene in polypropylene. Polym. Degrad. Stab. 98 (2013) 1495-1505. https://doi.org/10.1016/j.polymdegradstab.2013.04.009

[39] G. Huang, S. Wang, P.A. Song, C. Wu, S. Chen, X. Wang, Combination effect of carbon nanotubes with graphene on intumescent flame-retardant polypropylene nanocomposites. Compos. - A: Appl. Sci. Manuf. 59 (2014) 18-25. https://doi.org/10.1016/j.compositesa.2013.12.010

[40] M. Bhattacharya, Polymer nanocomposites-a comparison between carbon nanotubes, graphene, and clay as nanofillers. Materials. 9 (2016) 262. https://doi.org/10.3390/ma9040262

[41] J. Sandler, J.E. Kirk, I.A. Kinloch, M.S.P. Shaffer, A.H. Windle, Ultra-low electrical percolation threshold in carbon-nanotube-epoxy composites. Polymer. 44 (2003). 5893-5899. https://doi.org/10.1016/S0032-3861(03)00539-1

[42] X. Wen, Y. Wang, J. Gong, J. Liu, N. Tian, Y. Wang, T. Tang, Thermal and flammability properties of polypropylene/carbon black nanocomposites. Polym. Degrad. Stab. 97 (2012) 793-801. https://doi.org/10.1016/j.polymdegradstab.2012.01.031

[43] X. Wen, K. Szymańska, X. Chen, E. Mijowska, Nanosized carbon black as synergist in PP/POE-MA/IFR system for simultaneously improving thermal, electrical and mechanical properties. J. Therm. Anal. Calorim. 139 (2020) 1091-1098. https://doi.org/10.1007/s10973-019-08466-4

[44] X. Wen, Z. Liu, Z. Li, J. Zhang, D.Y. Wang, K. Szymańska, T. Tang, Constructing multifunctional nanofiller with reactive interface in PLA/CB-g-DOPO composites for simultaneously improving flame retardancy, electrical conductivity and mechanical properties. Compos. Sci. Technol. 188 (2020) 107988. https://doi.org/10.1016/j.compscitech.2019.107988

[45] J. Liu, Y. Zhang, Y. Guo, C. Lu, B. Pan, S. Peng, Q. Niu, Effect of carbon black on the thermal degradation and flammability properties of flame-retarded high impact polystyrene/magnesium hydroxide/microencapsulated red phosphorus composite. Polym. Compos. 39 (2018) 770-782. https://doi.org/10.1002/pc.23998

[46] L. Liu, X. Zhao, C. Ma, X. Chen, S. Li, C. Jiao, Smoke suppression properties of carbon black on flame-retardant thermoplastic polyurethane based on ammonium polyphosphate. J. Therm. Anal. Calorim. 126 (2016) 1821-1830. https://doi.org/10.1007/s10973-016-5815-x

[47] Y. Pan, Z. Guo, S. Ran, Z. Fang, Influence of fullerenes on the thermal and flame-retardant properties of polymeric materials. J. Appl. Polym. Sci. 137 (2020) 47538. https://doi.org/10.1002/app.47538

[48] Ü. Tayfun, M. Doğan, Flame-retardant properties of fullerene and nanodiamond-based polymer nanocomposites. In Flame Retardant Nanocomposites, Woodhead Publishing, pp. 263-286, 2024. https://doi.org/10.1016/B978-0-443-15421-8.00004-5

[49] Z. Fang, P. Song, L. Tong, Z. Guo, Thermal degradation and flame retardancy of polypropylene/C60 nanocomposites. Thermochim. Acta. 473 (2008) 106-108. https://doi.org/10.1016/j.tca.2008.04.019

[50] G.H. Yeoh, I.M. De Cachinho Cordeiro, W. Wang, C. Wang, A.C.Y. Yuen, T.B.Y. Chen, H.T. Chua, Carbon-based flame retardants for polymers: A bottom-up review. Adv. Mater. 36 (2024) 2403835. https://doi.org/10.1002/adma.202403835

Materials Research Forum LLC
https://doi.org/10.21741/9781644904077-5

[51] P.A. Song, H. Liu, Y. Shen, B. Du, Z. Fang, Y. Wu, Fabrication of dendrimer-like fullerene (C 60)-decorated oligomeric intumescent flame retardant for reducing the thermal oxidation and flammability of polypropylene nanocomposites. J. Mater. Chem. 19 (2009) 1305-1313. https://doi.org/10.1039/b815610g

[52] Z. Guo, Z. Wang, Z. Fang, Fabrication of 9, 10-dihydro-9-oxa-10-phosphaphenanthrene-10-oxide-decorated fullerene to improve the anti-oxidative and flame-retardant properties of polypropylene. Compos. B: Eng.183 (2020) 107672. https://doi.org/10.1016/j.compositesb.2019.107672

[53] R. Wang, L. Wu, D. Zhuo, Z. Wang, T.Y. Tsai, Fabrication of fullerene anchored reduced graphene oxide hybrids and their synergistic reinforcement on the flame retardancy of epoxy resin. Nanoscale Res. Lett. 13 (2018) 1-14. https://doi.org/10.1186/s11671-018-2678-z

[54] P. Song, L. Zhao, Z. Cao, Z. Fang, Polypropylene nanocomposites based on C 60-decorated carbon nanotubes: Thermal properties, flammability, and mechanical properties. J. Mater. Chem. A. 21 (2011) 7782-7788. https://doi.org/10.1039/c1jm10395d

[55] X. Zhou, S. Ran, H. Hu, Z. Fang, Improving flame-retardant efficiency by incorporation of fullerene in styrene-butadiene-styrene block copolymer/aluminum hydroxide composites. J. Therm. Anal. Calorim. 125 (2016) 199-204. https://doi.org/10.1007/s10973-016-5354-5

[56] T.Y. Tsai, N. Bunekar, C.C. Huang, Y.S. Huang, L.C. Chen, Novolac cured epoxy resin/fullerene modified clay composites: Applied to copper clad laminates. Rsc Adv. 5 (2015) 95649-95656. https://doi.org/10.1039/C5RA18073B

[57] B. Mazela, A. Batista, W. Grześkowiak, Expandable graphite as a fire retardant for cellulosic materials-A review. Forests, 11 (2020) 755-763. https://doi.org/10.3390/f11070755

[58] Q. Yang, Y. Geng, H. Dong, J. Zhang, X. Yu, L. Sun, Y. Chen, Effect of environmental regulations on China's graphite export. J. Clean. Prod. 161 (2017). 327-334. https://doi.org/10.1016/j.jclepro.2017.05.131

[59] H.D. Tuan Nguyen, H.T. Nguyen, T.T. Nguyen, A.K. Le Thi, T.D. Nguyen, Q.T. Phuong Bui, L.G. Bach, The preparation and characterization of MnFe2O4-decorated expanded graphite for removal of heavy oils from water. Materials. 12 (2019) 1913. https://doi.org/10.3390/ma12121913

[60] C. Alonso, A. Manich, A.D. Campo, P. Felix-De Castro, N. Boisseree, L. Coderch, M. Martí, Graphite flame retardant applied on polyester textiles: Flammable, thermal and in vitro toxicological analysis. J. Ind. Text. 51 (2022) 4424S-4440S. https://doi.org/10.1177/15280837211062056

[61] M.R. Nyden, R.H. Harris, Y.S. Kim, R.D. Davis, N.D. Marsh, M. Zammarano, Characterizing particle emissions from burning polymer nanocomposites. In Proceedings of the 21st BCC Conference on Flame Retardation (Vol. 1, pp. 717-719). (2010) Gaithersburg, MD, USA: National Institute of Standards and Technology-Nanotech.

[62] J.X. Bouillard, B. R'Mili, D. Moranviller, A. Vignes, O. Le Bihan, A. Ustache, D. Fleury, Nanosafety by design: Risks from nanocomposite/nanowaste combustion. J. Nanoparticle Res. 15 (2013) 1-11. https://doi.org/10.1007/s11051-013-1519-3

[63] R. Arvidsson, D. Kushnir, B.A. Sandén, S. Molander, Prospective life cycle assessment of graphene production by ultrasonication and chemical reduction. Environ. Sci. amp; Technol. 48 (2014) 4529-4536. https://doi.org/10.1021/es405338k

[64] S. Araby, B. Philips, Q. Meng, J. Ma, T. Laoui, C.H. Wang, Recent advances in carbon-based nanomaterials for flame retardant polymers and composites. Compos. B: Eng. 212 (2021) 108675. https://doi.org/10.1016/j.compositesb.2021.108675

[65] P. Song, Y. Shen, B. Du, Z. Guo, Z. Fang, Fabrication of fullerene-decorated carbon nanotubes and their application in flame-retarding polypropylene. Nanoscale. 1 (2009) 118-121. https://doi.org/10.1039/b9nr00026g

Materials Research Foundations 190 (2026) 153-193 | https://doi.org/10.21741/9781644904077-6

Chapter 6

Lignin: A Natural Flame-Retardant

Guocai Tian*

State Key Laboratory of Complex Non-Ferrous Metal Resource Clean Utilization, Faculty of Metallurgical and Energy Engineering, Kunming University of Science and Technology, Kunming, Yunnan Province, China, 650093

tiangc01@gmail.com, tiangc@kust.edu.cn

Abstract

Lignin, as a natural flame-retardant, has aroused great interest among researchers. The source, structure, and physical and chemical property of lignin were briefly introduced. The research status in the preparation and performance of lignin direct flame-retardant, physical synergy, and chemical modification flame-retardant were comprehensively reviewed and analyzed. The development prospects and challenges faced by lignin, a natural flame-retardant, were discussed in order to provide guidance and assistance for the advancement of eco-friendly, sustainable and effective flame-retardant materials.

Keywords

Flame-Retardant, Lignin, Chemically Modified Lignin, Lignin Blended Flame-Retardant, Lignin Synergistic Flame-Retardant, Nitrogen Modified Lignin, Nitrogen and Phosphorus Co-Modified Lignin, Silicon Modified Lignin, Nano-modified Lignin

Contents

1. Introduction

As the rapid advancement of the economy and society, the sharp rise in the demand for energy and materials gives rise to the over exploitation of fossil fuels, with the consequence that environmental problems become more and more serious [1,2]. Therefore, the replacement of petroleum-based raw materials with renewable resources has become one of the important topics of concern in the current academic community. Lignin, which takes the second place in terms of abundance among biomass components in the plant kingdom, has a global reserve of over 3.0×10^{11} tons and an annual regeneration capacity of 2×10^{10} tons, accounting for 30% of the Earth's non petrochemical organic carbon resources [3,4]. However, currently, less than 2% of the $(1.5\text{-}1.8) \times 10^{8}$ tons of industrial lignin produced globally each year is effectively utilized, mainly as low value fuels or chemical additives, which not only causes resource waste but also brings significant environmental pressure [5].

Lignin is widely present in the phloem and thin-walled tissues of plants, and a certain amount is also present in wood cells, forming the basic plants' skeleton with hemicellulose and cellulose. Lignin is cross-linked through hemicellulose and cellulose with the hydrogen and covalent bonds, and is a complex phenolic polymer formed from three alcohol monomers: coumarin, paclitaxel, and sinapine [5,6].

Lignin is an aromatic polymer with oxygen-riches that represents 15% to 30% of biomass. The lignin structure contains active functional groups such as alcohol hydroxyl, aromatic, phenolic hydroxyl, carbonyl, carboxyl, methoxy groups, and conjugated double bonds, which can occur many chemical reactions such as alcoholysis, reduction, oxidation, hydrolysis, acid hydrolysis, acylation, photolysis, halogenation, sulfonation, nitration, condensation, alkylation, or the graft copolymerization [7]. Lignin and its modified lignin were widely applied in fields such as electro-mechanical, agriculture, light industry, building materials, mining, environmental protection, petroleum, which becomes a hotpot of current research [8-10].

The function of flame-retardant is to change the reaction pathway of fiber combustion, promote fiber decomposition, produce non-combustible gases, or form a molten material cover on the surface of fibers, thereby preventing combustion from continuing [11,12]. Traditional flame-retardants have significant flame-retardant effects due to their low dosage, but they have potential toxicity issues [13,14]. Therefore, the application of lignin with high carbonization ability and rich hydroxyl groups has gradually attracted attention in the field of flame-retardant [15,16]. The latest research confirms that lignin's unique oxygen-rich aromatic structure gives it a unique advantage in the development of materials for flame-retardant. The lignin thermal degradation and the coke formation are usually completed through three ways: cleavages of the ether bond, dealkylation and cleavage of the C-C bond, as well as demethylation and carbonization. The pyrolysis and carbonization processes usually produce carbon layers with rough surfaces and high specific surface area (specific surface area>300 m^2/g), which exert flame-retardant effects through a dual mechanism of gas-phase quenching and condensed phase barrier, blocking the surface between the fire and the polymer matrix, which is beneficial for flame-retardant [16,17].

Due to the complexity of lignin molecular structure and its relatively poor thermal stability, as well as the elements absence of flame-retardant such as phosphorus and nitrogen, directly using lignin as a flame-retardant is not effective in enhancing the materials' flame-retardant performance. Usually, lignin is compounded with other flame-retardants or chemically modified to introduce flame-retardant elements or groups into the chemical structure of lignin, to gain a series of lignin based flame-retardants with superior performance. This chapter starts from the source, the structure, and physical and chemical properties of lignin, introduces the research status of lignin direct flame-retardant, physical synergy, and chemical modification flame-retardant, and comprehensively reviews and analyzes the recent progress in the preparation and performance research of lignin flame-retardants at home and abroad. At the same time, the development prospects and challenges faced by lignin, a natural flame-retardant, were discussed to provide guidance and assistance for advancing eco-friendly, sustainable and effective flame-retardant materials.

2. The source, structure, physical and chemical properties of lignin

2.1 Sources

Lignin, as a complex natural high molecular weight organic compound [18-26], is widely present in the xylem of most terrestrial plants, accounting for approximately 33% of their biomass. In gymnosperms (coniferous wood) and angiosperms (broad-leaved wood and grass), the lignin content ranges from 15% to 36%. Except for tropical ferns, lignin is distributed in all vascular plants. The distribution of lignin can be analyzed from different levels such as nature, the whole plant, and plant cell walls. Lignin with diverse structures and molecular weights exists in the roots, stems, leaves, bark, fruits, shells, and seeds of mature plants. The edible parts of plants such as millet, grains, wheat bran, beans, cabbage, plums, peanuts, cocoa, pears, strawberries, raspberries, etc. also contain lignin. Moreover, the more mature the plant, the higher its lignin content. After being ingested by the human body, these lignins can clear bile acids in the intestine, reduce cholesterol, and have definite effects on preventing rectal cancer and gallstones [18-20].

In plants, the distribution of lignin varies significantly in chemical composition being dependent on the plant species [20-22]. The lignin content varies among different tree species. For example, gymnosperms have a higher lignin content than angiosperms, while coniferous trees have an average lignin content of 29%, which is larger than the average lignin content of broad-leaved trees at 21%. The content of lignin in tropical wood is a little larger than that of temperate wood. The content of lignin in tree branches is larger than that of tree trunks. The bark is divided into outer bark and inner bark (phloem), with different chemical compositions. For example, the phloem of pine contains 17% lignin and the outer bark contains 44%. The phloem of spruce contains 16%, and the outer skin contains 27%. Birch has a phloem content of 25%, while poplar has a phloem content of 28%. There are different in lignin content between natural forests and

Materials Research Foundations 190 (2026) 153-193 https://doi.org/10.21741/9781644904077-6

artificial forests, as well as trees of different ages. The lignin content of coniferous artificial forests is greater than of broad-leaved artificial forests.

However, the difference in lignin content between populations is minimal, and it is not closely linked to the latitude, longitude, frost period, annual average temperature, annual rainfall, as well as a diameter at breast height and ring width of the source. Plants have significant differences in chemical composition at different stages of growth and development. The process of lignin filling the fiber framework within cell walls may result in uneven distribution within individuals. The higher the trunk, the lower the lignin content, exhibiting vertical distribution heterogeneity. In addition, the lignin composition of gymnosperms and angiosperms is different. The lignin of gymnosperms (such as coniferous wood) is mainly of the guaiacyl (G) type, while in angiosperms (such as broad-leaved wood), the lignin is mainly of the guaiacyl (G)-eugenol (S) type. The monocotyledonous plants in angiosperms (such as grasses) have lignin in the form of guaiacyl (G) -syringayl (S)-p-hydroxyphenyl (H) [20-22].

2.2 Structure units

Lignin is mainly comprised of three elements: carbon, hydrogen, and oxygen, with mass fractions of approximately 60%, 6%, and 30%, respectively. In addition, it also contains about 0.67% nitrogen element. It should be noted that the lignin elemental composition is not fixed and may vary according to plant species, origin, and separation methods. Natural lignin is usually classified into three categories: hardwood lignin, cork lignin, and herbaceous lignin. They have subtle differences in terms of structure and molecular weight. For example, compared to hardwood lignin, cork lignin has more hydroxyl groups [18-22].

At present, there is not any effective analytical method for precise characterization of lignin structure. However, with the help of biosynthesis or chemical degradation, combined with techniques of the Raman spectroscopy (Raman), the nuclear magnetic resonance spectroscopy (NMR), and Infrared spectroscopy (IR), some structural models of natural lignin can be derived. The basic lignin structural unit was shown in Figure 1. It can be clearly observed in Figure 1 that lignin is a very complex aromatic heteropolymer compound, mainly presents a branched network structure, which is composed of phenylpropane groups linked to one another via C-C bonds or ether bonds. Its basic structural units include p-hydroxyl phenol (H), syringyl (S), and guaiacyl (G). From the study of biosynthesis processes, it is known that these three basic structural units are first formed by aromatization of glucose to form shikimic acid, and then the basic structures of the three lignin types mentioned above are synthesized from shikimic acid.

Materials Research Foundations 190 (2026) 153-193 https://doi.org/10.21741/9781644904077-6

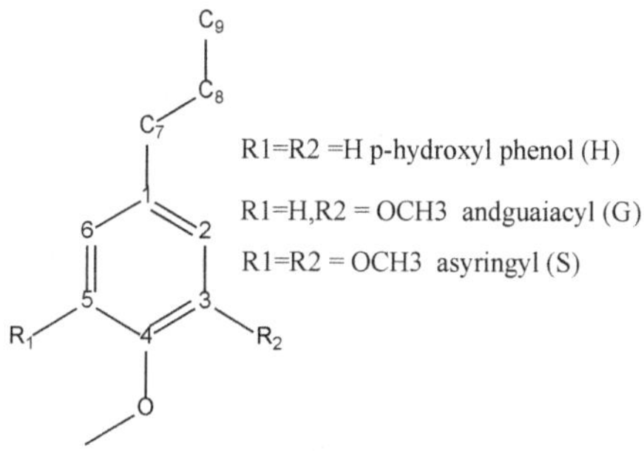

Fig. 1 The basic unit of lignin

2.3 Physical and chemical properties

The physical and chemical properties of lignin are influenced by various factors, including internal factors such as species, location, structure, and growth times of the plant, as well as external factors such as separation and extraction methods. In plants, the original body of lignin is related to the sugars, forming a structure with the larger spatial network. This complex carbohydrate lignin is insoluble in various solvents. So, the physical and chemical properties of lignin commonly mentioned are lignin with low molecular weight isolated and extracted through specific ways, other than the true lignin properties in plants [18-26].

(1) Color

The original lignin appears in a white or almost colorless state [18,19]. However, once separated and prepared, the color of lignin will change in depth due to different methods. For example, the spruce lignin obtained and named by Brauns' separation method has a creamy-white color. The colors ranging from light yellow brown to dark brown and darker for acid, copper ammonium, and periodate lignin. Through diazomethane methylation treatment, lignin color can be reduced and even turned white, which serves to extend the lignin application fields. Double bonds in unsaturated compounds such as vinyl and carbonyl groups exist in the structure of lignin, which can create conjugated systems in combination with benzene rings. In addition, the -NR (amino), -OH (hydroxyl), -OR (ether bond), -X (halogen), - COOH (carboxyl) group, and other color enhancing groups can darken the lignin color. In addition, the phenolic hydroxyl groups in lignin structural units are easily oxidized into colored substances such as quinines under the action of oxygen in the air (similar to the role of peroxidases inside plants), which can cause lignin to change color.

(2) Relative density

The density of lignin is an important parameter, which relative density is typically ranging from 1.35 to 1.50 $g.cm^{-3}$. The specific value may vary depending on the preparation method, raw material source, and measurement conditions. Taking pine wood as an example, the relative density of ethylene glycol lignin is 1.362 $g.cm^{-3}$, while the relative density of hydrochloric acid lignin is slightly lower, at 1.348 $g.cm^{-3}$, indicating that different separation processes have a certain impact on the density of lignin. In addition, the density of the solution used during the measurement process can also significantly affect the final results: for example, when using water as the measurement solution for pine wood sulfate lignin, the relative density is 1.451 g.cm^{-3}, but when using benzene for measurement, it drops to 1.436 g.cm^{-3}, which may be due to the different interactions between solvent molecules and lignin structure. Similarly, the measurement results of spruce lignin at 20 °C showed a relative density of 1.33 g.cm^{-3} when measured with water, while it increased to 1.391 g.cm^{-3} when measured with dioxane, further confirming the crucial role of solvent selection in density determination. These data not only reflect the complexity of lignin itself, but also provide important references for subsequent research on its structure performance relationship [18-24].

(3) Thermal properties

The thermal properties of lignin are closely related to its high carbon content (about 50% higher than cellulose), which makes it a highly promising biomass energy material. Due to carbon being the main contributor to combustion heat, lignin can release significant heat during combustion [18-24]. For example, the combustion heat of lignin with ash free spruce hydrochloric acid can reach 110 kJ/g, while the combustion heat of sulfuric acid lignin is about 109.6 kJ/g, which is close to the combustion efficiency of many traditional fossil fuels. Based on this characteristic, black liquor lignin in the paper industry is often recycled: after extracting lignin through concentration technology, it can be directly used as fuel, which not only reduces waste disposal costs but also achieves energy cycling. In addition, the thermal stability of lignin is also worth paying attention to, and its decomposition temperature is usually between 200-300 °C, depending on the degree of crosslinking and functional group type in the molecular structure. For example, sulfonated modified lignin may have a slightly lower thermal decomposition temperature than natural lignin due to the introduction of more polar groups. This controllable thermal behavior makes it have potential applications in fields such as biomass pyrolysis and carbon material preparation. In the future, by optimizing the lignin extraction process or chemical modification, its thermal utilization efficiency can be further improved, providing new ideas for the development of renewable energy.

(4) Solubility

Lignin, as the main component of plant cell walls, is significantly influenced by its chemical structure, separation methods, and environmental conditions such as pH and temperature for its solubility [18-24]. It has many polar groups, especially aggregates of a large number of hydroxyl groups, forming strong intermolecular and intramolecular hydrogen bonds. Natural lignin is usually insoluble in water and most organic solvents. However, some isolated lignin (such as Brauns lignin or organic solvent lignin) is soluble in pyridine, dioxane, methanol, ethanol, acetone, and dilute alkaline solutions, while acid lignin is almost insoluble in any solvent. The solubility of lignin can be significantly improved through chemical modification, such as sulfonation and alkali treatment. For example, alkaline lignin has enhanced solubility under alkaline conditions, while lignosulfonates (such as sodium lignosulfonate) have excellent water solubility due to the introduction of sulfonic acid groups. They can form stable solutions with high concentrations (up to 200 g/L) in neutral or alkaline solutions, but have lower solubility under strong acidic conditions. In addition, the solubility of lignin is closely related to the content and molecular weight of its functional groups (such as phenolic hydroxyl and carboxyl groups): an increase in polar groups or a decrease in molecular weight (such as a decrease in molecular weight from tens of thousands to thousands to tens of thousands after separation) can help improve solubility. In practical applications, lignosulfonates are widely used as anionic surfactants due to their good water solubility; Natural lignin needs chemical modification such as demethylation before it can

Materials Research Foundations 190 (2026) 153-193 https://doi.org/10.21741/9781644904077-6

be used in materials or energy fields. The solubility of lignin can be targeted through structural regulation and process optimization, providing key technical support for its high-value utilization.

(5) Electrochemical properties

Lignin, as a high molecular weight anionic electrolyte, has significant electrochemical properties in biomass resource utilization and paper industry. In an electric field environment, lignin molecules migrate towards the anode due to carrying negative charges, which provides a unique means for studying their structure [18-24]. Researchers can use glass fiber filter paper electrophoresis technology to deeply explore the chemical bond connection state or breaking mechanism between lignin and sugars, providing key information for understanding the decomposition process of plant cell walls. In addition, electrophoresis technology can also be used to separate lignin and sugars, which has potential application value in biomass refining processes. In the field of papermaking black liquor treatment, electrophoresis and electrodialysis technologies have shown significant advantages in efficiently separating lignin and its sulfonate derivatives. Lignosulfonate, as one of the main components of papermaking black liquor, has a different electrochemical behavior from natural lignin. Selective separation of the two can be achieved by regulating the electric field conditions. This separation not only helps to recover valuable lignin products, but also reduces the environmental load of black liquor treatment. It is worth noting that the electrophoretic behavior of lignin is influenced by multiple factors such as its molecular weight, degree of sulfonation, and solution pH value. Optimizing these parameters can improve separation efficiency. In the future, combining new electrode materials or pulsed electric field technology may further promote the industrial application of lignin electrochemical separation processes.

(6) Chemical properties

The molecular structure of lignin is extremely rich, containing lots of active groups of alcohol hydroxyl, phenolic hydroxyl, aromatic, carbonyl, carboxyl, methoxy, and the conjugated double-bonds, etc. These active functional groups endow lignin with strong chemical reactivity, enabling it to participate in various complex reactions such as hydrolysis, oxidation, acid hydrolysis, reduction, alcoholysis, acylation, sulfonation, photolysis, halogenation, condensation, alkylation, nitration, or graft copolymerization. Among numerous types of reactions, oxidation, sulfonation, acylation, condensation, and graft copolymerization is particularly critical and dominated. Among them, sulfonation reaction can be considered as the foundation and core prerequisite for the widespread application of lignin. Until now, lignin mostly functions in the form of lignin sulfonate in practical applications. In addition to participating in numerous chemical reactions, lignin also possesses special properties such as color reaction, adhesion, chelation, delayed effect, and adsorption, which further expand the application space of lignin in different fields [18-26]

Lignin, as a natural flame-retardant, has aroused great interest among people. In recent years, lignin has been widely and successfully used as a flame-retardant, and extensive research has been conducted. Figure 2 shows the current research field. Mainly including pure lignin, blended lignin, chemically modified lignin, etc. Below, we will introduce the current status and progress of the research one by one.

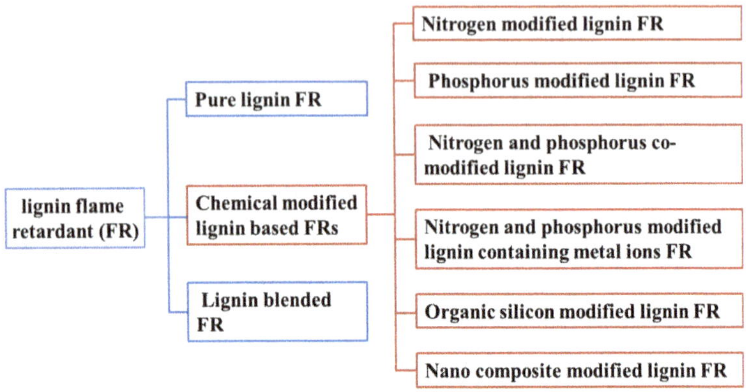

Fig. 2 The cassification of researched lignin flame-retardant (FR)

3. Lignin flame-retardant

3.1 Pure lignin flame-retardant

In flame-retardant systems, when lignin is present as a single component, it can exhibit flame-retardant effects on its own. As early as 1998, Gallina et al. conducted relevant research [27].They used lignin as an flame-retardant to test the resistance pf fire for polypropylene (PP) and found that lignin can largely decrease the release rate of heat for PP, prolonged the total combustion time, and increased carbon production. Subsequently, in 2011, Chen et al. discovered [28] that lignin can improve the PP flame-retardant properties. By alkylating lignin with bromododecane, the compatibility between PP matrix and lignin was significantly enhanced, and the heat resistance of PP/lignin composite material was increased by 25 ℃. Song et al. used styrene acrylonitrile butadiene copolymer (ABS) as a polymer matrix to study the effect of content of lignin and its matrix dispersibility on the polymer properties [29]. The research results indicate that increasing the lignin content can retard the process the ABS thermal degradation and enhance the residual carbon content. Cone calorimeter analysis further indicates that lignin can significantly decrease the rate of ABS heat release and delay the process of combustion. When 20 wt % lignin was added, the rate peak of heat release reduced by about 32% for ABS. The analysis of residual carbon indicates that the formed protective carbon layer with lignin is the key factors for enhancing flame-retardant.

In 2015, Torkelson et al. successfully prepared low-density polyethylene (LDPE) and PP hybrid materials containing 5% -30% (wt) lignin using solid-state shear crushing method [30]. Compared to pure polymers, hybrid materials with a lignin content of 30% (wt) have increased their Young's modulus about 81% and 62%, respectively. Lignin enhances the thermal stability and Vickers hardness of hybrid materials. De Chirico et al. prepared flame-retardant isotactic composites of PP/lignin using a melt-blending method [31]. It was found that when the lignin content increased from 5 wt % to the 15 wt %, the Young's modulus increased from 1.33 GPa to 1.64 GPa. Compared with pure PP, as adding a 15 wt % lignins, thermal decomposition temperature can increase from

Materials Research Foundations 190 (2026) 153-193 https://doi.org/10.21741/9781644904077-6

470 ℃ to 500 ℃ for PP. The carbon residual of the polymer increased from 7 wt % to 11 wt % at 700 ℃. Zhang et al. mixed hydride functionalized organosilicon with lignin to prepare elastomer and foam materials [32]. The cross-linked network formed by lignin and silicon endows lignin with the ability to scavenge free radicals and enhances the flame-retardant and thermal stability of foam.

Uddin et al. prepared lignin/polyvinyl alcohol (PVA) nanofiber membranes using electrospinning as separators for lithium-ion batteries (LIB) [33]. The PVA polymer provides the necessary skeleton structure for electrospinning, while lignin improves the porosity, wettability, and thermal stability of the membrane. Compared with ordinary polyolefin membranes, this membrane has higher flame-retardant and thermal stability. Mandlekar et al. investigated the effects of different lignin contents on the flame-retardant and thermal stability of bio-based polyamide 11 (PA11) [34]. Compared with pure PA11, sulfonated lignin composite materials have higher heat release rate and residual carbon content. The total heat releases are significantly reduced. In order to avoid the complex process of lignin-based fillers and extracting cellulose, the Eucommia ulmoides residue (ER) was directly used to enhance polyester properties and improve its flame-retardant by Wang et al. in 2020 [35]. By using the polyester matrix of polybutylene succinate (PBS) and adding 30 wt % ER for simple melt blending, the bending and tensile modulus of the polyester based composite materials were increased by 72% and 87%, respectively. Lignin and cellulose rich in six carbon rings further increased the content of residual carbon of PBS/ER composite materials, and ER also decreased the rate peak about 43% and the heat release rate minimized about 20% for PBS.

Lignin can be directly applied as an flame-retardant with many advantages, such as no need for chemical modification, low cost, and the ability to retain its natural carbonization properties (due to its benzene ring structure). It has significant flame-retardant effects on polar materials such as polyurethane (PU) and epoxy resin. However, it also has some limitations, such as low flame-retardant efficiency (limit oxygen index (LOI) value is usually < 25%) and poor compatibility with hydrophobic polymers such as PP and ABS. Usually, a higher addition amount (> 30%) is required, which can decrease the material mechanical properties and make it difficult to meet the V-0 level of flame-retardant requirements in UL-94 test.

3.2 Lignin blended flame-retardant

The usage of original lignin in the field of flame-retardant has certain limitations. However, sodium lignosulfonate, alkaline lignin and enzymolysis lignin can be used alone in the polyurethane foam system to enhance the flame-retardant and thermal stability of polyurethane foam. Chen et al. [28] conducted a study by adding alkylated sulfate lignin with a mass fraction of 20% to polypropylene, successfully increasing the LOI value of polypropylene to 25.2%. In subsequent experiments, when 20% hardwood sulfate lignin was in addition to polypropylene, the LOI is heightened to 24.6%. The lignin based flame-retardants have a significant effect on improving the flame-retardant of polypropylene materials. Song et al. [29] prepared composite materials using lignin, styrene grafted maleic anhydride copolymer, styrene ethylene butadiene, and ABS as raw materials. It indicated that lignin can ensure the protection of ABS by generating a carbon layer, thereby effectively improving its flame-retardant performance.

Lu et al. [36] studied the application of lignosulfonate flame-retardant system in polyurethane. During the experiment, as lignosulfonate was gradually added to the polyurethane system, its LOI

Materials Research Foundations 190 (2026) 153-193 https://doi.org/10.21741/9781644904077-6

value continued to rise, indicating that lignosulfonate can significantly enhance the flame-retardant properties of the polyurethane system. Zhang et al. [37] successfully synthesized Lig DOPO flame-retardant by connecting lignin with DOPO using diisocyanate as an intermediate medium. It was noted that the flame-retardant has certain combustion performance and thermal stability. When adding 25% flame-retardant, the LOI index can climb to 30.2%. Mandlekar et al. [38] used lignosulfonate lignin and kraft paper lignin as carbonization agents for polyamide phosphorus-based flame-retardants. The results show that the ternary blend containing sulfonated lignin can form carbon slag with a dense carbon layer, and achieve the flame-retardant effect through the condensed phase flame-retardant mechanism. In all ternary mixture systems, lignin degradation begins at lower temperatures, and the degradation process is extremely slow and continuous, leading to a minimization the temperature of initial decomposition. At the same time, a large amount of carbon residue is collected to achieve flame-retardant.

Zhang et al. [39] successfully prepared polyurethane foam coated with alkaline lignin or sodium lignosulfonate. The research data showed that the LOI of the polyurethane foam was up to 28.4%. The flame-retardant principle of the coating is that it can produce dense and expansive carbon layers during the combustion, capture-free radicals, release inert gas to dilute oxygen, and the condensed phase flame-retardant mechanism works together with the gas phase flame-retardant mechanism, significantly enhancing the flame-retardant performance of polyurethane foam. This means that the use of lignin alone can further enhance the flame-retardant of polyurethane foam and cotton fabrics without any phosphorylation modification or combined use with other flame-retardants.

Liu et al. [40] successfully prepared a novel flame-retardant of LMA by assembling lignin with amino trimethyl phosphate and melamine through electrostatic interactions. Owing to the high content of hydroxyl groups in LMA, it can undergo ring opening reactions with epoxy resin, resulting in excellent interfacial strength between LMA and epoxy resin. Epoxy resin LMA has successfully passed the V-0 level (UL-94) of vertical combustion test, with an LOI of 30.2%. Meanwhile, compared to pure epoxy resin, the smoke production rate and peak heat release rate of epoxy resin LMA decreased about 77.5% and 72.3%, respectively. This result clearly indicates that the excellent expanded carbon structure endows epoxy resin with outstanding flame-retardant properties. Cen et al. [41] introduced lignin as a structure enhancer and charring agent, and successfully prepared sodium alginate/sodium lignosulfonate/phytic acid super hydrophobic aerogel. The superhydrophobic material mainly uses the condensed phase flame-retardant mechanism to achieve flame-retardant: when heated, the material continues to absorb heat, and quickly forms a gel layer, effectively preventing the volatilization of combustible gases inside the combustor, thereby curbing the formation and spread of the flame.

Compared to traditional materials, nanomaterials exhibit significant advantages in thermodynamic properties. Among them, carbon nanomaterials of carbon nanotubes (CNT) and graphene nano-sheets (GnP) exhibit excellent flame-retardant properties among numerous polymer composite materials, attracting widespread attention from researchers. However, CNTs have hydrophobicity and are easy to aggregate under their strong van der Waals forces [42, 43]. And lignin take place at an ideal dispersant for CNTs. Therefore, compounding lignin with carbon nanomaterials can not only improve the dispensability of CNTs, but also further enhance the flame-retardant of the composite material [44].

Considering the synergistic flame-retardant effect between GnP and CNT [45,46], Song et al. [47] found that compared with paper coated separately, paper coated with a combination of GNP, CNTs, and lignin have a more uniform and dense structure, lower breathability, and more thermal conductivity paths, significantly improving the flame-retardant performance of paper. They also noted that adjusting the ratio of lignin to carbon nanomaterials can maintain good flame-retardant effects while reducing costs. Song et al. further prepared alkaline lignin pre adsorbed CNTs/GNP hybrid materials through an aqueous vacuum assisted impregnation method. After impregnating the hybrid material into heat-treated wood, a dense protective layer was created on the wood outer surface, effectively hindering heat and mass transfer during combustion, decrease the pyrolysis reaction rate, and greatly improving the flame-retardant properties of the modified wood [48].

Cellulose is widely distributed in nature as a biomass resource. The nano micro cellulose (NFC) prepared by cellulose is often used in the field of aerogel and other materials because of its superior flexibility [49]. However, the flammability of NFC severely restrains its application in high fire safety requirements fields [50,51]. Wang et al. [52] utilized lignin nanoparticles (LNP) to improve the stability, thermal conductivity, flexibility, and flame-retardant of boron nitride PVA/(BN)-OH composite films. The PVA/BN-OH/LNP composite membrane was successfully prepared by combining vacuum filtration with chemical crosslinking. When the amount of LNP added is 2.5% (vol), the maximum penetration thermal conductivity of the BN-OH/PVA/LNP composite film can reach 1.74 $W.m^{-1}.\ K^{-1}$, which is considerably larger than the composite film that no LNP is added. Song et al. prepared multi wall carbon nanotubes (MWCNTs)/lignin based phenolic foam composites by in-situ polymerization. The addition of MWCNTs dramatically increased the LOI of phenolic foam to 44.2%, while reducing the rate of heat release, mass loss and total heat release. This is because MWCNTs composite phenolic foam will form a dense and incompletely burned carbon layer on the surface during combustion, which can effectively isolate external heat and air, thus playing a role in flame-retardant.

In addition, studies have found that grafting a single boron element into the NFC chemical structure can improve its flame-retardant properties to a certain extent, but its carbon forming ability is limited, and it cannot form an effective barrier to hinder heat transfer. Lignin can be used as a carbonization promoter in NFC systems to replace expensive graphene oxide. Tong et al. [53] subjected kraft paper and unbleached bamboo pulp to quaternization pretreatment to obtain quaternized nanocellulose (Q-NFC), which was then mechanically decomposed and boronized to successfully prepare Q-NFC with flame-retardant properties. Research data shows that compared with B/Q-NFC containing only boron, Q-NFC containing boron and lignin (L-B/Q-NFC) has a lower total smoke production, only $(0.0013 \pm 0.0003)\ m^2/m^2$; Compared with untreated pure NFC, the rate of peak heat release $(41.1 \pm 2\ kW/m^2)$, total heat release, and total smoke generation of L-B/Q-NFC were significantly reduced, with reductions of about 79%, 56%, and 99%, respectively, demonstrating good flame-retardant performance.

3.3 Lignin synergistic flame-retardant

Lignin, rich in hydroxyl groups, numerous benzene rings and other active functional groups, can be used as a raw material for various polymer materials and partially replace petrochemical raw materials [54]. In flame-retardant systems, lignin can be used as an excellent carbonization agent to prepare of the outstanding-performance biomass based eco - friendly materials [55]. In the study of using pure lignin to prepare flame-retardant polypropylene (PP) composites, it was shown that the thermal degradation temperature of the composite material obtained by mixing compounds

Materials Research Foundations 190 (2026) 153-193 https://doi.org/10.21741/9781644904077-6

such as ammonium polyphosphate (APP) and polyvinyl alcohol (PVA) with lignin was higher than when lignin was used alone. This is explained by the fact that the generated conjugated system catalyzes the breaking of lignin carbon oxygen bonds, increasing the carbonization rate. Polylactic acid (PLA), as a biodegradable material, has a low price and considerable yield, but its poor heat resistance limits its wider application. Using industrial lignin as a carbon source and compounding it with APP to develop flame-retardant PLA fabric can significantly improve the fire resistance of PLA. Compared with PLA, the flammability of modified composite materials is relatively low: PLA fabrics containing 5% lignin (LIG) and 5% APP have lower total heat release, only (11 ± 1) MJ/m^2, and higher residual carbon content, reaching (36 ± 3)%, compared to PLA fabrics containing only 5% APP, showing better flame-retardant performance [56-59].

The synergistic flame-retardant system constructed by lignin and traditional flame-retardants such as ammonium dihydrogen phosphate, boric acid, ammonium polyphosphate (APP) and magnesium hydroxide, can replace pure lignin and achieve more efficient flame-retardant effects. Partial synergistic flame-retardants can catalyze the C-O bonds' breaking, char formation, and promote cross-linking by forming highly conjugated systems. For example, in flame-retardant PP systems, lignin and aluminum hydroxide PVA. The combination of melamine phosphate ester, mono-ammonium phosphate, and APP can increase the thermal degradation temperature, prolong the combustion time, and improve the carbonization rate of PP composite materials [60]. Although lignin content has a certain impact on flame-retardant, the specific effect depends on the actual application scenario and material combination.

Verdolotti et al. [61] analyzed the combustion behavior of thermoplastic corn protein bio-composites (TPZ) with added flame-retardants of APP and alkaline lignin (AL) under different fire scenarios. They found that composite materials containing both 3 wt % AL and 10 wt % APP showed an excellent flame-retardant performance under various fire conditions. In addition, phytic acid (PA) is also employed as a flame-retardant synergist for lignin/polylactic acid (PLA) composites. Cheng et al. [62] concluded that the better the dispersion of PA in the lignin/ PLA system, the better the flame-retardant effect. The mixture of PA and lignin showed a synergistic effect in enhancing the comprehensive performance of the resulting composite material. The heat release rate of all composite materials is lower than that of pure PLA, and in a composite containing with a10 wt % lignin and 10 wt % PA, the heat release rate decreases by 36%.

Réti et al. [63] incorporated APP and sulfate lignin into PLA to obtain a PLA/APP/LIG system (mass ratio of 60%: 30%: 10%). In the flammability test, comparing with pure PLA, the combustion level of this system has been raised to V-0 level, and the rate of peak heat release has changed from 325 kW/m^2 to 173 kW/m^2, dramatically improving the flame-retardant performance. The use of renewable lignin as a carbon source in PLA can replace traditional pentaerythritol, reducing petroleum resource consumption while maintaining a high flame-retardant effect, and even potentially decreasing the usage of APPs. Lu et al. [36] used lignin as a carbonization agent to explore the flame-retardant properties of lignin/PP expansion flame-retardant system. Among them, lignosulfonate was copolymerized with isocyanate instead of diethylene glycol (DEG) to prepare lignosulfonate rigid polyurethane (LRPU) foam. Lignosulfonate improved the thermal stability of LRPU foam, and the LRPU based on DEG and containing 15 wt % lignosulfonate had the best thermal stability. Mandlekar et al. [64] studied the combustion behavior and thermal decomposition of sulfonated lignin (LS) and zinc phosphate (ZnP) and in PA11 matrix. Adding ZnP and LS to PA11 can enhance the polymer matrix thermal stability. The vertical combustion test UL-94 denoted that LS alone had poor flame-retardant effect on PA11, but when used together

Materials Research Foundations 190 (2026) 153-193 https://doi.org/10.21741/9781644904077-6

with ZnP, it had a synergistic flame-retardant effect on PA11. When additive was 10 wt %, the flame-retardant effect was optimal.

Cayla et al. [65] studied the effect of APP/kraft lignin (KL) on the combustion behavior and thermal degradation of bio-based PA11. The KL addition can delay the thermal degradation of PA11, and compared with pure PA11, the heat release rate minimizes by 66%. When 13 wt % APP and 7 wt % KL added to the PA11 matrix, the best flame-retardant performance is exhibited. Mandlekar et al. [38] applied pure lignin (alkaline sulfate lignin (DL) or lignosulfonate) and hypophosphite (aluminum hypophosphite (AlP) or zinc hypophosphite) to the flame-retardant of polyamide 11 (PA11), and studied the effects of lignin type and hypophosphite on the combustion performance and thermal stability of PA11 by melt extrusion of a ternary mixture of PA11/LIG/P. It was found that both types of lignin can prompt the thermal stability of PA11 and produce stable carbon residue. The combination of AlP (PA11, with a ratio of 80%: 10%: 10% lignosulfonate to AlP) and lignosulfonate can largely minimize total heat release (22% reduction) of PA11, and the peak heat releases rate (74% reduction), and exhibiting better flame-retardant performance. This is because lignin forms a dense carbon layer with reacting with phosphate compounds, which are the high-temperature decomposition products of hypophosphite, thus achieving the purpose of flame-retardant. Wang et al. [66] prepared alkali lignin/ammonium polyphosphate intumescent flame-retardant polyurethane foam by adding refined alkali lignin and ammonium polyphosphate to polyurethane foam in different proportions to form an intumescent flame-retardant (IFR). It was shown that when the ratio of alkali lignin to the ammonium polyphosphate was set to 1:6 and the addition of IFR was 30%, the OLI of ammonium polyphosphate /alkali lignin intumescent flame-retardant polyurethane foam reached 26.3%. After the flame-retardant addition, a dense and continuous carbon layer was formed, which reduced the rate of thermal degradation of the material and promoted the residual carbon rate. The flame-retardant effect was attained by the condensed phase flame-retardant mechanism.

Jin et al. [67] used lignin as the initial raw material and grafted piperazine molecules onto lignin molecules through hydroxymethylation modification and Mannich reaction. The reverse co precipitation method was used to prepare piperazine modified lignin/magnesium hydroxide double coated red phosphorus flame-retardant. The flame-retardant exhibits a good synergistic effect among its components. Carretier et al. [68] introduced alkali lignin into the flame-retardant study of polyurethane elastomers and found that the lignin addition can largely diminish the temperature T_{max} and the rate of the peak thermal loss. Compared with the standalone APP, the mixture of lignin and APP can achieve a higher residual carbon rate after thermal degradation, which also confirms the interaction between the two during the thermal degradation process. Widsten et al. [69] compared the flame-retardant of chemically modified and unmodified and industrial lignin with the flame-retardant of an ammonium pentaerythritol/polyphosphate expansion system in the polypropylene matrix. Polypropylene is flammable and can generate a large amount of molten droplets and smoke when burned in air, which limits its further application. Research has demonstrated that compared to pure polypropylene, polypropylene/lignin composites reduce the percentage of heat reduction by 41% and total heat release by 36%. In terms of gas-phase flame-retardant mechanism, polypropylene/lignin composite materials with modified lignin have better flame-retardant properties. Liang et al. [70] investigated the flame-retardant effect of lignin containing flame-retardants. A renewable and eco-friendly lignin based flame-retardant was prepared by one-step reaction of melamine, ammonium polyphosphate, and aluminum lignin. It was noted that the flame-retardant performance of epoxy resin is optimal as adding 20% flame-retardant. The total heat release rate and smoke production can achieve 53.1 MJ/m^2 and 9.9 m^2, respectively. The gas-phase flame-retardant mechanism indicates that this lignin based flame-retardant has good smoke suppression and fire extinguishing properties.

Yan et al. [71] developed a functionalized layered double hydroxide (LDH) using lignosulfonate and nickel hydroxide ($Ni(OH)_2$) as raw materials- LS@$Ni(OH)_2$ flame-retardant. The results showed that the total heat release, peak heat release rate, and total smoke generation of this material were significantly reduced by 29.5%, 62.7%, and 46.2%, respectively. The flame-retardant mechanism of LS@$Ni(OH)_2$/LDH mainly relies on the physical barrier effect of hydrotalcite and the catalytic carbonization of transition metal elements and in condensed phase flame-retardant. Xue et al. [72] replaced polyether polyols in polyurethane raw materials with modified sodium lignosulfonate, and added nano silica and ammonium polyphosphate in different proportions to the rigid polyurethane material for synergistic flame-retardant, resulting in composite rigid polyurethane. It was found that when ammonium polyphosphate content was 25.0%, the content of nano silica was 1.0%, and the content of sodium hydroxymethyllignosulfonate was 4%, the maximum LOI of the composite material could reach 27.2%, and the maximum total smoke release and heat release and were reduced by 32.5% and 51.5%, respectively. Polyurethane materials with added synergistic flame-retardants have denser residual carbon and smaller pores, belonging to the condensed phase flame-retardant mechanism. Li et al. [73] synthesized a phosphorus containing lignin based flame-retardant wrapped in polysiloxane by reacting lignin, polysiloxane, and phosphate. The condensed phase flame-retardant mechanism was found that lignin, phosphate, and polysiloxane work together to form a cohesive and dense carbon layer, which hindered the oxygen and heat transfer and prevent the diffusion of combustible gases.

Lu et al. [74] produced a metal organic framework (P-MOF) containing phosphorus/nitrogen and cobalt components and added it to flame-retardant. Due to the smoke suppression (cobalt component) and fire extinguishing (phosphorus organic component) effects, the addition of P-MOF substantially reduced heat release and smoke production. The synergistic effect between P-MOF components and lignin inhibits the condensed phase combustion, leading to a higher degree of carbon residues graphitization. The flame-retardant effect is achieved through the combined action of gas-phase flame-retardant and condensed phase flame-retardant mechanism. Shi et al. [75] synthesized lignin-based bio-waterborne polyurethane using aqueous solutions of sodium lignosulfonate with different dispersants concentrations, and studied the effects of sodium lignosulfonate on the thermal stability and thermodynamic properties of polyurethane films. With the increase of sodium lignosulfonate, the ultimate LOI of lignin based bio-waterborne polyurethane can reach 24.7%, showing good flame-retardant effects in both gas-phase and condensed phase flame-retardant mechanisms. Shang et al. [76] formulated a composite expansion flame-retardant by compounding lignin and piperazine pyrophosphate in a mass ratio of 1:1, and utilized it for flame-retardant modified epoxy resin. It was found that the introduction of lignin and piperazine pyrophosphate enhanced the LOI of flame-retardant modified epoxy resin from 22.2% to 27.5%, and reduced total smoke release and the maximum heat release rate and by 22.25% and 60.11%, respectively. Research has found that the addition of composite expansion flame-retardants considerably promotes the carbonization rate of the system, making the residual carbon layer of epoxy resin flame-retardant materials denser after combustion. The combined effect of gas-phase flame-retardant and condensed phase flame-retardant mechanism and enhances the flame-retardant effect.

Materials Research Foundations 190 (2026) 153-193 https://doi.org/10.21741/9781644904077-6

From the previous research work, it can be observed that lignin complex as a flame-retardant has the behaviors of simple process and low cost. By compounding with inorganic flame-retardants (zinc borate, aluminum hydroxide,) or organic flame-retardants (phosphorus/nitrogen compounds) to produce synergistic effects, flame-retardant efficiency can be prioritized. However, the compatibility of compound flame-retardants is poor, resulting in unstable flame-retardant performance, and when the amount of flame-retardant added is high, and it will decrease the mechanical properties of the matrix material.

4. Chemical modified lignin-based flame-retardants

Table 1 Main chemical reactions for chemical modification of lignin

Chemical modification methods of lignin	**Reaction conditions and objectives**	**references**
Amination reaction	By utilizing the active amino groups on the benzene ring of lignin to condense with formaldehyde and acid, nitrogen-containing flame-retardant compounds are introduced onto the benzene ring to enhance the flame-retardant of lignin in plastics.	[77, 78]
esterification	Under the catalysis of 1-methylimidazole, react with succinic anhydride to optimize the molecular structure and improve thermal stability (more stable at high temperatures).	[79]
Hydroxymethylation reaction	React with formaldehyde in NaOH solution at 80 °C, then adjust the pH to 2 to introduce hydroxymethyl, increase the active group to introduce flame-retardant elements (phosphorus/nitrogen/metal).	[80]
Schiff Base reaction	Carbonyl groups undergo nucleophilic addition and dehydration condensation with triethylenetetramine (TETA) in aqueous solution, and phosphorus/nitrogen flame-retardant elements are introduced to suppress flame propagation.	[81]
Silanization reaction	Organosilicon reacts with lignin in a water/ethanol mixed solvent, introducing silicon elements to slow down the combustion rate and enhance flame-retardant and durability.	[82]

Even though lignin can impart polymers with a certain level of flame-retardant, the composites of lignin/polymer often struggle to meet the V-0 standard in the industrial UL-94 vertical combustion test. Its limited flame-retardant efficiency can be used as an enhancer for most flame-retardant systems existed, such as nitrogen phosphorus expanding flame-retardants (IFR). The common IFR system included three parts: carbonization agent, acid source, and foaming agent [83]. Lignin is an ideal choice for carbonization agent due to its high carbonization ability. Lignin with different chemical reaction sites with methoxy, hydroxyl, carboxyl and carbonyl groups, which can be chemically modified to introduce nitrogen or phosphorus elements. Compare with unmodified

Materials Research Foundations 190 (2026) 153-193 https://doi.org/10.21741/9781644904077-6

lignin, lignin modified with nitrogen or phosphorus exhibits superior flame-retardant properties in some composite materials [84]. Although lignin has environmental advantages in synergistic flame-retardant with other substances, there are shortcomings in terms of durability and stability. In recent years, scholars at home and abroad have been devoted to further optimizing the properties of lignin through chemical modification. Yang et al. [85] carried out hydroxymethylation, epoxidation and phenolic modification of lignin. It was found that the decomposition temperature of phenolic lignin based polyurethane foam was slightly larger than that of pure lignin based polyurethane foam, which means that phenolic lignin can enhance the heat resistance of foam.

By introducing elements or compounds such as nitrogen, phosphorus, and silicon, the flame-retardant properties of lignin can be considerably optimized. Chemical modification of lignin can not only customize its flame-retardant properties according to specific needs, but also bring additional benefits such as improved thermal stability and reduced smoke generation, making it a more advantageous choice in flame-retardant applications. Figure 3 shows the various chemically modified methods for lignin [24]. We will disscus them as follows.

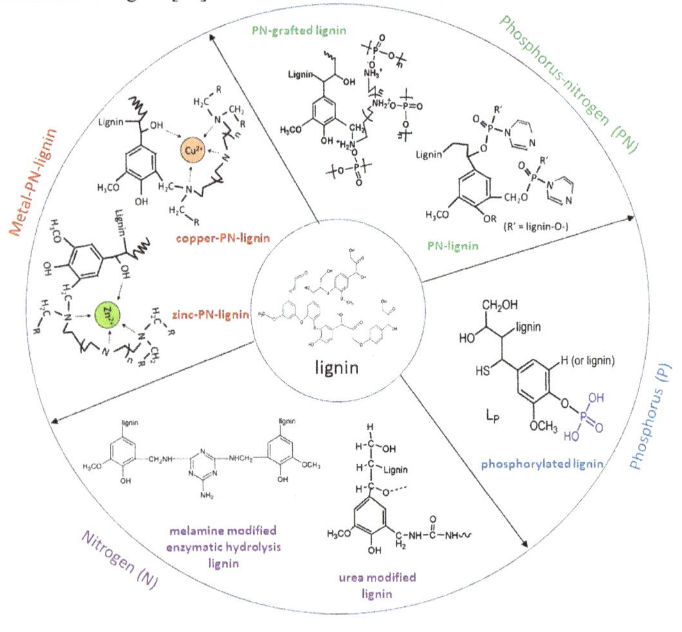

Fig. 3 Different chemically modified lignin using phosphorus, nitrogen, and metal ions. Reproduced with permission from [24] Copyright 2022, Springer Science Business Media, LLC, part of Springer Nature.

Materials Research Foundations 190 (2026) 153-193 https://doi.org/10.21741/9781644904077-6

4.1 Nitrogen or phosphorus modified lignin

4.1.1 Nitrogen modified lignin

Nitrogen plays a key role in flame-retardant systems. Generally speaking, nitrogen-containing lignin flame-retardants are produced by reacting amine and formaldehyde compounds with lignin such as resorcinol, urea, and melamine. This type of nitrogen-containing modified lignin flame-retardant has high thermal resistance and stability to thermal decomposition. Fu et al. [86] prepared a new expanding flame-retardant of melamine modified enzymatically hydrolyzed lignin (MEHL), and found that the MEHL decomposition temperature was larger than that of enzymatically hydrolyzed lignin (EHL). According to the vertical combustion UL-94 and LOI test, adding 50% EHLM and 12% microencapsulated red phosphorus and to rubber can successfully go through the V-0 level in the UL94 test, and the LOI attains 35%. Subsequently, Zhang et al. [87] synthesized urea modified lignin (UM Lig) using formaldehyde, lignin, and urea as raw materials through Mannich reaction. Pure PLA is flammable with an LOI of 21%. Adding 23 wt % APP, the LOI of PLA composite material promote to 30%. When the weight ratio of UM Lig to APP is 1: 4, the LOI can reach up to 34.5%. The PLA/UM Lig/APP blends can achieve the V-0 grade in UL-94 test, which both pure PLA/UM Lig and PLA failed the UL-94 test. In addition, cone calorimetry analysis noted that the heat release rate of PLA/APP/UM Lig was significantly lower than that of PLA/APP/lignin and PLA/APP systems. The TGA curve indicates that thermal stability of APP/UM Lig is higher than that of PLA.

Zhou et al. [88] produced nitrogen-containing lignin-based flame-retardants and added 7 wt% functionalized lignin. It was found that the carbon residue became denser. During the combustion process, nitrogen element releases ammonia, which dilutes oxygen and generates a dense layer of coke, blocking air from entering and achieving efficient flame-retardant in both condensed and gas-phase. The performance of phosphorus-based lignin-based flame-retardants modified by grafting is excellent. However, when the content increases to the critical value, simply increasing the amount of phosphorus-based flame-retardants no longer substantially improves the flame-retardant performance of the material [89]. However, phosphorus-based flame-retardants usually have satisfactory PN synergistic effects with nitrogen-containing compounds [90-91]. Wang et al. [92] synthesized a lignin based flame-retardant chemically grafted with polyethylene imine and polyphosphate. It was shown that phosphorus/nitrogen modified lignin conducted a stronger thermal stability and lower char formation temperature.

Phosphorus plays a key role in promoting the carbon formation in the composite, resulting in a thick and dense carbon layer in the resin matrix and achieving flame-retardant in a condensed phase. Wei et al. [93] developed a bio-based flame-retardant (Lig HCCP) through a one-step reaction between hexachlorocyclotriphosphazene and lignin. The flame-retardant mechanism is attributed to the synergistic effect of P-N, which helps to form an expanded and dense carbon layer. Weng et al. [94] prepared phosphorus and nitrogen doped lignin flame-retardant (NP-LMDL), and discussed the flame-retardant mechanism of phosphorus and nitrogen in foam. The results suggest that phosphorus and nitrogen in NP LMDLPFx work together to enhance the polymer carbonization, produce a rich phosphorus nitrogen residual carbon layer, and exert a condensed phase flame-retardant effect. Li et al. [95] prepared an innovative environmentally friendly silicon nitrogen phosphorus triple lignin flame-retardant (Lig-K-DOPPO) and found that the synergistic effect of N-P elements in Lig-K-DOPPO gives it excellent smoke suppression and flame-retardant. Zhong et al. [96] used sulfate lignin as raw material and introduced aldehyde groups from the outside through aldehyde reaction to prepare reactive lignin-based flame-retardants through nitrogen phosphorus modification. It was found that the modified phenolic foam had excellent flame-retardant when the substitution rate was 10%.

Chemical modification of lignin to prepare flame-retardants can significantly improve flame-retardant efficiency. Introducing a nitrogen source can promote the expansion into charcoal, increase the LOI to 27-29%, and also improve the smoke suppression performance of the material; But there are also disadvantages such as low grafting rate (<15%) and poor thermal stability (decomposition temperature < 250 ℃).

4.1.2 Phosphorus modified lignin

Pure lignin has limited effectiveness in polymer flame-retardant, while phosphorus containing lignin-based flame-retardants prepared by introducing phosphorus elements will generate polyphosphoric acid or phosphoric acid during combustion. These substances can remove moisture from lignin or polymer composite materials, thereby promoting carbonization reaction and achieving flame-retardant goals. Phosphorus functionalized lignin (P-lignin) combines the flame-retardant properties of phosphorus in the condensed and gas phase, as well as the condensed phase carbonization effect of lignin, thus exhibiting superior flame-retardant properties in polymer materials. Substances with volatile phosphorus can act as flame-retardants by catalyzing the free-radical recombination process. Compared with unmodified lignin, its flame-retardant is significantly improved [97]. When flame-retardants contain phosphorus groups, they not only have strong thermal stability, but also have relatively small adverse effects on the material [98,99].

In 2010, Bykov et al. [100] developed a lignin phosphorylation process, which resulted in materials with high phosphorus content of 16.6% (wt) and high adsorption performance. Research suggests that phosphorylation is more likely to take place at the carbonyl and phenolic hydroxyl sites of lignin. Subsequently, Xing et al. [101] extracted alkali lignin from wheat straw, and prepared phosphorus lignin from formaldehyde, phosphorus oxychloride and ethylene glycol through three steps of reaction, further enhancing the flame-retardant effect of phenolic encapsulated ammonium polyphosphate (PFAPP) on polyurethane foam (PUF). After adding lignin and PFAPP, the heat release rate of polyurethane foam reduced. Increasing the content of lignin and PFAPP, the average mass loss rate and total heat release are also decreased. Ferry et al. [102] used lignin as a flame-retardant for PBS biopolymers and successfully modified the surface of alkaline lignin by grafting macromolecular or molecular phosphorus compounds. Ammonium dihydrogen phosphate (DHAP) reduces the exothermic peak of the reaction, and when blended with PBS, the flame-retardant effect of the modified lignin is further enhanced.

Prieur et al. [103] used lignin phosphorylated adducts as flame-retardants for ABS and found that they could significantly increase the residual amount of polymers at high temperatures, enhancing the effect flame-retardant of polymers. When the P-lignin mass fraction is 30 wt %, it is well dispersed in ABS and markedly reduced the rate of peak heat release. Since P-lignin reacts with ABS during thermal decomposition to promote carbonization and form a protective layer, thereby reducing combustible gases released into the flame from ABS degradation. Mendis et al. [104] utilized pyridine catalyzed esterification reaction to phosphorylate lignin with diphenylphosphoryl chloride, thereby enhancing the carbonization ability of lignin. The results showed that the addition10% modified lignin to ABS decreased the total combustion heat by 20% and the rate of the peak heat release by 40%. Zhang et al. [105] prepared a novel lignin based flame-retardant of

LHD with a 9, 10-dihydro-9-oxa-10-phosphate-10-oxide, lignin, and hexamethylene diisocyanate. It was found that the residual residue of flame-retardant lignin based polyurethane carbon has a dense and continuous outer carbon layer on its surface, while the inner carbon layer has many expanding bubbles, demonstrating excellent performance in condensed phase flame-retardant.

In 2020, Hu et al. [106] synthesized lignin based flame-retardant LMD containing P. The results showed that the LOI of pure polylactic acid was 18.2%, and the LOI gradually increased with the addition of flame-retardant LMD. On this basis, adding 1, 3, 5-triglycidyl isocyanate (TGIC) can further enhance the flame-retardant performance of the system. Guo et al. [107] dissolved lignin and polyacrylonitrile (PAN) in DMSO, and made the LIG/PAN composite film, and then phosphorylated to gain the flame-retardant composite material flame-retardant-LIG/PAN. It was found that composite membranes can form a carbon layer with expansion effect during combustion, thereby delaying the PAN matrix thermal decomposition. The residual carbon content of flame-retardant-LIG/PAN increased to 15.9% as compared with pure PAN and LIG/PAN at 800 ℃. The rate of peak heat release decreased to 4.9 kW/m^2, and the flame-retardant and carbonization ability were significantly improved. Wang et al. [108] prepared a novel phosphorus containing lignin based flame-retardant composite material LMD and previously used as a component to develop a degradable flame-retardant polylactic acid/wood bio-composite material. The results indicate that flame-retardant originates from the phosphorus element and lignin in LMD.and Dai et al. [109] confirmed that LMD is a very good flame-retardant and reinforcing material for composite materials. The study concluded that composite materials with higher phosphorus content can achieve the best flame-retardant effect at the V0 level by UL-94 tested. Lee et al. [110] incorporated lignin fillers into polypropylene after phosphorylation modification to improve the flame-retardant of polypropylene based composites. The scanning electron microscope images after combustion testing, the solid carbon formed on the composite material surface by phosphorylated lignin can serve as a protective layer for polypropylene, so lignin phosphorylation can enhance its flame-retardant.

Researches have concluded that adjusting the grafting rate of flame-retardants can optimize the performance of lignin flame-retardants. Hajj et al. [111] grafted various phosphorus containing flame-retardants onto lignin rich Platycodon grandiflorus fibers and lignin poor flax fibers, and found that the thermal performance and flame-retardant were positively correlated with the grafting rate of phosphorus. Liu et al. [112] developed a lignin derived flame-retardant by grafting polyphosphate amide onto the lignin waste with graft polymerization methods. The obtained flame-retardant composite has achieved a satisfactory V-0 rating in UL-94 test. Zhang et al. [113] used 9, 10-dihydro-9-oxa-10-phosphaoyl-10-oxide and lignin, synthesized a composite material with exhibited excellent flame-retardant. Numerous studies have confirmed that introducing the phosphorus elements through esterification reaction for phosphorus (P) modification can achieve an LOI of over 32% and an increase in residual carbon rate to 40%. However, such reactions typically require high temperatures (>120 ℃) and catalysts, and the process is relatively complex.

4.2 Nitrogen and phosphorus co-modified lignin

Lignin modified by nitrogen phosphorus synergy can be applied extensively as a flame-retardant in various polymer material fields. Among them, the nitrogen source produces none combustible gases when heated or burned, which can dilute combustibles and isolate oxygen. During the combustion process, phosphorus sources will produce substances such as polyphosphate, which

Materials Research Foundations 190 (2026) 153-193 https://doi.org/10.21741/9781644904077-6

advances the phosphorus containing non-combustible carbon layer formation in lignin. Common nitrogen sources include imidazole, 3-amino-1, 2, 4-triazole, melamine, ammonium hydroxide, p-phenylenediamine, or 4, 4-diaminodiphenylmethane, Phosphorus sources include diethyl phosphite 9, 10-dihydro-9-oxa-10-phosphate phenanthrene oxide (DOPO), phosphorus trichloride, or polyphosphate. Nitrogen sources are generally used to form the chemical bonds with modified lignin or lignin, releasing NH_3 gas during combustion. The strategic role of phosphorus is to produce the carbon layer containing phosphorus, which usually includes the hypophosphite or phosphate compounds [116-130].

Yu et al. [114] performed methanol hydroxymethylation treatment on wheat straw alkali lignin through a three-step reaction, and chemically grafting phosphorus and nitrogen elements to improve the behavior of flame-retardant. The modified lignin (NP lignin) shows a stronger carbonization ability. The carbonization rate reached 61.4 wt %, while the carbonization rate of ungrafted lignin was only 40.7 wt % at 600 ℃, under a N_2 atmosphere. When 30% PN lig is added to PP composite materials, a thicker and denser carbon layer will be formed. Subsequently, Zhu et al. [115] grafted the lignin with functional groups containing phosphorus and nitrogen through liquefaction esterification salinization reaction, and prepared lignin-based melamine phosphate compounds (LPMC) for the first time. The compound replaced some copolymerized with isocyanate and polyols to obtain lignin modified polyurethane foam PU-LPMC. This kind of foamed plastic can produce a large amount of non-combustible gas during thermal degradation, and form a dense carbon layer $(C\text{-}P\text{-}N\text{-}O)_x$ on the foam surface. Through the synergistic effect of condensed and gas phase flame-retardant, the flame-retardant of the polymer is effectively improved.

Zhang et al. [116] prepared nitrogen phosphorus grafted lignosulfonate (PNLS) using lignosulfonate as a carbon source. After modification, the temperature of initial decomposition of lignin was substantially changed from 73 ℃ to 205 ℃. By incorporating it into a thermoplastic polyolefin elastomer (TPO), the ultimate LOI of the composite system reached 28.2%, which is greater than the TPU/LS system (22.6%). In the fire test, the TPO/PNLS system became the only system that reached the V-0 level, fully demonstrating the excellent thermal stability of this expanding flame-retardant. Hu et al. [117] synthesized phosphoramidite modified enzymatically hydrolyzed lignin (EHL-PM) for the preparation of novel flame-retardants. The test results demonstrate that when the amount of microencapsulated red phosphorus and EHL-PM added reaches 45%, the fire test can reach the highest level V-0, and a dense carbon layer with good flame-retardant effect can be formed. Liu et al. [118] modified lignin with N and P, and then co-bonded with metal ions to obtain modified lignin. Compared with unmodified lignin, modified lignin can more effectively improve the thermal stability of polypropylene. Meanwhile, the metal ions play a catalytic role, further optimizing the flame-retardant properties of the composite material. Costes et al. [119] used phosphorus and nitrogen modified lignosulfonates and organic solvent lignin as flame-retardants for PLA. It was found that untreated lignin can play a certain flame-retardant role in PLA, mainly because of the coke formation, but the thermal stability of PLA is considerably reduced. In contrast, lignin treated with nitrogen/phosphoruschemistry can reduce the thermal degradation of PLA, significantly improve flame-retardant, and achieve the V-0 level in the UL-94 test.

Lignin derivative vanillin can be coupled with diethyl phosphite and diamine, and then reacted with the epichlorohydrin to generate high-performance flame-retardant epoxy resin [120]. The epoxy resin obtained is flammable with an LOI of 24.6%, while cured bio-based epoxy resin, with good expansion and dense carbonization ability, the LOI increased to 31.4% -32.8%, reaching The V-0 level in UL-94 test. Matsushita et al. [121] introduced 3-amino-1, 2, 4-triazole into KL through Mannich reaction, then it reacted with phosphoryl chloride to introduce the phosphorus. Subsequently it reacted with α, ω-alkyl diols to form an expandable flame-retardant resin. This flame-retardant resin is non-flammability, and the heat release rate and heat release are significantly decreased as compared with commercial phenolic resins. Depending on the mass balance calculation, the volatile phosphorus containing substances generated during the combustion process plays a key flame-retardant role.

Reactive organic phosphorus, represented by DOPO and its derivatives, has the advantages of high flame-retardant and good thermal stability, which can form chemical bonds with the polymer matrix, endowing the material with a certain degree of flame-retardant. At present, research on its combined flame-retardant with lignin is continuously deepening. Examining the low compatibility with lignin, Zhang et al. [122] used hexamethylene diisocyanate (HDI) as a toughening agent to connect lignin and DOPO, and applied the resulting nitrogen phosphorus modified lignin LHD to enhance the flame-retardant properties of polyurethane. The carbon yield of LHD at 600 ℃ is greater than of pure DOPO, and the content of residual carbon increases with lignin content increasing. Moreover, the temperature at which LHD loses 5% weight (T_5) and 10% weight (T_1) are larger than that of pure lignin. It indicated that the modified lignin has a stronger thermal stability.

In 2019, Zhong et al. [123] encapsulated RP in a mixture of piperazine modified lignin (LP) and AlP to enhance the flame-retardant of ABS resin. It was found that the flame-retardant grade of ABS coated with RP was noticeably improved, reaching V-0 level. Zhou et al. [124] prepared EP/7-F-Lignin composite material (containing 7% modified lignin) by grafting polyphosphoric acid and polyethyleneimine (PEI) onto lignin for nitrogen phosphorus modification. Its maximum decomposition temperature drops to 364 ℃, and the char formation rate increases to 28% at 800 ℃. It was noted that epoxy resin with added P-N modified lignin was more prone to generating carbon layers. It can reach the V-0 level in the UL-94 test. Although inorganic phosphorus has abundant raw material sources and low prices, it has disadvantages such as large filling amounts and adverse effects on the material mechanical properties. With the continuous innovation and improvement of phosphorus based flame-retardant formulations, organic phosphorus with dual plasticizing and flame-retardant effects, combined with biomass such as lignin, is recognized as a substitute for halogen flame-retardants. According to different usage methods, organic phosphorus can be divided into reactive and additive types. Hu et al. [106] selected 1, 6-hexamethylene diisocyanate (MDI), which is cheaper than HDI, and used trichloroisocyanurate (TGIC) as a coupling agent to enhance the flame-retardant of wood/PLA materials, increasing the LOI value from 18.2% to 28.1%, significantly enhancing the flame-retardant of wood plastic composites. However, HDI and MDI themselves do not possess flame-retardant properties, which limits the polymer's ability to achieve higher flame-retardant performance.

Selecting piperazine (PA) and DOPO with excellentflame-retardant to prepare flame-retardant intermediates, and connecting the intermediates to lignin to form intermediate lignin. In addition, grafting DOPO onto intermediate lignin can obtain higher phosphorus content [78]. After introducing EP and burning it, the coke expands owing to the generation of non-combustible gases from nitrogen elements. Compared with materials containing the same proportion of initial lignin,

Materials Research Foundations 190 (2026) 153-193
https://doi.org/10.21741/9781644904077-6

modified lignin carbonized earlier, indicating that lignin has successfully achieved phosphorus nitrogen modification and improved the flame-retardant effect of EP. Reactive organic phosphorus endows the material with inherent flame-retardant properties, but the cost is high and only a few are put into actual production. Additive organic phosphorus is not difficult to process and has a wider range of applications. Yang et al. [125] synthesized lignin derivatives (TP-g-lignin) using PEPA and 3-isopropenyl-α, α-dimethylbenzyl isocyanate (TMI) and (substance ratio1.2:1), endowing PLA with good flame-retardant. At a high temperatures, the groups of phosphate in TP-g-lysin can be decomposed into the phosphorus containing substances, resulting in a residual carbon rate of over 26% for all different ratios of TP-g-lysin formulations. Regardless of the ratio, the rate of peak heat release for PLA composite materials will further increase with increasing the TP-g-lysin loading. Nevertheless, TMI components are flammable and release more heat during combustion.

The above research work shows that lignin can be co modified with nitrogen and phosphorus (N/P) by introducing phosphorus (such as DOPO) or nitrogen elements (such as melamine) through esterification, grafting and other reactions. The modified flame-retardant can dramatically improve its carbonization ability and flame-retardant efficiency, with excellent flame-retardant properties (such as an LOI value of over 30%), as well as smoke suppression and low toxicity. However, its disadvantage lies in the complex synthesis steps and harsh reaction conditions (usually requiring high temperatures or the addition of catalysts). In addition, modified lignin may lose some of its biodegradability.

4.3 Nitrogen and phosphorus modified lignin containing metal ions

By introducing metal ions, polymer dehydrogenation can be promoted and catalyzed into carbon in the condensed phase, thereby enhancing the flame-retardant effect of PN lignin. Research has found that introducing nickel acetate (cobalt or zinc) into the PP/PN lignin system can further enhance the thermal stability and flame-retardant of PN lignin [126]. For example, adding 2 wt % nickel acetate can produce a catalytic degradation effect on PP/PN lignin, while significantly increasing the residual carbon rate to 44%. The addition of $Ni(Ac)_2$ increased the LOI of PP/PN lignin from 22% (17.5% for pure PP) to 26%, demonstrating good flame-retardant. Liu et al. [127] added Zn^{2+} to PN/PBS lignin, and after adding 10 wt % PNZn lignin, total heat release rate and the rate of peak heat release of PBS were significantly reduced by 68% and 51%, respectively. In addition, the total smoke and dust emissions have considerably decreased by 50%. Adding PNZn lignin can form a dense, complete, and thick carbon layer, thereby enhancing carbonization performance. Subsequently, in 2016, they added copper (II) ions to PNCu lignin/PP composite materials [128]. Compared with unmodified lignin, PNCu lignin not only can reduce total heat releases rate and the combustion heat release rate, reduce the combustion process, but also can minimize the total smoke generation rate during combustion, effectively improving the flame-retardant and thermal stability of composite materials.

The introduction of aluminum ions not only improves the carbonization efficiency of combustibles, but also absorbs heat at high temperatures and reacts with oxygen to form an aluminum oxide layer, isolating combustible gases and playing a key role in both condensed and gas-phase flame-retardant. Song et al. [129] used lignin cracking products as the skeleton and grafted trivalent aluminum ions at the sites with reactive functional groups on their molecular chains. They also grafted various flame-retardant inorganic acid groups/amine groups to invent a lignin aluminum flame-retardant. Compared with traditional phenolic flame-retardant insulation

Materials Research Foundations 190 (2026) 153-193
https://doi.org/10.21741/9781644904077-6

materials, this flame-retardant extensively uses low-cost natural macromolecular lignin, making its terminal product lignin aluminum flame-retardant and its prepared flame-retardant insulation materials highly competitive in the market. Metal copper ions can catalyze the carbonization process of lignin at high temperatures, promoting the decomposition of lignin into carbonized products. Liu et al. [130] modified lignin with nitrogen, phosphorus, and zinc ions, using alkaline lignin as a bio-based additive for polybutadiene ester (PBS). Adding 10% PNZn lignin can significantly reduce total heat release AND the peak heat release rate and of PBS by 67% and 50%, respectively. The total smoke production is significantly decreased by 50%. Adding PNZn lignin can produce a complete, dense, and thick carbon layer, and PNZn lignin exhibits good flame-retardant behavior in the condensed phase flame-retardant mechanism. Compared to traditional flame-retardants, metal ions in metal ion modified lignin flame-retardants are usually less toxic, reducing the risk of harmful substances being released into the environment during material combustion. Guo et al. [131] prepared aminated polyacrylonitrile (A-PAN) using ethylenediamine treatment. Then sodium lignosulfonate was added and can adsorb on the interior and surface of A-PAN with strong ionic bonds and chelated with copper ions to obtain SLS/Cu/ A-PAN. Research has found that the synergistic effect of polyacrylonitrile and copper ions endows SLS/Cu/ A-PAN with a good flame-retardant behavior and high carbonization ability in condensed phase flame-retardant.

By introducing phosphorus, nitrogen, and copper through chemical grafting, functionalized lignin improves the flame-retardant behavior and thermal stability of wood plastic composites (WPC) compared to ordinary lignin due to the introduction of these elements [132-134]. It was reported that the copper addition had a synergistic effect with nitrogen and phosphorus elements [135], enhancing the dehydrogenation ability of WPC and catalyzing lignin carbonization. Pure alkaline lignin was selected to react with polyethyleneimine (PEI) in the presence of formaldehyde to get an intermediate, which was then reacted with diethyl phosphite (DEP) in the presence of formaldehyde. The resulting material was metalized with copper (II) acetate to form a lignin copper coordination compound [118]. In the process of thermal degradation of WPC, the addition of 10% F-lignin can substantially increase the maximum weight loss temperature of 5 ℃ and initial degradation temperature of 7 ℃ for WPC. When containing 0.33% copper ions, WP/PP/ 5-F-lignin exhibits the lowest total heat release with 60.5 ± 1.8 MJ/m^2, but excessive copper ions can damage the carbon layer and draw up a counterproductive effect.

From the above research, it can be concluded that the addition of metal ions (Cu^{2+}, Fe^{3+}, Zn^{2+}, Al^{3+}, etc.) can chelate and modify lignin. Modified lignin can dramatically enhance the catalytic conversion to charcoal, increasing the residual char rate of flame-retardants to 45-50%. Therefore, this type of flame-retardant is more suitable for high-temperature flame-retardant scenarios. However, the residue of these metals may further accelerate the aging of the material.

4.4 Organic silicon modified lignin

Silicon containing compounds have advantages such as environmental protection and good heat resistance, making them a high-quality flame-retardant additive for polymer materials. Organic silicon lignin flame-retardants have excellent heat resistance, oxidation resistance, and smoke suppression properties. However, due to its high cost and complex synthesis process, it is relatively rare in practical applications. During the combustion process, silicon presence can improve the thermal stability of the carbon layer, so silicon containing compounds can be incorporated into expansion flame-retardants to enhance the flame-retardant effect. Research has found that organosilicon can improve the biodegradation performance of lignin flame-retardants to a certain extent, which helps to reduce their long-term impact on the environment [136].

Zhang et al. [137] added LSH and APP to PLA to construct a new type of expandable flame-retardant system. Compared with PLA/PP and PLA/PP/lignin, PLA/PP/LSH composite material forms a continuous and dense expansion type carbonized layer, exhibiting excellent flame-retardant properties. Zhang et al. [138] first hydrolyzed tetraethoxysilane (ethyl orthosilicate) to obtain tetrahydroxysilane, which was further reacted and polymerized with lignin to obtain lignin silica hybrid composites (LSHs), which were then added to PLA together with ammonium polyphosphate. The LOI of PLA7 system (containing 15% APP and 5% LSH) amounts to 34%, the rate of peak heat release can reach 145 kW/m^2, and the residual carbon rate is 13.1% at 800 ℃ . Its flame-retardant behavior is better than that of PLA4 (containing 15% APP and 5% alkali lignin). This suggests that the stability of LSH samples at extreme temperatures is better than that of lignin samples, and more stable char can be generated after combustion. It was shown that the introduction of organosilicon can improve the flame-retardant properties of lignin and enhance the overall performance of flame-retardants [138]. Ma et al. [139] combined different types of organosilicon with lignin to synergistically flame-retardant epoxy resin, developing a green and environmentally friendly composite flame-retardant material. This study deeply analyzed the mechanism of lignin and organosilicon in synergistic flame-retardant, providing guidance for the subsequent development of lignin/organosilicon flame-retardants with excellent comprehensive performance. Lignin silica hybrid (LSH) was produced by the sol-gel method. After adding APP and LSH into PLA, LOI can reach to 34%, and can reach The V-0 level at UL -94 test. Compared with PLA/PP/lignin and PLA/PP, the presence of the silica results in a dense and continuous expansion of the char layer in the combustion. The combination of LSH and APP can significantly enhance the thermal stability and flame-retardant behavior of polymer composites [140].

Under the catalysis of three (pentafluorophenyl) borane [$B(C_6F_5)_4$], the surface of lignin reacts with the hydrosilylation functional groups on the organic silicon polymer, and the organic silicon crossed links, thereby obtaining a reinforced elastomer. Lignin acts as both a crosslinking agent and a reinforcing agent, and this composite material have superior high-temperature aging resistance [141]. Lu et al. [142] used wheat straw alkali lignin as the main raw material and modified it with γ-aminopropyltriethoxysilane to prepare N and Si containing lignin-based carbonization agents. The residual carbon rates of the carbonization agent at 500 ℃ and 800 ℃ reached 31.5% and 21.3%, respectively, substantially higher than the 2.0% and 1.9% of wheat straw alkali lignin. It indicates that the carbonization agent has excellent thermal stability and carbonization ability, which helps to reduce carbon emissions.

Introducing organic silicon into lignin molecules can improve the thermal insulation performance of materials. Song et al. [143] used γ - divinyl-3-aminopropyltriethoxysilane to introduce nitrogen and silicon elements into wheat straw alkali lignin, and CLig obtained was associated with APP as an expanding flame-retardant. Data analysis shows that PLA-6 (containing 18.4% APP and 4.6% CLig) has a higher LOI (30.5% and 25%, respectively) compared to PLA-8 (containing 18.4% APP and 4.6% Lig), with a remaining mass of 49.6%, much higher than the latter (19.5%). The cone calorimeter showed that APP and CLig reduced total heat release and the rate of peak heat release of PLA. It showed that CLig and APP enhanced the PLA thermal stability at high temperatures, and the flame-retardant effect of this expansion type flame-retardant was significant.

It showed that the introduction of organosilicon is used to generate a dense and uniform carbon layer, which can isolate the material surface from contact with flames and effectively delay the spread of fire [144]. Dai et al. [145] used the sol-gel method to prepare silicon modified lignin (SiO_2@Al) as the source of carbon in intumescent flame-retardant, and prepared intumescent flame-retardant composites of polypropylene. It indicates that the 4% SiO_2@Al polypropylene composite material can reached V-0 level through the UL-94 flame-retardant test. The addition of SiO_2@Al formed a more stable honeycomb like carbon layer in polypropylene, achieving flame-retardant effect in condensed phase flame-retardant.

The use of silane coupling agents can enhance the compatibility between polymer matrix and lignin, and the dispersibility of flame-retardants can be improved solely through organic silicon modification. The modified lignin has superior thermal stability and is suitable for high-temperature processing. But its disadvantage is that the cost of organic silicon sources is superior, and the improvement in flame-retardant efficiency is limited.

Table 2 the function of different modified methods for lignin

modification method of ligin	Flame-retardant mechanism
Nitrogen modification	After nitrogen modified lignin is thermally decomposed, non combustible gases such as water vapor, ammonia, and nitrogen are produced, which can dilute the concentration of oxygen and combustible gases and exert gas-phase flame-retardant effects.
phosphorus modification	Generate P-O-C bonds with high heat resistance and chemical stability. Under heating conditions, it will decompose to produce pyrophosphate or polyphosphoric acid, and release PO · and PO_2 · free radicals. Acidic products and free radicals can promote the dehydration and carbonization reaction of hydroxyl groups in lignin, thereby forming a dense oxygen and thermal insulation carbon layer, effectively protecting the substrate from further combustion.
Collaborative modification of P and N	Due to the synergistic effect of phosphorus and nitrogen elements in flame retardancy, they exhibit higher thermal stability and superior flame-retardant efficacy
metal modification	Introducing metal ions such as aluminum, copper, zinc, cobalt, etc. can catalyze the carbonization of lignin and improve the carbonization efficiency of combustibles. Ions absorb heat at high temperatures and react with oxygen to form a carbon layer that isolates flammable gases, exerting flame-retardant and smoke suppression effects.
Silicone Material	Organic silicon combustion generates SiO_2 with good thermal stability, which effectively isolates the contact between combustible surfaces and flames, delays the spread of fire, and improves flame retardancy.

4.5 Nano modified lignin

Lignin nanoparticles, since their large specific surface area and nanoscale size, can construct effective thermal barriers at high temperatures, thereby effectively slowing down flame propagation and heat conduction rates. By using chemical modification methods, grafting other elements onto lignin nanoparticles can further improve their flame-retardant properties. This composite material can precisely control the material flame-retardant properties from the micro and nano levels. Compared to traditional materials, nanomaterials exhibit significant optimization

Materials Research Forum LLC
https://doi.org/10.21741/9781644904077-6

in thermodynamic properties [146]. Using nanoscale lignin as raw material, as the number of oxide nanoparticles formed during combustion increases, the flammability and thermal stability of the polymer are improved. Based on this, Chollet et al. [147] used the dissolution precipitation method to prepare the lignin nanoparticles (LNP) with partially sulfated lignin particles (LMP), and functionalized them with diethyl 2- (triethoxysilyl) ethyl) phosphonic acid (SiP) and diethyl chlorophosphate. Subsequently, their flame-retardant effects were further investigated in the PLA system. At 700 ℃, LNP treated with SiP has a higher content of residual carbon that that of LMP treated with SiP. Moreover, the use of nanoscale lignin can reduce its dosage). Even if the content of LNP SiP is as low as 5%, it can keep the flame-retardant effect of 20% lignin dosage without reducing the rate of peak heat release.

Song et al. [148] successfully produced a lignin modified graphene/carbon nanotube hybrid coating with high flame-retardant using the Mayer rod method. The research results clearly demonstrate that the excellent flame-retardant of the coating is attributed to the synergistic effect between carbon nanotubes/graphene and lignin. Guo et.al [149] successfully synthesized nano silica using tetraethyl orthosilicate and lignin as raw materials, and introduced nano silica into phenolic resin through condensation reaction and intermolecular forces. The research revealed that when the mass ratio of nano silica in lignin phenolic foam was 0.5% (LPF -0.5), its heat release rate was 19.14 kW/m^2, effective combustion heat was 8.14 kJ/kg, and total heat release was 4.2 kW/m^2. The flame-retardant was better than other modified foams with nano silica content. Chollet et al. [150] prepared lignin nanoparticles through a dissolution precipitation process and functionalized them. As a result, it was found that functionalized lignin nanoparticles have higher residual carbon content, and the use of nano lignin not only reduces the amount of additives, but also maintains efficient flame-retardant effect throughout the condensed phase flame-retardant process. Liu et al. [151] elaborated on the preparation methods of lignin functional nanoparticles, mainly including self-assembly, mechanical method, polymerization assembly method, and freeze-drying carbonization method, providing valuable references for the related research of nano modified lignin. Meng et al. [152] gradually grafted phenyl phosphorus oxychloride and 1, 4-dimethoxyacetylene onto lignin, successfully synthesizing multifunctional lignin derivative nanoparticles (CP Lignin), and applied them to the preparation of thermoplastic polyurethane composites.

Compared with the control thermoplastic polyurethane, the smoke generation and the rate of peak heat release of the sample reduced by 50.0% and 53.8%, respectively. The reason behind this is the case that the degradation temperature of thermoplastic polyurethane is increased, while lignin or CP Lignin slows down the degradation process. CP Lignin is favorable to forming a uniformly distributed carbon layer, effectively enhancing the flame-retardant effect of thermoplastic polyurethane (TPU) in the condensed phase flame-retardant mechanism. Li et al. [153] prepared lignin nanoparticles using lignin as the starting material, mixed them with polyvinyl alcohol solution and chitosan solution, and successfully prepared lignin nano-particles/polyvinyl alcohol/chitosan ternary composite material through grafting modification. The ultimate LOI of the composite material reaches 27%, reaching the flame-retardant level.

From the above series of research works, it can be clearly seen that the use of nanotechnology (such as layered clay, carbon nanotubes) combined with lignin to form a nanocomposite lignin flame-retardant can generate a dense carbon layer, and its barrier effect is extremely significant. This kind of flame-retardant has a low addition amount (usually<5%) and has a little effect on the

Materials Research Foundations 190 (2026) 153-193 https://doi.org/10.21741/9781644904077-6

mechanical properties of the material. However, it has drawbacks such as difficulty in dispersing nanomaterials, difficulty in implementing industrial production, and high costs.

Table 3 Application and advantages and disadvantages of lignin based flame-retardants in flame-retardant plastics

The flame-retardant application of lignin		advantage	disadvantage
Unmodified lignin flame-retardant	Single lignin flame-retardant	LOI value is about 27% -28%	Large addition amount (>30 wt%), low flame-retardant/carbonization efficiency, high smoke production, poor durability
	Lignin physical synergistic flame-retardant	LOI value>30%, passed UL-94 The V-0 level	The addition amount is still relatively large (20-30 wt%), and the flame-retardant efficiency is limited by the material/ratio/dosage. The carbonization efficiency is low, the smoke production is high, and the durability is poor
Modified lignin flame-retardant	nitrogen or Phosphorus modified lignin flame-retardant	flame-retardant efficiency is improved (with an addition amount as low as 15-20 wt%), and the heat release is reduced by 20% -40%	Poor compatibility/ dispersibility/ durability
	Nitrogen and phosphorus modified lignin flame-retardant	Nitrogen phosphorus synergistic effect, flame-retardant effect better than single element modification (UL-94 V-0 pass rate)	
	Nitrogen phosphorus modification containing metal ions	Carbonization efficiency increases (residual carbon content>25 wt%), significantly suppressing smoke (smoke density 30% -50%)	
	Silicon modified lignin flame-retardant	Low toxicity and environmental protection, delaying the spread of fire, durability (maintaining 80% effectiveness after 100 washes)	
	Nano composite lignin flame-retardant	Forming a dense carbon layer with significant barrier effect; The addition amount is low (usually<5%) and has little effect on the mechanical properties of the material.	There are disadvantages such as difficulty in dispersing nanomaterials, difficulty in industrial production, and high cost.

Materials Research Foundations 190 (2026) 153-193
https://doi.org/10.21741/9781644904077-6

5. Conclusion and outlook

The application of lignin as a bio-based flame-retardant to prepare high-performance polymer materials has become a hot topic in research in recent years. This chapter briefly analyzes lignin and its structure and properties, and elaborates in detail on the lignin application in the field of flame-retardant. The chemical modification mainly introduces phosphorus nitrogen modification, nitrogen phosphorus modification containing metal ions, organic silicon modification, and nano modification. By delving into the research progress of lignin based flame-retardants, a theoretical basis is given for the advancement of efficient and environmentally friendly flame-retardant materials. Due to the limited flame-retardant properties of lignin, chemical modification of lignin with flame-retardants containing P, N, and Si is utilized to enhance the flame-retardant effect of polymers. Again, combining modified lignin with different metal ions is an effective measure to enhance the flame-retardant effect of modified lignin. Numerous methods can improve the flame-retardant performance of lignin based composite materials. These composite materials still cannot meet the industrial flame-retardant requirements. Therefore, the combination of conventional flame-retardants such as boric acid, ammonium polyphosphate, lignin flame-retardants, ammonium dihydrogen phosphate reduces the heat release rate of polymers while increasing the LOI and the level in UL-94 test.

At present, two challenges for lignin-based flame-retardants will be considered. One challenge is the interfacial and dispersion interactions of lignin in the polymer matrix, which usually cannot be easily mixed physically. In contrast, chemical modification of the lignin with various flame-retardant elements such as P, N and Si can improve its interfacial and dispersion interactions. However, the low addition of these elements often limits the improvement of polymer flame-retardant modification. Again, lignin is commonly scattered in the polymer matrix to form the aggregation, which mainly restricts the flame-retardant effect. The use of metal ions with excellent synergistic flame-retardant effect can further reduce the amount of lignin added while enhancing the flame-retardant performance of composite materials-based lignin. Another challenge is the case that the application of high levels of lignin derived flame-retardants often affects final polymer's mechanical properties. Efficient nanomaterials such as graphene, carbon nanotubes, and clay can be introduced into lignin composite materials to enhance the flame-retardant of the polymer while ensuring its mechanical properties. The future research directions are given below.

(1) Mechanism research and flame-retardant molecular design. By using in-situ TGA-FTIR/MS combined technology to track the staged pyrolysis gas of lignin flame-retardants, combined with Raman and XPS spectroscopy to examine the chemical structure evolution path of the carbon layer. Using molecular dynamics simulations to reveal the dynamic effects of silicon/phosphorus modification in the formation of carbon layer cross-linked networks (C-O-P/Si-O-Si), elucidating the mechanism of metal ion (Fe^{3+}/Zn^{2+}) catalyzed graphitization microcrystals orientation, regulating the morphology of the carbon layer, and revealing the thermal decomposition behavior and evolution mechanism of lignin flame-retardants. Exploring the structure-activity relationship between porosity, density, and flame-retardant performance (residual carbon strength, thermal insulation efficiency), and establish a dynamic model of "decomposition carbonization protection" to guide the optimization of flame-retardants. Using new methods such as machine learning to predict the relationship between the structure, active and performance of flame-retardants, accelerating the molecular design of new lignin flame-retardants. Construct a quantitative structure-activity model by integrating molecular structure (hydroxyl, substituent), thermal

Materials Research Foundations 190 (2026) 153-193 https://doi.org/10.21741/9781644904077-6

decomposition, and flame-retardant performance data. Screen phosphorus/nitrogen modified sites and predict performance such as LOI value and residual carbon rate. Through the closed-loop iteration of prediction synthesis testing, optimize the balance between flame-retardant efficiency and environmental friendliness, and decrease the risk of metal toxicity.

(2) Multi-functional composite system development. Explore the composite of lignin with new nanomaterials such as bio-char and MXene, and design a lignin based multi-component composite flame-retardant system. Constructing gas-solid dual phase flame-retardant function through phosphorus nitrogen synergistic modification, coupled with nano montmorillonite/ carbon nanotubes to construct a three-dimensional reinforced carbon layer structure. Studying the synergistic mechanism of chemical grafting (such as cyclic phosphate esters) and physical coating (bio-based plasticizers) on flame-retardant, mechanical properties and smoke suppression. Developing dynamic cross-linking technology to regulate the compatibility between flame-retardants and polymer interfaces, combined with micro combustion calorimetry, smoke density testing, and thermogravimetric infrared spectroscopy, to reveal the structure-activity relationship between multi-scale structures (microcrystals/pores/interfaces) and flame-retardant efficiency and smoke suppression. Exploring halogen-free minimal toxicity biodegradable composite formulas to construct a lignin composite flame-retardants with flame-retardant, smoke suppression, and antibacterial functions. Adapting to the application needs multiple scenarios such as plastics, rubber, coatings, etc. Therefore, developing lignin-based flame-retardants with multiple functions, and high added value is the future research direction.

(3) Green and efficient modification technology. The green and sustainable development of lignin flame-retardants is one of the keys focus for the future. Exploring more lignin sourced from renewable resources in terms of raw materials, continuously innovate technologies in terms of technology, reduce pollution from the source of preparation processes, develop environmentally friendly synthesis methods, reduce dependence on limited resources, and promote the sustainable development of the flame-retardant material industry. Focusing on the green and efficient modification technology of lignin, we will develop a solvent-free modification process that combines mechanochemical methods (such as ball milling shear induced hydroxyl activation) with bio-catalysis (laccase/peroxidase driven esterification crosslinking). Through force enzyme coupling, we will trigger the directed grafting of lignin aromatic rings with phosphorus/nitrogen reagents, replacing halogenated flame-retardant systems such as DOPO. Utilizing natural hydroxyl and methoxy groups of lignin to construct dynamic covalent bonds (Schiff base/disulfide bonds), designing a thermally triggered self-crosslinking flame-retardant network to simultaneously enhance the thermal stability and density of the carbon layer. Combining in-situ infrared and synchrotron radiation characterization to analyze the synergistic mechanism of mechanochemistry enzyme catalysis, integrating ultrasound assisted reaction enhancement and solvent circulation system, and creating a low-energy and low emission process. Evaluate the mechanical properties and flame-retardant efficiency of modified products in polylactic acid and epoxy resin, form a "natural functional group utilization green reaction enhancement self-assembly function regulation" technology chain, and promotes the large-scale green preparation of lignin flame-retardants.

(4) Optimization of lignin flame-retardant process and large-scale production. Study the thermal stability of lignin flame-retardants in melt processing and develop technical parameters suitable for industrial production such as injection molding and extrusion. Exploring the synergistic flame-retardant mechanism between lignin and bio-based polymers such as PLA and PHA, and promote

the application of all bio-based materials. Focusing on key technologies for continuous production of lignin flame-retardants, optimize modification process parameters (such as pH, temperature, reaction time), and promoting efficient solid-liquid separation and solvent recovery systems. By enhancing mass and heat transfer through micro-reactors, regulating the uniformity of substituent distribution, and combining online monitoring (such as near-infrared spectroscopy) to achieve precise process control. Exploring the melt blending compatibility of lignin and polyolefin substrates, and design a parameter window for twin-screw extrusion granulation process. Assessing the energy consumption, three waste emissions, and economy in large-scale production, integrate water-based reaction systems and nano-self-assembly technology, construct green and low-carbon process flows, and promote industrial applications. Establishing performance evaluation criteria (such as LOI, UL-94, and cone calorimeter) for the degradation and toxicity (EC50 value) of lignin flame-retardants. Developing industry standards and promoting the standardization of the industry.

Acknowledgement

This work is partially supported by Yunnan Fundamental Research Projects (No. 202501AS070127) and supported by the National Natural Science Foundation of China (No.52364042).

References

[1] Y. Liu, Y.Y. Liu, Y.J. Liu, X.M. Zhao, Research progress of lignin-based flame-retardants, Adv. Text. Technol. 32 (2024) 28-40. In Chinese

[2] M.C. Zhou, W. Liu, Carbon emission reduction and utilization of biomass resources, Chem. Eng. Des. 32 (2022) 11-14.

[3] Z. Li, N. Zhang, L.H. Hu, Y.H. Zhou, Research progress of lignin-based flame-retardants, J. Cellulose Sci. Technol. 29 (2021) 59-68. In Chinese.

[4] B.M. Upton, A.M. Kasko, Strategies for the conversion of lignin to high-value polymeric materials: Review and perspective, Chem. Rev. 116 (2016) 2275-2306. https://doi.org/10.1021/acs.chemrev.5b00345

[5] Y. Zhou, X. Zhang, L. Zhang, The structure of lignin and its modification status, Mod. Chem. Ind. 30 (2010) 63-66. In Chinese.

[6] Z. Li, Research on renewable biomass resource: Lignin, J. Nanjing For. Univ. (Nat. Sci. Ed.) 36 (2012) 1-7. In Chinese.

[7] S. Laurichesse, L. Averous, Chemical modification of lignins: Towards biobased polymers, Prog. Polym. Sci. 39 (2014) 1266-1290. https://doi.org/10.1016/j.progpolymsci.2013.11.004

[8] L. Costes, F. Laoutid, S. Brohez, P. Dubois, Bio-based flame-retardants: When nature meets fire protection, Mater. Sci. Eng. R Rep. 117 (2017) 1-25. https://doi.org/10.1016/j.mser.2017.04.001

[9] C. Rodríguez Correa, M. Stollovsky, T. Hehr, Y. Rauscher, B. Rolli, A. Kruse, Influence of the carbonization process on activated carbon properties from lignin and lignin-rich

biomasses, ACS Sustain. Chem. Eng. 5 (2017) 8222-8233. https://doi.org/10.1021/acssuschemeng.7b01895

[10] K.A.Y. Koivu, H. Sadeghifar, P.A. Nousiainen, D.S. Argyropoulos, J. Sipilä, Effect of fatty acid esterification on the thermal properties of softwood kraft lignin, ACS Sustain. Chem. Eng. 4 (2016) 5238-5247. https://doi.org/10.1021/acssuschemeng.6b01048

[11] Y. Zhou, Y. Qiu, C. Li, L. Jin, Present situation and development of flame-retardant fabrics, Silk Monthly 43 (2006) 70-73. In Chinese.

[12] D. Li, Q. Lü, H. Zhang, C. Li, L. Gai, Analysis on combustion performance of textile and apparel products and safety risk of banned/limited flame-retardant conten, Adv. Text. Technol. 22 (2014) 45-48, 52. In Chinese.

[13] Z. Zheng, X. Zhong, On the study of green environment-friendly flame-retardants, J. Zhejiang Fashion Inst. Technol. 6 (2007) 10-14. In Chinese.

[14] X. Liu, W. Ji, X. Wang, Q. Zhang, S. Zhong, Study on properties of polyester fabrics with dpap flame-retardant finishing, Adv. Text. Technol. 29 (2021) 89-94. In Chinese.

[15] V.K. Thakur, M.K. Thakur, P. Raghavan, M.R. Kessler, Progress in green polymer composites from lignin for multifunctional applications: A review, ACS Sustain. Chem. Eng. 2 (2014) 1072-1092. https://doi.org/10.1021/sc500087z

[16] K Klapiszewski, K. Bula, M. Sobczak, T. Jesionowski, Influence of processing conditions on the thermal stability and mechanical properties of PP/silica-lignin composites, Int. J. Polym. Sci. 2016 (2016) 1-9. https://doi.org/10.1155/2016/1627258

[17] J. Hu, D. Shen, S. Wu, H. Zhang, R. Xiao, Effect of temperature on structure evolution in char from hydrothermal degradation of lignin, J. Anal. Appl. Pyrolysis 106 (2014) 118-124. https://doi.org/10.1016/j.jaap.2014.01.008

[18] L. Dai, C.L. Si, Lignin and its functional materials, Science Press, Beijing, 2019. In Chinese.

[19] X.Q. Qiu, X.P. Ouyang, D.J. Yang, et al., Modification Theory and Technology for Efficient Utilization of Industrial Lignin, Science Press, Beijing, 2014. In Chinese.

[20] M.H. Liu, Y.F. Liu, Preparation and Application of Lignin Modified Materials, Chemical Industry Press, Beijing, 2020. In Chinese.

[21] J. Huang, S.Y. Fu, Lignin Chemistry and Modified Materials, Chemical Industry Press, Beijing, 2014. In Chinese.

[22] T.D. Jiang, Lignin (2nd ed.), Chemical Industry Press, Beijing, 2009. In Chinese.

[23] R. Zhong, Z.H. Ye, Transcriptional regulation of lignin biosynthesis in plants, Plant signaling & behavior, 4 (2009) 1028-1034 https://doi.org/10.4161/psb.4.11.9875

PMid:19838072 PMCid:PMC2819510

[24] N.N. Solihat, A..F Hidayat, M.N.A.M. Taib, M.H. Hussin, S.H. Lee, M.A.A. Ghani, S.S.A.O.A. Edrus, Recent developments in flame-retardant lignin-based biocomposite: manufacturing, and characterization. Journal of Polymers and the Environment 30 (2022) 4517-4537. https://doi.org/10.1007/s10924-022-02494-2

[25] I. Sapouna, M. Lawoko, Deciphering lignin heterogeneity in ball-milled softwood: Unravelling the synergy between the supramolecular cell wall structure and molecular events, Green Chemistry 23 (2021) 3348-3364.

[26] M.F. Devaux, F. Jamme, W. André, B. Bouchet, C. Alvarado, S. Durand, P. Robert, L. Saulnier,Synchrotron time-lapse imaging of lignocellulosic biomass hydrolysis: tracking enzyme localization by protein autofluorescence and biochemical modification of cell walls by microfluidic infrared microspectroscopy. Frontiers in Plant Science 9 (2018) 200.

[27] G. Gallina, E. Bravin, C. Badalucco, G. Audisio, M. Armanini, A. De Chirico, F. Provasoli, Application of cone calorimeter for the assessment of class of flame-retardants for polypropylene, Fire Mater. 22 (1998) 15-18.

[28] F. Chen, H. Dai, X. Dong, J. Yang, M. Zhong, Physical properties of lignin-based polypropylene blends, Polym. Compos. 32 (2011) 1019-1025. https://doi.org/10.1002/pc.21087

[29] P. Song, Z. Cao, S. Fu, Z. Fang, Q. Wu, J. Ye, Thermal degradation and flame retardancy properties of ABS/lignin: Effects of li.gnin content and reactive compatibilization, Thermochim. Acta 518 (2011) 59-65. https://doi.org/10.1016/j.tca.2011.02.007

[30] K.A. Iyer, J.M. Torkelson, Sustainable green hybrids of polyolefins and lignin yield major improvements in mechanical properties when prepared via solid-state shear pulverization, ACS Sustain. Chem. Eng. 3 (2015) 959-968. https://doi.org/10.1021/acssuschemeng.5b00099

[31] A. De Chirico, M. Armanini, P. Chini, et al., Flame-retardants for polypropylene based on lignin, Polym. Degrad. Stab. 79 (2003) 139-145. https://doi.org/10.1016/S0141-3910(02)00266-5

[32] J. Zhang, E. Fleury, Y. Chen, M.A. Brook Flame-retardant lignin-based silicone composites, RSC Adv. 5 (2015) 103907-103914. https://doi.org/10.1039/C5RA24093J

[33] M.J. Uddin, P.K. Alaboina, L. Zhang, S.J. Cho, A low-cost, environment-friendly lignin-polyvinyl alcohol nanofiber separator using a water-based method for safer and faster lithium-ion batteries, Mater. Sci. Eng. B 223 (2017) 84-90. https://doi.org/10.1016/j.mseb.2017.05.004

[34] N.K. Mandlekar, A. Cayla, F. Rault, S. Giraud, F. Salaun, G. Malucelli, J. Guan, Thermal stability and fire retardant properties of polyamide microcomposites containing different lignins, Ind. Eng. Chem. Res. 56 (2017) 13704-13714. https://doi.org/10.1021/acs.iecr.7b03085

[35] Y. Wang, J. Yue, R. Xie, C. Liu, L. Gan, J. Huang, High-value use of lignocellulosic-rich eucommia residue for promoting mechanical properties and flame retardancy of poly(butylene succinate), J. Appl. Polym. Sci. 137 (2020) 48543. https://doi.org/10.1002/app.48543

[36] W. Lu, Q. Li, Y. Zhang, S. Hirose, H. Hatakeyama, Z. Jin, Lignosulfonate/APP IFR and its flame retardancy in lignosulfonate-based rigid polyurethane foams, J. Wood Sci. 64 (2018) 287-293. https://doi.org/10.1007/s10086-018-1701-4

[37] Y. Zhang, Q. Zhao, L. Li, R. Yan, J. Zhang, J.C. Duan, B.J. Liu, Z.Y. Sun, M.Y. Zhang, W. Hu, Synthesis of a lignin-based phosphorus-containing flame-retardant and its application in polyurethane, RSC Adv. 8 (2018) 32252-32261. https://doi.org/10.1039/C8RA05598J

PMid:35547477 PMCid:PMC9086252

[38] N. Mandlekar, A. Cayla, F. Rault, S. Giraud, F. Salaün, J. Guan, Valorization of industrial lignin as biobased carbon source in fire retardant system for polyamide 11 blends, Polymers 11 (2019) 180. https://doi.org/10.3390/polym11010180

PMid:30960166 PMCid:PMC6401932

[39] D. Zhang, J. Zeng, W. Liu, X. Qiu, Y. Qian, H. Zhang, Y. Yang, M. Liu, D. Yang, Pristine lignin as a flame-retardant in flexible PU foam, Green Chem. 23 (2021) 5972-5980. https://doi.org/10.1039/D1GC01109J

[40] Y. Liu, B. Liu, Z. Sun, H Zhang, Y. Men, W. Hu, Bioinspired mono-component lignin endowing epoxy resin with simultaneously improving flame retardancy and mechanical properties, Compos. Commun. 35 (2022) 101306. https://doi.org/10.1016/j.coco.2022.101306

[41] Q. Cen, S. Chen, D. Yang, D. Zheng, X. Qiu, Full bio-based aerogel incorporating lignin for excellent flame retardancy, mechanical resistance, and thermal insulation, ACS Sustain. Chem. Eng., 11(2023) 4473-4484. https://doi.org/10.1021/acssuschemeng.2c07652

[42] R. Shvartzman-Cohen, Y. Levi-Kalisman, E. Nativ-Roth, R Yerushalmi-Rozen, Generic approach for dispersing single-walled carbon nanotubes: The strength of a weak interaction, Langmuir 20 (2004) 6085-6088. https://doi.org/10.1021/la049344j

[43] L. Gao, Y. Liu, Dispersion and surface modification of carbon nanotubes, Bull. Chin. Ceram. Soc. 24 (2005) 114-119.

[44] S.M. Goodman, N. Ferguson, A.B. Dichiara, Lignin-assisted double acoustic irradiation for concentrated aqueous dispersions of carbon nanotubes, RSC Adv. 7 (2017) 5488-5496. https://doi.org/10.1039/C6RA25986C

[45] A.B. Dichiara, T.J. Sherwood, J. Benton-Smith, J.C. Wilson, S.J. Weinstein, R.E. Rogers, Free-standing carbon nanotube/graphene hybrid papers as next generation adsorbents, Nanoscale, 6 (2014) 6322-6327. https://doi.org/10.1039/c4nr01028k

[46] W. Li, A. Dichiara, J. Bai, Carbon nanotube-graphene nanoplatelet hybrids as high-performance multifunctional reinforcements in epoxy composites, Compos. Sci. Technol. 74 (2013) 221-227. https://doi.org/10.1016/j.compscitech.2012.11.015

[47] K. Song, I. Ganguly, I. Eastin, A. Dichiara, Lignin-modified carbon nanotube/graphene hybrid coating as efficient flame-retardant, Int. J. Mol. Sci. 18 (2017) 2368. https://doi.org/10.3390/ijms18112368

[48] K. Song, I. Ganguly, I. Eastin, A. Dichiara, High temperature and fire behavior of hydrothermally modified wood impregnated with carbon nanomaterials, J. Hazard. Mater. 384 (2019) 121283. https://doi.org/10.1016/j.jhazmat.2019.121283

[49] N. Lavoine, I. Desloges, A. Dufresne, J. Bras, Microfibrillated cellulose-Its barrier properties and applications in cellulosic materials: A review, Carbohydr. Polym. 90 (2012) 735-764. https://doi.org/10.1016/j.carbpol.2012.05.026

[50] B. Wicklein, A. Kocjan, G. Salazar-Alvarez, F. Carosio, G. Camino, M. Antonietti, L. Bergström, Thermally insulating and fire-retardant lightweight anisotropic foams based on

 Materials Research Forum LLC
https://doi.org/10.21741/9781644904077-6

nanocellulose and graphene oxide, Nat. Nanotechnol. 10 (2015) 277-283. https://doi.org/10.1038/nnano.2014.248

[51]G. Zu, J. Shen, L. Zou, F. Wang, X. Wang, Y. Zhang, X. Yao, Nanocellulose-derived highly porous carbon aerogels for supercapacitors, Carbon 99 (2016) 203-211. https://doi.org/10.1016/j.carbon.2015.11.079

[52] X. Wang, S. Ji, X. Wang, H.Y. Bian, L.R. Lin, H.Q. Dai, H. Xiao, Thermally conductive, super flexible and flame-retardant BN-OH/PVA composite film reinforced by lignin nanoparticles, J. Mater. Chem. C 7 (2019) 14159-14169. https://doi.org/10.1039/C9TC04961D

[53] C. Tong, S. Zhang, T. Zhong, Z. Fang, H. Liu, Highly fibrillated and intrinsically flame-retardant nanofibrillated cellulose for transparent mineral filler-free fire-protective coatings, Chem. Eng. J., 419 (2021) 129440. https://doi.org/10.1016/j.cej.2021.129440

[54] C. Zhu, H. Zhang, R. Xiao, Y. Chen, D. Liu, F. Du, H. Ying, Research progress in catalytic valorization of lignocellulose, Sci. Sin. Chim. 45 (2015) 454-478. https://doi.org/10.1360/N032014-00280

[55] B. Liu, Application of lignin-based flame-retardant in composite materials, Changchun Univ. Technol., Changchun, 2021, pp. 13-23. In Chinese

[56] Y. Yang, X. Zhang, G. Hu, Research status of halogen-free flame-retardants, Shanxi Chem. Ind. 30 (1) (2010) 50-53. In Chinese

[57] X. Zhang, Y. Huang, G. You, H. He, Progresses in Research of Halogen-free Flame-retardants, Fine Chem. Intermed. 41 (2011) 1-8.

[58] Y. Bei, S. Weng, P. Jia, C. Bo, L. Hu, Y. Zhou, Research progress in lignin-based flame-retardants, Biomass Chem. Eng. 57 (2023) 71-78.

[59] A. Cayla, F. Rault, S. Giraud, F. Salaün, V. Fierro, A. Celzard, PLA with intumescent system containing lignin and ammonium polyphosphate for flame-retardant textile, Polymers 8 (2016) 331. https://doi.org/10.3390/polym8090331

[60] A. De Chirico, M. Armanini, P. Chini, G. Cioccolo, F. Provasoli, G. Audisio, Flame-retardants for polypropylene based on lignin, Polym. Degrad. Stab. 79 (2003) 139-145. https://doi.org/10.1016/S0141-3910(02)00266-5

[61] L. Verdolotti, M. Oliviero, M. Lavorgna, S. Iannace, G. Camino, P. Vollaro, A. Frache, On revealing the effect of alkaline lignin and ammonium polyphosphate additives on fire retardant properties of sustainable zein-based composites, Polym. Degrad. Stab. 134 (2016) 115-125. https://doi.org/10.1016/j.polymdegradstab.2016.10.001

[62] X. Cheng, J. Guan, R. Tang, K.Q. Liu, Phytic acid as a bio-based phosphorus flame-retardant for poly(lactic acid) nonwoven fabric, J. Clean. Prod. 124 (2016) 114-119. https://doi.org/10.1016/j.jclepro.2016.02.113

[63] C. Réti, M. Casetta, S. Duquesne, S. Bourbigot, R. Delobel, Flammability properties of intumescent PLA including starch and lignin, Polym. Adv. Technol. 19 (2008) 628-635. https://doi.org/10.1002/pat.1130

[64] N.K. Mandlekar, G. Malucelli, A. Cayla, F. Rault, S. Giraud, F. Salaün, J. Guan, Fire retardant action of zinc phosphinate and polyamide 11 blend containing lignin as a carbon source, Polym. Degrad. Stab., 153 (2018) 63-74. https://doi.org/10.1016/j.polymdegradstab.2018.04.019

[65] A. Cayla, F. Rault, S. Giraud, F. Salaün, R. Sonnier, Influence of ammonium polyphosphate/lignin ratio on thermal and fire behavior of biobased thermoplastic: The case of polyamide 11, Materials 12 (2019) 1146. https://doi.org/10.3390/ma12071146

[66] J. Wang, Y. Bian, Y. Yi, Y. Ma, Z. Ba, Z. Liu, Study on preparation and properties of puf intumescent flame retarded by alkali lignin/app, Plast. Sci. Technol. 47 (2019) 42-45.

[67] Y. Jin, R. Zhong, L. Xiong, W. Chen, Q. Lv, Synthesis of a pyrazine-modified lignin/magnesium hydroxide double-coated red phosphorus flame-retardant and its application in ABS, CN Patent 108912401B, 2020.

[68] V. Carretier, J. Delcroix, M. Pucci, P. Rublon. J.M. Lopez-Cuesta, Influence of sepiolite and lignin as potential synergists on flame-retardant systems in polylactide (PLA) and polyurethane elastomer (PUE), Materials 13 (2020) 2450.

[69] P. Widsten, T. Tamminen, A. Paajanen, T. Hakkarainen, T. Liitiä, Modified and unmodified technical lignins as flame-retardants for polypropylene, Holzforschung 75 (2021) 584-590. https://doi.org/10.1515/hf-2020-0147

[70] D.X. Liang, X.J. Zhu, P. Dai, X. Lu, H.Guo, H. Que, D. Wang, T. He, C. Xu, H.M. Robin, Z. Luo, X. Gu, Preparation of a novel lignin-based flame-retardant for epoxy resin, Mater. Chem. Phys. 259 (2021) 124101. https://doi.org/10.1016/j.matchemphys.2020.124101

[71] W.J. Yan, S. Xu, X.Y. Tian, J.J. Min, S.C. Liu, C.J. Ding, N.L. Wang, Y. Hu, Q.X. Fan, J.S. Li, H.Y. Zeng, Novel bio-based lignosulfonate and Ni(OH)2 nanosheets dual modified layered double hydroxide as an eco-friendly flame-retardant for polypropylene, Colloids Surf. A Physicochem. Eng. Asp. 655 (2022) 130195. https://doi.org/10.1016/j.colsurfa.2022.130195

[72] J.Y. Xue, K.Y. Xu, F.M. Meng, H. Lei, Phosphorus and Silicon Synergistically Improve the Flame-retardant Performance of Lignin-Based Polyurethane Materials, Plastics 51 (2022) 19-23.

[73] C.H. Li, B.K. Wang, L.J. Zhou, X Hou, S. Su, Effects of lignin-based flame-retardants on flame-retardancy and insulation performances of epoxy resin composites, Iran. Polym. J. 31 (2022) 949-962. https://doi.org/10.1007/s13726-022-01052-w

[74] X. Lu, A.F. Lee, X.L. Gu, Improving the flame retardancy of sustainable lignin-based epoxy resins using phosphorus/nitrogen treated cobalt metal-organic frameworks, Mater. Today Chem. 26 (2022) 101184. https://doi.org/10.1016/j.mtchem.2022.101184

[75] M.Q. Shi, X.W. Wang, J. Yang, Development of lignin-based waterborne polyurethane materials for flame-retardant leather application, Polym. Bull. 80 (2023) 5553-5571. https://doi.org/10.1007/s00289-022-04320-0

[76] X. Shang, X. Bi, H. Tan, Y. Zhang, Effect of lignin compounded with pyrophosphoric acid piperazine intumescent flame-retardant on flame-retardant properties of epoxy resin, J. Northeast For. Univ., 51 (6) (2023) 140-145. In Chinese.

[77] K. Komisarz, T.M. Majka, K. Pielichowski, Chemical and physical modification of lignin for green polymeric composite materials, Mater. 16 (2022) 16. https://doi.org/10.3390/ma16010016

[78] N.N. Solihat, A.F. Hidayat, M.N.A.M. Taib, M.H. Hussin, S.H. Lee, M.A.A. Ghani, S.S.A.O.A. Edrus, Recent developments in flame-retardant lignin-based biocomposite: Manufacturing, and characterization, J. Polym. Environ. 30 (2022) 4517-4537. https://doi.org/10.1007/s10924-022-02494-2

[79] L.J. Ding, H.T. Xu, L. Wu, Y. MaiMaiTi, Preparation of modified lignin and its effect on flame-retardant and antioxidant properties of polypropylene, Polym. Mater. Sci. Eng. 38 (2022) 66-71. In Chinese.

[80] W.Y. Xing, H.X. Yuan, P. Zhang, H. Yang, L. Song, Y. Hu, Functionalized lignin for halogen-free flame-retardant rigid polyurethane foam: Preparation, thermal stability, fire performance and mechanical properties, J. Polym. Res. 20 (2013) 234-243. https://doi.org/10.1007/s10965-013-0234-1

[81] Z. Li, S.X. Weng, F. Song, X. Ren, X. YAng, L. Hu, Synthesis and thermal analysis of lignin-based flame-retardant containing nitrogen and phosphorus, Chem. Ind. For. Prod. 41 (2021) 63-70. In Chinese

[82] X. Yang, Y. Song, K. Lin, J. Li, Preparation of lignin-based char-forming agent and its application in intumescent flame-retardant polypropylene, Polym. Mater. Sci. Eng. 38 (2022) 39-46. In Chinese.

[83] C. Lu, J. Wang, L. Chen, et al., The effect of adjuvant on the halogen-free intumescent flame-retardant ABS/PA6/SMA/APP blend, J. Appl. Polym. Sci., 118 (2010) 1552-1560. https://doi.org/10.1002/app.32527

[84] H. Ding, J. Wang, C. Wang, F. Chu, Synthesis of a novel phosphorus and nitrogen-containing bio-based polyols and its application in flame-retardant polyurethane sealant, Polym. Degrad. Stab. 124 (2016) 43-50. https://doi.org/10.1016/j.polymdegradstab.2015.12.006

[85] L. Yang, X. Wang, Y. Cui, Y. Tian, H. Chen, Z. Wang, Modification of renewable resources-lignin by three chemical methods and its applications to polyurethane foams, Polym. Adv. Technol. 25 (2014) 1089-1098. https://doi.org/10.1002/pat.3356

[86] R.L. Fu, X.S. Cheng, Synthesis and flame-retardant properties of melamine modified EH lignin, Adv. Mater. Res. 236-238 (2011) 482-485. https://doi.org/10.4028/www.scientific.net/AMR.236-238.482

[87] R. Zhang, X. Xiao, Q. Tai, H. Huang, Y. Hu, Modification of lignin and its application as char agent in intumescent flame-retardant poly(lactic acid), Polym. Eng. Sci. 52 (2012) 2620-2626. https://doi.org/10.1002/pen.23214

[88] S. Zhou, R. Tao, P. Dai, Z. Luo, M. He, Two-step fabrication of lignin-based flame-retardant for enhancing the thermal and fire retardancy properties of epoxy resin composites, Polym. Compos. 41 (2020) 2025-2035. https://doi.org/10.1002/pc.25517

[89] Y. Yu, X. Pu, B. Zheng, S. Zhang, Preparation and properties of phosphorus flame-retardant polyester, J. Zhejiang Sci-Tech Univ. (Nat. Sci. Ed.), 45 (2021) 205-211. In Chinese.

[90] Y. Wang, J. Zhou, X. Feng, H. Zhang, Preparation of new P-N synergistic flame-retardant and its application in PA66, J. Zhejiang Sci-Tech Univ. (Nat. Sci. Ed.) 47 (2022) 806-813. In Chinese.

[91] Y. Wang, Study on flame-retardant modification and properties of PA66 with P/N synergistic flame-retardant, Zhejiang Sci-Tech Univ., Hangzhou, 2022, pp. 1-60. In Chinese.

[92] S. Wang, S.Q. Ma, C.X. Xu, Y. Liu, J. Dai, Z. Wang, X. Liu, J. Chen, X. Shen, J. Wei, J. Zhu, Vanillin-derived high-performance flame-retardant epoxy resins: Facile synthesis and properties, Macromolecules 50 (2017) 1892-1901. https://doi.org/10.1021/acs.macromol.7b00097

[93] Y. Wei, S.Y. Zhu, Q.W. Qian, Q. Jiang, L. Zhang, K. Jin, W. Liu, Y. Qiu, Hexachlorocyclotriphosphazene functionalized lignin as a sustainable and effective flame-retardant for epoxy resins, Ind. Crops Prod. 187 (2022) 115543. https://doi.org/10.1016/j.indcrop.2022.115543

[94] S.X. Weng, Z. Li, C.Y. Bo, F. Song, Y. Xu, L. Hu, Y. Zhou, Design lignin doped with nitrogen and phosphorus for flame-retardant phenolic foam materials, React. Funct. Polym. 185 (2023) 105535. https://doi.org/10.1016/j.reactfunctpolym.2023.105535

[95] J.X. Li, Z.P. Yan, M. Liu, T. Lu, R. Liu, S. Zhao, Q. Lv, B. Li, S. Zhao, H. Wang, Triple silicon, phosphorus, and nitrogen-grafted lignin-based flame-retardant and its vulcanization promotion for styrene butadiene rubber, ACS Omega 8 (2023) 21549-21558. https://doi.org/10.1021/acsomega.3c00714

[96] L. Zhong, Design of reactive flame-retardant based on lignin and its application in flame-retardant phenolic foam, Yanzhou Univ., Yangzhou, 2023, pp. 32-45. In Chinese. https://doi.org/10.1080/09593330.2023.2176792

[97] X. Huo, C. Ding, Y. Wang, Research progress on lignin and its derivatives for flame-retardant modification of polymers, Shandong Chem. Ind., 51 (2022) 72-74. In Chinese.

[98] Q. Huang, Y. Shen, Z. Wang, The study of flame retardancy progress for grafted silk fabric with vinyl phosphorus flame-retardant, Silk Monthly, 50 (2013) 29-34. In Chinese.

[99] W. Hu, Q. Yan, C. Li, J. Shen, L. Wang, M. Sun, Progress of research on application of nitrogen and phosphorus flame-retardants, Chem. Ind. Times 33 (2019) 31-37.

[100] G.L. Bykov, B.G. Ershov, A sorbent based on phosphorylated lignin, Russ. J. Appl. Chem., 83 (2010) 316-319. https://doi.org/10.1134/S1070427210020254

[101]W. Xing, H. Yuan, P. Zhang, H. Yang, L. Song, Y. Hu, Functionalized lignin for halogen-free flame-retardant rigid polyurethane foam: Preparation, thermal stability, fire performance and mechanical properties, J. Polym. Res. 20(2013) 1-12. https://doi.org/10.1007/s10965-013-0234-1

[102] L. Ferry, G. Dorez, A. Taguet, B. Otazaghine, J.M. Lopez-Cuesta, Chemical modification of lignin by phosphorus molecules to improve the fire behavior of polybutylene succinate, Polym. Degrad. Stab. 113 (2015) 135-143.

[103] B. Prieur, M. Meub, M. Wittemann, R. Klein, S. Bellayer, G. Fontaine, S. Bourbigot, Phosphorylation of lignin to flame retard acrylonitrile butadiene styrene (ABS), Polym. Degrad. Stab. 127 (2016) 32-43. https://doi.org/10.1016/j.polymdegradstab.2016.01.015

[104] G.P. Mendis, S.G. Weiss, M. Korey, C.R. Boardman, Phosphorylated lignin as a halogen-free flame-retardant additive for epoxy composites, Green Mater. 4 (2016) 150-159. https://doi.org/10.1680/jgrma.16.00008

[105] Y. Zhang, Q. Zhao, L. Li, R. Yan, J. Zhang, J.C. Duan, B.J. Liu, Z.Y. Sun, M.Y. Zhang, W. Hu, Synthesis of a lignin-based phosphorus-containing flame-retardant and its application in polyurethane, RSC Adv. 8 (2018) 32252-32261. https://doi.org/10.1039/C8RA05598J

[106] W. Hu, Y.M. Zhang, Y.X. Qi, H. Wang, B. Liu, Q. Zhao, J. Zhang, J. Duan, L. Zhang, Z. Sun, B. Liu, Improved mechanical properties and flame retardancy of wood/PLA all-degradable biocomposites with novel lignin-based flame-retardant and TGIC, Macromol. Mater. Eng. 305 (2020) 1900840.

[107] Y.B. Guo, C.Z. Cheng, T.G. Huo, Y. Ren, X. Liu, Highly effective flame-retardant lignin/ polyacrylonitrile composite prepared via solution blending and phosphorylation, Polym. Degrad. Stab. 181 (2020) 109362. https://doi.org/10.1016/j.polymdegradstab.2020.109362

[108] Y.L. Wang, Y.M. Zhang, B.Y. Liu, Q. Zhao, Y. Qi, Y. Wang, A novel phosphorus-containing lignin- based flame-retardant and its application in polyurethane, Compos. Commun. 21 (2020) 100382. https://doi.org/10.1016/j.coco.2020.100382

[109] P. Dai, M.K. Liang, X.F. Ma, Y. Luo, M. He, X. Gu, Q. Gu, Highly efficient, environmentally friendly lignin-based flame-retardant used in epoxy resin, ACS Omega 5 (2020) 32084-32093. https://doi.org/10.1021/acsomega.0c05146

[110] J.H. Lee, D. Jang, I. Yang, S.M. Jo, S. Lee, Effect of phosphorylated lignin on flame retardancy of polypropylene-based composites, J. Appl. Polym. Sci. 139 (2022) e52519. https://doi.org/10.1002/app.52519

[111] R. Hajj, R.E. Hage, R. Sonnier, B. Otazaghine, S. Rouif, M.Nakhl, J.M. Lopez-Cuesta, Influence of lignocellulosic substrate and phosphorus flame-retardant type on grafting yield and flame retardancy, React. Funct. Polym. 153 (2020) 104612. https://doi.org/10.1016/j.reactfunctpolym.2020.104612

[112] L. Liu, B. Shi, A. Zhang, Y. Xue, J. Zhang, J. Dai, M. Hassanpour, L.C. Tang, Y. Shi, P. Song, A polyphosphoramide-grafted lignin enabled thermostable and fire-retardant polylactide with preserved mechanical properties, Compos. Part A Appl. Sci. Manuf. 160 (2022) 107028. https://doi.org/10.1016/j.compositesa.2022.107028

[113] A. Zhang, J. Zhang, L. Liu, J. Dai, X. Lu, S. Huo, M. Hong, X. Liu, M. Lynch, X. Zeng, P. Burey, Engineering phosphorus-containing lignin for epoxy biocomposites with enhanced thermal stability, fire retardancy and mechanical properties, J. Mater. Sci. Tech. 167 (2023) 82-93. https://doi.org/10.1016/j.jmst.2023.06.004

[114] Y. Yu, S. Fu, P. Song, X. Luo, Y. Jin, F. Lu, Q. Wu, J. Ye, Functionalized lignin by grafting phosphorus-nitrogen improves the thermal stability and flame retardancy of polypropylene, Polym. Degrad. Stab. 97 (2012) 541-546. https://doi.org/10.1016/j.polymdegradstab.2012.01.020

[115] H. Zhu, Z. Peng, Y. Chen, G. Li, L. Wang, Y. Tang, R. Pang, Z.U.H. Khan, P. Wan, Preparation and characterization of flame-retardant polyurethane foams containing phosphorus-nitrogen-functionalized lignin, RSC Adv. 4 (2014) 55271-55279. https://doi.org/10.1039/C4RA08429B

[116] H. Zhang, N. Feng, G.X. Zhang, Synthesis and char-forming flame retardancy of thermoplastic polyolefin elastomer grafted with phosphorus-nitrogen lignin sulfonate, Polym. Mater. Sci. Eng. 31 (2015) 11-16. In Chinese.

[117] H. Hu, Y.Liu, Y. Li, X. Hu, H. Zhou, Synthesis of a novel lignin-based intumescent flame-retardant and its performance in LLDPE/EVA blends, Met. Mater. Metall. Eng. 44 (2016) 3-7. In Chinese.

[118] L. Liu, M. Qian, P.Song, G. Huang, Y. Yu, S. Fu, Fabrication of green lignin-based flame-retardants for enhancing the thermal and fire retardancy properties of polypropylene/wood composites, ACS Sustain. Chem. Eng. 4 (2016) 2422-2431. https://doi.org/10.1021/acssuschemeng.6b00112

[119] L. Costes, F. Laoutid, M. Aguedo, A. Richel, S. Brohez, C. Delvosalle, P. Dubois, Phosphorus and nitrogen derivatization as efficient route for improvement of lignin flame-retardant action in PLA, Eur. Polym. J. 84 (2016) 652-667. https://doi.org/10.1016/j.eurpolymj.2016.10.003

[120] S. Wang, S. Ma, C. Xu, Y. Liu, J. Dai, Z. Wang, X. Liu, J. Chen, X. Shen, J. Wei, J. Zhu, Vanillin-derived high-performance flame-retardant epoxy resins: facile synthesis and properties, Macromolecules 50 (2017) 1892-1901. https://doi.org/10.1021/acs.macromol.7b00097

[121] Y. Matsushita, D. Hirano, D. Aoki, S. Yagami, Y. Takagi, K. Fukushima., A biobased flame-retardant resin based on lignin, Adv. Sustain. Syst. 1 (2017) 1700073. https://doi.org/10.1002/adsu.201700073

[122] Y.M. Zhang, Q. Zhao, L. Li, R. Yan, J. Zhang, J.C. Duan, B.J. Liu, Z.Y. Sun, M.Y. Zhang, W. Hu, Synthesis of a lignin-based phosphorus-containing flame-retardant and its application in polyurethane, RSC Adv. 8 (2018) 32252-32261. https://doi.org/10.1039/C8RA05598J

PMid:35547477 PMCid:PMC9086252

[123] R. Zhong, X. Lin, S. Huang, L. Xiong, Y. Jin, Preparation and flame retardancy of aluminum phosphate double-coated red phosphorus modified by piperazine-lignin, Polym. Mater. Sci. Eng. 35 (2019) 143-148. In Chinese.

[124] S. Zhou, R. Tao, P. Dai, Z. Luo, M. He, Two-step fabrication of lignin-based flame-retardant for enhancing the thermal and fire retardancy properties of epoxy resin composites, Polym. Compos. 41 (2020) 2025-2035. https://doi.org/10.1002/pc.25517

[125] H. Yang, B. Shi, Y. Xue, Z. Ma, L. Liu, L. Liu, Y. Yu, Z. Zhang, P.K. Annamalai, P. Song, Molecularly engineered lignin-derived additives enable fire-retardant, UV-shielding, and mechanically strong polylactide biocomposites, Biomacromolecules 22 (2021) 1432-1444. https://doi.org/10.1021/acs.biomac.0c01656

[126] Y. Yu, P. Song, C. Jin, S. Fu, L. Zhao, Catalytic effects of nickel (cobalt or zinc) acetates on thermal and flammability properties of polypropylene-modified lignin composites, Ind. Eng. Chem. Res. 51 (2012) 12367-12374. https://doi.org/10.1021/ie301953x

[127] L. Liu, G. Huang, P. Song, Y. Yu, S. Fu, Converting industrial alkali lignin to biobased functional additives for improving fire behavior and smoke suppression of polybutylene succinate, ACS Sustain. Chem. Eng. 4 (2016) 4732-4742. https://doi.org/10.1021/acssuschemeng.6b00955

[128] L. Liu, M. Qian, P. Song, G. Huang, Y. Yu, S. Fu, Fabrication of green lignin-based flame-retardants for enhancing the thermal and fire retardancy properties of polypropylene/wood composites, ACS Sustain. Chem. Eng. 4 (2016) 2422-2431.

[129] Y.J. Song, A lignin aluminum flame-retardant and its preparation method and application, Chinese Patent CN102220145A, 2011-10-19. In Chinese.

[130] L. Liu, G. Huang, P. Song, Y. Yu, S. Fu, Converting industrial alkali lignin to biobased functional additives for improving fire behavior and smoke suppression of polybutylene succinate, ACS Sustain. Chem. Eng. 4 (2016) 4732-4742. https://doi.org/10.1021/acssuschemeng.6b00955

[131] Y. Guo, C. Zuo, W. Tan, Y. Liu, L. Jiang, D. Yu, Y. Ren, X. Liu, Fabricating flame-retardant polyacrylonitrile fibers modified by sodium lignosulfonate and copper ions, Polym. Degrad. Stab. 206 (2022) 110176. https://doi.org/10.1016/j.polymdegradstab.2022.110176

[132] F. Chen, W. Liu, S.I. Seyed Shahabadi, J. Xu, X. Lu, Sheet-like lignin particles as multifunctional fillers in polypropylene, ACS Sustain. Chem. Eng. 4 (2016) 4997-5004. https://doi.org/10.1021/acssuschemeng.6b01369

[133] B. Tian, J. Li, Z. Li, W. Dong, N. Zhang, H. Zhao, Y. Liu, M. Di, Preparation of polypropylene with high melt strength by wet reaction blending of lignin, J. Appl. Polym. Sci. 139 (2022) 51224. https://doi.org/10.1002/app.51224

[134] Y.K. Recepoglu, A.Y. Goren, Y. Orooji, A. Khataee, Carbonaceous materials for removal and recovery of phosphate species: Limitations, successes and future improvement, Chemosphere 287 (2022) 132177. https://doi.org/10.1016/j.chemosphere.2021.132177

[135] Y. Yu, P. Song, C. Jin, S. Fu, L. Zhao, Catalytic effects of nickel (cobalt or zinc) acetates on thermal and flammability properties of polypropylene-modified lignin composites, Ind. Eng. Chem. Res. 51 (2012) 12367-12374. https://doi.org/10.1021/ie301953x

[136] S.Q. Ma, Study on the synergistic flame retardancy and mechanism of lignin/silicone epoxy resin, Chinese Patent CN103450636A, 2014-06-26. In Chinese.

[137] S. Zhang, N.J. Wu, M.J. Li, Research progress on organosilicon flame-retardants, Polym. Bull. 12 (2010) 72-77. In Chinese.

[138] R. Zhang, X. Xiao, Q. Tai, H. Huang, J. Yang, Y. Hu, Preparation of lignin-silica hybrids and its application in intumescent flame-retardant poly(lactic acid) system, High Perform. Polym. 24 (2012) 738-746. https://doi.org/10.1177/0954008312451476

[139] G. Wang, Research progress of flame-retardant silicone polymer materials, J. Shandong Ind. Technol. 13 (2014) 72. In Chinese.

[140] J. Zhang, Y. Chen, P. Sewell, M.A. Brook, Utilization of softwood lignin as both crosslinker and reinforcing agent in silicone elastomers, Green Chem. 17(2015) 1811-1819. https://doi.org/10.1039/C4GC02409E

[141] J. Zhang, E. Fleury, Y. Chen, M.A. Brook, Flame-retardant lignin-based silicone composites, RSC Adv. 5 (2015) 103907-103914.. https://doi.org/10.1039/C5RA24093J

[142] H. Lu, Y. Song, Z. Wang, D. Liu, Y. Ming, J. Zhou, Synthesis and characterization of a lignin-based charring agent containing nitrogen and silicon, Plast. Sci. Technol. 44 (2016) 90-94. In Chinese.

[143] Y. Song, X. Zong, N. Wang, N. Yan, X. Shan, J. Li, Preparation of γ-divinyl-3-aminopropyltriethoxysilane modified lignin and its application in flame-retardant poly(lactic acid), Mater. (Basel) 11 (2018) 1505. https://doi.org/10.3390/ma11091505

[144] Y.J. Huang, Preparation and properties of P-Si synergistic flame-retardant waterborne anticorrosion acrylic coating, MS Thesis, Zhejiang Sci-Tech University, Hangzhou, 2023, pp. 1-82. In Chinese.

[145] J. Dai, W.J. Chen, M.L. Pan, H. Ou, Study on preparation and properties of silicon modified lignin charring agent synergistic intumescent flame-retardant polypropylene, Ind. Saf. Environ. Prot. 49 (2023) 50-54. In Chinese.

[147] K.L. Song, I. Ganguly, I. Eastin, A. Dichiara, Lignin-modified carbon nanotube/graphene hybrid coating as efficient flame-retardant, Int. J. Mol. Sci. 18 (2017) 2368. https://doi.org/10.3390/ijms18112368

[148] F. Laoutid, L. Bonnaud, M. Alexandre, J.M. Lopez-Cuesta, P. Dubois, New prospects in flame-retardant polymer materials: From fundamentals to nanocomposites, Mater. Sci. Eng. R Rep. 63 (2009) 100-125. https://doi.org/10.1016/j.mser.2008.09.002

[149] Y.J. Guo, Preparation and properties of phenolic foam modified by nano silica and lignin, MS Thesis, Chinese Academy of Forestry, Beijing, 2018, pp. 1-67. In Chinese.

[150] B. Chollet, J.M. Lopez-Cuesta, F. Laoutid, L. Ferry, Lignin nanoparticles as a promising way for enhancing lignin flame-retardant effect in polylactide, Mater. (Basel) 12 (2019) 2132. https://doi.org/10.3390/ma12132132

[151] X. Liu, S. Li, S. Liu, J. Li, Z. Chen, Research progress in preparation and functional application of lignin-based nanoparticles, Biomass Chem. Eng. 54 (2020) 53-65. In Chinese.

[152] D. Meng, H.Wang, Y. Li,J. Liu, J. Sun, X. Gu, H. Wang, S. Zhang, Constructing lignin-based nanoparticles towards flame-retardant thermoplastic polyurethane composites with improved mechanical and oxidation resistant properties, Int. J. Biol. Macromol. 253 (2023) 126570. https://doi.org/10.1016/j.ijbiomac.2023.126570

[153] J. Li, S. Zhang, S. Zeng, M. Zhou, J. Cheng, Lignin nanoparticles/polyvinyl alcohol/chitosan ternary composites and its preparation method and application, Chinese Patent CN113652047B, 2023-02-24. In Chinese.

Chapter 7

Metal-Organic Frameworks as Flame-Retardant Materials for Polymers

T.C. Mokhena[1,2]*, S. Mapukata[1], M.J. Mochane[3], K. Shingange[4], M. Yılmaz[5]

[1]DSTI/Nanotechnology Innovation Centre, Advanced Materials Division Mintek, Randburg, South Africa

[2]School of Chemical and Physical Sciences, University of Mpumalanga, Mbombela 1200, South Africa

[3]Department of Chemistry, College of Science, Engineering, and Technology, University of South Africa, Roodepoort, Johannesburg, South Africa

[4]School of Physics, University of the Witwatersrand, 1 Jan Smuts Avenue, Braamfontein 2001, South Africa

[5]Department of Chemistry and Chemical Processing Technologies, Bahçe Vocational School, Osmaniye Korkut Ata University, Osmaniye, 80000, Türkiye

mokhenateboho@gmail.com

Abstract

Metal-organic frameworks (MOFs) are a novel class of nanostructured materials with multiple dimensions and intrinsic properties, enabling their versatility in various applications. They exhibit high porosity, large surface area, ease of functionalization, and suitable tailorability. These properties make MOFs exceptional nanofillers for creating high-performance multifunctional composites. Moreover, MOFs possess flame retardant qualities that can be enhanced by modifying metal ions or adding flame retardant components to their ligands. As such, recent research efforts on the flame-retardant effects and mechanisms of MOFs when incorporated into different polymers have intensified. This chapter reviews the use of MOFs and their hybrids as non-halogenated flame retardants (FRs). It also summarizes the synthesis and properties of MOFs and their impact on the flame retardancy and smoke suppression of various polymers. The chapter concludes with the challenges and future prospects of these materials as FRs.

Keywords

Metal-Organic Frameworks, Flame Retardants, Polymeric Materials, Polymer Composites, Nanomaterials

Contents

1. Introduction

Polymeric materials are integral to modern life, thanks to their unique properties and ease of processing [1-4]. They offer excellent electrical insulation, robust mechanical performance, and resistance to corrosion. However, their inherent flammability and smoke emission pose significant challenges, limiting their use in high-performance applications [4-6]. To expand their utility, it is essential to enhance their flame retardant and smoke suppression characteristics [7]. One promising approach is the inclusion of metal compounds into polymeric matrices [7-9]. This strategy can significantly improve fire resistance and reduce smoke production because metals can catalyze the conversion of CO to CO_2, while facilitating the formation of char through catalytic carbonization that protects the underlying material [1,2,10]. However, there is often poor compatibility between polymers and metal compounds, often serving as a major hurdle in their synergistic interactions. This incompatibility not only hampers the mechanical properties of the composite materials, but also prevents them from achieving the desired level of flame retardancy. Therefore, ongoing research is crucial to develop methods that enhance the compatibility and overall performance of these advanced materials [7,8,11].

A promising approach to incorporating metal ions into polymeric materials to enhance their fire resistance is through the use of metal-organic frameworks (MOFs) [5,12,13]. MOFs are porous, crystalline structures formed by the coordination of metal ions with organic linkers [14,15]. Since 1995, over 20,000 crystal structures have been reported, including ZIFs (e.g., ZIF-8 and ZIF-67) which are based on imidazolate as organic linker, MILs (e.g., MIL-53(Fe), MIL-101(Fe or Cr)), IRMOFs (e.g., IRMOF-1), HKUST (e.g., HKUST-1), and UiOs (e.g., UiO-66 and UiO-66-NH_2), which utilize benzenedicarboxylic acid or benzenetricarboxylic acid as organic linkers [15,16]. MOFs can be tailored for specific applications, making them versatile, hence they are widely used in catalysis, gas storage and separation, drug delivery, and chemical sensing, amongst others. They are known for their large surface area, high porosity, excellent mechanical properties, thermal stability, and tunable physicochemical properties. These characteristics, along with the presence of both organic and inorganic components, make MOFs ideal flame retardants (FRs). The organic linkers in MOFs act as compatibilizers, improving dispersion and interfacial adhesion between MOFs and the host polymeric materials. Additionally, the organic components can be modified to carry flame-retardant groups such as phosphorus, nitrogen, and aromatic derivatives. The inorganic components, or metal ions, serve as catalysts, facilitating the formation of continuous char and reducing the emission of flue gases. For example, some metal species like $Fe(OH)_3$ are effective catalysts for smog reduction [13].

MOFs, as hybrid organic-inorganic materials, offer a valuable approach to developing novel FRs [12,15]. Various strategies are employed to create FRs using MOFs as core materials, including in situ encapsulation, combining traditional FRs with MOFs, cladding to produce core-shell structures, etching, mixing different MOFs, and incorporating MOFs with other fillers. This chapter reviews the use of MOFs and thier hybrids as FRs of different polymeric materials. It highlights the challenges and future prospects of these materials as novel non-halogeneted FRs .

2. Metal-organic frameworks

2.1 Overview

MOFs are synthetic materials consisting of metal ions coordinated to organic linkers, forming a robust and versatile structure [16,17]. These materials share similarities with zeolites, particularly in terms of their high surface area and crystallinity. MOFs are highly customizable, allowing for the design of materials with specific properties tailored to various applications. Additionally, they can be cost-effective to produce, depending on the choice of metal source. A broad range of analytical techniques have been employed to characterize MOFs, including; X-ray diffraction (XRD) for determining crystal structure, solid-state nuclear magnetic resonance (NMR) for understanding the local chemical environment, as well as ultraviolet-visible (UV-vis) and infrared (IR) spectroscopies for probing electronic and vibrational properties [17]. To investigate the porosity and morphological characteristics of MOFs, techniques such as Brunauer-Emmett-Teller (BET) surface area analysis and various microscopy methods, including Scanning Electron Microscopy (SEM) and Transmission Electron Microscopy (TEM) are frequently utilized [16]. These analytical tools collectively provide comprehensive insights into the structural, chemical, and physical properties of MOFs, facilitating their development and optimization for a wide array of applications.

The fire resistance mechanisms of MOFs function in both the condensed and vapor phases [15]. During combustion, MOFs generate transition metals that act as Lewis acids with excellent catalytic effects, facilitating the carbonization of the host polymeric materials [12]. These metals promote the decomposition of the polymeric materials at relatively low temperatures, enhancing the charring process and producing a compact, stable char that reduces the production of flammable radicals. This char acts as an insulator, protecting the underlying material from heat and oxygen transmission, thereby delaying its decomposition. Additionally, the metals generated from MOFs can act as catalysts at higher temperatures, promoting the graphitization of the char. This results in a stable, high-quality char that further enhances fire resistance performance.

Some MOFs can release inert gases such as, ammonia, CO_2, and water at elevated temperatures, which dilute flammable gases in the combustion zone, thereby improving the flame retardancy of the sample. During combustion, MOFs also generate metal oxides that serve as smoke suppression agents [12]. These metal oxides can capture highly flammable products, preventing the formation of smog and soot particles. Due to their porosity and high specific surface area, MOFs can absorb highly flammable products as well as toxic gases generated during combustion [12]. Furthermore, well-known FRs can be incorporated into the MOF structure to develop FRs composed of essential molecules that further enhance fire resistance performance [12,18-20]. This combination of properties makes MOFs highly effective in improving the fire resistance of polymeric materials.

3. Preparation of flame-retardant metal-organic framework composites

3.1 Blending

The blending process can be emplyed for physical integration of FRs with MOFs using thermal processing techniques such as melt compounding [13,21]. This solvent-free approach is crucial for industrial viability. The integration leverages the porosity and high surface area of MOFs to facilitate the dispersion of FRs within their porous structure, thereby synergistically enhancing the fire resistance performance of the resulting composite material. Liu et al. [21] utilized iron-based MOFs in conjunction with phosphorus-based FRs to enhance the flame retardancy of high-density polyethylene (HDPE). The iron-based MOF was synthesized using a solvothermal technique, followed by roasting the MOF and phosphorus-based FRs at 80 °C for 6 hours in an oven. This was succeeded by melt blending using a torque rheometer at 160 °C for 20 minutes. The resulting composite was then heated at 165 °C for 10 minutes and subsequently melt-pressed at 10 MPa for 5 minutes. SEM and EDS analyses indicated that the additives were well-dispersed within the HDPE matrix. The study reported that the incorporation of both iron-based MOFs and phosphorus FRs significantly improved the overall flame-retardant properties of the composite material. While studies indicate that better dispersion of additives within polymeric materials can be achieved, this behavior often depends on the specific type of polymer and additive used. Therefore, a significant concern with the blending strategy is that the dispersion of FR additives is frequently compromised, resulting in uneven distribution of the additives. This uneven distribution can negatively impact both the mechanical performance and fire resistance of the composite material.

3.2 Grafting

Grafting techniques can be employed to incorporate functional groups containing FR elements onto MOFs to enhance their overall flame resistance performance [13,22]. These elements can be immobilized either internally or externally on MOFs, serving as additional materials that help

Materials Research Foundations 190 (2026) 194-226
https://doi.org/10.21741/9781644904077-7

improve the flame retardant properties of the host polymeric materials. Common methods for grafting different molecules onto MOFs include Click Chemistry, the Kabachnik-Fields method, Schiff base reactions, and many more. Wang et al. [23] prepared organophosphorus-modified MOFs using the Kabachnik-Fields reaction. Initially, the MOFs were synthesized in a hydrothermal reactor at 120°C for 24 hours, followed by vacuum drying at 80°C. To create bimetallic MOFs, $NaWO_4$ was dissolved in water with oxalic acid, and the resulting powder was ultrasonicated for 30 minutes before undergoing hydrothermal reaction at 180°C for 6 hours. After filtration and vacuum drying, the mixture was dispersed in tetrahydrofuran, then DOPO and polyoxymethylene were added and stirred at 50°C for 10 hours under an inert atmosphere to produce phosphorus-modified bimetallic MOFs, as illustrated in Figure 1. The presence of phosphorus-modified bimetallic MOFs enhanced fire resistance by promoting adsorption and catalytic charring of combustible products, effectively inhibiting the release of toxic smoke and heat. Additionally, the phosphorus element from organophosphorus played a crucial role in both condensed and gas phases. The combination of organophosphorus and bimetallic MOFs facilitated multi-coordination of metals, phosphorus elements, and hierarchical pores, synergistically improving the fire safety of the resulting composites. The bimetallic MOF contained NH_2 groups, complemented by phosphorus groups, which are known to promote charring while diluting combustible products, thereby enhancing overall fire resistance performance. This demonstrates that grafting can be a reliable and effective method to introduce crucial elements within the composite that enhance fire safety. It is worth mentioning that the elements must be chosen to synergistically work together to improve the fire resistance performance of the host polymeric materials. However, the main issue with grafting is that it is a complex process requiring precision in reaction conditions to achieve efficient grafting of rich FR elements into MOFs.

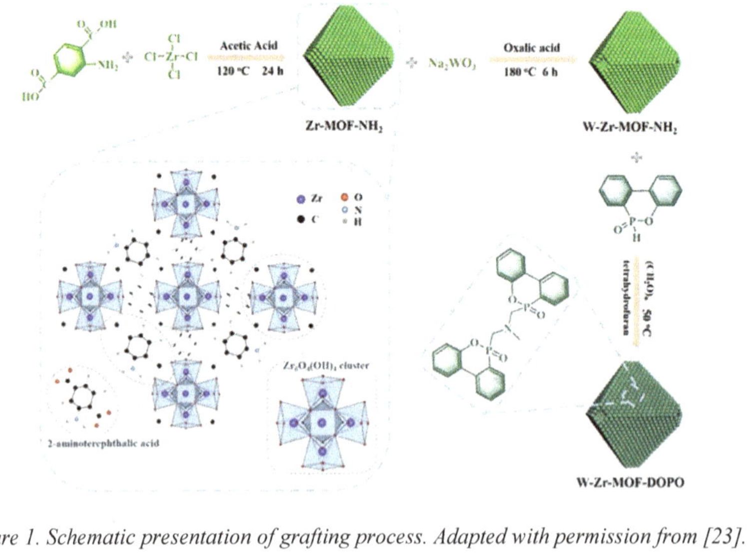

Figure 1. Schematic presentation of grafting process. Adapted with permission from [23].

3.3 Acid and Alkali Etching

The etching process can be used to enhance the number of active sites on MOFs [13,24]. It is essential to carefully control the etching conditions to avoid compromising the structural integrity of the MOF. If the etching is insufficient, the number of active sites remains limited, reducing the effectiveness of the MOF. Conversely, over-etching can cause the pore framework to collapse, leading to a loss of structural stability and functionality. Parameters such as etching duration, concentration of etching agents, and temperature therefore need to be meticulously optimised to ensure that the MOF retains its structural integrity while increasing the number of active sites. This delicate balance allows MOFs to maintain their high surface area and porosity, which are essential for their effectiveness in practical applications. Li et al. [24] employed a "one-stone-two-birds" strategy to etch zeolitic imidazolate framework (ZIF-67) using aluminum hypophosphite (AHP) as the acidic material. Carboxymethylcellulose (CMC) was used as a buffer and support to prevent the complete conversion of the MOF into amorphous cobalt phosphate hydrate[6]. The resulting material exhibited a surface area of 183 m^2/g and maintained its nanoporous structure due to the presence of the macromolecular CMC. The resulting etched MOF was identified as a core-shell nanomaterial. This structure was formed because the CMC did not infiltrate the smaller pores of the MOF, enabling the creation of a dual structure. The material comprised amorphous cobalt phosphate (ACP) as the shell and etched ZIF-67 as the core, with CMC wrapped around the surface like gauze.

3.4 In Situ Encapsulation

This method can be used to confine the guest FR within the porous MOFs to ensure its release at the appropriate time during combustion [13,25]. This method involves dissolving organic ligands and FR molecules in a solvent to achieve a homogeneous solution before introducing the metallic salt solution into the mixture. The mixture is then agitated under controlled temperature to facilitate the formation of the MOF structure while entrapping the FR molecules within its crystalline matrix. Triethyl phosphate (TEP) was encapsulated in MOF through an in situ process to produce phosphorus-containing MOFs as FRs for polylactic acid [25]. In this case, metal salts ($ZrCl_2$) and 2-aminoterephthalic acid (BCD-NH_2) as organic ligands were dissolved in TEP/DMF under sonication, followed by the addition of formic acid. The resulting mixture was heated at 120°C for 3 minutes until it turned yellow, then stirred further for 15 minutes. After filtration and washing, the obtained powder was oven-dried at 70°C for 12 hours to produce MOF@TEP FR. It was found that the presence of modified MOF improved mechanical performance and flame retardant properties, demonstrating that encapsulation can introduce FR elements into the cavity of MOFs to create novel FRs for various polymeric materials.

3.5 Adsorption

To enhance the flame retardant properties of MOF-based FRs, the guest FR can be adsorbed onto MOFs using mechanical methods [13,14,26,27]. Ultrasonication and stirring are commonly employed to ensure the adsorption of guest FRs onto MOFs. This process leverages the high surface area and porosity of MOFs to provide active sites for FR attachment. Triethyl phosphate (TEP) was used as a model FR and adsorbed onto MIL-53 [26]. In this case, the MOFs were synthesized in a hydrothermal reactor at 200°C for 4 hours, followed by washing and drying at 60°C for 8 hours. The obtained MOF was then dispersed in a TEP solution and stirred for 8 hours. The mixture was left to stand at 80°C for 12 hours, followed by filtering and drying at 70°C for 12 hours. XRD results showed a shift towards lower angles in the presence of TEP, indicating that TEP was adsorbed into the interior of the MOFs, creating larger interlayer spacing. XPS analysis also confirmed the successful adsorption of TEP onto MOFs, with characteristic peaks at 134 eV and 134.9 eV associated with P=O and P-O, respectively.

4. Fire Resistance properties of metal-organic frameworks in polymeric materials

4.1 Epoxy resin

Epoxy resin (EP) is renowned for its exceptional electrical insulation, excellent corrosion resistance, outstanding mechanical properties, and ease of processing, making it widely applicable across various industries, such as automotive, aerospace, electronics, and construction [2,28,29]. As one of the essential thermosetting resins, epoxy resin is also highly valued for its versatility and durability. However, its inherent high flammability and significant smoke production limit its use in high-performance applications. To address these limitations, researchers are exploring various methods to enhance the fire resistance and reduce smoke emissions of epoxy resin, such as incorporating fire resistant additives and developing advanced composite materials. These efforts aim to expand the potential applications of epoxy resin in fields requiring stringent fire safety standards. MOFs have been one of the most used FRs for EP as summarized in Table 1 [28,30-32].

In the work conducted by Wang et al. [23] for instance, they prepared organophosphorus-modified bimetallic organic frameworks, which were then introduced into EP to enhance its fire safety and smoke suppresion properties. The limiting oxygen index (LOI) value increased from ~ 25% to ~32%, achieving a UL94 V0 rating when 3% of the modified MOF was added to EP. The total smoke production (TSP) and peak heat release rate (pHRR) were reduced by approximately 36% and 58%, respectively, compared to neat EP. This improvement was attributed to the synergy between MOFs and grafted organophosphorus, which contributed in both gas and condensed phases. The resulting char was found to be intact and continuous when modified MOFs were used as FRs, whereas the char residue of EP and unmodified MOFs was fragile and cracked (see Fig. 2). The I_D/I_G ratio indicated that the modified MOF led to exceptional graphitization of the char layer, providing reliable protection for the underlying material. Analysis of evolved gases revealed that during early decomposition, the initial released gases contained phosphorus groups such as P=O, PO^{2-}, and P(O)-OH. At the maximum decomposition temperature, gases including aromatic compounds, hydrocarbons, water/phenol, and ester components were released. The release of these gases demonstrated the capability of modified MOFs in inhibiting the decomposition of EP.

Chen et al. [27] combined adenosine triphosphate (ATP) with boron nitride (BN) and Cu-MOF to develop a novel FR for EP. The initial pHRR and total heat release (THR) values of EP were approximately 989 kW/m^2 and 87 MJ/m^2, respectively. Introduction of the FR reduced these values to about 580 kW/m^2 and 59 MJ/m^2, representing reductions of 41% and 32%. Additionally, the peak smoke production rate (PSPR) and TSP decreased from 0.403 m^2/s and 32.7 m^2 to 0.226 m^2/s and 20.4 m^2, while the PCOPR and PCO_2PR dropped from 0.175% and 1.69% to 0.081% and 0.73%, respectively. These improvements indicate the flame retardancy of the FR, showcasing that careful selection of materials to produce FRs can create synergistic effects, delivering superior fire resistance and smoke suppression properties. In this study in particular, BN acts as a barrier during early combustion, preventing heat and flame penetration into the polymer, delaying its degradation, a phenomenon known as the labyrinth effect. As the temperature rises, ATP forms phosphoric acid, facilitating the dehydration and charring of the host polymer, leading to stable P-O-P or P=O structures. Phosphorus radicals generated by ATP trap reactive groups in the vapor phase, thereby inhibiting combustion. The carbon source from ATP also carbonizes, and adenine (a gas source) releases inert NH_3, promoting the formation of expanded char from molten char. This char, primarily composed of a P-C-N network, inhibits heat and combustible gases from entering the sample. Cu-MOF generates non-combustible CO_2 at higher temperatures, diluting combustible gases and providing a cooling effect. Additionally, Cu ions catalyze the conversion of combustible gases into residual carbon, enhancing char retention. Ultimately, BN and the resultant CuO remain in the condensed phase, reinforcing the char and acting as a shielding layer.

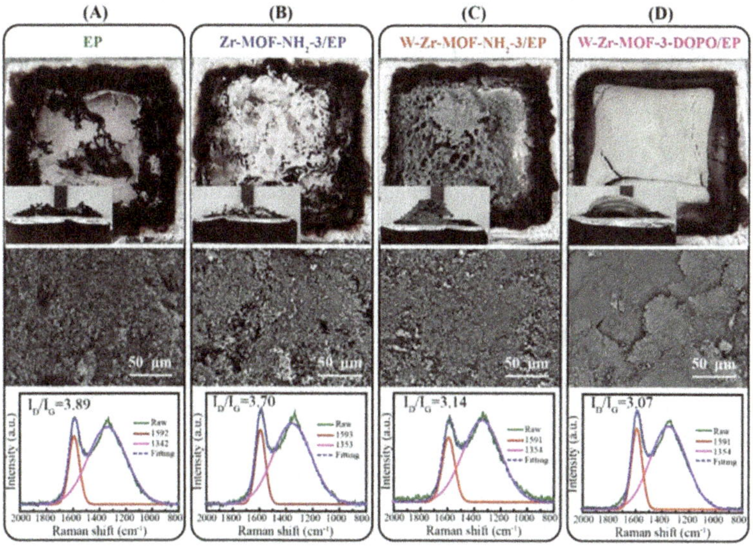

Figure 2. Digital images, SEM images and Raman spectra of external char residue of (A) neat EP, (B) unmodified MOFs/EP, and (C) modified MOFs/EP after cone colorimeter test. Adapted with permission from [23].

Xu et al. [28] developed a hybrid material by loading a rhombic dodecahedral MOF (ZIF-8) onto reduced graphene oxide (rGO), as a novel flame retardant for EP. The presence of just 2% ZIF-8 in the EP reduced the pHRR and THR values from 940 kW/m^2 and 41.2 MJ/m^2 to 395 kW/m^2 and 30.8 MJ/m^2, respectively. The smoke production rate (SPR) and TSP also decreased to 0.65 m^2/s and 48.6 m^2, respectively. This was attributed to the catalytic effect of ZIF-8 metal ions in facilitating charring. When both rGO and ZIF-8 were present, the pHRR and THR further decreased to 332 kW/m^2 and 27.3 MJ/m^2, while the SPR and TSP decreased to 0.53 m^2/s and 36.8 m^2, respectively, compared to neat EP. The LOI value reached approximately 27%, with a UL94 rating of V1. This enhanced performance was attributed to the synergistic effect between rGO and ZIF-8.

Materials Research Forum LLC
https://doi.org/10.21741/9781644904077-7

Lv et al. [29] developed a ternary intumescent flame retardant (IFR) system comprising of a core-shell structure of ZIF-8@polyphosphazene (PZN) and ammonium polyphosphate (APP) to enhance the flame retardancy and mechanical properties of EP. Ammonium salt FRs like AAP have attracted attention due to their ease of preparation, clean production, application and disposal with less harm to the environment. In their work, Lv and colleagues added ZIF-8@PZN and APP at a ratio of 1:6, observing an increase in the LOI value from 19% to 31% relative to EP, and achieving a UL94 V0 rating. EP possessed a PSPR of 0.347 m^2/s and PCOPR of 0.051 g/s. However, after the addition of 3% ZIF-8@PZN and 18% APP, these values decreased by ~73% and ~65%, respectively, suggesting an improvement in fire resistance. These enhancements are due to the high stiffness of the fillers, which restrict the movement of the polymer molecular chains. This restriction is attributed to the strong interfacial adhesion provided by the amine groups in ZIF-8@PZN.

Chai et al. [30] employed Cu-MOF and APP as FRs for EP. The primary goal was to explore the synergy between phosphorus-containing materials and metal ion-containing materials to enhance the flame retardancy of EP. They reported that the addition of 9% Cu-MOF/APP improved fire resistance performance, achieving a LOI value of approximately 29% and a UL94 V0 rating. The production of CO, CO_2, and the pHRR were reduced by ~25%, ~30%, and ~55%, respectively. These improved flame resistance was attributed to the decomposition of APP, which released phosphorus-containing radicals that acted as scavengers for highly flammable radicals. Moreover, the released phosphate promoted the charring process, along with the metal ions generated from the combustion of Cu-MOF [30].

Zhang et al. [31] were the first to successfully use Mo-based MOF as a FR for EP. The molybdenum oxides contributed to the catalytic effect of the MOF, leading to a high degree of graphitization and better thermal oxidation resistance, capable of inhibiting the transfer of heat, O_2, and combustible radicals, thereby hindering flame propagation.

4.2 Polyurethane

Polyurethanes (PU) are lightweight thermoplastic elastomers known for their excellent mechanical and processing properties [14,33-36]. They exhibit good wear resistance, and have excellent corrosion resistance. However, PU is highly flammable and releases significant amounts of toxic and harmful gases during combustion, leading to severe environmental pollution, casualties, and property damage, which limits its use in various fields [18]. To enhance the fire resistance and smoke suppression properties of PU, various FRs have been employed, including halogen-based, single-component organic or inorganic FRs, nano-based FRs, modified FRs, and hybrid FRs [37-39]. However, large quantities of these FRs are needed to achieve the desired flame retardant and smoke suppression properties, which can negatively impact the mechanical performance of the final product. MOFs, which combine organic and inorganic components, have shown promise as suitable FRs for PU. Introducing small amounts of these materials can impart flame retardant and smoke suppression properties without compromising the mechanical properties of the resulting PU composites. This improvement is attributed to the organic component enhancing compatibility with the host polymeric materials, while the transition metal inhibits smoke and acts as a catalyst for charring and the conversion of CO into CO_2 in the vapor phase. Table 2 lists some of the studies that have been conducted on the incorporation of MOFs in PU for enhanced flame resistance [33,40-44].

Table 1. Summary of studies of MOFs as FRs for epoxy resin

MOF-FR	UL94	LOI	pHRR reduction %	THR reduction %	TSP reduction %	Comments	Refs.
2% ZIF-67 + AHP	-	25.2	47.8	21.0	20.7	The MOF, with its high specific surface area, exposed a large number of active sites that acted as catalysts to convert harmful combustible gases that were adsorbed onto the MOF	[24]
2% ZIF-8	NR	25.6	57	25.2	22.9	Reduced graphene oxide acted as a protective shield, while the metal oxides produced from the combustion of ZIF-8 facilitated the charring process. This resulted in the formation of a compact and dense char, which effectively inhibited smoke release.	[28]
2%ZIF-8/RGO	V1	26.8	65	33.2	36.8		
9%Cu-MOF/APP	V0	29.4	34.6		54.7	Cu-MOF and APP facilitated the charring process, resulting in a denser char that inhibited the transfer of heat and the release of toxic gases from the sample	[30]
6% Mo-MOF	V2	32.8	44.7	2.5	25.1	The MOF enhanced the graphitization of the char, and the resulting inorganic layer of MoO_3 also contributed to shielding the underlying sample from heat and oxygen, thereby improving fire resistance performance	[31]
2% ZIF-8@MgAl-LDH	V1	24.7	51	30.4	31.1	Both FRs facilitate the charring process, resulting in high graphitization and thermal oxidation resistance. The char acts as a shield, protecting the underlying material from burning during combustion.	[32]
2% ZIF-8@MgAl-LDH	V1	25.5	62	39	37.5		
3%- ZIF-8-β-FeOOH (Z8Fe)	V1	27.3	52.0	34.9	60.9	Both FRs improved the flame retardancy and smoke suppression properties of the host polymeric material.	[68]
3% ZIF-67-β-FeOOH (Z67Fe)	V0	28.1	58.9	36.9	57.2		

Xu et al. [45] introduced three different types of two-dimensional plates of zinc, cobalt, and zinc-cobalt zeolite imidazole frameworks to reduce the fire risk of PU. With the addition of 3% of these MOFs, the LOI values increased from 23 to 25, 26, and 26 for Zn-ZIF-, Co-ZIF-, and Zn-Co-ZIF-based composites, respectively. The THR and pHRR decreased by 28.8%, 42.8%, and 43.8%, and 50%, 45%, and 51.5%, respectively. The presence of the MOFs acted as barrier layers that inhibited heat transfer into the sample, thereby delaying decomposition. Additionally, the MOFs produced ammonia and water, which diluted combustible gases and inhibited combustion of the composites. Finally, the decomposed MOFs generated metal oxides such as cobalt oxides and zinc oxides, which catalyzed and facilitated the graphitization of the char into a crosslinked layer that shielded heat transfer and reduced the release of heat and smoke. Wu et al. [46] synthesized a core-shell structured FRs composed of ZIF-8 and phytic acid by exploiting the electrostatic interaction between negatively charged phytic acid (PA) and positively charged hyperbranched polysiloxane (HBPSi), as shown in Fig. 3 [46]. The addition of 2% of this IFR significantly improved the flame resistance performance of the resulting composites. The pHRR decreased from 682 kW/m^2 to 442 kW/m^2, while the THR decreased from 73.4 kJ/m^2 to 54.9 kJ/m^2. The PSPR and TSP decreased to 0.055 m^2/s and 3.69 m^2, with reductions of 21.4% and 65.5%, respectively. Additionally, the tensile strength and elongation-at-break increased from 42.36 MPa and 2138.1% to 46.87 MPa and 2178.2%, respectively. These improvements were attributed to better interaction between the FR and PU, specifically between the amino groups of HBPSi and the carbonyl groups of PU.

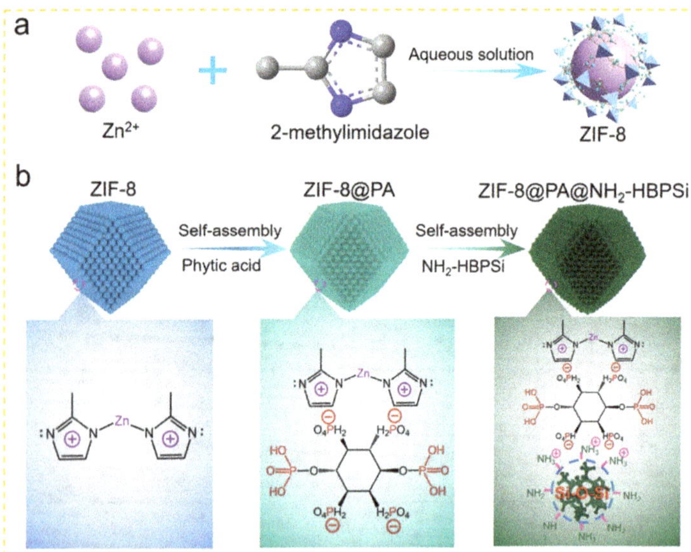

Figure 3. Schematic presentation of the preparation process of core-shell structure starting with (a) ZIF-8, (b) followed by inclusion of PA and then the hyperbranched polysiloxane (HBPSi). Reprinted with permsion from [46]

Materials Research Foundations 190 (2026) 194-226 https://doi.org/10.21741/9781644904077-7

Xu et al. [39] utilized a hybrid of zeolitic imidazolate framework-8 (ZIF-8)/α-zirconium phosphate (α-ZrP), created through the surface modification of α-ZrP with ZIF-8, to reduce the fire risk of PU. The prepared composites demonstrated excellent fire resistance performance compared to neat PU. The pHRR and THR were reduced by approximately 70% and 46%, respectively. Additionally, the smoke production rate (SPR) and TSP decreased by 59% and 41%, respectively. This improvement was attributed to two main factors: the physical barrier and catalytic carbonization provided by α-ZrP, which inhibited the entry of combustible gases and heat into the sample, and the metal oxide generated, which enhanced the charring of PU, resulting in a continuous, compact char that further prevented the release of smoke. Wang and colleagues from Hefei from China [47] introduced a hybrid of expanded graphite (EG) and α-ZrP into PU to enhance its smoke suppression and flame retardancy properties. The pHRR and THR decreased from 923 kW/m^2 and 41 MJ/m^2 to 153 kW/m^2 and 24 MJ/m^2, respectively, compared to neat PU. This improvement was attributed to the presence of EG, which formed a worm-like, fluffy char that protected the underlying material during combustion. Additionally, the decomposition of ZIF-8 generated ZnO, which facilitated charring and ensured the production of a compact char. This char inhibited the transfer of combustible gases and heat, thereby enhancing the fire retardant and smoke suppression properties of PU. Elsewhere, a multifunctional flame retardant composed of APP, SiO_2, and UiO-66-NH_2(Zr) was reported to provide excellent flame-retardant properties for PU. The authors began by preparing APP@SiO_2 using a one-pot synthesis method. APP and CTAB were dispersed in a suitable solvent, followed by the introduction of SiO_2 precursors, namely ethyl silicate (TEOS) and γ-(2,3-epoxypropoxy) propytrimethoxysilane (KH-560). $ZrCl_4$ was then dissolved in a suitable solvent, and APP@SiO_2 was added as the first solution. The second solution was prepared by mixing 2-aminoterephthalic acid and glacial acetic acid into a suitable solvent. This second solution was slowly added to the first solution while stirring, followed by refluxing at 120 °C for 24 hours. The sediments were collected and vacuum-dried at 80 °C to produce APP@SiO_2@UiO-66-NH_2(Zr). The composites were formed through melt mixing at 180-200°C, followed by melt pressing at 200°C for 10 minutes. Due to the FR system's essential elements for flame retardancy, the smoke suppression and flame resistance properties were significantly improved compared to neat PU. For instance, the pHRR declined by approximately 70%, 74%, and 75% for TPU/APP, TPU/APP@SiO_2, and PU/APP@SiO_2@UiO-66-NH_2(Zr) composites, respectively, with the heat release rate (HRR) decreasing by approximately 70%, 81%, and 88%, respectively. The LOI values increased from approximately 21% to 30% for TPU/APP@SiO_2@UiO-66-NH_2(Zr). Data plotted for THR, SPR, TSP, CO, and CO_2 yield curves confirmed the superior flame resistance performance with the addition of the FRs into PU. In Figures 4a and 4b, the pHRR and THR values drastically reduced, with the highest reduction recorded for PU/APP@SiO_2@UiO-66-NH_2(Zr) composites. The fire performance index (FPI) significantly increased by 141%, 179%, and 228% for PU/APP, PU/APP@SiO_2, and PU/APP@SiO_2@UiO-66-NH_2(Zr), respectively. The SPR composites declined by ~64%, ~66%, and ~64% for PU/APP, PU/APP@SiO_2, and PU/APP@SiO_2@UiO-66-NH_2(Zr) (see Figure 4c), while the CO and CO_2 yields were reduced by approximately 24%, 38%, and 52%, and 72%, 75%, and 77%, respectively, compared to the neat polymeric matrix. The highest reduction of approximately 88% in TSP was recorded for the PU APP@SiO2@UiO-66-NH_2(Zr) composite, demonstrating its exceptional smoke suppression properties. These enhancements were attributed to the synergy between the FR components. Early in the decomposition, UiO-66-NH_2(Zr) adsorbs a large amount of gases and reduces their release due to its porous structure and high specific area. After the MOF collapses, the resultant ZrO_2 facilitates the charring process to protect underlying materials. The silica-based FR produces a denser oxide film (SiO_2) on the surface of the composite during combustion, thereby inhibiting O_2 from entering the interior of the shielding layer and delaying PU decomposition. APP decomposes into phosphorus-containing compounds (phosphoric acid, metaphosphoric acid, polyphosphoric acid, and pyrophosphoric acid) that promote the charring process through dehydration of the host polymeric matrix. This leads to glassy phosphorus-rich materials protecting the surface of the polymer with a char layer that acts as a shield, inhibiting heat and smoke release. APP also releases scavenging radicals such as PO and PO_2, which quench flammable products like H, OH, and O in the combustion reaction, thereby breaking the combustion chain reaction. Additionally, H_2O and NH_3 produced during combustion contribute to the dilution of flammable products, thus improving the fire resistance performance of PU.

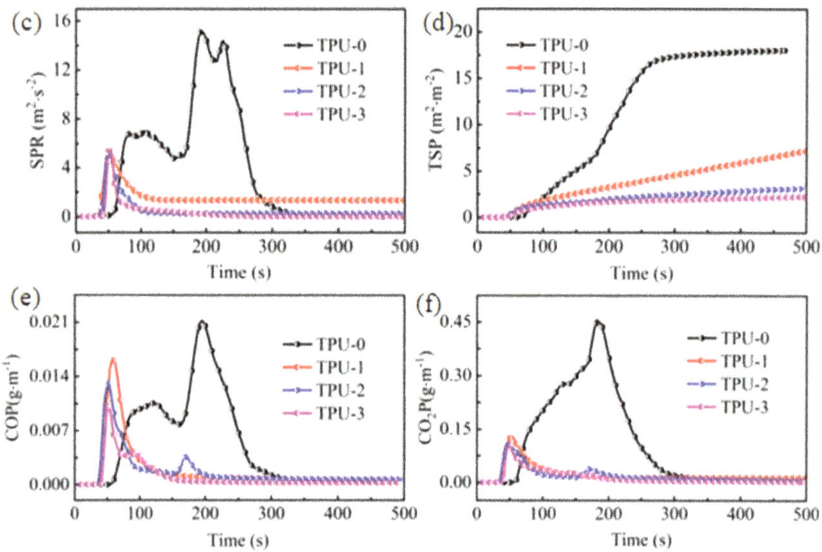

Figure 4. (a) HRR, (b) THR, (c) SPR, (d) TSP, (e) Carbon Monoxide (CO), and (f) Carbon Dioxide (CO_2) of PU-based composites. The composites include TPU-0 (neat PU), TPU-1 (PU/APP), TPU-2 (PU/APP@SiO_2), and TPU-3 (PU/APP@SiO_2@UiO-66-NH_2(Zr)). Reprinted with permsion from [42].

Wan et al. [14] designed Ni-MOF as a synergistic flame retardant for Mxene (Ti_3C_2Tx). The presence of 1% Ni-MOF@Mxene significantly enhanced the flame retardancy and smoke suppression properties of PU. The THR, smoke release rate (SRR), TSP, and the release of CO and CO_2 from the sample were reduced by 15%, 44%, 26%, 18%, and 8%, respectively, compared to neat PU. This behavior was attributed to the physical barrier provided by the highly thermally stable FR, catalytic carbonization due to the presence of five elements (C, O, N, Ni, and Ti), and the contribution of nitrogen (N) in releasing inert gases (N_2 or NH_3) in the vapor phase, which dilute combustible gases. Wang et al. [33] developed a hybrid flame retardant combining Co-MOF

Materials Research Foundations 190 (2026) 194-226
https://doi.org/10.21741/9781644904077-7

and APP to reduce the fire risk of PU. The Co-MOF compensated for the mechanical loss caused by APP. With a composition of 4.5% APP and 1.5% Co-MOF, the LOI value increased to approximately 28%, with no dripping and achieving a UL94 V0 rating. The pHRR and smoke production rate (SPR) decreased from 1359 kW/m^2 and 0.11 m^2/s to 257 kW/m^2 and 0.04 m^2/s, respectively, in the presence of the hybrid FR system. In addition to the type of MOF playing a critical role in the flame retardant and smoke suppression properties of polymeric materials, the shape of the MOFs was also reported to be crucial in reducing fire risks associated with these materials.

Geng et al. [48] prepared three types of ZIF-67 by adjusting the ratio between 2-methylimidazole and cobalt (II) nitrate hexahydrate, using CTAB as a surfactant. The resulting structures included dodecahedron (Fig. 5a,a1), cube (Fig. 5b,b1), and cross-shaped (Fig. 5c,c1), which were used separately as FRs for polyurethane (PU). Dodecahedral particles with a defined structure and size of 500 nm were achieved, while cubic structures with dimensions of 400 nm and heterotypic ZIF-67 with a cross-shaped structure were also obtained. Energy dispersive mapping of ZIF-67 cross-shaped particles confirmed the presence of Co, N, O, and C elements distributed across their surface (Figure 5d-g). Thermal degradation analysis under inert conditions revealed that the composites caused earlier degradation of PU compared to pure PU, with composites containing ZIF-cross-shaped particles exhibiting distinctive two-stage decomposition due to their different catalytic effects from varying crystal structures. Microcalorimeter tests (MCC) showed that all composites exhibited pHRR at around 280 to 400°C, significantly earlier than neat PU. However, the pHRR for PU/ZIF-67-cross-shaped was lower than its counterparts by approximately 35% and 23% for dodecahedral- and cube-shaped particles, respectively. The THR for cross-shaped particle composites was found to be quite higher than neat PU, attributed to a stronger catalytic oxidation effect that facilitated the complete combustion reaction, producing large quantities of CO_2 and H_2O. Both CO_2 and H_2O acted as inert gases in the vapor phase, diluting the concentration of combustible gases and hindering combustion. These results were confirmed using a cone calorimeter test, where the THR for PU was 146 MJ/m^2 and the pHRR was 1296 kW/m^2, significantly reduced to 708 kW/m^2 and approximately 131 MJ/m^2 with only 2% addition of cross-shaped ZIF-67. The FPI value increased from 0.0594 m^2·s/kW for neat PU to 0.0768 m^2·s/kW, while the fire growth index (FGI) was reduced from 5.204 kW/m^2·s to 4.926 kW/m^2·s. The PSPR declined from 0.178 m^2/s to 0.043 m^2/s, TSP reduced from 17 m^2 to 8 m^2 (a 50% drop), and the smoke factor decreased from approximately 22 to 5.9 MW/m^2, a 73% reduction compared to neat PU. The CO_2 release declined from 110 g to 100 g, a 9% drop. This demonstrates that the cross-shaped structured MOFs exhibited superior smoke suppression and fire resistance properties compared to other shapes. This was attributed to the particles' capability of inhibiting the decomposition of isocyanates into aromatic compounds, thus hindering the production of smoke precursors and reducing the generation and release of smoke.

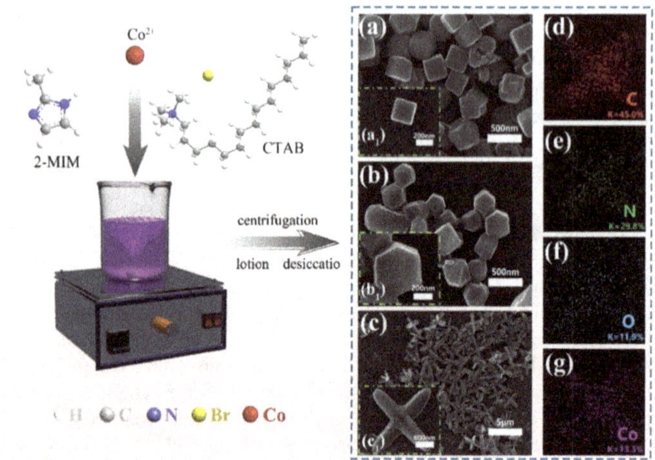

Figure 5. Synthetic route for ZIF-67 (on the left side). SEM micrographs of (a, a1) ZIF 67 nanocages, (b, b1) ZIF-67 cube-shaped and (c,c1) ZIF-67-cross-shaped and (d-g) EDS mapping for ZIF-cross-shaped. Adapted with permission from [48].

Table 2. Summary of studies of MOFs as FR for PU

FR	UL94	LOI	pHRR reduction %	THR reduction %	TSP reduction %	Comments	Refs.
3% Zn-ZIF-L	-	25.3	28.8%	7.8%	18.5%	All the nanosheets exhibited excellent fire resistance performance	[45]
3% Co-ZIF-L	-	26.0	42.8%	11.4	18.5		
3% Co-Zn-ZIF-L	-	25.7	43.8%	9.3	22.9		
3% ZIF-8 + EG	V1	30.2	83.4	42.6	22.4	Synergy between ZIF-8 and EG enhanced the flame retardant and smoke suppression properties	[47]
2% ZIF-8	-	-	18.3	19.8	-	The IFR adsorbed combustible products and promoted charring, generating a char that acted as a protective shield.	[46]
2% ZIF-8@PA@HBPSi	-	-	35.2%	25.2%	65.5%		
4% MH@MOF-P	-	-	45.8		21.4	MH@MOF can be utilized as adsorbents for phosphates to serve as FRs	[40]
Co-ZIF-L@RGO (3.0/27.0)	V0	36.6	84.4	70.1	62.5	Co-ZIF-L was effectively loaded onto rGO, resulting in excellent fire resistance performance	[69]
6% APP/Co-MOF(4.5/1.5)	V0	28	81.1	19.2	31.1	The inclusion of Co-MOF compensated for the loss of mechanical performance in	[33]

Materials Research Foundations 190 (2026) 194-226
https://doi.org/10.21741/9781644904077-7

						the resulting composites cuased by APP	
6.5%APP/0.5%ZIF-67@GO	V0	27.4	81.2	29	23.4	A tensile strength of approximately 18 MPa and an elongation-at-break of around 1026% were achieved, with a prolonged burning time exceeding 600 seconds, compared to 200 seconds for neat PU	[41]
2% GO@NH_2-UIO-66(Zr)	V0	28.4	74.6	35.0	-	The resultant ZrO_2 from the combustion of the FR, along with the formed char layer and noncombustible gases, significantly contributes to excellent fire resistance performance.	[43]
APP@SiO_2@UiO-66-NH_2(Zr)	-	29.9	75.1	87.8	87.7	The fire resistance performance was attributed to each component within the FR system, namely UiO-66-NH_2(Zr), silicon FR, phosphorus FR, and nitrogen FR.	[42]
0.0625% Cu-MOF	V2	23	26.0	1.3	-	Both Cu_MOF and APP decompose early, producing copper compounds, phosphorus-containing, and amino compounds, which promote the dehydration of PU to form polyurea and release CO_2. These components also contribute to the formation of high-quality char with significant expansion, capable of shielding the matrix. Copper-containing compounds migrate to the surface, enhancing the graphitization of the char, which hinders heat and flammable products from entering or leaving the sample, thereby reducing the generation of toxic and dangerous gases.	[44]
Cu-MOF (0.0625)/APP (7.9375)	V0	27	19.2	63.4	-		

4.3 Polyethylene

Polyethylene (PE) is one of the most prevelant polymers in different indutries ranging from packaging to high end indutries [21]. Similar to any polymer, PE is also flammable, it generates a large amount of heat that is crucial for sustenance of the the combustion process and promotes the propagation of flame. Hence, a wide variety of FRs are often introduced into PE to broaden its application while reducing the risk of fire.

Lui et al. [21] synthesized FeMOF (MIL-53(Fe)) using a solvothermal method and fabricated HDPE/ADP/Fe-MOF and HDPE/AHP/Fe-MOF composites through melt compounding and melt pressing with high-density polyethylene (HDPE), aluminum diethyl hypophosphite (ADP), and aluminum hypophosphite (AHP). Incorporation of 0.75 wt% MIL-53(Fe) and 14.5 wt% ADP increased the LOI value by approximately 49%, while the combination of 0.75 wt% MIL-53(Fe) and 14.5 wt% AHP resulted in a 35% increase compared to neat HDPE. Additionally, the pHRR was reduced by approximately 22% and 2.3% when compared to ADP-only and AHP-only samples, respectively. The synergy between ADP and MIL-53(Fe) proven by analysis of char from cone calorimeter tests (CCT), which revealed a continuous structure without visible voids. This is because the presence of MIL-53(Fe) enhances char quality due to its ability to facilitate char formation, thereby inhibiting heat and oxygen exchange to the underlying material and leading to an overall reduction in combustibility of the composite.

4.4 Polypropylene

Polypropylene (PP) is one of the most widely used thermoplastic polymers [49, 50]. However, reducing its fire risk is particularly challenging due to its non-charring behavior and high heat release rate. Despite these challenges, a variety of FRs have been employed to mitigate its fire risk [49-51]. Recently, MOFs have emerged as alternative FRs for PP [49]. Xie and colleagues from Nanjing, China, prepared PP composites using nano ZIF-8 and dried grains with solubles from distillers as fillers to enhance flame retardancy and mechanical properties [49]. The PP was melt-blended with the fillers using a twin-screw extruder, followed by injection molding. ZIF-8 was synthesized at room temperature for an hour, then dried, while the grains with solubles were dried, crushed, and sieved through a 60 mesh. The resulting ZIF-8 particles were hexagonal with an average diameter of 80 nm. The presence of ZIF-8 improved the dispersion of the grains with solubles due to the interaction of organic linkers with polymer chains and the coordination of metal ions with amino or hydroxyl groups of the grains with solubles, leading to enhanced mechanical properties. The composite containing 19% ZIF-8 and 1% grains with solubles achieved a tensile strength of approximately 34 MPa and an elongation-at-break of around 25%. PP exhibited LOI value of approximately 18 and a horizontal burning velocity of 31 mm/min, indicating its fire risk. The introduction of 0.5%, 1%, and 3% ZIF-8 resulted in LOI values of ~20%, ~23%, and ~25%, and burning velocities of ~25, ~20, and ~15 mm/min, respectively. Additionally, the composites achieved a UL94 V2 rating. The reduced fire risk was attributed to the catalytic effect of ZIF-8, which facilitated the charring process and crosslinking of the char. The char layer shielded the sample from heat and prevented the transfer of combustible gases. Both fillers also released inert gases (N_2, NH_3, and H_2O) in the vapor phase, which diluted combustible radicals and had a cooling effect on the sample.

4.5 Polystyrene

Polystyrene (PS) is a widely used thermoplastic known for its unique features, including easy processing, excellent electrical insulation, exceptional corrosion resistance, and good optical properties [26]. It is primarily used in packaging, decoration, and electrical insulation. However, PS is flammable and produces large quantities of dangerous smoke during combustion, limiting its applications beyond traditional uses. During combustion, PS generates cyclopentadienyl, which readily forms chain reactions with other hydrocarbons, resulting in significant smoke production. To expand the applications of PS, it is essential to introduce FRs that can enhance its smoke suppression and flame retardency properties [52]. Hou et al. [52] investigated the use of Fe-MOF and Co-MOF as FRs to enhance the fire resistance performance of PS. They synthesized the MOFs through a solvothermal method and incorporated them into PS using a solution blending technique. The pHRR and THR values decreased by ~14% and ~28%, and ~4% and ~18%, respectively, when 2% Fe-MOF and 2% Co-MOF were added to PS. The total CO production was reduced by ~3% and ~8% for PS/Fe-MOF and PS/Co-MOF composites compared to PS, demonstrating improved fire safety. During combustion, gaseous products from the MOFs dilute volatile compounds and reactive oxygen, while metal oxides act as thermal and fuel barriers, enhancing the flame retardancy of PS. Chen et al. [53] also synthesized a Zr-based MOF (UiO-66) using a solvothermal method to enhance the fire resistance performance of PS. They reported a 27% and 15% reduction in pHRR and THR, respectively, with the inclusion of Zr-MOF in PS. Additionally, the TSP was reduced by more than 35% with the addition of 5% Zr-MOF.

Hou et al. [54] synthesized flower-like Ni-MOF using a solvothermal method to enhance the FR of PS. Graphene oxide, obtained via the Hummers method, was functionalized with nickel 2-methylimidazole MOF (Ni-MOF-GO) using the same solvothermal technique. The Ni-MOF-GO demonstrated superior FR properties compared to other samples. The pHRR and THR decreased by ~33% and ~18%, respectively, compared to neat PS. The PSPR decreased from 0.45 m^2/s for PS to 0.39 and 0.36 m^2/s for GO and Ni-MOF composites, respectively. Ni-MOF-GO further reduced PSPR to 0.07 m^2/s, confirming the synergistic behavior of GO and Ni-MOF. The enhancements in flame retardance were attributed to the layered construction and nickel-containing structure of Ni-MOF-GO. Moreover, the presence of vertically aligned channels within the Ni-MOF sheets created large pore volumes and specific surface areas for absorbing flammable products, as well as numerous active sites for CO catalytic oxidation and smoke reduction. Therefore, absorption, physical barriers, and catalytic effects played crucial roles in enhancing the fire resistance performance of PS.

Zhao et al. [26] incorporated MIL-53/TEP into PS to reduce its fire risk using melt compounding process. The authors used cone calorimeter to evalute the fire risk performance of PS and its composites. The pHHR of PS was found to be ~1071 kW/m^2 with the loading of MIL-53/TEP 3% reducing it to ~806 kW/m^2, which is ~25%. FGI was also reduced from 5.2 to 4.3, while PCOPR and the PCO_2PR decreased from 0.155 g/s and 2.807 g/s to 0.092 g/s and 0.875 g/s after the addtion of 3% MIL-53/TEP, i.e. ~41% and ~69% redustions respectively. This indicate that this FR inhibit the CO and CO_2 production and thus reducing risk of asphyxiation in the event of fire. The efficiency of the FR was attributed to TEP releasing phosphorus-based radical when the MOF structure collapses. These radicals act as scavenger for $HO^·$ and $H^·$ radicals and thus inhibit fire propagation.

Xu et al. [55] studied the flame retardant performance of PS with the incorporation of a cobalt-based MOF. In their study, the MOF was modified with phosphonitrilic chloride trimer (PCT) post-synthesis to create the PCT@MOF-NH_2 flame retardant system. Their findings indicated that the THR value decreased by 12.9%, 16%, and 32% for PS/MOF-NH_2-3.0%, PS/PCT-3.0%, and PS/PCT@MOF-NH_2-3.0%, respectively, compared to neat PS. Additionally, PS/PCT@MOF-NH_2 showed a significant decrease in pHRR, SPR, and CO release compared to other composites and neat PS. These enhancements, greater than those observed in PS composites containing either MOF-NH_2 or PCT alone, demonstrate the synergistic FR effect of MOF-NH_2 and PCT.

 Materials Research Forum LLC
https://doi.org/10.21741/9781644904077-7

A recent study by Yuan et al. [56] reported the immobilization of bis(diphenylphosphino)ferrocene (BDF) into porous Fe-MOF to design a novel FR (D-Fe-MOF) for PS. The porous Fe-MOF was produced using ammonia as an etching agent, followed by the introduction of BDF. The untreated Fe-MOF was diamond-shaped, with a more pronounced porous structure due to the disruption of carboxyl-metal bonds by the base. BDF was found to be attached to D-Fe-MOF, with some adsorbed within the MOF. A 49% reduction in pHRR was achieved with the addition of 3% D-Fe-MOF in the PS. The PCOPR and PCO_2PR were also reduced from 0.126% and 3.276% to 0.090% and 1.575%, respectively, indicating reductions of 56% and 48% compared to neat PS. The results depict the synergy between phosphorus-containing FRs and MOFs, which work together in both condensed and vapor phases. Phosphorus-containing FRs release radicals such as PO that dilute and scavenge flammable products during combustion. The presence of metal ions in MOFs also catalyzes the charring process, forming denser char that traps flammable products and prevents air from reaching the burning zone, thereby reducing fire hazards associated with PS.

4.6 Polyethylene terephthalate

Polyethylene terephthalate (PET) is known for its excellent chemical stability, mechanical performance, and low air permeability, making it ideal for textiles and packaging [22]. However, its high flammability limits its applications. Enhancing the flame retardancy properties of PET could expand its use into electronics, military equipment, safety protection, and transportation. MOFs have been used to reduce PET's fire hazard [22]. Ma et al. [22] synthesized Fe-, Al-, and Cu-based MOFs as FRs for PET. The LOI values increased from 22% to 28%, 25%, and 24% with Fe-MOF, Al-MOF, and Cu-MOF, respectively. The ignition times for MOF-PET composites were shorter than for PET alone, indicating the catalytic effect of MOFs in converting polymer melt into char residue. All MOF-PET composites achieved a UL94 V2 rating.

In another study, Fe-MOF (MIL-88B) was used as a FR for PET, which had been functionalized with a phosphorus-containing compound [57]. The presence of the phosphorus-containing compound (P-PET) and the inclusion of MOF (MIL-P-PET) increased the LOI values to 29% and 30%, respectively, both achieving a UL94 V0 rating. Compared to P-PET, MIL-P-PET reduced the pHRR by 22%. However, the catalytic effect of ferric ions from Fe-MOF was inactive in the presence of phosphorus-containing compounds, leading to incomplete carbonization and less dense char, which resulted in increased smoke release. This was due to the quenching effect from the P=O and ferric ion complexation during combustion. This demonstrates the importance of carefully selecting materials in the flame retardant system to avoid antagonistic effects that could lead to unexpected fire resistance performance.

4.7 Polycarbonate

Polycarbonate (PC) is a widely used engineering polymer with limited flame resistance and a relatively high heat distortion temperature due to its rigid main chain backbone [58,59]. However, to expand its applications beyond traditional uses, there is a need to improve its flame retardancy and the incorporation of MOFs has garnered attention [58,59]. For instance, Sai et al. [58] prepared PC/Zr-based MOF composites through melt compounding by melt-pressing. The incorporation of 4% and 6% Zr-MOF effectively improved the LOI value from 25% to 28.2% and 28.3%, respectively, achieving a UL94 V0 rating. Digital images of the samples after the UL94 vertical burning and LOI tests are shown in Figure 7. The pHRR and THR decreased by up to 47% and 21%, respectively, compared to PC. The FPI decreased by 185% with the addition of 4% Zr-

MOFs. To furter impreve fire resistence perfoamnce of the PC additional mofication on MOFs is often employed. For instance, Sai et al. [59] synthesized Zr-MOFs on cerium phenylphosphonate lamella using a hydrothermal method. The inclusion of the binary hybrid FR Zr-MOF@CeHPP into PC significantly prolonged the pHRR and time to ignition (TTI) to 145 seconds and 18 seconds, respectively. The composites exhibited suppressed fire hazard, with reductions in THR, pHRR, peak specific extinction area (PSEA), and total smoke release (TSR) by 20%, 45%, 74%, and 18%, respectively. The composites achieved a UL94 V0 rating, with the LOI value increasing from 25% to 28% with the addition of only 2% Zr-MOF@CeHPP into PC.

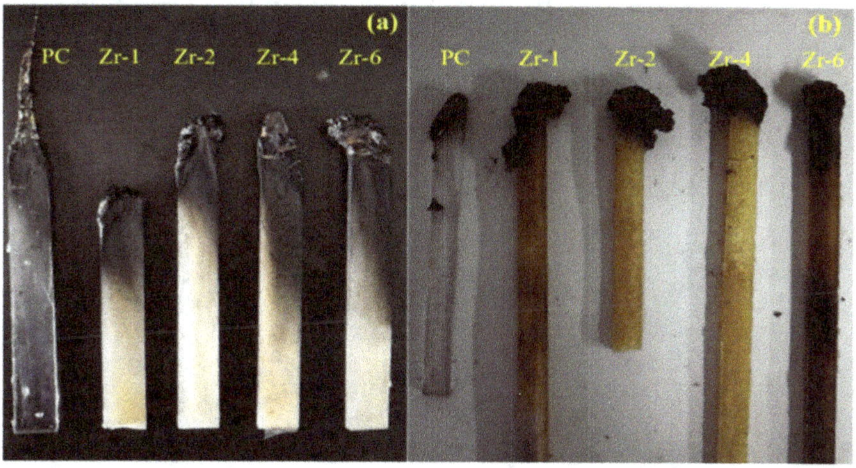

Figure 6. Digital photos of PC composites samples after (a) UL 94 and (b) LOI test. Adapted with permission from [58].

Qian et al. [60] investigated the use of black phosphorus (BP) modified with MIL-53(Al) through a hydrothermal method as a FR for PC. The LOI value reached approximately 31 with a UL94 V0 rating when 1% of MIL-53 functionalized BP was incorporated into the PC matrix. The THR decreased by approximately 8%, 14%, and 19% with the inclusion of 0.3%, 0.5%, and 1.0% of the flame retardant, respectively. Additionally, the pHRR was reduced by approximately 33%, 47%, and 49% when the flame-retardant amount was 0.3%, 0.5%, and 1.0%, respectively.

4.8 Polylactic acid

Polylactic acid (PLA) is one of the most extensively researched biopolymers as a replacement for oil-based synthetic polymers due to its ease of processability, relatively excellent mechanical performance, and biodegradability [25,61]. However, like other polymers, PLA has an LOI value of about 19%, indicating that it is highly flammable, posing a fire hazard for real-life applications [61]. Over the past decades, various additives have been studied to enhance not only the flame retardancy but also to enhance the durability, stabilty and mechanical strength [25]. With focus on the MOF-based additives, Zhang et al. [61] investigated the use of β-cyclodextrin-MOF@Schiff base as a suitable flame retardant for PLA. As shown in Figure 8, β-cyclodextrin was dispersed in methanol and mixed with the MOF precursor, also dissolved in methanol, followed by washing and drying to obtain. The Schiff base was then inserted into the hydrophobic cavity of β-cyclodextrin through a diffusion process. The LOI value reached 25 with a 3% loading of the composite on PLA, and further loading increased the LOI value to 29%, making the material non-flammable. With the addition of 5% and 7%, the materials qualified for a UL94 V1 rating. During early combustion, PLA generates highly flammable molecular segments responsible for the rapid spread of fire. However, the inclusion of the FR with host-guest structures releases guest molecules and Co-containing radicals. The silanol hydroxyl groups from the Schiff base and the OH groups from β-cyclodextrin contribute to char formation in the condensed phase, leading to crosslinked stable char. The cobalt oxide from the MOF also facilitates the aromatization process through catalytic carbonization.

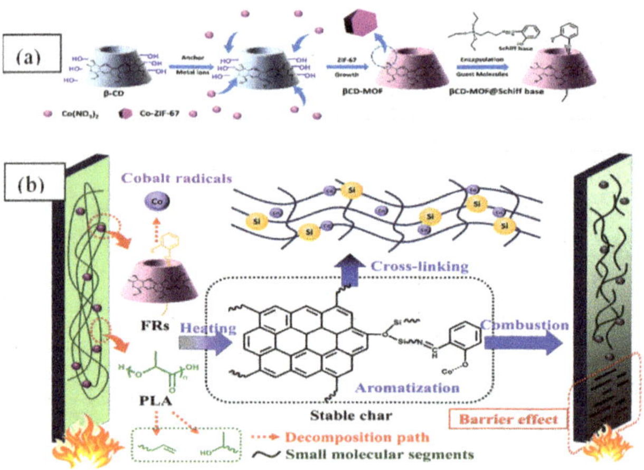

Figure 7. Schematic presentation of the synthesis of (a) β-cyclodextrin-MOF@Schiff base. (b) Proposed flame retardant mechanism for PLA/β-cyclodextrin-MOF@Schiff base composites. Adapted with permission from [61].

Jin et al. [62] enhanced the flame retardancy of PLA by using a cobalt-based MOF (ZIF-67) combined with piperazine pyrophosphate (PAPP). The addition of 4.9% PAPP and 0.2% MOF increased the LOI value from 21% to 33% and achieved a UL94 V0 rating. The pHRR decreased by 28%, while the FPI increased by 67%. The improvements in flame retardency were attributed to the phosphate species generated by PAPP, which promoted the charring process, and the cobalt oxides from the MOF, which stabilized the char, effectively shielding the underlying material. Similarly, Hou et al. [63] incorporated a Co-MOF synthesized using organic ligands modified with 9,10-dihydro-9-oxa-10-phosphaphenanthrene-10-oxide (DOPO) into PLA. The synergistic effect between DOPO and Co-MOF enhanced its flame retardancy properties. The HRR decreased by 27%, peak smoke production by 56%, and total CO yield by 20%, confirming the effectiveness of Co-MOF modified with DOPO as a flame retardant. Zhao et al. [64] reported a significant reduction in CO release by 80% and a 50% reduction in PSPR . The authors synthesized Ce-based MOFs by coordinating binary acid and amino structures with cerium oxides, making them suitable

FRs for PLA. They further introduced phytic acid to strengthen the hydrogen bonding within the network and improve flame retardancy by incorporating phosphorus into the system. The pHRR decreased by approximately 28%.

Zhang et al. [65] synthesized a nickel-based MOF (Ni-MOF) using an eco-friendly and cost-effective method for use as a FR for PLA. The Ni-MOF, modified with phosphate, was prepared through a hydrothermal method using an autoclave at 160°C for 15 hours. To prepare the composites, PLA, wood fiber, and APP were dried and then mixed with phosphate-Ni-MOF using melt compounding followed by melt-pressing. The addition of the filler significantly increased the LOI value for neat PLA, achieving a UL94 V0 rating. The significant improvement in fire resistance performance was further substantiated by cone calorimeter test findings, where PLA/APP/Ni-PO/WF exhibited a decrease in pHRR and a decline in THR compared to neat PLA. The data curves shown in Figure 9 indicate that the addition of Ni-PO and APP can effectively reduce the fire risk of PLA. The HRR curves had multiple peaks associated with two scenarios during combustion. The first main peak is attributed to the formation of char that acts as a thermal insulator and separates O_2 from the sample, thereby reducing the heat release rate. However, the second peak is due to the degradation and breakage of the char structure, which increases the burning area due to exposure of the sample to air. The smoke production rate was also found to decrease to 0.64 m^2, 0.55 m^2, and 0.41 m^2 when 1%, 3%, and 5% of Ni-PO were loaded into the PLA-WF/APP composite. The enhanced performance is due to APP and Ni-PO promoting charring and providing active sites for flammable radicals, retaining them within the char.

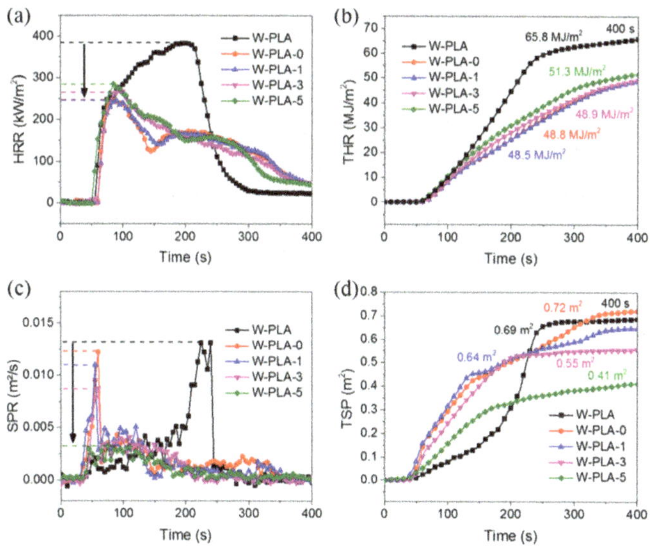

Figure 8. HRR, THR, SPR, and TSP of PLA composites indicate that the presence of APP and Ni-PO in PLA/WF composites reduces the fire risk of PLA. Adapted with permission from [65].

Materials Research Foundations 190 (2026) 194-226
https://doi.org/10.21741/9781644904077-7

4.9 Other polymers

MOFs have also been successfully incorporated into a range of other polymers with the aim of enhancing their flame resistance properties and showcasing their possible applications. For instance, Zhou et al. [66] prepared a hybrid of cellulose nanofibers (CNF)@MOF aerogel with high thermal insulation and excellent flame retardancy using a stepwise assembly approach. The authors immersed sealed vessels containing CNFs into liquid nitrogen, followed by freeze drying to create a self-standing aerogel. They then infiltrated the aerogel with precursors to form CNF-Al-MIL-53 through another freeze drying process. The inclusion of the MOF (Al-MIL-53) imparted flame retardancy to the CNF aerogels, such that when subjected to an alcohol lamp, the CNF-Al-MIL-53 aerogel self-extinguished after removing the flame. This is contracry to the CNF aerogel, which ignited and completely burned within 3 seconds. Similar findings were observed in a study by Nabipour [67], who produced ZIF-8@cellulose aerogels by in situ growth of the MOFs on the surface of cellulose fibers. The presence of the MOF (ZIF-8) enhanced the fire resistance performance of cellulose, with the pHRR decreasing from 140 kW/m^2 to 48 kW/m^2 and the THR decreasing from 13 MJ/g to 9.9 MJ/g.

In the work done by Wang et al. [68], they employed a phosphorylated Zn-based MOF (ZIF-8-P) as an effective FR to reduce the fire risk of poly(methyl methacrylate) (PMMA). Encouragingly, the as-prepared PMMA/ZIF-8-P composite demonstrated efficient fire safety properties (e.g., lower surface resistance; a decrease of ca. 73.1% in the t-PHRR; a lower amount of total pyrolysis products), surpassing those of pure PMMA and the composite containing unphosphorylated MOFs (PMMA/ZIF-8). In order to explore the flame retardance mechanism, the researchers used SEM to examine the compositions and structures of residual carbon generated from the PMMA composites after combustion as shown in Figure 10 a,b. Different from pure PMMA with no residual carbon, both PMMA/ZIF-8 and PMMA/ZIF-8-P could form char after combustion. The remnant char of the PMMA/ZIF-8 sample possessed lots of fragile holes, while a compact and continuous char morphology could be obtained for the PMMA/ZIF-8-P sample. This is because the phosphate groups in the latter composite can act as a highly efficient catalysts toward the formation of graphitized char and can self-decompose into a protective phosphorus-rich surface layer during a burning process. Raman spectroscopy was further used to analyze the residual char structures (Fig. 10c), revealing the obvious G-band and D-band peaks around 1585 and 1350 cm^{-1}, respectively. The intensity ratio (I_G/I_D) of PMMA/ZIF-8-P exceeded that of PMMA/ZIF-8 (0.4177 vs. 0.4001), demonstrating the higher graphitization level of char from the PMMA/ZIF-8-P composite, which is beneficial to blocking heat and gas transfer. From analysis above, the researchers put forward a possible fire safety mechanism for the ZIF-8-P in PMMA as shown in Fig. 10d. They elucidated that the ZIF-8-P decomposes to generate metal-containing residue as well as amino and phosphoric acid compounds, which could form graphitized char during combustion and build a phosphorus-rich protective layer. This thus impedes heat and mass transmission, oxygen permeation, and the release of combustible gases.

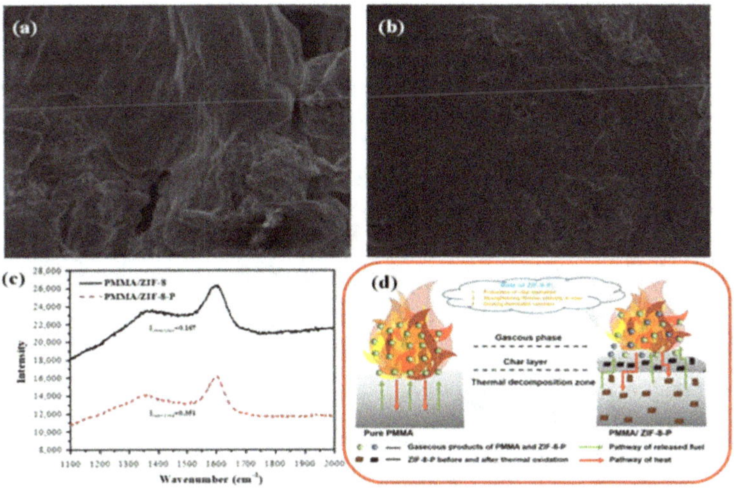

*Figure 9. SEM of the char, including the (**a**) PMMA/ZIF-8 and (**b**) PMMA/ZIF-8-P composites. (**c**) Raman spectra for the carbon residue of the PMMA/ZIF-8 and PMMA/ZIF-8-P composites. (**d**) Schematic diagram of improving the flame retardance of the PMMA/ZIF-8-P composite. Adapted with permission from [68].*

In the same breath, Lin et al. [69] developed a nanohybrid system using oxygen vacancy defect-LA-MOF@MXene composite to enhance the flame resistance of acrylonitrile-butadiene-styrene (ABS). The presence of 0.5% La-MOF@MXene improved the tensile strength and elongation-at-break of ABS. The addition of 1% La-MOF resulted in an LOI value of around 19%, while 1% MXene led to approximately 20%. The inclusion of 1% La-MOF@MXene resulted in an LOI value of around 22%. Cone calorimeter tests revealed that 1% La-MOF@MXene reduced the pHRR from 1013 kW/m^2 to 714 kW/m^2 and the THR from 118 MJ/m^2 to 96 MJ/m^2. The PSPR and TSP decreased by approximately 47% and 29%, respectively, while the PCOP and peak PCO$_2$P declined by around 19% and 36%. Additionally, the smoke formation rate decreased from approximately 2322 kW/m^2 to about 1337.3 kW/m^2, a reduction of around 43% with the addition of 1% La-MOF@MXene. These improvements were attributed to the barrier effect of the flame retardant, the denser char layer, and the catalytic effect of oxygen vacancies.

With the aim of enhancing the flame resistance properties of the highly durable, abrasion resistant polyamide 66 (PA66), Liu et al. [70] incorporated aluminum-based MOFs (Al-MOFs) flame-retardant additives into it. The MOF was attained by converting post-consumer PET waste into functional materials. At an optimal 4 wt% loading, the PA66/Al composite achieved a 40 % increase in LOI (from 25 % to 35 %) and reduced pHRR by 33.04 % (from 727.61 kW·m^{-2} to 487.24 kW·m^{-2}), while enhancing tensile strength by 19.74 % (from 57.42 MPa to 68.76 MPa).

From the reported studies, it is evident that most polymers can easily absorb heat and release pyrolysis products without limitation under the action of heating. Large amounts of heat, dense smoke, toxic volatile gases, and suffocating gases (CO and CO$_2$) are released in the process [71].

In the presence of MOFs however, it can be elucidated that they mainly enhance the thermal stability and influence the fire retardancy of polymers in two aspects: (1) by catalytically promoting the formation of a dense and thermally stable char, which acts as a barrier to protect the polymer from the heat and oxygen; and (2) suppressing the production of smoke and toxic CO through its catalytic oxidation behavior [70-73]. This mechanism is depicted in Fig. 11 [71].

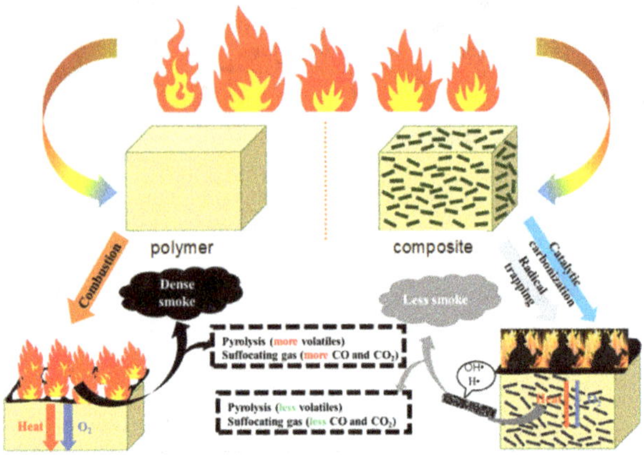

Figure 10. Schematic illustration of the proposed mechanism of MOF-polymer composites. Adapted with permission from [71].

As listed in Table 3, in addition to the polymers discussed in this chapter, there is a wide catalogue of polymers that have been explored and can still be studied for incorporating MOFs (and other FR), thereby enhancing their applicability and value. The evolution towards eco-friendly polymer composites represents a pivotal advancement in fire retardant technology, with plenty of room for future innovation and widespread implementation.

Table 3. List of polymers used for making fire-retardant composite polymers, along with their melting or boiling points (MP/BP). Adapted with permission from [74].

Polymer or Adhesive Polymer	Chemical Formula	MP/BP (°C)
Polyurethane (PU)	$(C_3H_8N_2O)_n$	136
Acrylic resins	$(C_5H_8O_2)_n$	160
Epoxy resins	$(C_{21}H_{25}ClO_5)n$	120
Polyvinyl chloride (PVC)	$(C_2H_3Cl)_n$	100
Polyethylene (PE)	$(C_2H_4)_n$	115–135
Polypropylene (PP)	$(C_3H_6)_n$	130–170
Polyester resins	$(C_{14}H_{22}O_6)_n$	170–172
Silicone polymers	$(SiO_2)_n$	1414
Ethylene-vinyl acetate (EVA) copolymers	$(C_2H_4)_n(C_4H_6O_2)_m$	90
Polyvinyl alcohol (PVA)	$(C_2H_4O)_n$	200
Ethylene propylene diene monomer (EPDM)	$(C_8H_{16})_n$	100–160
Polyvinyl acetate (PVAc)	$(C_4H_6O_2)_n$	60

Polycarbonate (PC)	$(C_{16}H_{14}O_3)_n$	220–230
Acrylonitrile butadiene styrene (ABS)	$(C_8H_8 \cdot C_4H_6 \cdot C_3H_3N)_n$	105
Polyimides	$(C_{22}H_{10}O_4)_n$	360
Polyamide (Nylon)	$(C_{12}H_{22}N_2O_2)_n$	220
Phenolic resins	$C_8H_6O_2$	90–150
Urea-formaldehyde resins	$C_2H_6N_2O_2$	130
Melamine-formaldehyde resins	$C_4H_8N_6O$	354
Polylactic acid (PLA)	$(C_3H_4O_2)n$	150–160
Polyhydroxy alkanoates (PHA)	$(C_6H_{10}O_2)_n$	170
Starch-based polymers	$(C_6H_{10}O_5)_n$	200–220
Cellulose acetate	$(C_{10}H_{16}O_8)_n$	230
Lignin-based polymers	Varies	108–150

5. Challenges and future prospects

Significant progress has been made in the development of MOFs and their polymer hybrids as alternative non-halogenated FRs because halogen-based FRs are now banned. The studies presented indicate that flame resistance varies for each host polymeric material. Therefore, this chapter summarizes the observations reported for each polymer to provide guidance for those interested in pursuing research on novel MOFs as effective FRs. The organic linkers in MOFs act as compatibilizers, improving the dispersion and interfacial interaction of the FRs with the host polymeric materials. This has resulted in many FRs having no adverse effect on the mechanical performance of some polymers. Depending on the metals used, the fabrication of MOFs can be quite expensive. Moreover, MOFs are typically synthesized in small quantities in a laboratory setting due to the intricate nature of their synthesis which requires meticulous control of reaction conditions, including temperature, pressure, and solvent composition, requiring a significant amount of time [71]. Despite these challenges, the rate at which research output is being produced on MOFs as FRs clearly indicates that these materials are the future of non-halogenated FRs. They can become integral to novel FRs that possess the desired properties for reducing fire risk associated with polymeric materials.

The unique features of MOFs, such as their high porosity and specific surface area, make them excellent templates for introducing various materials containing flame retardant elements, resulting in multifunctional FRs. Furthermore, MOFs facilitate the charring process, leading to high graphitization and thermal oxidation resistance. The char acts as a shield, protecting the underlying material from burning during combustion. The transition metals in MOFs serve as oxygen-catalytic and char-forming agents, facilitating the oxidation of carbon oxides and trapping flammable volatiles from the polymeric material. This process results in a compact and denser char that effectively inhibits the transmission of heat and flammable products, thereby reducing the release of toxic and dangerous gases.

6. Conclusions

In summary, this chapter reviewed the use of MOFs as FRs for various polymers. The synthesis methods and flame-retardant effects of MOFs and their hybrids were discussed. MOFs and their hybrids have proven to be effective materials in reducing the fire risk associated with polymeric materials. The unique structure of MOFs, which contains both organic components and metal ions, allows for enhancement of the flame retardancy and smoke suppression of the resulting composite materials. The organic components can serve as a carbon source and release inert products that dilute flammable compounds and oxygen, thereby reducing the heat generated from the polymer. Transition metals, acting as Lewis acids, serve as oxygen-catalytic and char-forming agents that facilitate the oxidation of carbon oxides and trap flammable volatiles from the polymeric material. This process leads to the formation of a compact and denser char that effectively inhibits the transmission of heat and flammable products in or out of the sample, resulting in a decrease in the release of toxic and dangerous gases. Additionally, MOFs possess high porosity and specific surface areas, making them excellent templates for introducing various materials containing flame retardant elements. These features also contribute to the absorption of flammable products, thereby improving the fire resistance performance of the resulting materials. Despite the promising results, research on the use of MOFs as FRs is still at its infancy, and extensive work still needs to be done to optimise the process and enhance the commercial value of these materials. As more researchers design novel FRs using MOFs, there is significant potential for them to become central to the flame retardant market in the near future. The ongoing research and development in this field will likely lead to innovative solutions that enhance fire safety and broaden the applications of polymeric materials.

Acknowledgements

The authors are thankful to the Department of Science, Technology and Innovation/Nanotechnology Innovation Centre Mintek for the financial support.

References

[1] B.-W. Liu, H.-B. Zhao, L. Chen, L. Chen, X.-L. Wang, Y.-Z. Wang, Eco-friendly synergistic cross-linking flame-retardant strategy with smoke and melt-dripping suppression for condensation polymers, Compos. B Eng. 211 (2021) 108664. https://doi.org/10.1016/j.compositesb.2021.108664

[2] X. Feng, W. Xing, L. Song, Y. Hu, In situ synthesis of a MoS2/CoOOH hybrid by a facile wet chemical method and the catalytic oxidation of CO in epoxy resin during decomposition, J. Mater. Chem. A. 2 (2014) 13299-13308. https://doi.org/10.1039/C4TA01885K

[3] T.C. Mokhena, E.R. Sadiku, M.J. Mochane, S.S. Ray, Mechanical properties of fire retardant wood-plastic composites: A review, Express Polym. Lett. 15 (2021) 744-780. https://doi.org/10.3144/expresspolymlett.2021.61

[4] A. Toldy, G. Harakály, B. Szolnoki, E. Zimonyi, G. Marosi, Flame retardancy of thermoplastics polyurethanes, Polym. Degrad. Stab. 97 (2012) 2524-2530. https://doi.org/10.1016/j.polymdegradstab.2012.07.015

[5] K. Song, H. Zhang, Y.-T. Pan, Z. Ur Rehman, J. He, D.-Y. Wang, R. Yang, Metal-organic framework-derived bird's nest-like capsules for phosphorous small molecules towards flame retardant polyurea composites, J. Colloid Interface Sci. 643 (2023) 489-501. https://doi.org/10.1016/j.jcis.2023.04.047

[6] L. Theys, M.J. Mochane, T.G. Mofokeng, M.T. Motloung, M.P. Motloung, T.C. Mokhena, The effect of expandable graphite and montmorillonite (MMT) clay on the morphology, thermal stability, and flammability properties of the maize stalk/PBS bio-composite, J.

Thermoplast. Compos. Mater. (2025) 08927057251329245. https://doi.org/10.1177/08927057251329245

[7] M.J. Mochane, S.I. Magagula, J.S. Sefadi, E.R. Sadiku, T.C. Mokhena, Morphology, Thermal Stability, and Flammability Properties of Polymer-Layered Double Hydroxide (LDH) Nanocomposites: A Review, Crystals, 2020. https://doi.org/10.3390/cryst10070612

[8] T.C. Mokhena, E.R. Sadiku, S.S. Ray, M.J. Mochane, K.P. Matabola, M. Motloung, Flame retardancy efficacy of phytic acid: An overview, J. Appl. Polym. Sci. 139 (2022) e52495. https://doi.org/10.1002/app.52495

[9] M.J. Mochane, T.H. Mokhothu, T.C. Mokhena, Synthesis, mechanical, and flammability properties of metal hydroxide reinforced polymer composites: A review, Polym. Eng. Sci. 62 (2022) 44-65. https://doi.org/10.1002/pen.25847

[10] S. Royer, D. Duprez, Catalytic Oxidation of Carbon Monoxide over Transition Metal Oxides, ChemCatChem 3 (2011) 24-65. https://doi.org/10.1002/cctc.201000378

[11] T.C. Mokhena, E.R. Sadiku, S.S. Ray, M.J. Mochane, T.E. Motaung, The effect of expanded graphite/clay nanoparticles on thermal, rheological, and fire-retardant properties of poly(butylene succinate), Polym. Compos. 42 (2021) 6370-6382. https://doi.org/10.1002/pc.26304

[12] J. Hu, Y.-T. Pan, K. Zhou, P. Song, R. Yang, A new way to improve the fire safety of polyurethane composites with the assistance of metal-organic frameworks, RSC Appl. Polym. 2 (2024) 996-1012. https://doi.org/10.1039/D4LP00257A

[13] J. Yuan, Y.-T. Pan, Y. Lin, W. Zhang, R. Du, Y. Liu, R. Yang, Introduction of Phosphorous to Metal-Organic Frameworks for Fire-Safe Polymers: From Synthesis to Application, Polym. Sci. Technol. (2025). https://doi.org/10.1021/polymscitech.5c00003

[14] M. Wan, C. Shi, X. Qian, Y. Qin, J. Jing, H. Che, F. Ren, J. Li, B. Yu, Interface assembly of flower-like Ni-MOF functional MXene towards the fire safety of thermoplastic polyurethanes, Compos. - A: Appl. Sci. Manuf. 163 (2022) 107187.. https://doi.org/10.1016/j.compositesa.2022.107187

[15] X. Sun, W. Miao, Y.-T. Pan, P. Song, S. Gaan, L.H. Ibarra, R. Yang, Metal-organic frameworks: a solution for greener polymeric materials with low fire hazards, Adv. Sustain. Syst. 9 (2025) 2400768. https://doi.org/10.1002/adsu.202400768

[16] M.U. Farooq, R.R. Zairov, M.K. Kadirov, O.G. Sinyashin, B. Arkook, M. Harb, A step towards rational design of hierarchical porous MOFs architectures for emerging practical implementations, Chem. Eng. J. 515 (2025) 163604. https://doi.org/10.1016/j.cej.2025.163604

[17] S. Carrasco, Metal-organic frameworks for the development of biosensors: A current overview, Biosensors. 8 (2018). https://doi.org/10.3390/bios8040092

[18] F. Yu, X. Yu, P. Jia, X. Zhou, H. Wang, Y. Hou, B. Wang, L. Song, Y. Hu, Synergic effect between La-MOF and MXene: Revolutionizing TPU with improved mechanical strength and fire safety, Chem. Eng. J. 499 (2024) 156145. https://doi.org/10.1016/j.cej.2024.156145

[19] F. Tian, Y. Wu, C. Zhu, Y. She, Y. Jin, B. Wang, H. Chen, H. Xu, Y. Liu, S. Wang, X. Xu, Preparation of ZIF67-modified phosphate compounds for enhancing fire safety of strandboards, Chem. Eng. J. 483 (2024) 149393. https://doi.org/10.1016/j.cej.2024.149393

[20] X. Bi, K. Song, Z. Zhang, T. Lin, Y.-T. Pan, W. Fu, P. Song, J. He, R. Yang, Joint exfoliation of mxene by dimensional mismatched sic/zif-67 toward multifunctional flame retardant thermoplastic polyurethane, Small 20 (2024) 2403375. https://doi.org/10.1002/smll.202403375

[21] H. Liu, L. Deng, Z. Zhao, X. Sun, R. Zhou, C.-M. Shu, J. Jiang, Effects of iron-based metal-organic framework-MIL-53(Fe) and phosphorus-based flame retardants on the flame retardancy of high-density polyethylene, J. Therm. Anal. Calorim. 149 (2024) 4505-4516. https://doi.org/10.1007/s10973-024-13062-2

[22] C. Li, Y. Long, Y. Lou, X. Huo, L. Ma, Y. Tang, C. Hao, X. Wen, Three-in-one novel flame retardant through grafting organophosphorus onto Fe-MOF functionalized Mg(OH)2 for improving fire safety and mechanical properties of PBS composites, Sustain. Mater. Technol. 40 (2024) e00906. https://doi.org/10.1016/j.susmat.2024.e00906

[23] X. Wang, T. Wu, J. Hong, J. Dai, Z. Lu, C. Yang, C. Yuan, L. Dai, Organophosphorus modified hollow bimetallic organic frameworks: Effective adsorption and catalytic charring of pyrolytic volatiles, Chem. Eng. J. 421 (2021) 129697. https://doi.org/10.1016/j.cej.2021.129697

[24] Q. Li, X. Song, Y.-T. Pan, J. Sun, A. Bifulco, R. Yang, Dual function of carboxymethyl cellulose scaffold: A one-stone-two-birds strategy to prepare double-layer hollow ZIF-67 derivates for flame retardant epoxy composites, J. Colloid Interface Sci. 674 (2024) 445-458. https://doi.org/10.1016/j.jcis.2024.06.189

[25] X.G. Wang, P. Qi, S.J. Zhang, S.L. Jiang, Y.C. Li, J. Sun, B. Fei, X.Y. Gu, S. Zhang, A novel flame-retardant modification strategy for UiO66-NH2 by encapsulating triethyl phosphate: preparation, characterization, and multifunctional application in poly (lactic acid), Mater. Today Chem. 30 (2023) 101550. https://doi.org/10.1016/j.mtchem.2023.101550

[26] H. Zhao, B. Yuan, Y. Zhan, F. Yang, J. Zhou, C. Qi, C. Lei, Y. Li, Upgrading the pore-size scale of MIL-53 from microporous to macroporous for adsorbing triethyl phosphate and reducing the fire risk of polystyrene, Compos. - A: Appl. Sci. Manuf. 159 (2022) 107003. https://doi.org/10.1016/j.compositesa.2022.107003

[27] C. Chen, B. Wang, G. Xiao, M. Cao, F. Zhong, Z. Yang, J. Zhou, M. Wang, R. Zou, Tri-source integrated adenosine triphosphate loaded BN in synergy with Cu-MOF to improve the fire safety of epoxy resin, Constr. Build. Mater. 394 (2023) 132258. https://doi.org/10.1016/j.conbuildmat.2023.132258

[28] B. Xu, W. Xu, G. Wang, L. Liu, J. Xu, Zeolitic imidazolate frameworks-8 modified graphene as a green flame retardant for reducing the fire risk of epoxy resin, Polym. Adv. Technol. 29 (2018) 1733-1743. https://doi.org/10.1002/pat.4278

[29] J. Wang, Y. Wei, Z. Wang, X. He, C. Wang, H. Lin, Y. Deng, MOFs-derived self-sacrificing template strategy to double-shelled metal oxides nanocages as hierarchical interfacial catalyst for suppressing smoke and toxic gases releases of epoxy resin, Chem. Eng. J. 432 (2022) 134328. https://doi.org/10.1016/j.cej.2021.134328

[30] J. Chai, Y. Hou, Y. Zhang, Y. Xia, S. Ran, S. Wang, N. Chen, Z. Guo, J. Li, Z. Fang, A Cu-MOF-decorated ammonium polyphosphate for epoxy resins with reduced smoke release and enhanced char layer, J. Appl. Polym. Sci. 141 (2024) e55918. https://doi.org/10.1002/app.55918

[31] F. Zhang, X. Li, L. Yang, Y. Zhang, M. Zhang, A Mo-based metal-organic framework toward improving flame retardancy and smoke suppression of epoxy resin, Polym. Adv. Technol. 32 (2021) 3266-3277. https://doi.org/10.1002/pat.5338

[32] A. Li, W. Xu, R. Chen, Y. Liu, W. Li, Fabrication of zeolitic imidazolate frameworks on layered double hydroxide nanosheets to improve the fire safety of epoxy resin, Compos. - A: Appl. Sci. Manuf. 112 (2018) 558-571. https://doi.org/10.1016/j.compositesa.2018.07.001

[33] H. Wang, H. Qiao, J. Guo, J. Sun, H. Li, S. Zhang, X. Gu, Preparation of cobalt-based metal organic framework and its application as synergistic flame retardant in thermoplastic polyurethane (TPU), Compos. B Eng. 182 (2020) 107498. https://doi.org/10.1016/j.compositesb.2019.107498

[34] Y.N. Kim, Y.-M. Ha, J.E. Park, Y.-O. Kim, J.Y. Jo, H. Han, D.C. Lee, J. Kim, Y.C. Jung, Flame retardant, antimicrobial, and mechanical properties of multifunctional polyurethane nanofibers containing tannic acid-coated reduced graphene oxide, Polym. Test. 93 (2021) 107006. https://doi.org/10.1016/j.polymertesting.2020.107006

[35] C. Liu, K. Xu, Y. Shi, J. Wang, S. Ma, Y. Feng, Y. Lv, F. Yang, M. Liu, P. Song, Fire-safe, mechanically strong and tough thermoplastic Polyurethane/MXene nanocomposites with exceptional smoke suppression, Mater. Today Phys. 22 (2022) 100607. https://doi.org/10.1016/j.mtphys.2022.100607

[36] Y. Shi, C. Liu, L. Fu, Y. Feng, Y. Lv, Z. Wang, M. Liu, Z. Chen, Highly efficient MXene/Nano-Cu smoke suppressant towards reducing fire hazards of thermoplastic polyurethane, Compos Part A Appl Sci Manuf. 150 (2021) 106600. https://doi.org/10.1016/j.compositesa.2021.106600

[37] S.-C. Huang, C. Deng, Z.-Y. Zhao, H. Chen, Y.-Y. Gao, Y.-Z. Wang, Phosphorus-containing organic-inorganic hybrid nanoparticles for the smoke suppression and flame retardancy of thermoplastic polyurethane, Polym. Degrad. Stab. 178 (2020) 109179. https://doi.org/10.1016/j.polymdegradstab.2020.109179

[38] X. Ji, D. Chen, Q. Wang, J. Shen, S. Guo, Synergistic effect of flame retardants and carbon nanotubes on flame retarding and electromagnetic shielding properties of thermoplastic polyurethane, Compos. Sci. Technol. 163 (2018) 49-55. https://doi.org/10.1016/j.compscitech.2018.05.007

[39] B. Xu, W. Xu, Y. Liu, R. Chen, W. Li, Y. Wu, Z. Yang, Surface modification of α-zirconium phosphate by zeolitic imidazolate frameworks-8 and its effect on improving the fire safety of polyurethane elastomer, Polym. Adv. Technol. 29 (2018) 2816-2826. https://doi.org/10.1002/pat.4404

[40] B. Tu, K. Zhou, Q. Zhou, K. Gong, D. Hu, Waste to resource: preparation of an efficient adsorbent and its sustainable utilization in flame retardant polyurethane composites, RSC Adv. 11 (2021) 9942-9954. https://doi.org/10.1039/D0RA10873A

[41] Q. Liu, H. Wang, H. Li, J. Sun, X. Gu, S. Zhang, Constructing a novel synergistic flame retardant by hybridization of zeolitic imidazolate framework-67 and graphene oxide for thermoplastic polyurethane, Polym. Adv. Technol. 33 (2022) 2374-2385. https://doi.org/10.1002/pat.5694

[42] M. Wan, C. Shi, X. Qian, Y. Qin, J. Jing, H. Che, F. Ren, J. Li, B. Yu, K. Zhou, Design of novel double-layer coated ammonium polyphosphate and its application in flame retardant thermoplastic polyurethanes, Chem. Eng. J. 459 (2023) 141448. https://doi.org/10.1016/j.cej.2023.141448

[43] Q. Yao, S. Li, J. Ban, B. Yang, T. Xu, H. Ren, Y. Huang, Functionalized Zr-MOFs for enhancing flame retardancy and smoke suppression performance on TPU, ChemistrySelect 9 (2024) e202401227. https://doi.org/10.1002/slct.202401227

[44] X. Chen, X. Chen, S. Li, C. Jiao, Copper metal-organic framework toward flame-retardant enhancement of thermoplastic polyurethane elastomer composites based on ammonium polyphosphate, Polym. Adv. Technol. 32 (2021) 2829-2842. https://doi.org/10.1002/pat.5260

[45] W. Xu, Z. Cheng, D. Zhong, Z. Qin, N. Zhou, W. Li, Effect of two-dimensional zeolitic imidazolate frameworks-L on flame retardant property of thermoplastic polyurethane elastomers, Polym. Adv. Technol. 32 (2021) 2072-2081. https://doi.org/10.1002/pat.5237

[46] W. Wu, W. Huang, Y. Tong, J. Huang, J. Wu, X. Cao, Q. Zhang, B. Yu, R.K.Y. Li, Self-assembled double core-shell structured zeolitic imidazole framework-8 as an effective flame retardant and smoke suppression agent for thermoplastic polyurethane, Appl. Surf. Sci. 610 (2023) 155540.. https://doi.org/10.1016/j.apsusc.2022.155540

[47] G. Wang, W. Xu, R. Chen, W. Li, Y. Liu, K. Yang, Synergistic effect between zeolitic imidazolate framework-8 and expandable graphite to improve the flame retardancy and smoke suppression of polyurethane elastomer, J. Appl. Polym. Sci. 137 (2020) 48048. https://doi.org/10.1002/app.48048

[48] Y. Geng, Z. Zhao, R. Li, X. Liu, G. Li, M. Wan, L. Liu, X. Chen, C. Jiao, Facile-controllable synthesis of 3D cross-shaped ZIF-67 to achieve highly efficient flame retardancy and smoke suppression for TPU composites, Polym. Degrad. Stab. 227 (2024) 110845. https://doi.org/10.1016/j.polymdegradstab.2024.110845

[49] J. Xie, X. Shi, M. Zhang, X. Dai, X. Wang, Improving the flame retardancy of polypropylene by nano metal-organic frameworks and bioethanol coproduct, Fire Mater. 43 (2019) 373-380. https://doi.org/10.1002/fam.2709

[50] S. Zhang, B. Li, M. Lin, Q. Li, S. Gao, W. Yi, Effect of a novel phosphorus-containing compound on the flame retardancy and thermal degradation of intumescent flame retardant polypropylene, J. Appl. Polym. Sci. 122 (2011) 3430-3439. https://doi.org/10.1002/app.34428

[51] X. Chen, C. Jiao, Flame retardancy and thermal degradation of intumescent flame retardant polypropylene material, Polym. Adv. Technol. 22 (2011) 817-821. https://doi.org/10.1002/pat.1583

[52] Y. Hou, W. Hu, Z. Gui, Y. Hu, preparation of metal-organic frameworks and their application as flame retardants for polystyrene, Ind. Eng. Chem. Res. 56 (2017) 2036-2045. https://doi.org/10.1021/acs.iecr.6b04920

[53] W. Chen, Y. Jiang, R. Qiu, W. Xu, Y. Hou, Investigation of UiO-66 as flame retardant and its application in improving fire safety of polystyrene, Macromol. Res. 28 (2020) 42-50. https://doi.org/10.1007/s13233-019-7165-6

[54] Y. Hou, W. Hu, X. Zhou, Z. Gui, Y. Hu, Vertically Aligned Nickel 2-Methylimidazole Metal-Organic Framework Fabricated from Graphene Oxides for Enhancing Fire Safety of Polystyrene, Ind. Eng. Chem. Res. 56 (2017) 8778-8786. https://doi.org/10.1021/acs.iecr.7b01906

[55] Z. Xu, W. Xing, Y. Hou, B. Zou, L. Han, W. Hu, Y. Hu, The combustion and pyrolysis process of flame-retardant polystyrene/cobalt-based metal organic frameworks (MOF) nanocomposite, Combust. Flame. 226 (2021) 108-116. https://doi.org/10.1016/j.combustflame.2020.11.013

[56] B. Yuan, S. Wu, H. Zhao, Y. Niu, Z. Gao, Y. Zhu, Y. Kong, H. Jin, P. Wang, Controlled release of phosphorus-containing free radical from the porous channels of metal-organic frame to reduce the combustion heat and smoke releases of polystyrene, J. Therm. Anal. Calorim. 149 (2024) 8027-8039. https://doi.org/10.1007/s10973-024-13244-y

[57] T. Ma, W. Wang, R. Wang, Thermal Degradation and Carbonization Mechanism of Fe-Based Metal-Organic Frameworks onto Flame-Retardant Polyethylene Terephthalate, Polymers, 2023. https://doi.org/10.3390/polym15010224

[58] T. Sai, S. Ran, Z. Guo, Z. Fang, A Zr-based metal organic frameworks towards improving fire safety and thermal stability of polycarbonate, Compos. B Eng. 176 (2019) 107198. https://doi.org/10.1016/j.compositesb.2019.107198

[59] T. Sai, S. Ran, Z. Guo, H. Yan, Y. Zhang, P. Song, T. Zhang, H. Wang, Z. Fang, Deposition growth of Zr-based MOFs on cerium phenylphosphonate lamella towards enhanced thermal stability and fire safety of polycarbonate, Compos. B Eng. 197 (2020) 108064. https://doi.org/10.1016/j.compositesb.2020.108064

[60] Z. Qian, B. Zou, Y. Xiao, S. Qiu, Z. Xu, Y. Yang, G. Jiang, Z. Zhang, L. Song, Y. Hu, Targeted modification of black phosphorus by MIL-53(Al) inspired by "Cannikin's Law" to achieve high thermal stability of flame retardant polycarbonate at ultra-low additions, Compos. B Eng. 238 (2022) 109943. https://doi.org/10.1016/j.compositesb.2022.109943

[61] J. Zhang, X. Zhang, R. Wang, W. Wang, H. Zhao, S. Yang, Z. Dong, D.-Y. Wang, Y.-T. Pan, Cyclodextrin-based host-guest hierarchical fire retardants: Synthesis and novel strategy to endow polylactic acid fire retardancy and UV resistance, Carbohydr. Polym. 341 (2024) 122313. https://doi.org/10.1016/j.carbpol.2024.122313

[62] X. Jin, J. Zhang, Y. Zhu, A. Zhang, R. Wang, M. Cui, D.-Y. Wang, X. Zhang, Highly efficient metal-organic framework-based intumescent poly(L-lactic acid) towards fire safety, ignition delay and UV resistance, Int. J. Biol. Macromol. 250 (2023) 126127. https://doi.org/10.1016/j.ijbiomac.2023.126127

[63] Y. Hou, L. Liu, S. Qiu, X. Zhou, Z. Gui, Y. Hu, DOPO-modified two-dimensional co-based metal-organic framework: preparation and application for enhancing fire safety of poly(lactic acid), ACS Appl. Mater. Interfaces. 10 (2018) 8274-8286. https://doi.org/10.1021/acsami.7b19395

[64] K. Zhao, W. Hu, Y. Hou, Nanoconfinement-enhanced fire safety and mechanical properties of polylactic acid with nanocerium metal-organic frameworks, ACS Appl. Mater. Interfaces. 16 (2024) 46750-46760. https://doi.org/10.1021/acsami.4c09184

[65] L. Zhang, S. Chen, Y.-T. Pan, S. Zhang, S. Nie, P. Wei, X. Zhang, R. Wang, D.-Y. Wang, Nickel metal-organic framework derived hierarchically mesoporous nickel phosphate toward smoke suppression and mechanical enhancement of intumescent flame retardant wood fiber/poly(lactic acid) composites, ACS Sustain. Chem. Eng. 7 (2019) 9272-9280. https://doi.org/10.1021/acssuschemeng.9b00174

[66] S. Zhou, V. Apostolopoulou-Kalkavoura, M.V. Tavares da Costa, L. Bergström, M. Strømme, C. Xu, Elastic aerogels of cellulose nanofibers@metal-organic frameworks for thermal insulation and fire retardancy, Nano-Micro Lett. 12 (2019) 9. https://doi.org/10.1007/s40820-019-0343-4

[67] H. Nabipour, S. Nie, X. Wang, L. Song, Y. Hu, Highly flame-retardant zeolitic imidazole framework-8@cellulose composite aerogels as absorption materials for organic pollutants, Cellulose 27 (2020) 2237-2251. 860-9. https://doi.org/10.1007/s10570-019-02860-9

[68] L. Wang, X. Hu, Z. Mao, J. Wang, X. Wang, phosphorylated metal-organic framework for reducing fire hazards of poly(methyl methacrylate), Polymers 14 (2022) 4871. https://doi.org/10.3390/polym14224871

[69] Y. Lin, W. Dong, S. Li, S. Zhang, X. Chen, B. Wang, L. Du, Constructing surface oxygen vacancies defect-La-MOF@MXene for enhanced fire safety and smoke suppression properties in robust ABS nanocomposites, Polym. Degrad. Stab. 234 (2025) 111251. https://doi.org/10.1016/j.polymdegradstab.2025.111251

[70] B. Liu, Y. Huang, Z. Dong, Y. Wu, H. Huang, L. Bing, Z. Zhu, J. Wei, R. Wang, Upcycling PET waste into Al-MOFs with one-step hydrothermal method and its application in flame retardant of PA66, Polym. Degrad. Stab. 241 (2025) 111611. https://doi.org/10.1016/j.polymdegradstab.2025.111611

[71] X. Li, F. Zhang, M. Zhang, X. Zhou, H. Zhang, Comparative Study on the Flame Retardancy and Retarding Mechanism of Rare Earth (La, Ce, and Y)-Based Organic Frameworks on Epoxy Resin, ACS Omega 6 (2021) 35548-35558. https://doi.org/10.1021/acsomega.1c05088

[72] J. Zhang, Z. Li, X. Qi, W. Zhang, D.-Y.Wang, Size-tailored bimetallic metal-organic framework (MOF) on graphene oxide with sandwich-like structure as functional nano-hybrids for improving fire safety of epoxy, Compos. B Eng. 188 (2020) 107881 https://doi.org/10.1016/j.compositesb.2020.107881

[73] B. Hou, Y.-T. Pan, P. Song, Metal-organic frameworks as promising flame retardants for polymeric materials, Microstructures 3 (2023) 2023039. https://doi.org/10.20517/microstructures.2023.37

[74] H. Kolya, C.-W. Kang, Eco-friendly polymer nanocomposite coatings for next-generation fire retardants for building materials, Polymers 16 (2024) 2045. https://doi.org/10.3390/polym16142045

Materials Research Foundations 190 (2026) 227-247
https://doi.org/10.21741/9781644904077-8

Chapter 8

Nitrogen-Based Flame Retardants

G.G. Flores-Rojas[1,2,3], O. Farias-Elvira[1], R. Vera-Graziano[3], L. A. Perez-Carrillo[1], E. Bucio[2], and E. Mendizábal[1]

[1]Department of Chemistry and Chemical Engineering, University Center for Exact Sciences and Engineering, University of Guadalajara, Guadalajara 44430, Jalisco, Mexico

[2]Department of Radiation Chemistry and Radiochemistry, Institute of Nuclear Sciences, National Autonomous University of Mexico, Mexico City 04510, Mexico

[3]Institute of Materials Research, National Autonomous University of Mexico, Mexico City 04510, Mexico

ggabofo@hotmail.com

Abstract

The increasing use of flammable materials in technological and domestic applications has generated the need to increase the safety of these materials against fire risks by using flame retardant compounds. Non-halogenated compounds have stood out for their advantages, such as their low toxicity and lower smoke density. Nitrogen-based flame retardants have stood out for allowing their application in places with limited access and ventilation. However, despite the advantages, effectiveness, and advances in developing nitrogen-based flame retardants, these continue to be combined with other types of retardants, such as metal hydroxides or phosphorus-based flame retardants.

Keywords

Retardants, N-based Flame, Melamine, Intumescent, Nitrogen-based Radical, Phosphazenes

Contents

Materials Research Forum LLC
https://doi.org/10.21741/9781644904077-8

1. Introduction

Flammable materials have several applications in materials focused on construction, technology, aerospace, textile, and automotive, among others [1], has generated a greater need for standard safety requirements for these materials in extreme conditions that can cause their combustion. The most used compounds as retardants are those based on halogens, mainly bromine, which are the most important and used compounds. However, these have begun to be gradually replaced by other types of friendlier retardants that are not halogen-based [2]. Some of the compounds are those based on metal hydroxides [3] or those based on phosphorus [4], silicon [5], and borates [6], the latter presenting a synergistic effect with nitrogen-based retardants [7,8]. In the case of metal hydroxides such as aluminum or magnesium hydroxide, these can decompose into water after combustion, making them friendly to the environment. However, these have low activity since they require high concentrations (up to 60%) that are capable of affecting some of the properties of the matrix, for example the mechanical properties [9].

Nitrogen-based flame retardants are a small group that has grown rapidly and has attracted widespread interest as they are more environmentally friendly and have a similar efficiency to halogen-based compounds and metal hydroxides [2]. These retardants are mainly used in polyolefins, intumescent paints, textiles, and wallpapers, among others [10–12]. Unlike other flame retardants, nitrogen-based flame retardants do not show interference with other additives added to the matrix and, on the contrary, this group of retardants presents numerous advantages, such as their easy handling as they are solid compounds and their lower toxicity [13] and corrosion than those based on halogens that release gases such as hydrogen chloride or hydrogen bromide, as well as the absence of dioxins [14].

The most commonly used inorganic and organic compounds are ammonium polyphosphate and melamine, which are incorporated as intumescent agents in most cases in polyurethanes and various recycled polymers since these types of flame retardants have high decomposition temperatures [15,16]. In this context, nitrogen-based flame retardants are environmentally friendly due to their non-toxic emissions and residues, since they do not require the addition of other elements that are not already present in the polymer [17–19], making the disposal of their residues similar to that of fertilizers due to having a similar elemental composition in nitrogen and phosphorus, and some application are described in Table 1.

Materials Research Foundations 190 (2026) 227-247 https://doi.org/10.21741/9781644904077-8

Table 1. Main applications of N-based flame retardants

Application	Description	Flame retardant	Advantages	Limitations
Polyurethane foams [20]	Use in furniture to improve fire resistance by forming a protective layer	Melamine, ammonium polyphosphate	High thermal stability, low emission of toxic fumes	It can affect mechanical properties if not controlled
Intumescent paints [21]	Formulation for coatings that expand and protect during fires	Melamine, melamine phosphates	Effective protection by forming intumescent layers	High cost and specialized application
Papers, textiles and coatings [22]	Fire protection in fibrous materials with retardants based on guanidine and phosphates	Guanidine phosphates, sulfonamides	Protection in natural substrates	Variable compatibility
Electrical cables and wires [23,24]	Fire retardation and smoke emission reduction in electrical cables	Halogen-free flame retardant polymers, such as melamine polyphosphate	Increased safety, lower toxicity	Balance between electrical properties and retardance
Coatings on composite materials [21]	Combined use to improve thermal resistance and char formation in composites	Mixtures with melamine and phosphates	Greater durability and efficient retardation	Compatibility and formulation costs

2. Types of nitrogen-based flame retardants (NFRs)

NFRs include salts derived from ammonia, melamine, and guanidine compounds such as melamine phosphates, ammonium polyphosphate, guanidine phosphates, guanidine sulfamate, among others [23,25], which are used in materials such as textiles and paints, and these compounds are the most used [13]. In addition, these compounds are characterized by an intense endothermic decomposition with a release of non-flammable nitrogen and ammonia gases, which are capable of mitigating flames [26,27]. Other compounds used as NFRs are N-alkoxy compounds and phosphazenes [28,29]. These compounds present strong interactions with the polymeric base, leading to the formation of a condensed phase that decomposes by decarbonization, which limits their use to specific polymers such as polyolefins and epoxy resins [30,31].

 Materials Research Forum LLC
https://doi.org/10.21741/9781644904077-8

2.1 Ammonia flame retardants

Ammonia flame retardants are used as salts, particularly those of ammonium polyphosphate, ammonium sulfamate, and ammonium pentaborate [32–34]. These types of retardants undergo endothermic decomposition, which causes the release of ammonia gas, leading to rapid carbonization of the material through the formation of an acidic residue that is initially generated in the combustion process [35,36].

2.1.1 Ammonium polyphosphate

Ammonium polyphosphate compounds can be classified into two groups according to their crystalline phase, which are I (APPI) and II (APPII) (Fig. 1). The mode of action of APP compared to halogenated retardants is that they act in the gaseous phase, producing dense smoke by incomplete combustion, while APP acts in the solid phase, resulting in lower smoke emission comparable to that of the original material with complete combustion. APPI and APPII compounds present significant differences in their chemical structure and molecular weight, and these differences affect their solubility in water and thermal behavior, influencing their performance as flame retardants [37,38].

APPI compounds have a linear chain length with a number of phosphate units (n) generally less than 100, which is lower compared to the APPII type. This type of compound has a greater solubility in water compared to the APPII type and a decomposition temperature that begins at approximately 150 ° C, making it ideal for use where a rapid response to thermal exposure is required. However, this characteristic also limits their use in materials focused on constant exposure to high-temperature processes [39,40].

In the case of APPII-type compounds, these present a branched or cross-linked chemical structure with a repeating number of units greater than 1000, providing a higher molecular weight and conferring high thermal stability with a decomposition temperature that oscillates around 300 °C and a lower solubility in water. These characteristics of APPII make it ideal in intumescent systems because they offer better behavior as flame retardants, greater durability in humid conditions, and better compatibility with thermoplastic and thermosetting polymers. Furthermore, recent studies have shown that APP II generates greater carbon formation during combustion, which contributes to protecting polymeric materials against fire [39,41]. Other research has explored the modification of APP compounds, such as the surface modification of APPII or the obtaining of APPV [42,43]. In the case of surface modification of APPII with organic compounds or silane coatings, improving its dispersion in polymeric matrices and reducing the migration of additives, thus optimizing its performance in advanced industrial applications, including high-performance polymers, intumescent paints, and composite materials [44,45]. While APPV-type compounds exhibit improved characteristics, such as greater water insolubility and thermal stability between 300–580 °C, which broaden their applications, these characteristics, which contribute to the improved properties of APPV, could be due to its cross-linked structure, obtained from APPII compounds by heat treatments at 410 °C [46].

The effectiveness of APP as a flame retardant is enhanced by combining it with synergistic compounds, which are carbon-forming agents or foaming agents. The combination of these compounds with APP allows reducing the amount of APP required to achieve an acceptable flame retardant effect in polyolefins. For example, commercial products such as Exolit® AP 760 are synergistic blends that have been shown to increase polypropylene's limiting oxygen index (LOI) almost linearly with increasing loading [47–49].

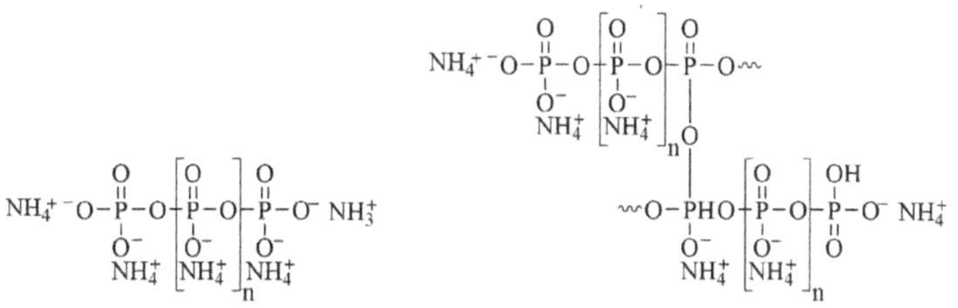

APPI, Linear APP, n<100 APPII, Branched APP, n>1000

Figure 1. Types of ammonium polyphosphate based on their chemical structure and repeating units.

2.1.2 Ammonium pentaborate and sulfamate

Ammonium pentaborate and sulfamate are inorganic compounds employed as flame retardants due to their ability to act in both the solid and gaseous phases of the combustion process. They contribute to thermal inhibition and the suppression of free radicals.

2.1.2.1 Ammonium pentaborate

Ammonium pentaborate is a borated salt that acts as a retarder in cellulosic materials and polymers such as epoxies, polyamides, and polyurethanes. It is primarily used when an easily soluble alkaline retardant is required or when alkali metals are not possible [34,50]. Ammonium pentaborate is used by spraying or by immersing the material. Ammonium pentaborate solutions are applied in most cases by spraying or through immersion to produce a flame retardant product [23], where reduction of thermal stability is the primary flame inhibition mechanism, through the formation of carbon residues and a protective layer of nitrous oxide that presents the following processes [34,50,51]:

- Endothermic decomposition of the flame retardant with a possible temperature range of 100–450°C, absorbing heat in the combustion zone [34,50].
- Formation of a coating of carbonized material rich in boron, which acts as a barrier that retains decomposition products, inhibiting the passage of oxygen and heat transfer.
- Ammonia is released during its thermal decomposition, which displaces oxygen in the combustion zone and dilutes flammable gases [34].

Additional studies have shown that ammonium pentaborate can improve the fire resistance of wood-plastic composites and flame-resistant textiles, mainly when used in conjunction with other synergistic additives such as phosphates [34,50,52].

2.1.2.2 Ammonium sulfamate

Ammonium sulfamate has similar applications to ammonium pentaborate and is notable for its use in cellulosic and some polymeric materials [17,53,54]. Its mechanism of action includes [55,56]:

Materials Research Foundations 190 (2026) 227-247 https://doi.org/10.21741/9781644904077-8

- Endothermic decomposition, which reduces the temperature of the combustion zone [54].
- Release of water vapor and ammonia gas, which act as diluents of the oxygen available in the combustion zone [57].
- Inhibition of the free radical chain reaction in the gas phase (similar to some halogenated mechanisms but without generating toxic or corrosive smoke) [58].

This compound is especially useful in formulations seeking to comply with strict environmental regulations, as it does not emit halogenated compounds or heavy metals during combustion. Its use has also been tested in mixtures with other nitrogenous compounds, such as urea, to enhance its flame retardant properties [59,60].

2.2 Melamine-based flame retardants

Melamine-based compounds used as flame retardants are widely used in polymer and composite materials. Melamine ($C_3H_6N_6$), due to its high nitrogen content (~66 wt%), acts as an effective thermal inhibition agent, undergoing endothermic decomposition leading to quenching during combustion, release of water, CO_2, and ammonia, providing a layer of char that also contributes to combustion inhibition (Fig. 2) [17,23,61,62]. These compounds are generally used as melamine salts, where their acid counter ion also presents a retarding effect, and the selection of this acid can be crucial in determining in most cases, not only the thermal stability of the retarding compound but also ensuring the effectiveness in promoting carbonization of the polymeric material and other polymeric interactions, where some of the factors to consider in the selection of the acid is the susceptibility of the matrix polymer to hydrolysis [62–65]. In addition to melamine and its salts used as flame retardants, other derivatives are used, taking advantage of their synergistic effect with other phosphorus-based flame retardants as well as some metal hydroxides and halogenated retardants [66–68]. Some of the disadvantages of this type of retardant are their low solubility, which limits their applications, in addition to requiring high loads in the base polymers, which diminishes their mechanical properties [69,70].

The mechanism of action of melamine and its derivatives as flame retardants have a similar mechanism of action caused by the reactive amino groups of melamine, which can be considered partial and is based on the following synergistic processes:

- Endothermic decomposition: this property of melamine causes a decrease in the temperature of the combustion zone, slowing the decomposition of the base material [26].
- Release of non-flammable gases: when melamine reaches its decomposition temperature, it releases many non-flammable gases such as ammonia, CO_2, and water. These gases disrupt the combustion zone, hindering combustion by forming an inert atmosphere that suppresses oxidation reactions [40].
- Formation of a carbonized layer when melamine acts as a foaming agent, classified as an intumescent system that promotes expansion, generating carbonization in the form of foam, acting as a thermal and physical barrier against fire [23].

Materials Research Forum LLC
https://doi.org/10.21741/9781644904077-8

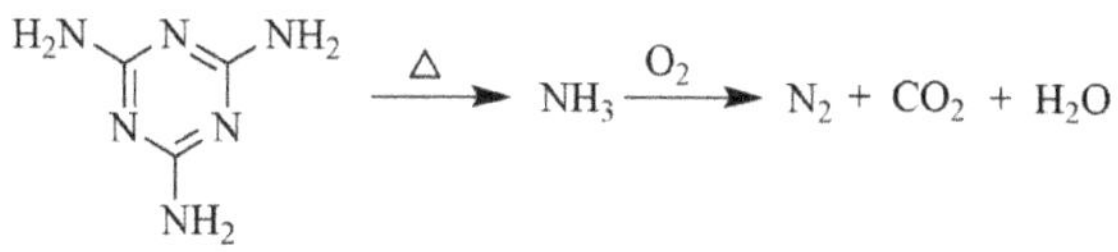

Figure 2. Melamine combustion products.

2.2.1 Melamine salts

Melamine salts are obtained by combining melamine with organic or inorganic acids, producing some of the following salts such as melamine cyanurate, melamine phosphate/melamine pyrophosphate, melamine formaldehyde, among others [71–73], which are used in polymers such as polyamides (melamine cyanurate), polyurethane foams, polyesters (melamine phosphate/melamine pyrophosphate) or used as flame retardant materials such as a resin, as is the case for melamine formaldehyde [26,74–76]. These salts present greater thermal stability compared to melamine, which has a low sublimation temperature of around 250-350 C [21,77–79], therefore making it impossible to apply them in polymers that are processed at a temperature higher than this, such as polypropylene, polyesters, and polyamines, where melamine cyanurate salts or melamine (poly)phosphate are used, which have a decomposition temperature of 300 and 350 C respectively [80,81].

2.2.1.1 Melamine cyanurate

Melamine cyanurate salt (MCA) is synthesized by the stoichiometric reaction between melamine and cyanuric acid, forming a complex with a molecular structure that presents a high organization by hydrogen bonds, conferring high thermal stability and low solubility (Fig. 3) [82,83]. This salt stands out for its thermal stability and compatibility with various polymers due to its crystalline morphology that allows an easy and homogeneous dispersion in the polymer matrix without significantly compromising the mechanical properties by requiring effective concentrations lower than 10%, as is the case with nylon-6, where the nitrogen-rich structure of melamine inhibits fire spread in the base polymer [83–85]. These characteristics of melamine cyanurate distinguish it from other additives that can affect the processability or mechanical resistance of the polymer due to its high concentration in the base polymer [86].

Some of the key advantages of melamine cyanurate are its possible application in materials that may be in contact with food where a tolerance of up to 3.2 mg/kg may exist, as well as in sensitive environments [87,88]. Other advantages are its affinity with polyamides, polyvinyl alcohol, cellulose, polypropylene, polyacrylics, and polyesters, maintaining their mechanical resistance and transparency properties in the corresponding cases [89], making this salt preferred for materials that require maintaining their aesthetic and structural properties at low costs because smaller adequate quantities are required [90]. Despite having good efficacy, melamine cyanurate has a lower performance than phosphorus salts, which also form a protective carbon layer. Furthermore, its low solubility makes it challenging to use in liquid coatings, where more polar derivatives such as melamine formaldehyde are preferred. Therefore, its effectiveness depends mainly on the polymeric matrix, where polystyrene or polypropylene, for example, has shown good efficiency [72,91–93].

 Materials Research Forum LLC
https://doi.org/10.21741/9781644904077-8

The applications of melamine cyanurate are mainly in the electrical, automotive, textile, construction materials (thermal insulation) and food utensils (used in films and plastic containers that require high heat resistance, such as those used in baking, microwave ovens, and sterilization). In the textile, automotive and electrical sectors, it is used in components such as connectors, battery housings and printed circuits, as well as high-security upholstery, which is used in car and airplane cabins because it has a delayed ignition time without generating toxic and corrosive fumes that damage devices and people, complying with some regulations such as UL94, UL94 V0 and IEC 60695-11-10 [23,88,94–97]. In simple terms, melamine cyanurate salt can be considered an effective and economical solution for engineering applications requiring fireproofing properties. However, this compound can exhibit erratic or unreliable performance depending on its polymer structure, so an appropriate formulation is required.

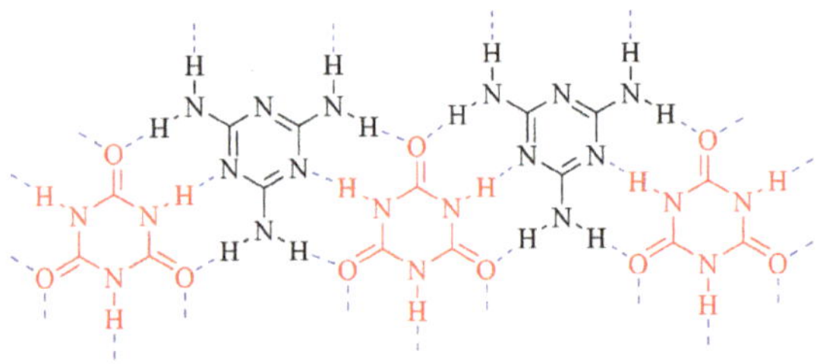

Figure 3. Structure of meleamine cyanurate.

2.2.1.2 Melamine polyphosphate

This compound is considered a flame retardant obtained by combining melamine and phosphoric acid (Fig. 4), and its thermal stability increases according to the acid used, where the polyphosphate has the highest thermal stability thanks to the number of repeating units, followed by the pyrophosphate and finally the phosphate has the lowest thermal stability due to having only one phosphate unit in its structure, where the phosphate, when reaching a temperature above 200 °C, undergoes a condensation reaction to form melamine pyrophosphate and finally, when it reaches a temperature above 260 ° C, it gives rise to the formation of melamine polyphosphate which undergoes an endothermic decomposition above 350 °C [98,99]. This compound offers increasing thermal stability with a decomposition that releases water, phosphoric and polyphosphoric acid, promoting heat dissipation, forming a protective layer of carbon residue due to its intumescent effectiveness, which is benefited by the melamine that causes a carbon foamy residue as a layer that decreases the oxygen available in the combustion process, protecting the polymer. Despite its compatibility with various polymers, such as polyamides and polyurethanes, it can be less efficient in polymers with a low charring tendency; however, melamine phosphate can be combined with other compounds to optimize the formation of a protective layer of carbon residue [100–104].

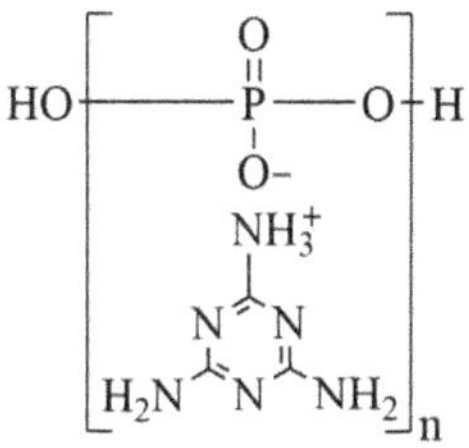

n=1, melamine phosphate
n=2, melamine pyrophosphate
n=3, melamine polyphosphate

Figure 4. Structure of melamine phosphates.

2.2.1.3 Melamine condensates

Melamine condensates are compounds formed through the controlled thermal condensation of melamine and other precursors, such as cyanamide, dicyanamide, and various acids, including phosphoric acid, polyphosphoric acid, or methanesulfonic acid. This process leads to increasingly complex chemical structures derived from melamine, which decomposes at around 350°C. These structures include melam salts, which also decompose in a similar temperature range of 350-370°C. In contrast, more complex compounds like melem and melon have higher decomposition temperatures of 460°C and 600°C, respectively, due to their intricate structures. [105–107]. These compounds are part of a family of nitrogen-based chemicals characterized by triazine and heptazine rings. They are becoming increasingly important as flame retardants because of their high nitrogen content, their ability to create protective layers during combustion, and their superior thermal stability compared to melamine, which sublimates from 250 °C to 350 °C [21,79,108]. The mechanism by which melamine condensates function is akin to that of their precursor compounds. They undergo endothermic decomposition, releasing non-combustible gases and forming charred residues that provide a fire-protective layer over the underlying material [23].

Melamine condensates offer several advantages over other salts, particularly due to their high capacity for forming intumescent carbon. They create dense and stable carbonized layers, which surpass those produced by alternatives like cyanurate or melamine phosphate. When combined with these other salts, melamine condensates can enhance performance in specific applications due to their synergistic effects. These characteristics allow it to expand its industrial applications in technical polymers such as biobased rigid polyurethanes, reinforced polyamides, and thermoplastic polyesters, as well as intumescent coatings in the form of paints and construction materials such as MDF boards, decorative laminates, thermal insulation or in automotive plastic housings and heat-sensitive electronic connectors. However, some of its disadvantages are its low solubility, which limits its processing, lower commercial availability, and production costs [106,109].

In summary, melamine condensates represent an advanced solution as flame retardants because they can form protective char layers, inhibit fire propagation through endothermic mechanisms, and release inert gases. Also, the propensity of melamine condensates to be chemically functionalized gives them substantial promise in areas such as the flame retardant polymer sector, where further studies on chemical functionalization are needed to broaden the chemicals as a material and establish them as a substantial part of today's modern fire design. However, these materials have issues with production costs and complex processing [110,111].

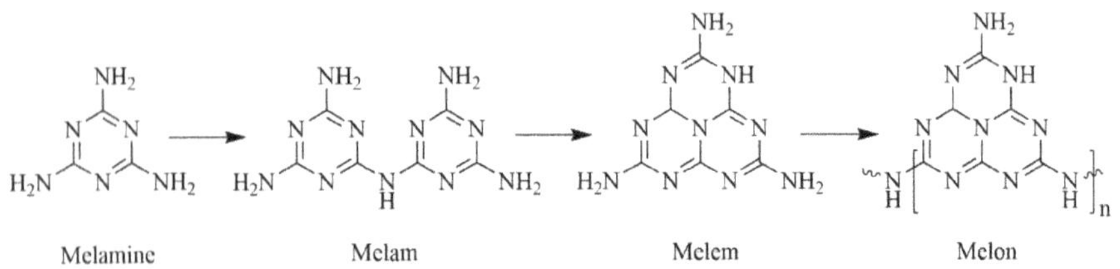

Figure 4. Condensation products of melamine.

2.3 Nitrogen-based radical generators

Nitrogen-based radical generators mainly include compounds such as N-alkoxyamines (NOR), azoalkanes (AZO), azoxys, hydrazones, amidines, azines, triazenes and oxidized cyclic amines. Their chemical structure consists of N–O or N–N bonds, which are susceptible to homolytic cleavage at high temperatures, inhibiting combustion by forming inert gases and the formation of free radicals that interfere with the free radical propagation reactions that sustain the flame. The thermal decomposition of these compounds releases stable or semi-stable free radicals such as nitroxyl ($R_2NO\cdot$), aminol ($R_2N\cdot$), or alkyl ($R\cdot$) radicals, which reduce the spread of fire through the Denisov cycle, which consists of interrupting or inhibiting the chain reactions of free radicals such as $H\cdot$ and $OH\cdot$, which propagate oxidation and thermal degradation in polymers and organic materials, thus contributing to protection against fire and degradation (Fig. 5) [28,112].

These retardants can be used in solid or semi-solid matrices, depending on the polymer and manufacturing process, showing high compatibility with technical polymers, as well as with other retardant compounds such as hydrazones and azines. In contrast, other compounds offer a hybrid retardant mechanism, such as tetrapotassium azo diphosphonate (INAZO), which combines action mechanisms in the gaseous and solid phases due to its chemical composition with nitrogen and phosphorus [113,114]. These characteristics position them as an advanced alternative to traditional retardants such as melamine salts and their condensates, characterized by their high retardant efficiency at low concentrations that can be even less than 1% by weight [28,115].

The thermal stability of nitrogen radical generators is adequate for their use in most engineering polymers. For example, azoalkanes and N-alkoxyamines have a decomposition temperature between 200 and 300 °C, releasing active radicals in the temperature range where the polymers begin to thermally degrade [115]. This timing ensures that the retardant effect is activated just when needed, maximizing fire protection without compromising stability during processing or normal use of the material. However, some of these types of retardants can have a very low stability temperature, limiting their use, as is the case with AIBN, which has a stability range of 65-70 °C [116]. The main applications of this class of retardants are based on thermoplastic polymers such as polypropylene (PP), as well as paints [112,117].

Materials Research Forum LLC
https://doi.org/10.21741/9781644904077-8

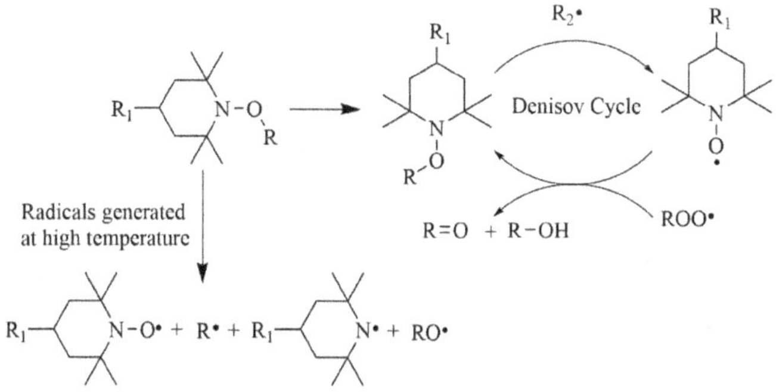

Figure 5. Mode of action of N-alkoxy radicals and Denisov Cycle.

2.4 Phosphazenes, phosphame and phosphoroxynitride

Phosphazene, phosphame, and phosphoroxynitride compounds are important flame retardants based on a composition of nitrogen and phosphorus [10,118,119]. Phosphame and phosphazane type compounds are compounds that present a polymeric structure crosslinked by P-N or –P=N– bonds, while the phosphoroxynitride compound presents a P–N–O bond [120]. These compounds offer advanced solutions for fire protection in various materials with better effectiveness compared to phosphates [121]. The main mechanism of action of these compounds is the development of a carbonized surface on the material that protects it during the combustion process (intumescence), release of non-combustible gases, and capture of free radicals in the flame, diluting the flammable gases and cooling the affected area, mechanisms that slow down or stop the spread of fire [10]. Some of the advantages of this type of retardant are their high thermal and chemical stability as well as their high efficiency, which allows their use in polymers, textiles, and composite materials with a low ratio between the retardant and the matrix [121,122], besides to presenting lower toxicity in combustion gases [122]. However, these can have some disadvantages, such as presenting a relatively high cost compared to other retardants due to their complexity of synthesis and formulation. In addition, their incorporation can affect the mechanical or aesthetic properties of the matrix if its formulation is not optimized [118].

Conclusion

Nitrogen-based flame retardants present a wide variety of uses, either individually or in combination with other compounds that optimize their effectiveness. These compounds decompose during combustion, releasing volatile agents that suppress flames and therefore the spread of fire through intumescence and the formation of protective layers. Additionally, they reduce the temperature in the combustion zone by dissipating heat through endothermic processes. Some of these compounds directly interfere with polymer combustion by acting as potent halogen-free carbonizing agents, making them attractive for applications requiring low emission of toxic and corrosive fumes.

Acknowledgment

The authors would like to thank the following institutions for their support: University of Guadalajara for PROSNI 2022 (E.M.), Dirección General de Asuntos del Personal Académico at the National Autonomous University of Mexico for Grant IN204223 (E.B.), and SECIHTI for the postdoctoral fellowship (CVU 407270) awarded to G.G.F.R.

References

[1] T.-M. Nguyen, S. Chang, B. Condon, J. Smith, Fire self-extinguishing cotton fabric: Development of piperazine derivatives containing phosphorous-sulfur-nitrogen and their flame retardant and thermal behaviors, MSA 05 (2014) 789-802. https://doi.org/10.4236/msa.2014.511079

[2] N. Arastehnejad, M.R. Sulaiman, R.K. Gupta, Nitrogen-Based Ecofriendly Flame retardants for Polyurethane Foams, in: R.K. Gupta, P.K. Kahol (Eds.), ACS Symposium Series, American Chemical Society, Washington, DC, 2021, pp. 167-185. https://doi.org/10.1021/bk-2021-1380.ch006

[3] T. Adschiri, Y. Hakuta, K. Sue, K. Arai, Hydrothermal synthesis of metal oxide nanoparticles at supercritical conditions, J. Nanoparticle Res. 3 (2001) 227-235. https://doi.org/10.1023/A:1017541705569

[4] S.V. Levchik, E.D. Weil, A review of recent progress in phosphorus-based flame retardants, J. Fire Sci. 24 (2006) 345-364. https://doi.org/10.1177/0734904106068426

[5] M. Kilinc, Silicon Based Flame retardants, in: A.B. Morgan, C.A. Wilkie (Eds.), Non-Halogenated Flame retardant Handbook, 1st ed., Wiley, 2014, pp. 169-199. https://doi.org/10.1002/9781118939239.ch5

[6] S. Bourbigot, M.L. Bras, R. Leeuwendal, K.K. Shen, D. Schubert, Recent advances in the use of zinc borates in flame retardancy of EVA, Polym. Degrad. Stab. 64 (1999) 419-425. https://doi.org/10.1016/S0141-3910(98)00130-X

[7] J.E. Hendrix, J.E. Bostic, E.S. Olson, R.H. Barker, Pyrolysis and combustion of cellulose. I. Effects of triphenyl phosphate in the presence of nitrogenous bases, J. Appl. Polym. Sci. 14 (1970) 1701-1723. https://doi.org/10.1002/app.1970.070140705

[8] C.Q. Yang, X. Qiu, Flame-retardant finishing of cotton fleece fabric: part I. The use of a hydroxy-functional organophosphorus oligomer and dimethyloldihydroxylethyleneurea, Fire Mater. 31 (2007) 67-81. https://doi.org/10.1002/fam.926

[9] M.J. Mochane, T.H. Mokhothu, T.C. Mokhena, Synthesis, mechanical, and flammability properties of metal hydroxide reinforced polymer composites: A review, Polym. Eng. Sci. 62 (2022) 44-65. https://doi.org/10.1002/pen.25847

[10] A.M. Palve, J.S. Suroshe, R.K. Gupta, Recent developments in nitrogen- and phosphorous-based flame retardants for polyurethanes, in: R.K. Gupta (Ed.), ACS Symposium Series, American Chemical Society, Washington, DC, 2021, pp. 189-212. https://doi.org/10.1021/bk-2021-1400.ch011

Materials Research Forum LLC
https://doi.org/10.21741/9781644904077-8

[11] A. Castellano, C. Colleoni, G. Iacono, A. Mezzi, M.R. Plutino, G. Malucelli, G. Rosace, Synthesis and characterization of a phosphorous/nitrogen based sol-gel coating as a novel halogen- and formaldehyde-free flame retardant finishing for cotton fabric, Polym. Degrad. Stab. 162 (2019) 148-159. https://doi.org/10.1016/j.polymdegradstab.2019.02.006

[12] H. Sun, W. Wang, Y. Liu, Q. Wang, A highly efficient, colorless phosphorus-nitrogen synergistic flame retardant for durable flame retardancy in wood pulp paper, Polym. Degrad. Stab. 215 (2023) 110468. https://doi.org/10.1016/j.polymdegradstab.2023.110468

[13] H. Horacek, R. Grabner, Advantages of flame retardants based on nitrogen compounds, Polym. Degrad. Stab. 54 (1996) 205-215. https://doi.org/10.1016/S0141-3910(96)00045-6

[14] H. Feuchter, F. Poutch, A. Beard, The impact of halogen-free phosphorus, inorganic and nitrogen flame retardants on the toxicity and density of smoke from 10 common polymers, Fire Mater. 47 (2023) 1003-1023. https://doi.org/10.1002/fam.3145

[15] Z.-B. Shao, C. Deng, Y. Tan, M.-J. Chen, L. Chen, Y.-Z. Wang, Flame retardation of polypropylene via a novel intumescent flame retardant: Ethylenediamine-modified ammonium polyphosphate, Polym. Degrad. Stab. 106 (2014) 88-96. https://doi.org/10.1016/j.polymdegradstab.2013.10.005

[16] A. Konig, U. Fehrenbacher, T. Hirth, E. Kroke, Flexible polyurethane foam with the flame retardant melamine, J. Cell. Plast. 44 (2008) 469-480. https://doi.org/10.1177/0021955X08095502

[17] D. Price, Y. Liu, G.J. Milnes, R. Hull, B.K. Kandola, A.R. Horrocks, An investigation into the mechanism of flame retardancy and smoke suppression by melamine in flexible polyurethane foam, Fire Mater. 26 (2002) 201-206. https://doi.org/10.1002/fam.810

[18] J. Gao, Y. Wu, J. Li, X. Peng, D. Yin, H. Jin, S. Wang, J. Wang, X. Wang, M. Jin, Z. Yao, A review of the recent developments in flame retardant nylon composites, Compos. C: Open Access 9 (2022) 100297. https://doi.org/10.1016/j.jcomc.2022.100297

[19] Y. Ding, L. Guo, J. Yang, Y. Li, Z. Cao, G. Lu, Z. Zhao, Functionalisation of multiwalled carbon nanotubes with melamine phosphate and their influence on morphology, thermal stability, flame retardancy and mechanical properties of ABS, Plast. Rubber Compos. 50 (2021) 92-103. https://doi.org/10.1080/14658011.2020.1840200

[20] M. Chen, Y. Yuan, W. Wang, L. Xu, Recent advances in flame retardant flexible polyurethane foams, Fire 8 (2025) 90. https://doi.org/10.3390/fire8030090

[21] R.G. Puri, A.S. Khanna, Intumescent coatings: A review on recent progress, J. Coat. Technol. Res. 14 (2017) 1-20. https://doi.org/10.1007/s11998-016-9815-3

[22] R. Nazir, S. Gaan, Recent developments in P(O/S)-N containing flame retardants, J. Appl. Polym. Sci. 137 (2020) 47910. https://doi.org/10.1002/app.47910

[23] C.A. Wilkie, A.B. Morgan, Fire Retardancy of Polymeric Materials, 3rd ed., CRC Press, Boca Raton, 2024. https://doi.org/10.1201/9781003380689

[24] Z. Liu, B. Li, K. Huang, Y. Lu, S. Yu, Z. An, Z. Ma, J. Ye, Y. Chen, Chen, Research on preparation and performance tests of the fire retardant coatings for special functionalized wires and cables, ACE, 65 (2024) 342-351. https://doi.org/10.54254/2755-2721/65/20240533

Materials Research Forum LLC
https://doi.org/10.21741/9781644904077-8

[25] C. Wan, G. Zhang, F. Zhang, A novel guanidine ammonium phosphate for preparation of a reactive durable flame retardant for cotton fabric, Cellulose 27 (2020) 3469-3483. https://doi.org/10.1007/s10570-020-03003-1

[26] Y. Liu, Q. Wang, The investigation on the flame retardancy mechanism of nitrogen flame retardant melamine cyanurate in polyamide 6, J. Polym. Res. 16 (2009) 583-589. https://doi.org/10.1007/s10965-008-9263-6

[27] H.-H. Kim, M.-J. Sim, J.-C. Lee, S.-H. Cha, The effects of chemical structure for phosphorus-nitrogen flame retardants on flame retardant mechanisms, J. Mater. Sci. 58 (2023) 6850-6864. https://doi.org/10.1007/s10853-023-08414-6

[28] C.-E. Wilén, R. Pfaendner, Design and utilization of nitrogen-containing flame retardants based on N-alkoxyamines, azoalkanes and related compounds, in: Polymer green flame retardants, Elsevier, 2014, pp. 267-288. https://doi.org/10.1016/B978-0-444-53808-6.00008-1

[29] X. Zhou, J. Wang, L. He, S. Qiu, R. Pan, Preparation of an integrated P-N-Si tube containing phosphazene-triazine-silica structure: An efficient flame retardant agent for epoxy resin, Compos. Commun. 38 (2023) 101504. https://doi.org/10.1016/j.coco.2023.101504

[30] S. Qiu, C. Ma, X. Wang, X. Zhou, X. Feng, R.K.K. Yuen, Y. Hu, Melamine-containing polyphosphazene wrapped ammonium polyphosphate: A novel multifunctional organic-inorganic hybrid flame retardant, J. Hazard. Mater. 344 (2018) 839-848. https://doi.org/10.1016/j.jhazmat.2017.11.018

[31] S.M. Seraji, H. Gan, S.R. Swan, R.J. Varley, Phosphazene as an effective flame retardant for rapid curing epoxy resins, React. Funct. Polym.164 (2021) 104910. https://doi.org/10.1016/j.reactfunctpolym.2021.104910

[32] L. Duan, H. Yang, L. Song, Y. Hou, W. Wang, Z. Gui, Y. Hu, Hyperbranched phosphorus/nitrogen-containing polymer in combination with ammonium polyphosphate as a novel flame retardant system for polypropylene, Polym. Degrad. Stab. 134 (2016) 179-185. https://doi.org/10.1016/j.polymdegradstab.2016.10.004

[33] L. Xia, H. Wang, B. Shentu, Z. Weng, Ammonium sulfamate flame retarded polyamide 6: Morphology and thermal degradation, Chin. J. Polym. Sci. 28 (2010) 753-759. https://doi.org/10.1007/s10118-010-9116-8

[34] R.E. Myers, E.D. Dickens, E. Licursi, R.E. Evans, Ammonium pentaborate: an intumescent flame retardant for thermoplastic polyurethanes, J. Fire Sci. 3 (1985) 432-449. https://doi.org/10.1177/073490418500300604

[35] F. Fang, X. Zhang, Y. Meng, Z. Gu, C. Bao, X. Ding, S. Li, X. Chen, X. Tian, Intumescent flame retardant coatings on cotton fabric of chitosan and ammonium polyphosphate via layer-by-layer assembly, Surf. Coat. Technol. 262 (2015) 9-14. https://doi.org/10.1016/j.surfcoat.2014.11.011

[36] G. Camino, L. Costa, L. Trossarelli, Study of the mechanism of intumescence in fire retardant polymers: Part V-Mechanism of formation of gaseous products in the thermal degradation of ammonium polyphosphate, Polym. Degrad. Stab. 12 (1985) 203-211. https://doi.org/10.1016/0141-3910(85)90089-8

[37] M. Watanabe, M. Sakurai, M. Maeda, Preparation of ammonium polyphosphate and its application to flame retardant, Phosphorus Res. Bull. 23 (2009) 35-44. https://doi.org/10.3363/prb.23.35

[38] M. Watanabe, H. Suzuki, S. Kawasaki, M. Maeda, M. Sakurai, Preparation of ammonium polyphosphate forms II and V and their Use as flame retardants, Phosphorus Res. Bull. 15 (2004) 95-99. https://doi.org/10.3363/prb1992.15.0_95

[39] I. Mihajlović, Recent Development of Phosphorus Flame retardants in Thermoplastic Blends and Nanocomposites, in: P.M. Visakh, Y. Arao (Eds.), Flame retardants, Springer International Publishing, Cham, 2015, pp. 79-114. https://doi.org/10.1007/978-3-319-03467-6_4

[40] A.B. Morgan, M. Klatt, Nitrogen-Based Flame retardants, in: A.B. Morgan (Ed.), Non-Halogenated Flame retardant Handbook 2 Nd Edition, 1st ed., Wiley, 2021, pp. 236-270. https://doi.org/10.1002/9781119752240.ch5

[41] K. Kishore, K. Mohandas, Action of phosphorus compounds on fire-retardancy of cellulosic materials: A review, Fire Mater. 6 (1982) 54-58. https://doi.org/10.1002/fam.810060203

[42] G. Liu, X. Liu, J. Yu, A Ammonium polyphosphate with crystalline form V by ammonium dihydrogen phosphate process, Ind. Eng. Chem. Res. 49 (2010) 5523-5529. https://doi.org/10.1021/ie100453a

[43] J. Yan, P. Xu, P. Zhang, H. Fan, Surface-modified ammonium polyphosphate for flame retardant and reinforced polyurethane composites, Colloids Surf. A: Physicochem. Eng. Asp. 626 (2021) 127092. https://doi.org/10.1016/j.colsurfa.2021.127092

[44] S. Wang, R. Ma, Effect of phosphorus-containing silane coupling agents modified ammonium polyphosphate on the flame retardancy and mechanical properties of epoxy resin, 142 (2025) e56807. https://doi.org/10.1002/app.56807

[45] Z. Wang, Y. Jiang, X. Yang, J. Zhao, W. Fu, N. Wang, D.-Y. Wang, Surface modification of ammonium polyphosphate for enhancing flame retardant properties of thermoplastic polyurethane, Materials 15 (2022) 1990. https://doi.org/10.3390/ma15061990

[46] D. Pan, J. Zhu, Y. Wu, K. Chen, B. Wu, L. Ji, Study on the crystal transformation of ammonium polyphosphate crystalline form V, Phosphorus Sulfur Silicon Relat. Elem. 191 (2016) 1306-1312. https://doi.org/10.1080/10426507.2016.1192621

[47] L. Sun, Y. Qu, S. Li, Co-microencapsulate of ammonium polyphosphate and pentaerythritol in intumescent flame retardant coatings, J. Therm. Anal. Calorim. 111 (2013) 1099-1106. https://doi.org/10.1007/s10973-012-2494-0

[48] C.-H. Ke, J. Li, K.-Y. Fang, Q.-L. Zhu, J. Zhu, Q. Yan, Y.-Z. Wang, Synergistic effect between a novel hyperbranched charring agent and ammonium polyphosphate on the flame retardant and anti-dripping properties of polylactide, Polym. Degrad. Stab. 95 (2010) 763-770. https://doi.org/10.1016/j.polymdegradstab.2010.02.011

[49] D. Enescu, A. Frache, G. Gualtieri, Combustion behavior of polypropylene-based composites used in industrial plasticollar, Compos. Interfaces 20 (2013) 241-253. https://doi.org/10.1080/15685543.2013.793099

[50] K.K. Shen, R. O'Connor, Flame retardants: borates, in: G. Pritchard (Ed.), Plastics Additives, Springer Netherlands, Dordrecht, 1998, pp. 268-276. https://doi.org/10.1007/978-94-011-5862-6_30

[51] K.K. Shen, Boron-Based Flame retardants in Non-Halogen Based Polymers, in: A.B. Morgan (Ed.), Non-Halogenated Flame retardant Handbook 2 Nd Edition, 1st ed., Wiley, 2021, pp. 309-336. https://doi.org/10.1002/9781119752240.ch7

[52] X. Zhang, Z. Wang, S. Zhou, F. You, D. Li, C. Zhou, Yu Pan, J. Wang, Enhanced flame retardancy level of a cotton fabric treated by an ammonium pentaborate doped silica-KH570 sol, J. Ind. Text. 52 (2022) 15280837221116590. https://doi.org/10.1177/15280837221116590

[53] W. Tang, S. Zhang, J. Sun, X. Gu, Flame Retardancy and Thermal stability of polypropylene composite containing ammonium sulfamate intercalated kaolinite, Ind. Eng. Chem. Res. 55 (2016) 7669-7678. https://doi.org/10.1021/acs.iecr.6b01722

[54] J.B. Dahiya, B.K. Kandola, A. Sitpalan, A.R. Horrocks, Effects of nanoparticles on the flame retardancy of the ammonium sulphamate-dipentaerythritol flame-retardant system in polyamide 6, Polym. Adv. Technol. 24 (2013) 398-406. https://doi.org/10.1002/pat.3095

[55] T. Hirata, T. Nishimoto, DSC, DTA, and TG of cellulose untreated and treated with flame retardants, Thermochim. Acta 193 (1991) 99-106. https://doi.org/10.1016/0040-6031(91)80177-K

[56] Y. Hu, X. Wang, eds., Flame retardant Polymeric Materials: A handbook, 1st ed., CRC Press, Boca Raton : CRC Press, 2019. https://doi.org/10.1201/b22345

[57] P.E. Gagnon, J.L. Boivin, C. Haggart, A Study of the reactions of ammonium sulphamate with amides and ureas, Can. J. Chem. 34 (1956) 1662-1670. https://doi.org/10.1139/v56-214

[58] B.K. Kandola, A.R. Horrocks, D. Price, G.V. Coleman, Flame retardant treatments of cellulose and their influence on the mechanism of cellulose pyrolysis, J. Macromol. Sci., Polym. Rev. 36 (1996) 721-794. https://doi.org/10.1080/15321799608014859

[59] A. Pal, A.K. Samanta, T.R. Kar, Eco-friendly fire-retardant finishing of cotton fabric with mixture of ammonium sulfamate and sodium Stannate with and without zinc acetate as external reagent, Cellulose 30 (2023) 11813-11828. https://doi.org/10.1007/s10570-023-05543-8

[60] A. Pal, A.K. Samanta, T.R. Kar, Eco-friendly fire-retardant finishing of Cotton fabric with mixture of Ammonium Sulfamate and Sodium Stannate, (2022). https://doi.org/10.21203/rs.3.rs-2081673/v1

[61] S.V. Levchik, A.I. Balabanovich, G.F. Levchik, L. Costa, Effect of melamine and its salts on combustion and thermal decomposition of polyamide 6, Fire Mater. 21 (1997) 75-83. https://doi.org/10.1002/(SICI)1099-1018(199703)21:2<75::AID-FAM597>3.0.CO;2-P

[62] T. Li, S. Li, T. Ma, Y. Zhong, L. Zhang, H. Xu, B. Wang, X. Sui, X. Feng, Z. Chen, Z. Mao, Flame-retardant poly (ethylene terephthalate) enabled by a novel melamine polyphosphate nanowire, Polym. Adv. Technol. 31 (2020) 795-806. https://doi.org/10.1002/pat.4815

[63] J. Lu, P. Jiang, Z. Chen, L. Li, Y. Huang, Flame retardancy, thermal stability, and hygroscopicity of wood materials modified with melamine and amino trimethylene phosphonic acid, Constr. Build. Mater. 267 (2021) 121042. https://doi.org/10.1016/j.conbuildmat.2020.121042

[64] Q. Jiang, P. Li, Y. Liu, P. Zhu, Flame retardant cotton fabrics with anti-UV properties based on tea polyphenol-melamine-phenylphosphonic acid, J. Colloid Interface Sci. 629 (2023) 392-403. https://doi.org/10.1016/j.jcis.2022.09.084

[65] P. Gijsman, R. Steenbakkers, C. Fürst, J. Kersjes, Differences in the flame retardant mechanism of melamine cyanurate in polyamide 6 and polyamide 66, Polym. Degrad. Stab. 78 (2002) 219-224. https://doi.org/10.1016/S0141-3910(02)00136-2

[66] A.D. Naik, G. Fontaine, F. Samyn, X. Delva, Y. Bourgeois, S. Bourbigot, Melamine integrated metal phosphates as non-halogenated flame retardants: Synergism with aluminium phosphinate for flame retardancy in glass fiber reinforced polyamide 66, Polym. Degrad. Stab. 98 (2013) 2653-2662. https://doi.org/10.1016/j.polymdegradstab.2013.09.029

[67] H. Gao, S. Hu, H. Han, J. Zhang, Effect of different metallic hydroxides on flame-retardant properties of low density polyethylene/melamine polyphosphate/starch composites, J. Appl. Polym. Sci. 122 (2011) 3263-3269. https://doi.org/10.1002/app.34398

[68] P.J. Davies, A.R. Horrocks, A. Alderson, Possible phosphorus/halogen synergism in flame retardant textile backcoatings, Fire Mater. 26 (2002) 235-242. https://doi.org/10.1002/fam.801

[69] Y. Shi, Z. Wang, J. Zhou, Facile synthesis of a flame retardant melamine phenylphosphate and its epoxy resin composites with simultaneously improved flame retardancy, smoke suppression and water resistance, RSC Adv. 8 (2018) 39214-39221. https://doi.org/10.1039/C8RA08696F

[70] W. Hou, Y. Fu, C. Zeng, N. Liu, C. Yin, Enhancement of flame retardancy and mechanical properties of polyamide 6 by incorporating melamine cyanurate combined with attapulgite, J. Appl. Polym. Sci. 137 (2020) 47298. https://doi.org/10.1002/app.47298

[71] K. Salasinska, K. Mizera, M. Celiński, P. Kozikowski, M. Borucka, A. Gajek, Thermal properties and fire behavior of polyethylene with a mixture of copper phosphate and melamine phosphate as a novel flame retardant, Fire Saf. J. 115 (2020) 103137. https://doi.org/10.1016/j.firesaf.2020.103137

[72] A. Dorieh, M. Farajollah Pour, S. Ghafari Movahed, A. Pizzi, P. Pouresmaeel Selakjani, M. Valizadeh Kiamahalleh, H. Hatefnia, M.H. Shahavi, R. Aghaei, A review of recent progress in melamine-formaldehyde resin based nanocomposites as coating materials, Prog. Org. Coat. 165 (2022) 106768. https://doi.org/10.1016/j.porgcoat.2022.106768

[73] P. Zhao, C. Guo, L. Li, Exploring the effect of melamine pyrophosphate and aluminum hypophosphite on flame retardant wood flour/polypropylene composites, Constr. Build. Mater. 170 (2018) 193-199. https://doi.org/10.1016/j.conbuildmat.2018.03.074

[74] M.-J. Chen, Z.-B. Shao, X.-L. Wang, L. Chen, Y.-Z. Wang, Halogen-free flame-tetardant flexible polyurethane foam with a novel nitrogen-phosphorus flame retardant, Ind. Eng. Chem. Res. 51 (2012) 9769-9776. https://doi.org/10.1021/ie301004d

[75] T. Nguyen, D. Hoang, J. Kim, Effect of ammonium polyphosphate and melamine pyrophosphate on fire behavior and thermal stability of unsaturated polyester synthesized from poly(ethylene terephthalate) waste, Macromol. Res. 26 (2018) 22-28. https://doi.org/10.1007/s13233-018-6004-5

[76] O. Zybina, M. Gravit, Intumescent coatings for fire protection of building structures and materials, Springer International Publishing, Cham. 2020. https://doi.org/10.1007/978-3-030-59422-0

[77] C. Hoffendahl, S. Duquesne, G. Fontaine, F. Taschner, M. Mezger, S. Bourbigot, Decomposition mechanism of fire retarded ethylene vinyl acetate elastomer (EVA) containing aluminum trihydroxide and melamine, Polym. Degrad. Stab. 113 (2015) 168-179. https://doi.org/10.1016/j.polymdegradstab.2014.09.016

[78] H. Yan, Y. Chen, S. Xu, Synthesis of graphitic carbon nitride by directly heating sulfuric acid treated melamine for enhanced photocatalytic H2 production from water under visible light, Int. J. Hydrog. Energy 37 (2012) 125-133. https://doi.org/10.1016/j.ijhydene.2011.09.072

[79] B.V. Lotsch, W. Schnick, New Light on an Old Story: Formation of melam during thermal condensation of melamine, Chem. - Eur. J. 13 (2007) 4956-4968. https://doi.org/10.1002/chem.200601291

[80] P. Zhao, C. Guo, L. Li, Flame retardancy and thermal degradation properties of polypropylene/wood flour composite modified with aluminum hypophosphite/melamine cyanurate, J. Therm. Anal. Calorim. 135 (2019) 3085-3093. https://doi.org/10.1007/s10973-018-7544-9

[81] K. Zhang, K. Wu, Y.-K. Zhang, H.-F. Liu, M.-M. Shen, W. Hu, Flammability characteristics and performance of flame-retarded epoxy composite based on melamine cyanurate and ammonium polyphosphate, Polym.-Plast. Technol. Eng. 52 (2013) 525-532. https://doi.org/10.1080/03602559.2012.762520

[82] X.-J. Li, C. Liu, C.-Y. Zhang, Z.-B. Shao, B. Zhao, Self-assembly of triazolyl-based cyclomatrix polyphosphazene and melamine cyanurate for flame retardant, smoke-suppressing, and mechanically robust epoxy resin, Polym. Degrad. Stab. 235 (2025) 111268. https://doi.org/10.1016/j.polymdegradstab.2025.111268

[83] W. Wang, Y. Liu, Q. Wang, Synthesis of melamine cyanuric based flame retardant via hydrogen bond self-assembly and in-situ dispersion strategies for improving comprehensive performance of epoxy resin, Compos. - A: Appl. Sci. Manuf. 176 (2024) 107826. https://doi.org/10.1016/j.compositesa.2023.107826

[84] X. Lu, X. Qiao, T. Yang, K. Sun, X. Chen, Preparation and properties of environmental friendly nonhalogen flame retardant melamine cyanurate/nylon 66 composites, J. Appl. Polym. Sci. 122 (2011) 1688-1697. https://doi.org/10.1002/app.34035

[85] A. Casu, G. Camino, M. De Giorgi, D. Flath, V. Morone, R. Zenoni, Fire-retardant mechanistic aspects of melamine cyanurate in polyamide copolymer, Polym. Degrad. Stab. 58 (1997) 297-302. https://doi.org/10.1016/S0141-3910(97)00061-X

[86] Y. Chu, Z. Chen, Y. Yu, T. Chen, Z. Chen, Q. Zhang, C. Li, H. Yang, J. Jiang, Synthesis of a novel phosphorus-containing melamine cyanurate derivative to enhance the fire resistance and mechanical properties of epoxy resin, Colloids Surf. A: Physicochem. Eng. Asp. 652 (2022) 129789. https://doi.org/10.1016/j.colsurfa.2022.129789

[87] Y.-C. Tyan, M.-H. Yang, S.-B. Jong, C.-K. Wang, J. Shiea, Melamine contamination, Anal. Bioanal. Chem. 395 (2009) 729-735. https://doi.org/10.1007/s00216-009-3009-0

[88] B.G. Hammond, S.J. Barbee, T. Inoue, N. Ishida, G.J. Levinskas, M.W. Stevens, A.G. Wheeler, T. Cascieri, A review of toxicology studies on cyanurate and its chlorinated derivatives., Environ. Health Perspect. 69 (1986) 287-292. https://doi.org/10.1289/ehp.8669287

[89] M. Yusuf, A review on flame retardant textile finishing: Current and future trends, CSM 3 (2019) 99-108. https://doi.org/10.2174/2405465803666180703110858

[90] A.R. Horrocks, Flame retardant challenges for textiles and fibres: New chemistry versus innovatory solutions, Polym. Degrad. Stab. 96 (2011) 377-392. https://doi.org/10.1016/j.polymdegradstab.2010.03.036

[91] Z. Wu, W. Xu, Y. Liu, J. Xia, Q. Wu, W. Xu, Preparation and characterization of flame-retardant melamine cyanurate/polyamide 6 nanocomposites by in situ polymerization, J. Appl. Polym. Sci. 113 (2009) 2109-2116. https://doi.org/10.1002/app.30022

[92] S. Zhang, W. Ji, Y. Han, X. Gu, H. Li, J. Sun, Flame-retardant expandable polystyrene foams coated with ethanediol-modified melamine-formaldehyde resin and microencapsulated ammonium polyphosphate, J. Appl. Polym. Sci. 135 (2018) 46471. https://doi.org/10.1002/app.46471

[93] F. Salaün, M. Lewandowski, I. Vroman, G. Bedek, S. Bourbigot, Development and characterisation of flame retardant fibres from isotactic polypropylene melt-compounded with melamine-formaldehyde microcapsules, Polym. Degrad. Stab. 96 (2011) 131-143. https://doi.org/10.1016/j.polymdegradstab.2010.10.009

[94] W.K. Patrick Lim, M. Mariatti, W.S. Chow, K.T. Mar, Effect of intumescent ammonium polyphosphate (APP) and melamine cyanurate (MC) on the properties of epoxy/glass fiber composites, Compos. B: Eng. 43 (2012) 124-128. https://doi.org/10.1016/j.compositesb.2011.11.013

[95] J. Zhang, H. Zhang, X. Wang, M. Zhang, Modified hemp fiber combined with melamine cyanurate for enhancing fire safety and mechanical properties of unsaturated polyester resins for automobile parts, AIP Advances 11 (2021) 125329. https://doi.org/10.1063/5.0076024

[96] A.B. Morgan, C.A. Wilkie, eds., Non-halogenated flame retardant handbook, 1st ed., Wiley, 2014. https://doi.org/10.1002/9781118939239

[97] J. Shen, J. Liang, X. Lin, H. Lin, J. Yu, S. Wang, The flame retardant mechanisms and preparation of polymer composites and their potential application in construction engineering, Polymers 14 (2021) 82. https://doi.org/10.3390/polym14010082

[98] T. Zhou, W. Chen, W. Duan, Y. Liu, Q. Wang, In situ synthesized and dispersed melamine polyphosphate flame retardant epoxy resin composites, J. Appl. Polym. Sci. 136 (2019) 47194. https://doi.org/10.1002/app.47194

[99] M. Nowak, B. Cichy, E. Kuźdżał, Kinetics of melamine phosphate thermal decomposition in DSC studies, J. Therm. Anal. Calorim. 126 (2016) 277-285. https://doi.org/10.1007/s10973-016-5705-2

[100] H. Yang, L. Song, Q. Tai, X. Wang, B. Yu, Y. Yuan, Y. Hu, R.K.K. Yuen, Comparative study on the flame retarded efficiency of melamine phosphate, melamine phosphite and melamine hypophosphite on poly(butylene succinate) composites, Polym. Degrad. Stab. 105 (2014) 248-256. https://doi.org/10.1016/j.polymdegradstab.2014.04.021

[101] D. Xu, H. Lu, Q. Huang, B. Deng, L. Li, Flame retardant effect and mechanism of melamine phosphate on silicone thermoplastic elastomer, RSC Adv. 8 (2018) 5034-5041. https://doi.org/10.1039/C7RA12865G

[102] L. Zhong, K.-X. Zhang, X. Wang, M.-J. Chen, F. Xin, Z.-G. Liu, Synergistic effects and flame retardant mechanism of aluminum diethyl phosphinate in combination with melamine polyphosphate and aluminum oxide in epoxy resin, J. Therm. Anal. Calorim. 134 (2018) 1637-1646. https://doi.org/10.1007/s10973-018-7699-4

[103] U. Braun, B. Schartel, M.A. Fichera, C. Jäger, Flame retardancy mechanisms of aluminium phosphinate in combination with melamine polyphosphate and zinc borate in glass-fibre reinforced polyamide 6,6, Polym. Degrad. Stab. 92 (2007) 1528-1545. https://doi.org/10.1016/j.polymdegradstab.2007.05.007

[104] P. Mali, N. Sonawane, Nilesh Pawar, V. Patil, UV-curing of novel tri-functional melamine-phosphate oligomer: An effect of coating with polyurethane acrylate toward mechanical, thermal, and flame retardant properties, Polym. Polym. Compos. 30 (2022) 09673911221079116. https://doi.org/10.1177/09673911221079116

[105] E. Wirnhier, M.B. Mesch, J. Senker, W. Schnick, Formation and characterization of melam, melam hydrate, and a melam-melem adduct, Chem. - Eur. J. 19 (2013) 2041-2049. https://doi.org/10.1002/chem.201203340

[106] V.M. Karlik, Derivatives of carbamide, symmetrical triazines, and heptazines: Ecologically clean fire retardants for different thermostable compositions, Polym. Sci. Ser. D 2 (2009) 170-173. https://doi.org/10.1134/S1995421209030095

[107] H. Jiang, M. Bi, D. Ma, B. Li, H. Cong, W. Gao, Flame suppression mechanism of aluminum dust cloud by melamine cyanurate and melamine polyphosphate, J. Hazard. Mater. 368 (2019) 797-810. https://doi.org/10.1016/j.jhazmat.2019.02.001

[108] S. Lu, W. Zhou, M. Yang, G. Chen, W. Hong, D. Yu, Z. Zheng, X. Chen, Preparation and flame retardant mechanism of polyheptazine/PA6 nanocmposites, Polymer 182 (2019) 121810. https://doi.org/10.1016/j.polymer.2019.121810

[109] R. Vijayarangan, M. Sakar, R. Ilangovan, Stabilization of melon phase during the formation of g-C3N4 from melamine and its structure-property relationship towards photocatalytic degradation of dyes under sunlight, J. Mater. Sci: Mater. Electron. 33 (2022) 9057-9065. https://doi.org/10.1007/s10854-021-07108-6

[110] E. Rezvani Ghomi, F. Khosravi, Z. Mossayebi, A. Saedi Ardahaei, F. Morshedi Dehaghi, M. Khorasani, R.E. Neisiany, O. Das, A. Marani, R.A. Mensah, L. Jiang, Q. Xu, M. Försth, F. Berto, S. Ramakrishna, The flame retardancy of polyethylene composites: From fundamental

concepts to nanocomposites, Molecules 25 (2020) 5157. https://doi.org/10.3390/molecules25215157

[111] G.L. Nelson, ed., Fire and Polymers: Hazards Identification and Prevention, American Chemical Society, Washington, DC, 1990.. https://doi.org/10.1021/bk-1990-0425

[112] B. Wang, J. Li, X. Lai, H. Li, Y. Chen, X. Zeng, Synthesis of a novel N -alkoxyamine containing macromolecular intumescent flame retardant and its synergism in flame-retarding polypropylene, Polym. Adv. Technol. 32 (2021) 2452-2464. https://doi.org/10.1002/pat.5275

[113] X. Liu, H. Jia, H. Li, Flame-retarding quasi-solid polymer electrolytes for high-safety lithium metal batteries, Energy Storage Mater. 67 (2024) 103263. https://doi.org/10.1016/j.ensm.2024.103263

[114] T. Tirri, M. Aubert, C.-E. Wilén, R. Pfaendner, H. Hoppe, Novel tetrapotassium azo diphosphonate (INAZO) as flame retardant for polyurethane adhesives, Polym. Degrad. Stab. 97 (2012) 375-382. https://doi.org/10.1016/j.polymdegradstab.2011.12.005

[115] M. Aubert, R.C. Nicolas, W. Pawelec, C. Wilén, M. Roth, R. Pfaendner, Azoalkanes-novel flame retardants and their structure-property relationship, Polym. Adv. Technol. 22 (2011) 1529-1538. https://doi.org/10.1002/pat.1642

[116] C. Li, C. Ma, J. Li, Highly efficient flame retardant poly(lactic acid) using imidazole phosphate poly(ionic liquid), Polym. Adv. Technol. 31 (2020) 1765-1775. https://doi.org/10.1002/pat.4903

[117] J.P. Candlin, Chapter 3 Polymeric Materials: Composition, Uses and Applications, in: Comprehensive Analytical Chemistry, Elsevier, 2008, pp. 65-119. https://doi.org/10.1016/S0166-526X(08)00403-0

[118] G. Yang, W.-H. Wu, Y.-H. Wang, Y.-H. Jiao, L.-Y. Lu, H.-Q. Qu, X.-Y. Qin, Synthesis of a novel phosphazene-based flame retardant with active amine groups and its application in reducing the fire hazard of Epoxy Resin, J. Hazard. Mater. 366 (2019) 78-87. https://doi.org/10.1016/j.jhazmat.2018.11.093

[119] J. Hu, F. Xin, K. Feng, Y. Chen, Y. Kang, A novel flame retardant based on Kabachnik-Fields reaction applied to fire retardant the unsaturated polyester resin, Eur. Polym. J. 228 (2025) 113832. https://doi.org/10.1016/j.eurpolymj.2025.113832

[120] E.D.D. Weil, N.G. Patel, Phospham-A stable phosphorus-rich flame retardant, Fire Mater. 18 (1994) 1-7. https://doi.org/10.1002/fam.810180102

[121] J.C. Markwart, A. Battig, L. Zimmermann, M. Wagner, J. Fischer, B. Schartel, F.R. Wurm, Systematically controlled decomposition mechanism in phosphorus flame retardants by precise molecular architecture: P-O vs P-N, ACS Appl. Polym. Mater. 1 (2019) 1118-1128. https://doi.org/10.1021/acsapm.9b00129

[122] G.F. Levchik, Y.V. Grigoriev, A.I. Balabanovich, S.V. Levchik, M. Klatt, Phosphorus-nitrogen containing fire retardants for poly(butylene terephthalate), Polym. Int. 49 (2000) 1095-1100. https://doi.org/10.1002/1097-0126(200010)49:10<1095::AID-PI405>3.0.CO;2-B

Materials Research Foundations 190 (2026) 248-267 https://doi.org/10.21741/9781644904077-9

Chapter 9

Nitrogen-Based Flame Retardants: Chemistry, Mechanisms, Classes and Applications

Ernest Debbarma, Mithun Roy and Susanta Ghanta*

Department of Chemistry, National Institute of Technology Agartala, Agartala-799046, Tripura, India

susanta.ghanta82@gmail.com

Abstract

Many chemicals are used to retard the burning rate or can reduce the chances of spreading of fire, are known as flame retardants. They are found in a wide range of consumer goods, such as electronics, textiles, plastics and many more. The basic requirements of flame retardants are enhanced by the ongoing growth of plastics used in building and construction materials, in electronic and electric applications. Several industries are looking towards non-halogenated flame-retardant products throughout the decade. Non-halogenated flame retardants are classified in three distinct categories as metal hydroxides, phosphorus-based molecules and N_2-related molecules. N_2-related flame retardants are often used in conjunction with phosphorus- related molecules but it can work on their own and ought to be considered a distinct separate class of flame retardants. The several uses and applications of N_2-related flame retardants are discussed along with ammonium and melamine related flame retardants such as N_2 molecules related to cyanuric acid, azoalkanes, phosphazenes, phospham, and N-alkoxy hindered amines, in this chapter.

Keywords

N_2-Related Flame Retardants, Non-Halogenated Flame Retardants, Azo-Alkanes, Melamine($C_3N_6H_6$), Phosphazenes, Phospham

Contents

Materials Research Forum LLC
https://doi.org/10.21741/9781644904077-9

1. Introduction

Various chemicals known as flame retardants (FRs) are sprayed on objects to stop them from burning or to delay the spread of fire. The term applies to the function, not a specific composition, of such chemicals [1,2], achieved through diverse chemistry involving elements such as bromine, chlorine, phosphorus, nitrogen, and aluminum. The definition can also be "a chemical which, when added to a combustible material, delays ignition and reduces flame spread of the resulting material when exposed to flame impingement" **[3]**. Among these, nitrogen-based flame retardant (FR) is one of the important safer alternatives to harmful halogenated retardants that are used in the majority before the emergence of non-halogenated retardants like nitrogen-based FRs. However, FRs do not make molecules completely fireproof, and their effectiveness varies depending on the polymer and fire scenario. Their usage is ever growing with fire incident occurring worldwide, making their impacts in environment and health including persistent, bio-accumulative and toxic substances (PBTs), of major concern. Halogenated FRs, widely used since the 1930s, were initially favored for their strong flame-suppressing properties. However, their chemical stability, once considered an advantage, allowed them to persist in the environment, reach the stratosphere, and contribute to ozone depletion. Additionally, when activated by fire, they release corrosive acid gases (HF, HCl, HBr), which can irritate eyes and lungs and worsen the toxicity of carbon monoxide in fire scenarios. Due to their small, lipophilic nature, they tend to bio-accumulate in fatty tissues and have been linked to toxicity, leading to restrictions or bans on their use in many applications.

Therefore, more eco-friendly alternatives to flame retardants were recommended and nitrogen-based flame retardants (NBFRs) have gained prominence as environmentally preferable alternatives to conventional halogenated FRs due to their superior environmental and toxicological profiles [3]. These compounds are increasingly used in polymers, textiles, coatings, and electronics as sustainable back-ups to Br_2 and chlorinated FRs, which face regulatory restrictions due to their bio-accumulative and ecotoxicological risks [4]. NBFRs split into subgroups, including ammonium-related, melamine ($C_3N_6H_6$) and melamine-related, N_2- related radical($R^{\cdot}$) generators, organophosphorus-nitrogen molecules and cyanuric-acid-based FRs. The following groups will be discussed in detail, with further subdivisions to elucidate their distinct chemical characteristics.

2. Principal Types of FRs Based on N_2

NBFRs [5-7] may be categorized into two principal groups according to their mechanism of FR action: mechanisms in the condensed and vapor phases. The more profound and well-celebrated class of NBFRs is extracted primarily from ammonia or melamine, despite derivatives of urea and guanidine have also been developed. These compounds operate mainly via endothermic decomposition, releasing large volumes of non-flammable gases like N_2 and ammonia, thereby diluting flammable gases and suppressing combustion in the vapor phase. To enhance thermal stability and reduce volatility under processing or operational conditions, these compounds are often converted into salt forms through reactions with inorganic acids. Typical examples include the use of melamine in polyurethane stretchable foams; melamine phosphates or ammonium polyphosphate-penta-erythritol networks in poly-olefins; melamine, melamine phosphates, or dicyandiamide in swollen paints; guanidine phosphates and ammonium salts in textiles; guanidine sulfamate in wallpapers. These NBFRs are extensively used due to their adaptability, non-toxic gas release profiles, and compatibility with a broad range of thermoplastics and coatings.

The second class of NBFRs operates predominantly in the condensed phase and consists of powerful interaction with the polymer matrix during thermal degradation. Rather than volatilizing, these retardants promote either extensive char formation, which acts as a thermal barrier, or facilitate polymer decomposition that encourages dripping away from the flame. As this mechanism relies on specific chemical interactions with the degrading polymer, its application is restricted to particular polymer systems. Notable examples include N-alkoxy overcrowded amines in polyolefins, melamine cyanurate in polyamides, and phosphazene compounds in epoxy resins, the latter of which have been the focus of considerable research in China. While NBFRs are commonly grouped by their predominant mechanism, their actual performance can vary significantly depending on the chemical composition of the host polymer. Indeed, vapor-phase active NBFRs may also promote char formation under certain conditions, while some condensed-phase additives may display vapor-phase activity depending on the matrix polymer. These findings highlight the critical importance of not only considering the intrinsic properties of the flame retardant, but also the reactivity and decomposition pathways of the substrate. The synergistic or antagonistic interactions between the flame retardant and polymer determine the overall effectiveness and mode of flame-retardant action.

3. FRs Based on ammonia

By volume, ammonia-based flame retardants make up the biggest group of flame retardants that contain nitrogen. Because ammonia exists in a gaseous state upto a temperature of -33°C, it must be employed in salt form when incorporated into solid materials for flame retardancy. Ammonia-based flame retardants usually undergo endothermic decomposition (as it absorbs heat, cooling the material and slowing combustion), releasing ammonia (which will dilute flammable gases and reduce oxygen availability) and leaving behind an acidic residue that promotes char. This char acts as an intumescent barrier: it provides thermal insulation, blocks oxygen and flammable gases from reaching the surface, and stabilizes the material's surface, thereby preventing dripping or melting. There are lean monetary ammonium-related FRs exercised, of which the best spectacular is aluminum polyphosphate (APP) [8] after aluminum pentaborate and ammonium sulfamate. Instead of being combined undeviatingly to molecules, aluminum phosphate, which comes in mono-basic, di-basic, and tri-basic forms is mostly acted as a dry powder extinguishing agent. However, an exception exists when it is incorporated into cellulosic fibres to serve as flame-retardant barriers, such as those used in mattresses. Similarly, ammonium sulfate is employed for flame-retarding cellulosic fibres in barrier applications, although it is not typically used in dry powder fire extinguishing systems.

Materials Research Foundations 190 (2026) 248-267 https://doi.org/10.21741/9781644904077-9

3.1 Ammonium Poly-phosphate

One of the most prominently used NBRFs is ammonia poly-phosphate (APP) [8], which is an inorganic salt of poly-phosphoric acid and ammonia, identified by CAS number 68333-79-9 and having both linear and branched chain forms. Chemical formula of APP is $H(NH_4PO_3)_nOH$, showing that each monomer consists of an orthophosphate radical of a phosphorus atom with three oxygen and one negative charge neutralized by an ammonium, leaving two bonds free to polymerize. Some monomers in the branching situations bond to three other monomers instead of the ammonium anion. It is a non-toxic flame retardant, available primarily in two crystalline forms, APP Phase I (linear, n < 100), and APP Phase II (branched, n > 1000) both APP Phase I and II figure is depicted below in Fig. 1 (A & B) respectively.

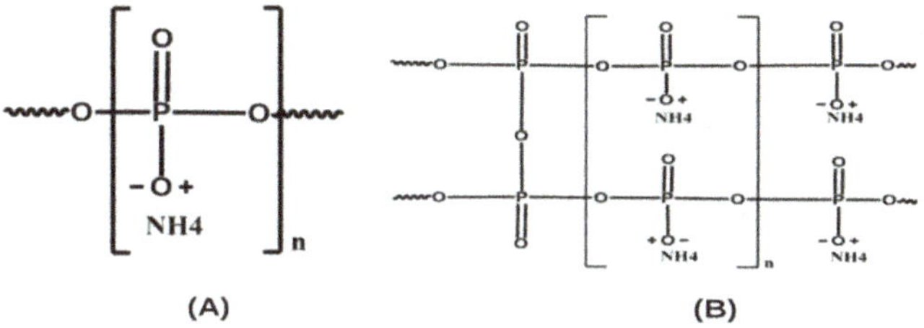

Fig. 1: The APP Phase I (A) and Phase II (B) figures are depicted here.

APP functions by dissolving into ammonia and polymeric phosphoric acid. As indicated before, endothermic decomposition eliminates heat, and ammonia and water are then produced, which dilutes the flammable gasses. An unstable phosphate ester is created when polyphosphoric acid combines with the -OH groups in polymer matrix or an additional synergist. The phosphate ester is then dehydrated in the following step. Surface facing the H origin (charring) develops a foamed char coating. The foamed char layer then serves as insulation, stopping the polymer from breaking down any further. APP's flame-retardant effectiveness is greatly enhanced by the insertion of synergistic chemicals like as pentaerythritol by-products, carbohydrates, and foaming delegates (melamine, etc.). A model for intumescent systems has been established by studying the mechanism of breakdown [9-11], and reference [12] provides an example of a reaction scheme (see Scheme 1).

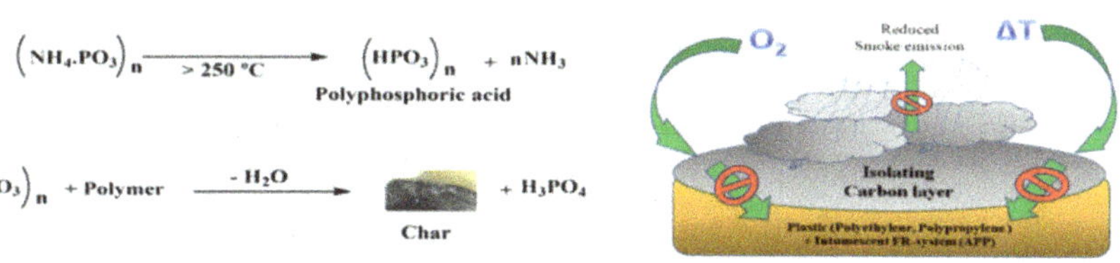

Scheme 1: The char formation and intumescence reaction scheme by APP is shown here.

As char-forming agents, polyhydric alcohols like pentaerythritol or starch are added to improve efficiency and enable lower loadings. The use of additional gas-forming chemicals, such melamine, can significantly amplify the blowing effect. However, 20% to 30% of this synergistic APP mixture is required to give polyolefins a satisfactory level of flame retardancy. Charring synergists and intumescent flame retardants like APP prevent the polymer from flaming but do not obstruct the gaseous flame. Consequently, less flammable gas is emitted, which results in a flame that is not sustained. The density of the smoke is similar to that of the virgin material since there is no interaction with the gaseous flame. The primary shortcoming of APP is its ability to absorb water over time, limiting it to only non-water-sensitive appliances. Thus, electrical appliances like switches and other electronic applications cannot use APP. Another shortcoming of APP is the temperature limits at which it can be safely used; it depends upon the category and crystal phase unoccupied; the elevated temperature will differ from 150 °C to 200 °C, though only unique categories of APP will have a 200 °C temperature limit.

3.2 Other Ammonia Salts

There aren't many other ammonium-based commercial flame retardants. Ammonium pentaborate, ammonium sulfamate, and ammonium sulfate have all been described in the literature. The regulated interaction of NH_3, H_2O, and H_3BO_3 produces ammonium pentaborate ($NH_4B_5O_8$), a white crystalline alkaline salt outcome. It is employed when alkali metals cannot be used or when an alkali borate that is easily soluble is required. When cellulosic materials burn, borate alters the oxidation reactions that result in the generation of carbon residue. The burned material keeps the breakdown products and creates a barrier against combustion. To create a fire-hindered outcome, ammonium pentaborate solutions can be sprayed or dipped onto paper. In cotton textile industry, ammonium pentaborate solutions be served as a component in flame proofing formulations. Ammonium pentaborate has also worked as a FR in thermoplastic polyurethane (TPU), epoxy, and urethane foam, among other polymers. The H_2O solubility of the ammonium pentaborate prohibits its applications as a launder durable FR for textiles, and this may not repellent to humidity/ H_2O exposure that can clean the ammonium pentaborate off paper or any other materials it was registered to. According to Myers et al. [13], ammonium pentaborate creates a multicellular glassy structure that shields the TPU below. The uses of ammonium pentaborate in polyamide 6 (PA6) has been investigated by Levchik et al. [14]. Ammonium pentaborate was demonstrated in this study to decrease PA6's thermal stability, which increases the production of carbonated residue upon combustion. Ammonium pentaborate has a fire-retardant effect that is comparable to that of ammonium polyphosphate, although it reduces thermal stability more. Ammonium pentaborate is useful for a component of multiflame retardant solutions in stiff polyurethane foam [15]. The flame resistance of PA6 dealt with diammonium imidobisulfonate ($NH(SO_3NH_2)_2$) and ammonium sulfamate ($NH_4SO_3NH_2$) has been studied [16–19]. It was demonstrated that when paired with trace levels (1%) of another char-forming synergist, like dipentaerythritol, 2% ammonium sulfamate was all that was needed to reach a UL94 V0 in PA6. They asserted that a protective char layer is formed as a result of a CP sulfating reaction between the primary PA6 -NH_2 groups and the dipentaerythritol's -OH groups. Wilkie [20] narrated a diminution in the rate of H release (HRR) by cone calorimeter assessment of a polystyrene molecules having as small as 5% of ammonium sulfamate as FR. He noticed a diminution of HRR by nearly 50%. The 10% increment by amount of ammonium sulfamate didn't produce in any extra enhancement of flame inhibiting. Finally, ammonium sulfate served as a FR remedy for cellulosic fiber calcine hurdle in mattresses and quite useful for this from 1970s [21]. It didn't demonstrated efficacy apart from wood or cotton materials. Similarly, these ammonium salts try to gain the thermally decomposable temperatures(T) which will restrict what polymers/compounds they can be utilized with.

Furthermore, the ammonium salts have issues with H_2O absorption (H_2O solubility too) that would extremely extremity their applications in electrical/electronic uses and anyone H_2O responsive end-use utilization or else these ammonium salts be provide FR assistance during the right end-use applications that won't have H_2O/humidity manifestation factors. The summery of other ammonium salts has been arranged in Table 1.

Table 1: The summery of other ammonium salts in N_2 related FR are tabulated here.

Flame Retardant	Chemical Formula	Applications	Mechanism	Advantages	Limitations	Related Paper Authors
Ammonium Pentaborate	$NH_4B_5O_8$	-Paper (spraying/dipping) - Cotton textiles - Polymers (epoxy, TPU, urethane foam, PA6)	- Borates alter oxidation reactions - Forms carbon residue and barrier char layer	- Effective for cellulosic and polymeric materials - Enhances char formation	- Water-soluble (not durable for laundering) - Not humidity-resistant - Reduces PA6 stability	Myers et al. [13], Levchik et al. [14], [15]
Ammonium Sulfamate	$NH_4SO_3NH_2$	- PA6 (with dipentaerythritol) - Polystyrene	- Sulfating reaction with PA6 and dipentaerythritol to form protective char	- As little as 2% effective in PA6 - 50% HRR reduction in polystyrene	- No added benefit beyond 5–10% loading	[16–19], Wilkie [20]
Ammonium Sulfate	$(NH_4)_2SO_4$	- Cellulosic materials (e.g., wood, cotton in mattress burn barriers)	- Decomposes to act as fire barrier in cellulosic materials	- Effective in cellulosic substrates	- Ineffective in non-cellulosic materials - Sensitive to heat, humidity, and water	[21]
General Limitations	–	- Not suitable for electronic/electrical applications	- Flame retardant effect via thermal decomposition and char formation	- Effective when dry/stable environment maintained	- Water solubility/absorption - Thermal decomposition limits polymer compatibility	–

4. FRs Based on Melamine($C_3N_6H_6$)

Another widely used flame retardant is melamine and its condensation products and salts. Much of its uses are observed to be in conjunction with phosphorus-related FRs, much like the ammonium-based variants and also combined with metal hydrates and halogenated FRs. Melamine undergoes endothermic decomposition, which results in the thermal of the polymer matrix and liberate non-combustible gases (H_2O, CO_2 and NH_3) [22]. Thus, it dilutes the burnable breakdown outcomes of the polymer matrix and thereby hinders the combustion. Melamine cannot be broadly applied as FR, but the efficiency for flame retardancy can be increased in select polymer classes even though the endothermic reaction is majorly independent of the polymer matrix. In the case of melamine derivative salts, its selection heavily depends on the thermal conditions during the treatment process of the polymer matrix and its susceptibility towards hydrolysis. When these melamine salts when used, the acid component often enhances flame retardancy through charring or interaction with the polymer matrix, making selection of acid a very crucial to efficiency and thermal stability, leading to a wide range of chemical solutions with melamine salts as FRs.

Materials Research Foundations 190 (2026) 248-267 https://doi.org/10.21741/9781644904077-9

4.1 Melamine as FR

Melamine or 1,3,5-triazine-2,4,6-triamine is a white crystalline powder characterized by a molecular mass of 126.12, molecular formula $C_3H_6N_6$, liquefaction point roughly 354°C with ρ =1.573 g/cm^3 and CAS: 108-78-1 [23]. At around 200°C melamine starts to undergo sublimation and thereby diminish the fuel gases and O_2 near the flaming origin. Melamine begins to sublimate at around 200°C, a process that plays a significant role in its flame-retardant behavior. During combustion, sublimation reduces the concentration of fuel gases and O_2 near the ignition origin, thereby inhibiting flame propagation. Moreover, melamine's decomposition is markedly endothermic approximately 470 kcal/mole allowing it to absorb substantial thermal energy and act as an effective heat sink during fire scenarios. This thermal behavior is complemented by the formation of a protective char layer, which includes intermediate compounds such as melam and melem. This char barrier is essential to melamine's flame-retardant efficacy [24]. In applied contexts, melamine is extensively utilized as an additive in flexible polyurethane foams, commonly found in upholstered furniture and mattresses. In these systems, it is typically dispersed within polyols. To comply with stringent fire safety standards, a high melamine loading ranging from 20 to 30 parts by weight relative to the polyol is required. It is frequently employed in conjunction with tris(1-chloro-2-propyl) phosphate (TCPP) to enhance performance [25]. From a solubility perspective, *melamine* dissolves little in cold water but dissolves more in hot water, a characteristic that may influence its utility across diverse environmental conditions [26]. Its interaction with polymers varies depending on the matrix. In polyurethanes, for instance, the amino (-NH_2) groups of melamine may interact with urethane linkages, potentially modifying the material's thermal decomposition pathways. Conversely, in polymers such as polyolefins, where such interactions are unlikely, melamine is less effective in altering condensed-phase behavior or decomposition chemistry. The molecular structure of melamine is illustrated in Scheme 1.

4.2 Melamine Salts

Melamine's relatively low sublimation temperature (~250°C) significantly limits its suitability for high-temperature polymer processing, as it begins to decompose and volatilize before many engineering polymers reach processing temperatures [27]. To address this limitation, melamine is often converted into thermally stable salts such as melamine cyanurate and melamine phosphate, which improve its performance as a FR, particularly in polymers like polypropylene, polyamide, and polyesters [28]. It has been observed that thermo gravimetric analysis (TGA) reveals a clear thermal stability hierarchy among melamine and its salts: Melamine < Melamine Cyanurate < Melamine Phosphate < Melamine Polyphosphate. It's interesting to note that the salts made from cyanuric acid and polyphosphoric acid exhibit far higher thermal stability than melamine itself, but melamine phosphate just slightly improves thermal resistance. Melamine cyanurate begins to degrade around 300°C, while melamine polyphosphate shows an onset of decomposition at approximately 350°C, affirming its potential for high-temperature applications [29]. Despite structural similarities among these salts, their modes of FR action differ significantly. Melamine cyanurate primarily functions via gas-phase flame inhibition (FI), releasing non-flammable gases such as NH_3 and H_2O, whereas melamine polyphosphate contributes to condensed-phase protection by uplifting the formation of a robust, thermally insulating char layer. This distinction underlines the importance of tailoring melamine derivatives to specific polymer systems and thermal requirements.

Materials Research Forum LLC
https://doi.org/10.21741/9781644904077-9

4.3 Melamine Cyanurate

Melamine cyanurate (MC) (see Fig. 2), having CAS: 37640-57-6, also known as the melamine cyanuric acid adduct or complex, is a crystalline material formed from a 1:1 molar ratio of melamine and cyanuric acid [30]. The crystal structure of the melamine cyanuric acid adduct reveals that the complex is stabilized by a vast 2D hydrogen bonding network, proving that the compound is not a salt but a supramolecular assembly held together by hydrogen bonds [31]. This unique supramolecular structure imparts significant thermal stability and renders the complex insoluble in most common solvents, which is beneficial in applications requiring thermal resistance and minimal migration from polymer matrices. MC is used in polymers such as polyamides, thermoplastic polyurethanes, and polyesters, which are processed at T upto 300 °C. At temperatures exceeding 320 °C, it encounters endothermic breakdown into melamine and cyanuric acid, which further degrade into NH_3, H_2O, and CO_2. Since the 1970s, melamine cyanurate has been widely employed as a FR in polyamide 6 (PA6) and its copolymers, such as polyamide 6,6 (PA66). However, MC is unsuitable for use in PA66 alone due to its lower thermal stability; it begins to decompose at around 300 °C, while PA66 typically requires processing temperatures up to 330 °C [32-36].

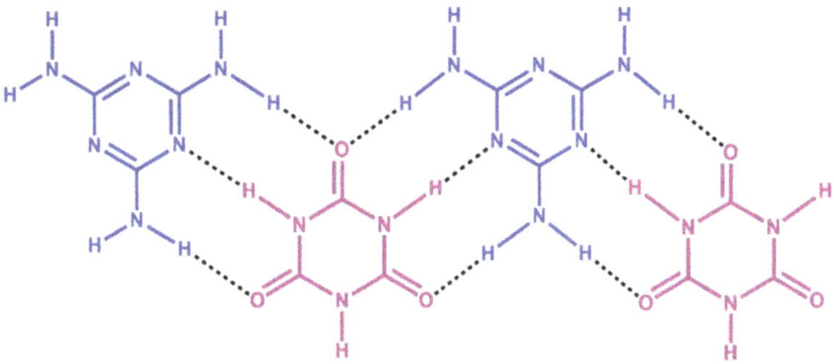

Fig. 2: The 2D Structure of melamine cyanurate is shown here.

In polyamides, MC acts through a "runaway" mechanism by significantly reducing the polymer's melt viscosity during thermal decomposition. This causes the solid to rapidly liquefy and drip away from the ignition source, interrupting sustained pyrolysis and enabling a UL94 V-0 flammability rating at additive loadings as low as 6-10% and thicknesses down to 0.4 mm (Yoon, 2022). This mechanism is highly specific to polyamides due to the interaction of cyanuric acid with the polymer matrix, which induces chain scission and accelerates melt flow [37]. However, while effective in flame retardancy, this accelerated degradation can increase total heat release, as the polymer decomposes and pyrolyses more rapidly than non-retarded counterparts. Furthermore, the dripping effect is not suitable for all fire safety applications, particularly where falling molten material poses risks to people evacuating.

Nevertheless, the scenario is different in the case of glass fiber reinforced polyamides as glass fiber entangles and hinders melt flow, resulting in larger, more energetic droplets that MC alone cannot extinguish under UL94 conditions. Consequently, MC's efficacy diminishes in such systems due to insufficient heat dissipation before the droplets make contact. Moderate FR performance (e.g.,

Materials Research Foundations 190 (2026) 248-267
https://doi.org/10.21741/9781644904077-9

UL94 V-2) can still be achieved using short fibres like milled glass or fillers such as Wollastonite. Outside of polyamides, MC is less effective alone and is typically used in synergistic formulations. In thermoplastic polyurethanes (TPUs), combinations of MC with organic phosphates (e.g., 25% MC + 7.5% phosphate ester) have achieved UL94 V-0 ratings. Synergism with nanoclays has also been observed in TPUs, although this combination tends to be antagonistic in polyamides due to the opposing effects on melt viscosity [38,39]. In polyesters, MC is often combined with phosphinates, where the primary flame-retardant effect arises from the gas-phase activity of the phosphinate, while MC confers via a H sink and reduction mechanism [40]. However, such systems struggle to meet stringent fire safety standards (e.g., UL94 V-0) at low thicknesses without additional cross-linking agents.

4.4 Melamine Polyphosphate

Chemically known as 1,3,5-triazine-2,4,6-triamine phosphate (CAS No. 41583-09-9), melamine polyphosphate (MPP) is commonly found as an odorless, white powder. It is structurally derived from melamine and phosphoric acid, forming melamine (mono)phosphate, a salt illustrated in Fig. 3(A). When subjected to thermal treatment, melamine phosphate undergoes a series of transformations: at temperatures above approximately 200 °C, it converts to melamine pyrophosphate, and upon further heating beyond 260 °C, it polymerizes to form melamine polyphosphate (MPP). These transformations are associated with the release of H_2O, thereby functioning as a heat sink through an endothermic effect. The thermal stability of these compounds increases progressively in the order: melamine **phosphate < melamine pyrophosphate < melamine polyphosphate,** reflecting the enhanced condensation and reduced volatility associated with each successive structure. At temperatures exceeding 350 °C, MPP decomposes endothermically, releasing polyphosphoric acid. This acid forms a protective coating on the polymer substrate, acting as a barrier against combustion by shielding the underlying material and promoting the formation of a char layer. This char significantly reduces oxygen availability at the combustion site and impedes further degradation of the polymer. Simultaneously, the melamine released during decomposition acts as a blowing agent, expanding the char into a protective, foamed barrier [27]. Thanks to its high thermal stability, MPP is primarily utilized in thermoplastic and thermosetting applications. Commercial formulations are available from companies such as BASF (e.g., Melapur® 200 series) and Budenheim (e.g., Budit® 3114). Despite its extensive investigation in polyamide systems, MPP demonstrates limited flame-retardant efficiency when used alone. Nevertheless, in highly glass-fibre-filled polyamides, relatively low MPP concentrations can achieve UL94 V-0 flammability classifications [28].

Materials Research Foundations 190 (2026) 248-267
https://doi.org/10.21741/9781644904077-9

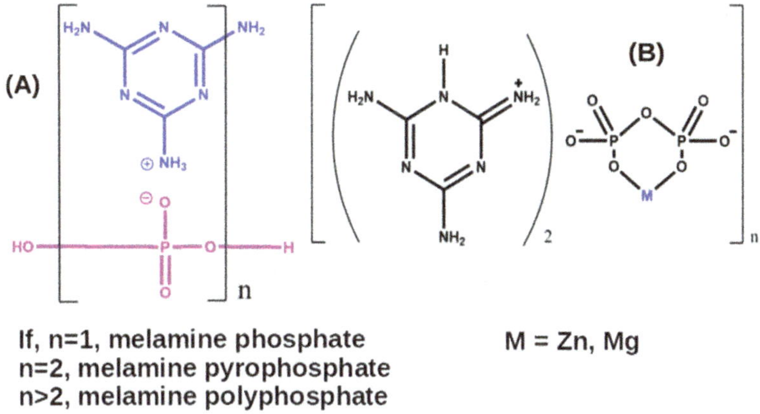

***Fig.** 3: (A) The structure of MPP, (B) Metal incorporated MPP, are shown here.*

MPP is especially effective in glass fibre-reinforced polyamide 6 and 66. When combined with metal phosphinates, MPP exhibits synergistic flame-retardant behavior. In a significant advancement, Wehner and Dave pioneered the development of a multi-component FR system by integrating melamine, polyphosphate, and alluminium, resulting in melamine-poly(alluminium phosphate). This compound, patented under the trade name *Safire® 200*, represents a major milestone in the evolution of halogen-free flame-retardant technologies for engineering polymers. Following this innovation, related derivatives *Safire® 400* and *Safire® 600* were synthesized by substituting alluminium with zinc and magnesium, respectively, yielding melamine-poly(zinc phosphate) and melamine-poly(magnesium phosphate) [35,41,42]. Illustration for possible structure of *Safire® 400* and *Safire® 600 is depicted in Fig. 3(B).*

4.5 Melamine condensates and its Salts

Melamine, being volatile, starts to sublime at approximately 220 °C, whereas its condensation products show greater thermal stability. Upon thermal treatment, melamine produces melam, melem, and melon (see Scheme 2) and their salts, which are reviewed extensively by Schwarzer et al. [43]. Gavrilova et al. [44] provide a precise method for preparation of melam, which is generally synthesized by heating melamine or its salts at around 300 °C. In more recent developments, a commercial synthesis route has been disclosed, wherein melamine is thermally treated in the presence of an acid, preferably *p*-toluene sulfonic acid, to produce melam and its salts [45]. Melam, melem, and their salts are widely employed as halogen-free FRs for thermoplastics, particularly polyamides. The Dutch State Mines (DSM), a global chemical company, has reported the effective use of approximately 7% melam as a FR in Polyamide 6 (PA 6, or Nylon 6), Polyamide 66 (PA 66, or Nylon 66), and Polyamide 46 (PA 46), a high-temperature-resistant polyamide [46]. The use of salts derived from phosphoric acid, polyphosphoric acid, or methanesulfonic acid of melam and melem as FR agents has been reported. These salts are directly synthesized by heating melamine, followed by neutralisation with the appropriate acid at temperatures exceeding 300 °C [47]. For instance, calcination of a melamine and polyphosphoric

Materials Research Foundations 190 (2026) 248-267 https://doi.org/10.21741/9781644904077-9

acid mixture at 300-400 °C for several hours results in the formation of melem polyphosphate [48]. As the tendency shifts towards more thermally stable polyamides in electrical and electronic applications, more products of melamine and their salts can be expected.

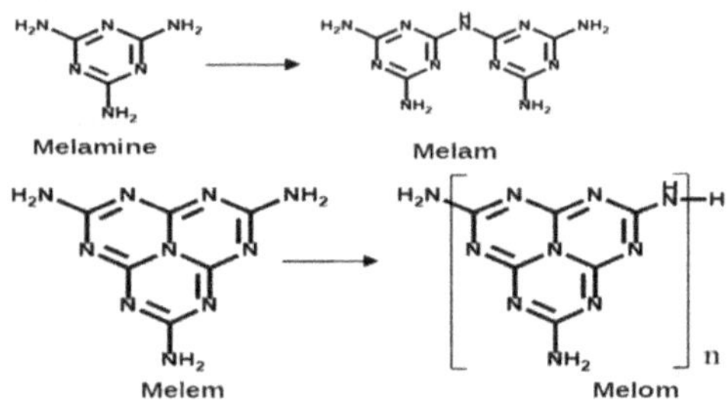

Scheme 2: Condensation products of melamine

5. N_2-Based Radical Generators

N_2-related radical generators are a distinct class of FRs used in polyolefins. These compounds produce radicals during combustion, triggering polymer chain scission and causing the material to melt and drip away from the flame. This dripping removes fuel from the fire, effectively helping to suppress it. Since only low levels of additive are needed, the polymer's mechanical properties are largely preserved [49]. Like melamine cyanurate, their action is driven by a self-accelerating degradation process. Among these, N-alkoxy overcrowded amines (NOR-type) and azoalkanes are the most studied. NOR FRs, such as Flamestab® NOR® 116 (see Fig. 4), are already used commercially in polyolefin films and fibres [50,51]. Scheme 3 outlines the two key modes of action for NOR compounds. At lower temperatures (below 150 °C), they act as stabilizers through the Denisov Cycle a radical-scavenging mechanism [52-54]. In this cycle, hindered amines react with peroxy radicals to form aminoxyl radicals, which then neutralize other reactive species. This closed-loop process protects the polymer from oxidative degradation and regenerates itself for repeated action. As temperatures rise during combustion, the chemistry shifts. NOR compounds decompose through cleavage of the N-O or O-R bonds, releasing radicals that promote rapid depolymerization. The resulting low-molecular-weight fragments are not volatile, so they drip from the flame without igniting similar to how molten candle wax behaves.

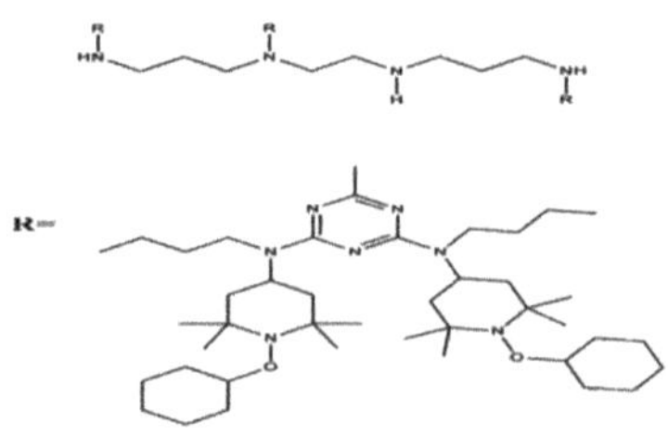

***Fig. 4**: Overcrowded N-alkoxy amine counterbalance Flamestab NOR®116*

Materials Research Foundations 190 (2026) 248-267
https://doi.org/10.21741/9781644904077-9

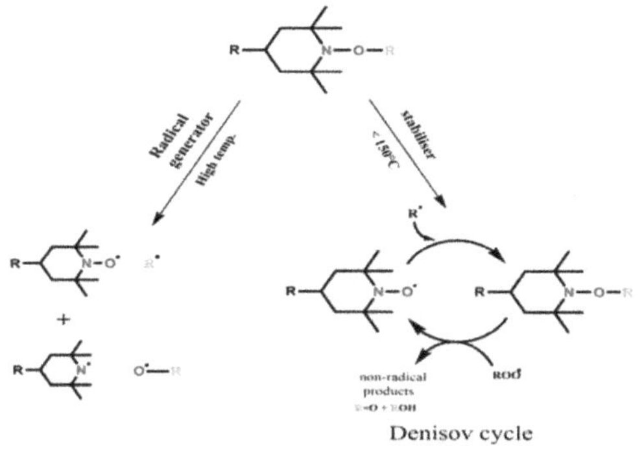

Scheme 3: N-alkoxy hindered amines reaction scheme is shown here.

However, this faster breakdown also increases the rate of heat release. For already high H freeing materials like polyolefins, NOR additives can worsen flammability under certain fire conditions. Therefore, careful consideration of fire risk is essential when using these compounds. To mitigate this, NORs are often combined with phosphorus-based synergists. Blends with compounds like cyclic dimethylspirophosphonate or melamine methane phosphonate have proven effective, particularly in polyolefins over 100 µm thick [55].

Other nitrogen-based radical generators, including azo and triazine derivatives, also show flame retardant potential. These release nitrogen gas and radicals upon decomposition, enabling polymer breakdown and dripping. While currently limited to laboratory use, they may be viable in applications where flame tests permit drip-based suppression.

6. Phosphazenes, Phospham and Phosphoroxynitride

Phosphorus–nitrogen systems like phosphazenes, phospham, and phosphoroxynitride (PON) are known for their thermally stable P-N frameworks. In these molecules, phosphorus is covalently bonded to two N-atoms typically through one single and one double bond. In phosphazenes, pentavalent structures are also possible, while phospham and PON exhibit highly cross-linked networks, with PON also containing P=O bonds and variable hydrogen content [56]. The chemistry of phosphazenes dates back to the 19th century, with Liebig first describing their synthesis in 1832 through the reaction of phosphorus pentachloride (PCl_5) and ammonium chloride (NH_4Cl). This reaction produces cyclic chlorophosphazenes like $(NPCl_2)_3$, depending on the sublimation conditions [57] (Scheme 5.1). Phosphazenes have found use in flame retardants, particularly in epoxy resins for printed circuit boards. A prominent example is hexaphenoxycyclotriphosphazene (HPCTP), valued for its good solubility in $(CH_3)_3COCH_3$ and its ability to achieve UL-94 V-0 ratings at loadings of 18% or more. Synergistic effects have also been reported when HPCTP is combined with alluminium diethyl phosphinate, enhancing flame retardant performance [58]. In polyesters such as polybutylene terephthalate (PBT), both phospham and PON have been evaluated. Phospham is extremely thermally stable, losing only about 10% of its weight at 500 °C, but it does not contribute to char formation or improve flammability ratings. PON, while increasing char yield, does not enhance flame retardancy. In fact, the residue it forms may increase the emission of combustible gases, making it counterproductive in some systems [59,60]. Conversely, in polyamides, phospham has shown more promising results. In PA46, for instance, its addition raised the limiting oxygen index (LOI) from 21 to 28. The inclusion of novolak resins further improved flame retardant performance.

7. FRs Based on Cyanuric Acid

Another class of N_2-related FRs is derived from cyanuric acid chemistry. Cyanuric chloride reacts sequentially with alcohols and amines to form esters and amides. When one mole of morpholine reacts with cyanuric chloride, it yields a bifunctional intermediate that can be further polymerized using bifunctional amines or esters. One such polymer, produced by MCA Technologies [61], combines morpholine and piperazine, forming a compound with a molecular weight of approximately 2800 Da and a nitrogen content of around 31%. This material is marketed under the service mark ppm-Triazine HF™ (see Fig. 5). These triazine($C_3N_3H_3$)-based polymers are primarily char-forming, and this property allows them to act synergistically with other FRs such as ammonium polyphosphate (APP) or phosphinates. Notably, a 1:3 blend of ppm- triazine HF™ and APP in polypropylene was found to function effectively as an intumescent system. With a 20% loading, the limiting oxygen index (LOI) increased from 19 (in neat polypropylene) to 31, and the material achieved a UL-94 V0 rating. Cone calorimetry results were particularly promising: a 10% loading of the system reduced the peak HRR (pHRR) by a factor of five, while a 20% loading achieved a tenfold reduction. Additionally, total smoke emission was significantly reduced at 20% storing. However, the 10% storing did not yield a major enrichment in smoke suppression, which was also reflected in its flame spread behavior.

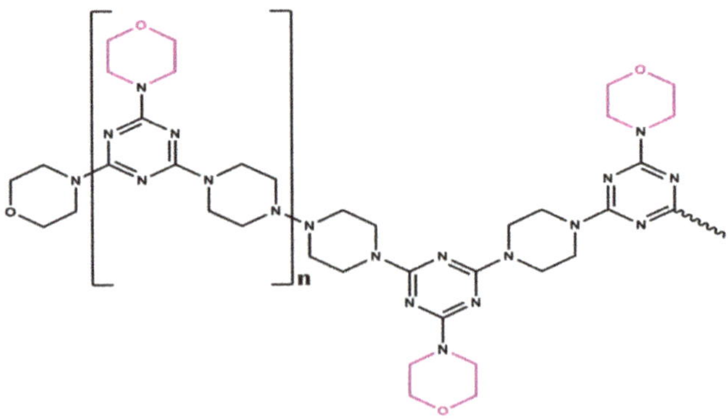

Fig. 5: Piperazine and cyanuric acid polymer as per MCA Technologies represented here.

8. Evaluating FR Mechanisms of Action and Fire Behavior

The most crucial factor in the research and development of FRs and flame-retarded polymers is the quantitative evaluation of fire hazards. Moreover, among the most widely used fire inspection

Materials Research Foundations 190 (2026) 248-267 https://doi.org/10.21741/9781644904077-9

just provide a liberty or declining result, providing no information on the mechanisms of action. However, determining the primary modes of action may be crucial to scientifically grounded development and study, and it also helps with their quantitative evaluation. Cone calorimeter measurements enable a quantitative evaluation of the modes of action as well as the fire dangers, such as HRR (heat release rate), pHRR (peak HRR), and THE (total heat evolved), resulting from the burning material in growing fires. The cone calorimeter that compute heat (H) and mass(M), therefore, effective H of combustion (EHOC) and char/residue harvest, be a potent device for computing the impact of fuel dilution and FI in the gas phase, besides the improvement of inorganic surplus and carbonaceous char (CC) in the condensed phase (CP). Additionally, the overall effect of intumescence or the development of a surplus protective layer can be evaluated in relation to lowering the rate of heat emission. The EHC is computed by the cone calorimeter consequently the flame's combustion efficiency (χ) and the EHOC (h_c^0) of the volatiles. Since the cone calorimeter mimics a fire that is well-ventilated, χ= 1 for the majority of the polymers. h_c^0 is decreased by fuel dilution, while χ is inhibited by flame. The THE is proportional to χh_c^0 since it represents the entire M deficit multiplied by the EHC. Here $(1 - \mu)$, (μ represent the char harvest from 0 to 1) accounts for dilution of fuel and charring in the condensed phase. So the decrease total M is the sample magnified by $(1 - \mu)$ and THE is given by

$$THE \sim \chi h_c^0(1-\mu)m_0$$

When scrutinizing the outcomes of a FR polymer to the results of the polymer itself, this THE may serve to measure the FR effects of charring, fuel dilution, and FI [62]. A FRs ability to inhibit flames in terms of χ can also be calculated and evaluated by comparing the THE outcomes of the Microscale Combustion Calorimeter and Cone Calorimeter tests [63]. The values of THE or EHC and μ, respectively, may serve to calculate the safeguarding layer impact of the fire surplus as empirical factor $\theta(t)$, assuming the HRR as: $HRR(t) \sim \theta(t)\ \chi h_c^0(1-\mu)$. The processes and mechanisms of flame retardancy be ascertained by means of further fire tests, analytical methods, and other cone calorimeter parameters also. For instance, the hike in smoke and CO production the whole time of the flaming moment indicates inadequate combustion and lower EHC. However, CO and smoke are produced as a result of both the polymer's burning and pyrolysis. To determine the routes of action of flame retardancy, pyrolysis often thermo gravimetric tests combined with produced gas scrutiny or surplus scrutiny be frequently used. This is hardly true, though, because these analytical techniques are not fire tests; rather, they only provide information about pyrolysis, like phosphorus discharge or CC, and consequently the mechanisms underlying the modes of action of FRs. Extreme caution is needed, as well as multi-methodical approaches and accurate quantification. For example, the behavior and fate of the entire amount of phosphorus should be examined and measured in both the gas phase and CP. The numbers and ratios of volatile outcomes frequently undergo significant variations, but only when this is accompanied by a similar change in carbonaceous char does this suggest a reduction in heat evolve therefore the conservancy of atoms equals the conservancy of EHOC. All fire surplus also functions as safeguarding layer, frequently referred to as a "barrier to H and M transport," depending considerably more on its characteristics than its quantity. Although this is undoubtedly the case, it should be highlighted that, in the majority of instances, very little significant M be trapped as volatiles that cause the fire surplus to inflate as balloon. The change in H transmission is always the primary consequence. Because of the development of an inorganic-carbonaceous residual safeguarding layer, layered silicate polymer nanocomposites have been found to exhibit a lower HRR and pHRR in the cone calorimeter. For certain systems, it has been demonstrated in detail that the sole significant factor is H shielding, or hike in reradiation by the heated surface [64]. Additional investigation of fire residues is done to look at flame retardant processes that occur in the CP and start the accumulation of CC or inorganic surplus [65]. Scanning electron microscopy (SEM) images are now routinely used to highlight morphological characteristics; SEM combined with energy dispersive X-ray spectroscopy (SEM/EDX) provides additional message on the chemical makeup of surplus. To comprehend the chemistry underlying the fire surplus and safeguarding layer, the development of CC, metal phosphates, silica, borates, or other glasses can be successfully addressed using elemental analysis, attenuated total reflection (ATR)-FTIR, Raman spectroscopy, and solid state NMR. Analyzing the residue's quantity, shape, and chemical makeup enables us to illustrate the potential of the residual safeguarding layer as well as how much fuel is prevented from burning in the flame [62]. The combination of the gas phase flame with the CP pyrolysis has a significant impact on the physico-chemical process of burning a polymer. Fire behavior and ignition characteristics are significantly influenced by a number of physical processes, including the materials' heat absorption. The reflectance-absorptance property of the polymer surface and its dependence on the radiation origins unavoidable be taken into account for ignition. Linteris employed four distinct sample platelets of varying thicknesses (0.4, 0.6, 0.9, and 1.4 mm) put under the cone heater at 5.0 kW/m2 irradiation in order to quantify the absorption of radiative H flux in the cone calorimeter [66]. To measure the amount of radiation that passed through the sample, a H flux meter was placed beneath the platelets. At deeper depths, carbon nanoparticles, including carbon nanotubes, have been shown to exhibit significant changes in heat absorption, lowering the tig. Increased thermal conductivity causes the surface temperature to rise more slowly, decompose and release volatile fuel later, and ignite later. Accordingly, it has been suggested that in carbon nanoparticle polymer nanocomposites, the higher thermal conductivity, which raises the tig, competes with the enhanced heat absorption in the top layer, which lowers the tig. Due to polymeric materials' propensity for easy ignition and widespread flame spread, coatings have been tested as a potent flame-retardant method to move the ignition time to a higher order of magnitude. In addition to sacrificial and intumescent coatings, heat reflective coatings have also been proposed. Three-layer systems, which include an adhesive, an infrared mirror, and a safeguard coating, increase the duration of ignition significantly by reducing H absorption by two fold of magnitude. The MARHE is also effectively reduced by shifting tig by a factor of five to ten.

Dripping and melt flow are important phenomena that affect fire characteristics, especially in response to fire assessments like the vertical UL 94. Melt flow and soaking generally have both positive and negative effects. Flaming soaking can cause pool fires, enhance the spread of flames, and provide as an extra source of ignition. However, because they take heat and mass out of the main pyrolysis zone, melt flow and soaking are employed as effective FR modes of action. It turns out that passing a lot of fire tests requires properly calibrated melt flow and soaking behavior. For example, effective retail molecules categorized as V-0 in UL 94 use both prompt unwatered self-extinguishing role and prompt non-flaming soaking self-extinguishing role. The interplay and clash between gasification, charring, and soaking in UL 94 fire role have been better understood. Special polytetrafluoroethylene fibers are frequently mixed with FRs in thermoplastics to introduce flow restrictions at lowering shear rates and accomplish V-0 unwatered. To accomplish the UL 94 classifications V-2 or non-flaming soaking V-0, non-charring foams like PS foams, foils, and thermoplastics are flame-retarded using radical generators that enhance decomposition to liquid outcomes in the CP, either alone or in combination with FRs that cause **FI**.

8. Conclusion

FRs based on nitrogen are a unique class of materials with a wide range of uses. By intumescence and heat removal, the majority of them endothermically break down under circumstances of combustion into inert volatiles, preventing a fire from spreading. To increase charring, these substances are frequently used with phosphorus acid generators. Here N_2 related FRs are pretty idle supporters of the intumescent process by producing the foamy gasses and, as a result of endothermic breakdown, acting as a H sink. N_2 related molecules can, depending on properties of polymer these supplements are added to, either help the polymer drain exterior H origin (NOR or melamine cyanurate) or aid in the development of char (triazine). If not, the more popular method is to think about combining phosphorus and N_2 to give a polymer the desired flame-retardant property. N_2 related FRs be appropriate non-halogenated molecules to look at, and they are probably going to be in use as long as it is appropriate and they offer a significant fire safety asset. All FR science is of governmental surveillance, together with N_2 related FRs. This method should be applied when the FR has the potential to react with the polymer. The majority of N_2 related FRs, with the exception of melamine in polyurethane foam, are non-reactive supplements, indicating that their application be restricted in some contexts (crystalline thermoplastics) but pacific permissible in thermosetting polymers. Protocols are changing, though, and this class of FR which until to now has shown some promising persistence, bioaccumulation, and toxicity profiles may escape examination and find greater application in the future. Examples of this include some ammonium salts.

References

[1] S. Kemmlein, O. Hahn & O. Jann, Emissions of organophosphate and brominated flame retardants from selected consumer products and building materials, Atmos. Environ. 37 (2003) 5485-5493. https://doi.org/10.1016/j.atmosenv.2003.09.025

[2] N. Surak, National Institute of Environmental Health Sciences, EHPonline: http://ehis. niehs. nih. Gov.

[3] E. D. Weil and S. V. Levchik, Flame Retardants for Plastics and Textiles: Practical Applications. 2nd ed., Hanser Publishers, Munich, Germany: Carl HanserVerlag, 2016. https://doi.org/10.1007/978-1-56990-579-1

[4] I. Van der Veen & J. de Boer, Phosphorus flame retardants: properties, production, environmental occurrence, toxicity and analysis, Chemosphere. 88 (2012) 1119-1153. https://doi.org/10.1016/j.chemosphere.2012.03.067

[5] A. B. Morgan & M. Klatt, Nitrogen-based flame retardants. Non-halogenated flame retardant handbook, 236-270, 2021. https://doi.org/10.1002/9781119752240.ch5

[6] M. Klatt, Nitrogen-based flame retardants. Non-Halogenated Flame Retardant Handbook; AB, Morgan, CA, Wilkie, Eds, pp.143-168, 2014. https://doi.org/10.1002/9781118939239.ch4

[7] H. Horacek & R. Grabner, Advantages of flame retardants based on nitrogen compounds. Poly. Deg. Stab. 54 (1996) 205-215. https://doi.org/10.1016/S0141-3910(96)00045-6

[8] E.D. Weil, S.V. Levchik, Flame retardants for plastics and textiles, p. 16. Hanser Publishers, Munich, Germany, 2009. https://doi.org/10.3139/9783446430655.fm

[9] G. Camino, N. Grassie, I. C. McNeill, Influence of the fire retardant, ammonium polyphosphate, on the thermal degradation of poly(methyl methacrylate). J. Poly. Sci. Polym. Chem. Ed., 16 (1978) 95 - 106. https://doi.org/10.1002/pol.1978.170160110

[10] G. Camino, L. Costa, L. Trossarelli, Study of the mechanism of intumescence in fire retardant polymers: Part III-Effect of urea on the ammonium polyphosphate-pentaerythritol system, Poly. Deg. Stab. 6 (1984) 221-229. https://doi.org/10.1016/0141-3910(84)90098-3

[11] G. Camino, L. Costa, L. Trossarelli, F. Costanzi, A. Pagliari, Study of the mechanism of intumescence in fire retardant polymers: Part VI-Mechanism of ester formation in ammonium poly phosphate penta erythritol mixtures. Poly. Deg. Stab. 12 (1985), 213-228. https://doi.org/10.1016/0141-3910(85)90090-4

[12] Flame Retardants for Fireproof Plastics, understanding its mechanism, SpecialChem, https://polymer-additives.specialchem.com/selection-guide/flame-retardants?src=Beyable, 2024, (accessed on 25/5/2025).

[13] R. E. Myers, E. D. Dickens Jr., E. Licursi, R. E. Evans, Ammonium penta borate: an intumescent flame retardant for thermoplastic polyurethanes. J. Fire Sci. 3 (1985) 432-49. https://doi.org/10.1177/073490418500300604

[14] G. F. Levchik, S. V. Levchik, A. F. Selevich, A. I. Lesnikovich, Influence of ammonium penta borate on combustion and thermal decomposition of polyamide 6, VESTSI-AKADEMIIA NAVUK BELARUSI SERYIA KHIMICHNYKH NAVUK (1995):34-38.

[15] E. Akdogan, M. Erdem, M. Erdem Ureyen, M. Kaya, Synergistic effects of expandable graphite and ammonium pentaborate octahydrate on the flame-retardant, thermal insulation, and mechanical properties of rigid polyurethane foam, Polym. Compos. 41 (2020) 1749-1762. https://doi.org/10.1002/pc.25494

[16] M. Lewin, J. Brozek, M. Martens, The system polyamide/sulfamate/dipentaerythritol: flame retardancy and chemical reactions, Polym. Adv. Technol. 13 (2002) 1091-1102. https://doi.org/10.1002/pat.243

[17] M. Lewin, J. Zhang, E. Pearce, J. Gilman, Flammability of polyamide 6 using the sulfamate system and organo-layered silicate, Polym. Adv. Technol. 18 (2007) 737-745. https://doi.org/10.1002/pat.964

[18] M. Coquelle, S. Duquesne, M. Casetta, et al., Investigation of the decomposition pathway of polyamide 6/ammonium sulfamate fibers, Poly. Deg. Stab. 106 (2014) 150-157. https://doi.org/10.1016/j.polymdegradstab.2014.02.007

[19] J. B. Dahiya, B. K. Kandola, A. Sitpalan, A. R. Horrocks, Effects of nanoparticles on the flame retardancy of the ammonium sulphamate-dipentaerythritol flame-retardant system in polyamide 6, Polym. Adv. Technol. 24 (2013) 398-406. https://doi.org/10.1002/pat.3095

[20] A. Wilkie, J. He, Some new sulfur-containing materials as putative fire retardants, in: 24th Annual Conference on Recent Advances in Flame Retardancy of Polymeric Materials, BCC Research, 2013.

[21] S. L. LeVan, Chemistry of Fire Retardancy, in: The Chemistry of Solid Wood, R. Rowell (Ed.), Chapter 14, pp. 531-574, American Chemical Society, Washington, DC, USA, 1984.

[22] L. Costa, G. Camino, Thermal behaviour of melamine, J. Therm. Anal. 34 (1988) 423-429.

[23] Melamine. CAS Common Chemistry. CAS, a division of the American Chemical Society, n.d. https://commonchemistry.cas.org/detail?cas_rn=108-78-1 (accessed on 25-5-2025) (CAS RN: 108-78-1). Licensed under the Attribution-Noncommercial 4.0 International License (CC BY-NC 4.0).

[24] J. König, U. Braun, H. Bahr, B. Schartel, Thermal decomposition of melamine cyanurate and melamine polyphosphate in fire retardancy of polymers, Fire Mater. 33 (2009) 63-78. https://doi.org/10.1002/fam.984

[25] E. D. Weil, V. Choudhary, Flame retardants for flexible polyurethane foams, J. Fire Sci. 13 (1995) 297-317. https://doi.org/10.1177/073490419501300202

[26] R. Chapman, P. R. Averell, Solubility of melamine in water, Ind. Eng. Chem. 35 (1943) 137-138. https://doi.org/10.1021/ie50398a003

[27] A. R. Horrocks, D. Price, Advances in Fire Retardant Materials, Elsevier, Amsterdam, 2008.

[28] E. D. Weil, S. V. Levchik, Flame Retardants for Plastics and Textiles: Practical Applications, 2nd ed., Carl Hanser Verlag GmbH & Co. KG, Munich, 2015.

[29] M. Lewin, S. M. Atlas, E. M. Pearce (Eds.), Flame-Retardant Polymeric Materials, Springer, New York, NY, 1990.

[30] V. Sangeetha, N. Kanagathara, R. Sumathi, N. Sivakumar, G. Anbalagan, Spectral and thermal degradation of melamine cyanurate, J. Mater. 7 (2013) 262094. https://doi.org/10.1155/2013/262094

[31] T. J. Prior, J. A. Armstrong, D. M. Benoit, K. L. Marshall, The structure of the melamine-cyanuric acid co-crystal, Cryst. Eng. Comm. 15 (2013) 5838-5843. https://doi.org/10.1039/c3ce40709h

[32] S. V. Levchik, A. I. Balabanovich, G. F. Levchik, L. Costa, Effect of melamine and its salts on combustion and thermal decomposition of polyamide 6, Fire Mater. 21 (1997) 75-83. https://doi.org/10.1002/(SICI)1099-1018(199703)21:2<75::AID-FAM597>3.0.CO;2-P

[33] Jr. J. Kersies, C. Fuerst, Flame retardant mechanistic aspects of melamine cyanurate in polyamide 6 and 66, Interflam '99, conference paper, 1998.

[34] A. Casu, G. Camino, M. De Giorgi, D. Flath, V. Morone, R. Zenoni, Fire-retardant mechanistic aspects of melamine cyanurate in polyamide copolymer, Poly. Deg. Stab. 58 (1997) 297-302. https://doi.org/10.1016/S0141-3910(97)00061-X

[35] A. B. Morgan, M. Klatt, Nitrogen-based flame retardants, in: A. B. Morgan (Ed.), Non-halogenated Flame Retardant Handbook, 2nd ed., Wiley, 2021. https://doi.org/10.1002/9781119752240.ch5

[36] A. I Balabanovich, The effect of melamine on the combustion and thermal decomposition behaviour of poly(butylene terephthalate), Poly. Deg. Stab. 84 (2004) 451-458. https://doi.org/10.1016/j.polymdegradstab.2003.12.003

[37] P. Gijsman, R. Steenbakkers, C. Fürst, J. Kersjes, Differences in the flame retardant mechanism of melamine cyanurate in polyamide 6 and polyamide 66, Poly. Deg. Stab. 78 (2002) 219-224. https://doi.org/10.1016/S0141-3910(02)00136-2

[38] D. Tabuani, F. Bellucci, A. Terenzi, G. Camino, Flame retarded Thermoplastic Polyurethane (TPU) for cable jacketing application, Poly. Deg. Stab. 97 (2012) 2594-2601.

[39] R. K. Gupta (Ed.), Materials and Chemistry of Flame-Retardant Polyurethanes. Volume 1: A Fundamental Approach, ACS Symp. Ser., American Chemical Society, Washington, DC, 1399 (2021). https://doi.org/10.1016/j.polymdegradstab.2012.07.011

[40] U. Braun, B. Schartel, Flame retardancy mechanisms of aluminium phosphinate in combination with melamine cyanurate in glass-fibre-reinforced poly(1,4-butylene terephthalate), Macromol. Mater. Eng. 293 (2008) 206-217. https://doi.org/10.1002/mame.200700330

[41] A. D. Naik, G. Fontaine, F. Samyn, X. Delva, J. Louisy, S. Bellayer, Y. Bourgeois, S. Bourbigot, Mapping the multimodal action of melamine-poly(aluminium phosphate) in the flame retardancy of polyamide 66. RSC Adv. 4 (2014), 18406-18418. https://doi.org/10.1039/C4RA02005G

[42] A. D. Naik, G. Fontaine, F. Samyn, X. Delva, Y. Bourgeois, S. Bourbigot, Melamine integrated metal phosphates as non-halogenated flame retardants: Synergism with aluminium phosphinate for flame retardancy in glass fiber reinforced polyamide 66, Poly. Deg. Stab. 98 (2013) 2653-2662. https://doi.org/10.1016/j.polymdegradstab.2013.09.029

[43] A. Schwarzer, T. Saplinova, E. Kroke, Tris-s-triazines (s-heptazines) - From a mystery molecule to industrially relevant carbon nitride materials, Coord. Chem. Rev. 257 (2013) 2032-2062. https://doi.org/10.1016/j.ccr.2012.12.006

[44] N. K. Gavrilova, V. A Gal'perin, A. I. Finkel'shtein, A. G. Koryakin, Synthesis of melam from melamine, Zh. Org. Khim. 13 (1977) 669-70.

[45] Melamine condensates prodn. for flame-retardant mouldings - by heating melamine (salt) in presence of organic acid or ammonia or melamine saltof organic acid under suitable reaction conditions,WO9616948.,Patentapplication(inactive)https://patents.google.com/patent/WO1996016948A1/zh(accessed 28/5/25).

[46] Melam-based flame retardant polyamide composition, WO9617013 A, Patentapplication (inactive) https://pubchem.ncbi.nlm.nih.gov/patent/WO-9617013-A1(accessed 28/5/25).

[47] Methanesulfonic acid melam as a flame-retardant for polyamide resins, comprises reacting melamine with methanesulfonic acid and baking product at a specific temperature, WP2002121816, BASF internal patent application, expired.

[48] Resin composition comprising polyamide resin, EP0994156 A, https://patents.google.com/patent/EP0994156A1/en (accessed 28/5/25).

[49] C. E. Wilen, R. Pfaendner, Non-halogen radical generators for enhanced flame retardancy of plastics, in: A. B. Morgan, C. A. Wilkie (Eds.), Fire Retardancy of Polymeric Materials, 3rd ed., CRC Press, 2024. https://doi.org/10.1201/9781003380689-11

[50] P. Arnaboldi, L. Davis, S. B. Samuels, J. M. Zenner, I. Vulic, A revolutionary light stabilizer system for polyolefins and other resins, Polym. Polym. Compos. 10 (2002) 93-100.

[51] M. Aubert, R. C. Nicolas, W. Pawelec, C. E. Wilen, M. Roth, R. Pfaendner, Azoalkanes-novel flame retardants and their structure-property relationship, Polym. Adv. Technol. 22 (2011) 1529-1538. https://doi.org/10.1002/pat.1642

[52] J. L. Hodgson, M. L. Coote, Clarifying the mechanism of the Denisov cycle: How do hindered amine light stabilizers protect polymer coatings from photo-oxidative degradation?, Macromolecules, 43 (2010) 4573-4583. https://doi.org/10.1021/ma100453d

[53] P. P. Klemchuk, M. E. Gande, Stabilization mechanisms of hindered amines, Poly. Deg. Stab. 22 (1988) 241-274. https://doi.org/10.1016/0141-3910(88)90014-6

[54] P. Bortolus, S. Dellonte, A. Faucitano, F. Gratani, Photostabilizing mechanisms of hindered-amines light stabilizers: interaction with electronically excited aliphatic carbonyls, Macromolecules, 19 (1986) 2916-2922. https://doi.org/10.1021/ma00166a007

[55] Flame-retardant composition comprising a phosphonic acid derivative, WO2010026230 A, https://patentscope.wipo.int/search/en/detail.jsf?doc Id=WO2010026230 (accessed 28/5/25).

[56] H. N. Stokes, On the chloronitrides of phosphorus, Amer. Chem. 17 (1985) 275-291.

[57] C. W. Allen, The use of phosphazenes as fire resistant materials. J. Fire Sci. 11 (1993) 320-328. https://doi.org/10.1177/073490419301100404

[58] M. Gleria and R. De Jaeger (Eds.), Phosphazenes: A Worldwide Insight, Nova Science Publishers Inc, Hauppauge, NY USA, 2004.

[59] G.F. Levchik, Y.V. Grigoriev, A.I. Balabanovich, S.V. Levchik, M. Klatt, Phosphorus-nitrogen containing fire retardants for poly(butylene terephthalate), Polym. Int. 49 (2000) 1095-1100. https://doi.org/10.1002/1097-0126(200010)49:10<1095::AID-PI405>3.0.CO;2-B

[60] E. D. Weil, N. G. Patel, Phospham - A stable phosphorus-rich Flame Retardant. J. Fire Sci. 18 (1994) 1-7. https://doi.org/10.1002/fam.810180102

[61] MCA Technologies - Commercial Publications. https://www.mcatechnologies.com/products/fire-retardants/index.html, (accessed on 28/5/25).

[62] S. Brehme, S. Bernhard, G. Jürgen, O. Fischer, D. Pospiech, Y. Bykov, and M. Döring. Phosphorus polyester versus aluminium phosphinate in poly (butylene terephthalate)(PBT): flame retardancy performance and mechanisms, Poly. Deg. Stab. 96(2011) 875-884. https://doi.org/10.1016/j.polymdegradstab.2011.01.035

[63] B. Schartel and T. R. Hull, Development of fire-retarded materials-interpretation of cone calorimeter data. Fire Mater. 31 (2007) 327-354. https://doi.org/10.1002/fam.949

[64] B. Schartel, & A. Weiß, Temperature inside burning polymer specimens: Pyrolysis zone and shielding. Fire Mater. 34 (2010) 217-235. https://doi.org/10.1002/fam.1007

[65] S. Levchik, & C. A.Wilkie, Char formation. Fire retardancy of polymeric materials, 589, 2000.

[66] G. Linteris, M. Zammarano,B. Wilthan & L. Hanssen, Absorption and reflection of infrared radiation by polymers in fire-like environments. Fire Mater. 36 (2012) 537-553. https://doi.org/10.1002/fam.1113

Materials Research Foundations 190 (2026) 268-294 https://doi.org/10.21741/9781644904077-10

Chapter 10

Electrospun Flame Retardant Polymer Materials

T.C. Mokhena[1,2]*, M.J. Mochane[3], T.A. Nhlapo[4], V.C Agbakoba[5], K. Shingange[6], M. Ntwaeaborwa[7], T.S. Motsoeneng[8], S. Mapukata[1]

[1]DSTI/Nanotechnology Innovation Centre (NIC), Advanced Materials Division, Mintek, Randburg, South Africa

[2]School of Chemical and Physical Sciences, University of Mpumalanga, Mbombela 1200, South Africa

[3]Department of Chemistry, College of Science, Engineering, and Technology, University of South Africa, Roodepoort, Johannesburg, South Africa

[4]Department of Medical Physics, Sefako Makgatho Health Sciences University, Medunsa, South Africa

[5]Department of Physics, Nelson Mandela University, Gqeberha, South Africa

[6]School of Physics, University of the Witwatersrand, Department of Physics, Braamfontein, 2001

[7]Faculty of Natural and Applied Sciences, Sol Plaatjie University, South Africa

[8]Department of Chemistry, University of Zululand, KwaDlangezwa Campus, 3886

mokhenateboho@gmail.com

Abstract

Over the past decades, there has been significant interest in producing polymer nanofibers from various polymeric materials, with applications ranging from flexible devices to biomedical materials. Electrospinning is a flexible processing technique that allows the processor to produce polymer nanofibers with diameters ranging from a few microns to hundreds of nanometres. This technique offers the advantage of modifying polymers at the nanoscale. Recently, a wide variety of flame-retardant (FR) additives have been incorporated into electrospun polymeric materials to enhance their fire resistance performance. This interest is driven not only by the desire to broaden the applications of electrospun nanofibers but also by the need to address the number of fire accidents that result in loss of life and property. This chapter describes recent advances in the use of FR additives in electrospun polymeric materials. It discusses the preparation of nanofibrous membranes, their flame resistance properties, and their applications.

Keywords

Polymers, Electrospinning, Flame Retardants, Fire Resistance Performance, Nanofibers, Applications

Materials Research Foundations 190 (2026) 268-294 https://doi.org/10.21741/9781644904077-10

Contents

1. Introduction

Fire resistance has become a crucial aspect of polymeric materials and technologies from a safety perspective [1-3]. This focus is not only to expand the applications of polymeric materials but also to address fire accidents that result in significant loss of life and property. Fire poses a major hazard due to the high number of fatalities and severe burns. Flame-retardant (FR) materials have recently gained prominence in polymeric materials due to international attention on safety issues [1]. Naturally, polymeric materials are flammable, making the inclusion of FRs essential. The addition of FR to polymeric materials is crucial for reducing flammability, thereby broadening their application. Flame-retardants are materials that can stop or reduce the spread of fire [2,3]. FRs act by reducing the concentration of combustible components within a material to levels that are insufficient to maintain combustion. These materials operate in both gas and condensed phases. In the condensed phase, they act as barriers that inhibit the transmission of oxygen and heat, thereby limiting the spread of fire. FRs can reduce the concentration of combustible components within a product to levels below those necessary to sustain burning. In the gas phase, flame retardants catalyze volatile molecules. When exposed to heat, they release radicals that act as scavengers for flammable organic molecules (e.g., H and OH). This results in the production of non-combustible molecules that can extinguish the fire. Additionally, flame retardants can promote crosslinking reactions or create cohesive and continuous char, which acts as an insulator to protect the underlying material and prevent further deterioration. Using different flame retardants together can enhance flame resistance performance beyond what a single flame retardant can achieve [2]. For example, one flame retardant might release radical scavengers to produce non-combustible materials that extinguish the fire, while another facilitates charring to protect the underlying material. Phosphorus-based flame retardants are known to release phosphorus radicals that scavenge H and OH radicals [3]. The presence of montmorillonite can encourage the production of continuous char that shields the underlying materials. At the same time, montmorillonite acts as a nano-barrier, protecting the underlying materials while limiting the passage of oxygen and heat.

Electrospinning is one of the versatile technologies for producing polymeric fibres with diameters from micro- to nano-scale [4-6]. It has the potential to produce fibres that can be used in different fields such as biomedical, protective clothing, drug delivery, biosensor, filtration, catalysis, energy and tissue engineering [4,7-10]. Research has intensified in using electrospinning to produce polymeric materials containing flame retardants, enabling their application in various fields such as textiles, energy, and more. A classic electrospinning setup consists of three main components: a voltage supply, a spinneret, and a grounded collector. Electrospinning has the potential as a manufacturing process to produce functionalized fibres with multifunctional properties from various materials. It is a reliable and effective method for creating high-performance materials. Both natural and synthetic polymers are commonly used as base materials to create nanofibers through electrospinning techniques. These techniques can effectively disperse FRs within polymeric materials, thereby enhancing their overall fire resistance performance. The paradigm shift towards flame-retardant polymer materials is justified by an increase in the production of commercial FRs, with a growing number of patents and research articles based on flame-retardant polymer materials. Therefore, this chapter aims is detail the preparation of polymer nanofibers filled with flame retardants (FRs). It highlights the challenges and applications of these electrospun materials.

2. Flame-retardant polymer materials

Fire resistance performance has become a critical property of polymeric materials. There are various families of flame retardants, including halogens, minerals, phosphorus, and nitrogen-based materials. The incorporation of FRs into polymeric materials can be approached in three main ways, depending on the processing technique used: (i) the blending approach which involves mixing FR additives with polymeric materials before processing [11]. (ii) Chemical grafting approach is the method that involves chemically grafting FR molecules or monomers onto polymeric matrix chains. The FRs are attached to the polymer through polymerization, forming covalent bonds. This method is often referred to as reactive processing because it involves bond formation to ensure that FRs are attached onto the polymeric materials. (iii) The surface coating approach involves FRs applied as a coating on the surface of already processed polymeric materials [12]. This coating protects the material in case of fire exposure. There are various families of flame retardants, including halogens, minerals, phosphorus, and nitrogen-based materials.

In general, several methods are used to determine the fire resistance of polymers, including cone calorimeter, pyrolysis combustion flow calorimeter (PCFC), Microscale Combustion Calorimeter or MCC, limiting oxygen index (LOI) and [13], and Underwriters Laboratories 94 (UL94) test [3]. MCC and LOI are frequently used techniques for fibrous materials, especially electrospun nanofibers. The principle of LOI is to study the least amount of O_2 in a mixture of O_2 and N_2 required to ensure protection of the materials against combustion. According to ISO 4589, LOI must be conducted on rectangular-shaped samples with dimensions of $80 \times 10 \times 4$ mm^3. During the test, the sample is subjected to a flame for 3 minutes or until it burns a length of 5cm while positioned vertically [3].

Materials Research Foundations 190 (2026) 268-294 https://doi.org/10.21741/9781644904077-10

MCC is often used to evaluate the flammability of materials. ASTM D7309 for MCC allows the analysis of samples of relatively low amounts, usually 2-4 mg, undergoing pyrolysis in a small chamber under an inert atmosphere [3]. According to the standard, the sample is subjected to heat at a rate of 0-4°C/s until reaching a maximum temperature of 750°C. Pyrolysis releases gases from the sample, which are then burnt in a combustor to determine the amount of O_2 used and the heat release rate of the combustion process. The principle of this technique relies on Huggett's relation, meaning 1 kg of oxygen used in combustion is equivalent to 13.1 MJ of energy released in the process. MCC provides essential parameters including peak heat release rate (PHRR) and total heat release (THR). Lower PHRR and THR values are associated with excellent flame resistance.

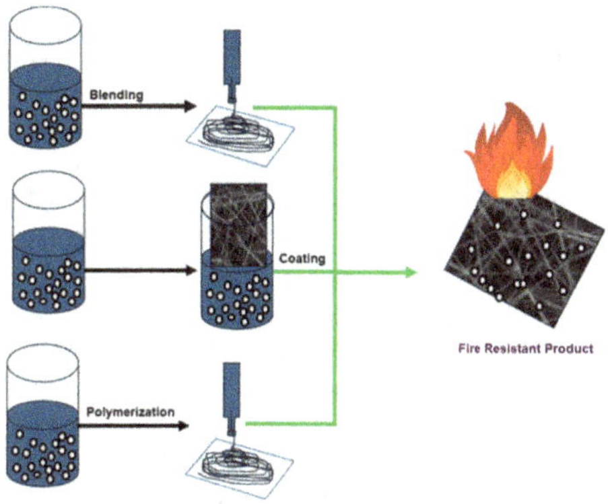

Figure 1. A visual representation of the three main approaches to introducing flame retardants into polymeric materials: blending additives, chemical grafting, and surface coating.

Cone calorimetry is also used to measure the flammability of polymers [14]. Its basic principle involves measuring the reduction of oxygen concentration of the sample when subjected to a certain heat flux (10-100 kW/m^2) [3]. Measurements are often performed according to ASTM E 1354 and ISO 5660. The technique also uses Huggett's relation. Parameters such as heat release rate and total heat release can be quantified from determination of gas flow and O_2 concentration. Additional parameters include time-to-ignition (TTI), time to flame out, mass loss rate, total smoke release (TSR), and levels of carbon dioxide, oxygen, and carbon monoxide.

PCFC is one of the fire test methods for quantifying the combustibility of small samples in milligrams (5-50 mg) [3]. This test has been standardized and classified as ASTM D7309-07. PCFC is regarded as more fire calorimetry parallel rather than a thermal analysis. It uses an oxygen combustion method to quantify heat release and heat release rate. It can separately measure the combustion of volatile and non-volatile components of the sample.

Underwriters Laboratories 94 (UL94) is a fire testing method for polymers serve as components in devices and appliances [3]. The testing procedure includes several tests, such as the radiant panel flame spread test, horizontal (H) tests, and vertical (V) tests. However, the most utilised test is UL94V, which measures the ignitability and flame spread of vertical bulk materials when exposed to a flame. This method has been standardised under IEC 60695-11-10 for small flames (50 W) and ASTM D3801-10. The flame is applied to the base of a sample, approximately 127 mm in length, for 10 seconds. Once the flame is removed, the time taken for the flame to cease is recorded as t1. Afterwards, the sample is subjected to the flame again for another 10 seconds, and the after-flame time (t2) is recorded along with the afterglow time (t3) [3,11]. During testing, cotton is placed beneath the sample to detect whether any burning droplets fall ignite the cotton. Based on the average results from 5 samples, the material is classified as V0, V1, or V2, with V0 being the most flame-retardant, followed by V1, and V2 being the least flame-retardant.

3. Electrospun flame-retardant polymeric materials

The basic electrospinning technique consists of three main components: a high voltage supply, a syringe, and a grounded collector. In principle, the polymer solution is loaded into the syringe and subjected to high voltage to produce polymeric nanofibers. The high voltage supply imparts an electrical charge to the polymer solution, resulting in the formation of a conical-shaped polymer droplet at the tip of the syringe, known as the Taylor cone. The charged jet then protrudes from the tip of the Taylor cone and travels towards the grounded collector. During this journey, the solution is stretched, leading to the deposition of fibres with diameters ranging from nano- to micro-scale onto the collector.

This technology also offers the opportunity to incorporate nanofillers, thereby fabricating multifunctional materials for various applications. Research has increasingly focused on employing the electrospinning method to design flame-retardant polymer materials [15]. In the case of flame-retardant polymeric materials, a wide variety of polymers have been electrospun alone or in the presence of fillers to enhance their fire resistance performance. This improvement allows their application in lithium-ion battery membranes, protective textiles, and many other uses, as detailed in Table 1. Advances in electrospinning methods have shown that it is possible to enhance production and include other constituents. This inclusion often addresses issues related to the spinnability of polymeric materials when combined with other substances and/or due to their ultimate application. For example, Wang et al. [16] prepared a PVDF-co-chlorotrifluoroethylene (CTFE) nanofibrous membrane filled with Sb_2O_3 nanoparticles for use as a separator in lithium-ion batteries. They employed the sequential electrospinning method. The resulting nanofibrous composite membrane demonstrated remarkable electrolyte uptake, high ionic conductivity, thermal stability, and enhanced fire resistance performance, leading to exceptional performance in Li||$LiFePO_4$ cells. This is due to the choice of technology used in the preparation, which produces a fibrous membrane composed of materials that synergistically enhance the overall properties of the resulting nanofibers. This improvement enables their application as safe separators in lithium-ion batteries. Elsewhere, the co-axial electrospinning technique was employed to fabricate a core-shell structure for a lithium-ion battery separator. This approach aimed to overcome the spinnability issues associated with polyetherimide, which possesses excellent heat resistance and mechanical strength to support the stable skeleton of the resultant nanofibers. Polyvinylidene fluoride-hexafluoropropylene (PVDF-HFP) was preferred as the shell material due to its high electrolyte affinity, facilitating the rapid transmission of Li^+ ions. The high porosity and three-dimensional structure of the resultant nanofibrous membrane further contributed to the performance of the separator by ensuring the uniform distribution of lithium ions on the battery electrode, thus aiding in the deposition of lithium ions. Consequently, the membrane maintained high discharge capacity of 178 $mAhg^{-1}$ at 0.2C and a capacity retention rate of ~90% after 200 cycles. In addition, the membrane exhibited high thermal stability, allowing its use at elevated temperatures (65°C) while maintaining excellent performance with stable charging/discharging.

Materials Research Foundations 190 (2026) 268-294 https://doi.org/10.21741/9781644904077-10

Table 1. Summary of studies on flame retardant/polymeric nanofibers and anticipated applications

Polymer	Additive	Electrospinning technology	Optimal loading percentage and conditions	Diamaters (nm)	Highlights	Applications	Refs.
PVDF	Silica	Single-needle	37-64wt% Silica TCD=15 cm	200-900	Flexible, super-insulating, and smokeless PVDF/SiO_2 nanofibers demonstrated excellent fire-retardant properties.	Personal Protective equipment (e.g., cloves, boots, clothing	[15]
Melamine	PVA	Single-needle	5-9wt% PVA Voltage= 15-21kV	0.3-0.6*	PVA enhanced the spinnability of melamine-based nanofibers, with their diameters varying based on processing parameters and solution properties.	High temperatures filters	[17]
Melamine (melamine (MA), melamine phosphate (MPP), melamine cyanurate (MCA))	PAN	Solution blowing spinning	10wt% PAN Flow rate=50mLh-1	500	PAN enhanced the spinnability of melamine-based materials, resulting in a highly flame-resistant material with improved softness, making it suitable for textile applications	Flame-retardant textiles	[18]
PVDF/PMIA/PVDF	-	Sequential spinning	15wt% PMIA 14% PVDF Flow rate=2mLh-1 25 cm 30kV Temp: 23±2°C Humidity: 45±3%	PVDF=367 PMIA=100	The sandwich nanofibrous structured materials achieved suitable mechanical and thermal properties, qualifying them for lithium-ion battery applications.	Battery separator	[19]
Polyimide/PVDF/polyimide	-	Sequential spinning	16%PVDF Flow rate=100μlh-1 for PI 125μlh-1 for PVDF Voltage: 10 kV TCD: 10cm	PI: 212 PVDF:317	The membrane demonstrated a shutdown function at high operational temperatures, which is crucial for preventing the thermal runaway issues that lithium-ion batteries face.	Battery separator	[20]
PAN	-	Single-needle	Feeding rate: 0.8 ml/h TCD:15 cm Voltage:15 kV Rotating drum speed: 400 rpm	265±64	High fire resistance performance reaching LOI value of 49%	-	[21]

PAN	Boron nitride and Triphenyl phosphate	Single needle	Flow Rate: 0.15 mm min^{-1} Voltage: 18kV TCP: 20 cm	-	The resultant membrane did not ignite upon contact with flame but burned for 6.9 seconds before self-extinguishing	Battery separator	[22]
PAN	DOPO	Single needle	Flow rate: 2 mLh^{-1} Voltage: 20kV Temp: 25°C TCD: 15cm	-	The inclusion of DOPO enhanced the flame-retardant properties of PAN, resulting in a discharge capacity of 129 mAh g^{-1} and a retention rate of 83% after 100 cycles at a C-rate of 0.6C.	Battery separator	[23]
PAN	CNTs & Ochre	Single needle	Voltage: 16 kV Flow rate: 0.004 mL min^{-1} TCD: 14 cm Take-up velocity: 9.8 ms^{-1}	155-176	High fire resistance performance and mechanical properties due to synergy between CNTs and ochre	Automobile or textile	[24]
PAN	CNTs & APP	tri-axial electrospinning system	Flow rate: 0.35/0.35/0.3 $mL.h^{-1}$ Conc: PAN/PAN&CNTs/PAN&APP 12%/12%&3%/12%&2% Temp: 25±5°C RH: 50±10% TCD 12 cm Voltage: 15kV	595	The membrane demonstrated excellent particulate capture efficiency and exhibited no signs of burning during the vertical burning test, achieving a Limiting Oxygen Index [13] value of approximately 31%.	Vehicle interior, firefighting, or fresh air filtration	[25]
Polyurethane (PU)	MgO & AgNPs	Single needle	Flow rate: 1 mLhr-1 Voltage: 13 kV TCD: 15 cm	106	Nanofibers exhibited excellent flame retardant properties and demonstrated antibacterial activity against both gram-negative and gram-positive bacteria.	Protective clothing	[26]

**=micrometer*

3.1 Electrospun melamine resin-based nanofibers

Melamine nanofibers are often prepared from melamine formaldehyde resin for high-performance applications [17,27]. These materials are preferred for heat and flame resistance applications such as firefighting apparel, drapery material, upholstery, etc. [17]. The maximum temperature at which melamine fibre can be used is 370°C. In their study, Zu-Sheng et al. [17] investigated the maximum temperature at which electrospun melamine microfibers can be used as high-temperature resistant filters. The authors used thermogravimetric analysis to evaluate the maximum degradation temperature. Since polyvinyl alcohol (PVA) was used in the solution to enhance the spinnability of melamine fibres, the TG curve showed two mass loss peaks. The first mass loss, observed gradually from 200°C, was associated with PVA, while the second sharp mass loss at 370°C was associated with triazine-based units, as shown in Fig. 2.

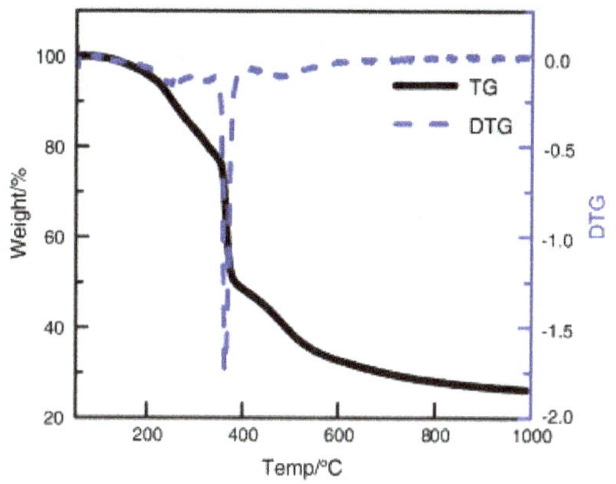

Figure 2. TG and DTG curves of electrospinning melamine nanofibers [17].

It is important to note that melamine-based materials are challenging to electrospin. Therefore, electrospinnable polymers are often incorporated into the system to enhance the electrospinnability of the resulting materials [17,28]. Elsewhere, polyacrylonitrile (PAN) was used as a co-spinning polymer for melamine-based nanofibers prepared using solution blow spinning [18]. In this case, different melamine-based raw materials, *viz.* melamine, melamine urate, and melamine phosphate, were used to prepare the electrospun nanofibers with the aid of PAN. The resulting nanofibers exhibited diameters within 500 nm, demonstrating the effectiveness of SBS technology as a reliable method for scaling up the production of electrospun nanofibers. The cone calorimeter test indicated that PAN nanofibers exhibit the highest total heat release of approximately 189 $MJ/m^2/g$ when compared to melamine-based electrospun nanofibers. Melamine (MA), melamine urate (MU), and melamine polyphosphate (MPP) reduced the total heat release of PAN to 1.22, 1.11, and 0.10 $MJ/m^2/g$, respectively. However, the heat release rate increased rapidly for MA- and MU-based nanofibers within 20 seconds, compared to MPP-based materials. This suggests that MPP enhances the overall flame resistance performance of the resulting fibers. The study on the flame retardant capabilities of MPP-based nanofibers for blankets or suits involved subjecting them to a 1200°C flame, as depicted in Fig. 3. The fibres were observed to extinguish quickly once removed from the flame source.

Materials Research Forum LLC
https://doi.org/10.21741/9781644904077-10

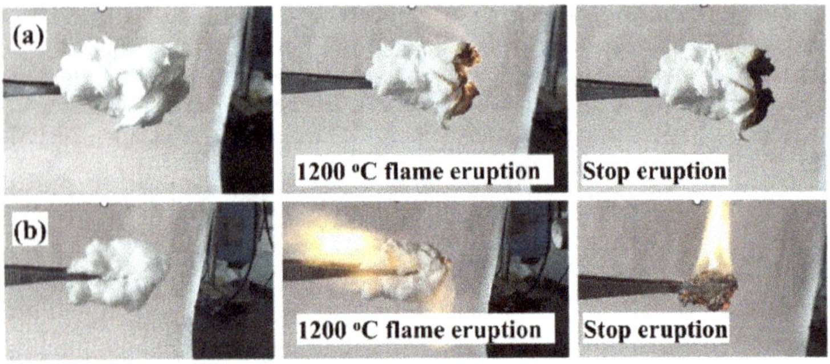

Figure 3. Flame retardant performance of (a) MPP and (b) PAN-based nanofibers under 1200°C acetylene flame [18].

3.2 Electrospun polyamide

A diverse range of fire-resistant fillers and nanofillers have been incorporated into polyimide to enhance its fire resistance and expand its applications in areas where fire safety is critically important [29,30]. In general, the presence of these fillers enhances fire resistance by either catalyzing the formation of char that shields the underlying material, acting as a heat nano-barrier layer, or generating free active radicals that scavenge combustible radicals. These processes collectively improve fire resistance by removing heat from the polymer matrix, thereby delaying combustion. It's worth noting that, despite the different technologies used to prepare these nanofibrous membranes, the presence of these fillers enhances fire resistance performance. The choice of filler plays a crucial role in determining the extent of the flame-retardant properties of the resulting materials.

Xiao et al. [29] studied a core-shell electrospun nanofibrous structure with enhanced fire resistance performance, using PA66 as the shell and graphene/red phosphorus (NG-RP) as the core. The authors reported that the presence of NG-RP improved the thermal stability of PA66 and increased the residue content after thermogravimetric analysis. Micro combustion calorimetry tests showed that the presence of NG-RP in the core reduced the Total Heat Release (THR) and Peak Heat Release Rate (PHRR) by approximately 51% and 27%, respectively, compared to neat PA66 nanofibers. Additionally, when exposed to flame for 12 seconds according to ASTM D6413-11a, NG-RP containing fibres took longer to ignite and sustained less damage compared to neat PA66 fibres. The samples self-extinguished the flame, leaving most of the membrane intact, indicating that NG-RP effectively improved the fire resistance performance of the resultant core-shell nanofibrous structure. Thum et al. [30] incorporated a nanohybrid of tannic acid into PA66 to enhance the flame-retardant properties of the resulting fibres. Micro combustion calorimetry (MCC) results showed that heat capacity and total heat release were reduced with the addition of tannic acid. This improvement was attributed to tannic acid acting as an intumescent material within PA66, facilitating charring that eliminates flaming melt dripping and allows the fibres to self-extinguish when removed from the flame.

Montmorillonite was added to Nylon 6 (polyamide 6) to enhance the flame-retardant properties of the resulting electrospun nanofibers [31]. The presence of montmorillonite was found to reduce

Materials Research Foundations 190 (2026) 268-294 https://doi.org/10.21741/9781644904077-10

the Peak Heat Release Rate (PHRR) during combustion. It was concluded that the dispersion of nanofillers is critical for improving the overall properties of the composite material, especially fire resistance. Highly dispersed nanofillers act as a nano-barrier, coating the underlying polymeric material and delaying the release of combustible substances, thereby enhancing the overall fire resistance performance.

3.3 Electrospun polyacrylonitrile (PAN)

Polyacrylonitrile (PAN) is one of the most extensively studied electrospun polymeric materials due to its unique features, such as high thermal stability and excellent mechanical properties [6,32,33]. It is often used for producing carbon fibers through a process involving stabilization and carbonization [6,32-34]. Typically, the polymer is first spun into nanofibers and then stabilized through heat treatment in an air- or oxygen-containing environment to facilitate cyclization and crosslinking. This process inhibits melting and assists in carbonization in an inert atmosphere, removing all non-carbon components to yield fibers with a carbon content exceeding 50% of the original PAN mass. The stabilization process has been used to promote crosslinking between nanofibers, producing heat- and fire-resistant nanofibrous materials. Ismar and Sarac pre-carbonized electrospun PAN nanofibers to eliminate the need for adding flame-retardant additives [21]. They electrospun PAN and then subjected it to thermal oxidation treatment at 300°C for 1 hour, resulting in fibers with an average diameter of 265 nm. The thermogravimetric curve for the oxidized PAN nanofibers showed no weight reduction over the investigated temperature range, demonstrating that the material is highly thermally stable. During fire testing, the fibers did not burn, melt, or drip, achieving a limiting oxygen index [13] value of approximately 49%. This high thermal stability was ascribed to the resultant chemical structure, where thermal oxidation transformed PAN into a ladder structure with partial carbon content. However, there is growing interest in further improving the heat and fire resistance of PAN without compromising its flexibility. This has led many researchers to investigate the inclusion of flame retardants and other substances to enhance the overall heat and fire resistance of PAN-based nanofibers. For instance, Yusuf et al. [23] introduced 9,10-dihydro-9-oxa-10-phosphaphenanthrene-10-oxide (DOPO) into electrospun PAN to enhance its flame-retardant properties. The authors indicated that the presence of DOPO reduced the peak heat release rate (PHRR) by 18%, 33%, and 49% when 10%, 15%, and 20% of DOPO were introduced into PAN, respectively. This behaviour was attributed to the high phosphorus content in DOPO, which quenches radicals released during combustion. During combustion, DOPO releases PO radicals that can capture hydrogen (H) and hydroxyl (HO) radicals, thereby reduce the heat release rate and interrupting the combustion chain.

In addition to incorporating flame-retardant fillers, nanoparticles were introduced into PAN nanofibers to enhance their fire resistance performance. The presence of nanoparticles not only improves fire resistance but also enhances the overall properties of the resultant nanofibrous membranes. Bai et al. prepared a PAN nanofibrous membrane with excellent fire resistance and thermal conductivity using the electrospinning method [22]. To enhance thermal conductivity and fire resistance, modified boron nitride (FBN) and triphenyl phosphate (TPP) were added to the PAN spinning solution. The PAN membrane burned relatively slowly, completely combusting after 3.67 seconds. However, the presence of FBN and TPP delayed ignition under the flame, with the burning process taking approximately 6.9 seconds before self-extinguishing. The peak heat release rate (PHRR) of PAN was found to be 382 W/g, while the nanofibrous composite had a PHRR of around 329 W/g. The total heat release (THR) was reported to be approximately 67%

compared to a commercial separator, and the heat release capacity showed a reduction of 73%, indicating that the resultant membrane had better flame-retardant properties. This behaviour was attributed to the heat resistance of PAN and the presence of both FBN and TPP fillers. During combustion, FBN blocks the transfer of decomposed volatiles and fixes them in the condensed phase, while TPP promotes the dehydration and carbonization of the volatiles, facilitating the formation of char that acts as an insulator, protecting the underlying material. Additionally, TPP decomposes into PO radicals that capture and trap the flammable radicals (viz. H and OH) generated from internal chemical reactions within the battery. All these processes collectively contribute to the improvements in the thermal stability and fire resistance of separators, thereby preventing the occurrence of battery thermal runaway.

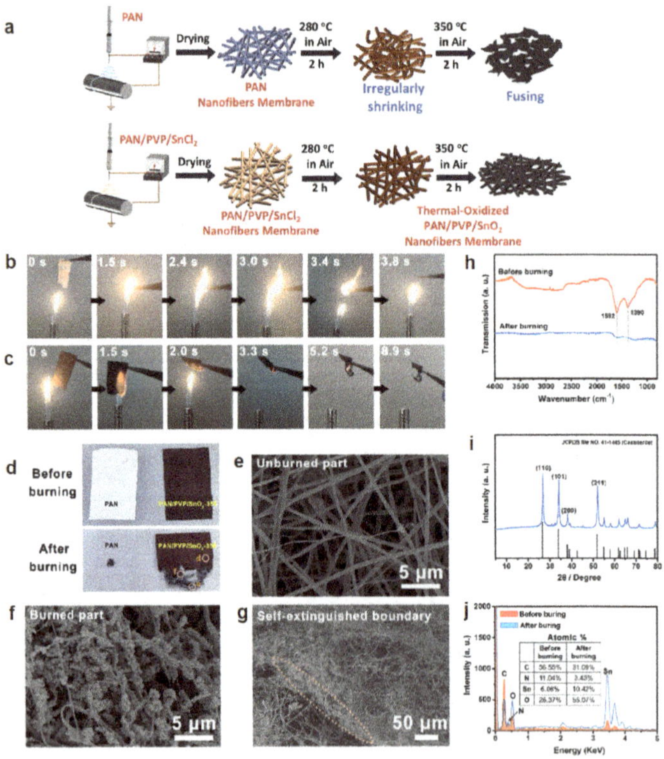

Figure 4. (a) schematic representation of preparation process of stabilized PAN-based nanofibers. (b) Fire ignition behaviour of neat PAN, and (c) Nanofibrous composite membrane. (d) Digital images of neat PAN and nanofibrous PAN-based membrane before and after burning. SEM micrographs of (e) unburned, (f) burned and (g) self-extinguished boundary. (h) IR spectra, (i) XRD, and EDX of post-burned nanofibrous PAN-based membrane. Reproduced with permission from [35].

Materials Research Foundations 190 (2026) 268-294 https://doi.org/10.21741/9781644904077-10

Moon and Emrick produced flexible, colourful, and environmentally benign PAN nanofibers by incorporating carbon nanotubes (CNTs) and ochre (Oc) nanofibers through the electrospinning technique [24]. The resultant composite nanofibrous membrane had an average diameter of 155-176 nm. The tensile strength reached approximately 80 MPa, which further increased to around 177 MPa after the stabilization process, due to increased orientation and partial crosslinking caused by cyclization. The heat release capacity decreased from 143 to 24 J/g·K, and the total heat release reduced from 8.7 to 2.1 kJ/g. Additionally, the LOI values increased from 22.1% to 34.5% after stabilization. This improvement was attributed to the formation of a thermally stable aromatic ladder structure resulting from the stabilization process. Similarly, Kang et al. [35] used the stabilization approach of PAN-based fibers to enhance the heat and fire resistance of the resultant composite materials for particulate filtration. The authors introduced $SnCl_2$ as a precursor for SnO_2 nanoparticles in the PAN/polyvinylpyrrolidone (PVP) blend, followed by spinning and thermal oxidation to create a highly porous nanofibrous membrane, as shown in Fig. 4a. The inclusion of $SnCl_2$ prevented the membrane from collapsing during thermal treatment, resulting in a porous membrane with a particulate removal efficiency of 99.89%, a pressure drop of 42 Pa, and a quality factor of 0.101. Additionally, the membrane can operate at high temperatures up to 350°C, indicating its high thermal stability due to the stabilization process. The stabilized nanofibrous composite membrane exhibited superior fire resistance compared to neat PAN. Neat PAN ignited immediately upon exposure to flame and burned completely within 3.8 seconds (Fig. 4b). In contrast, the stabilized nanofibrous composite membrane (OPAN/PVP/SnO_2-350) merely shrank and self-extinguished within 6.9 seconds after removal from the flame (Fig. 4c). SEM images showed a clear boundary between the burned area and the self-extinguished area (Fig. 4g). The burned area revealed a large number of particles decorating the fibres (Fig. 4f), indicating that PAN was not completely burned, as confirmed by FTIR results showing characteristic absorption bands at 1592 and 1390 cm^{-1} (see Fig. 4h). XRD results demonstrated no changes in the structural nature of the attained rutile SnO_2, while energy dispersive spectroscopy (EDX) analysis showed an increase in Sn and O with a decrease in C and N after burning. The proposed mechanism involves SnO_2 and a graphitic layer formed during the combustion process, coating the underlying materials and preventing the diffusion of volatiles and heat. Zhao et al.[25] prepared a multifunctional air filter with excellent fire-resistant and fire-warning capabilities using a tri-channel concentric spinneret to create a membrane composed of PAN/CNTs/ammonium polyphosphate (APP). In this case, the authors did not apply post-treatment; however, the presence of both APP and CNTs enhanced the fire-resistant properties. The as-prepared nanofibrous composite exhibited a high Limiting Oxygen Index value of approximately 31%, compared to neat PAN with an LOI value of around 18% [13]. Exposure to flame demonstrated that the composite is hardly flammable, and continuous exposure to flame led to the formation of char. The enhanced fire resistance performance was attributed to the decomposition of APP, which releases PO radicals that act as scavengers for combustible H and OH radicals. Cone calorimeter tests indicated that Peak Heat Release Rate (PHRR), Total Heat Release (THR), and Total Smoke Production (TSP) were further reduced when both CNTs and APP were used as fillers. The presence of CNTs provides an additional barrier, shielding the polymer matrix from degradation and reducing its heat release. Regarding fire warning performance, the composite triggers the warning system within 5 seconds due to changes in electrical resistance. The electrical resistance decreased from 20 MΩ to 0.45 MΩ within 2 seconds upon encountering flame and further reduced to 0.1 MΩ within 5 seconds, instantly triggering the warning system. This effect is attributed to APP's ability to catalyze PAN, thereby facilitating charring. The resulting char promotes conductive pathways

Materials Research Forum LLC
https://doi.org/10.21741/9781644904077-10

together with CNTs and thus enhance sensing behaviour. This indicates that the choice of electrospinning technology can help construct multifunctional structures with excellent fire resistance. By using PAN as the outer shell, PAN/CNT as the middle layer, and PAN/APP as the inner layer, it is possible to improve the mechanical properties, electrical conductivity, and fire resistance of the final product without the need for post-treatment. Additionally, the selection of electrospinning technology can be tailored to suit the specific application.

The addition of other materials into PAN to improve the flame-retardant property was also reported [36]. Cai et al. [36] fabricated ferric chloride-containing PAN nanofibers using the electrospinning technique. $FeCl_3$ was introduced into the PAN solution, followed by electrospinning to produce fibres with diameters ranging from 50 to 300 nm, depending on the $FeCl_3$ content. Higher $FeCl_3$ content resulted in finer fibres with smaller diameters (i.e., 50-200 nm with 5% $FeCl_3$ versus 150-300 nm with 1% $FeCl_3$). This was attributed to changes in solution properties, particularly increased solution conductivity. Fire resistance performance was studied using a micro combustion calorimeter (MCC). It was found that the peak heat release rate (PHRR) was reduced by 2.8%, 10.4%, and 17.4% with 1%, 3%, and 5% $FeCl_3$ loading, respectively, in comparison with the neat PAN. This reduction was ascribed to metal chlorides delaying the escape of combustible radicals, thereby promoting radical recombination and intermolecular crosslinking. The presence of Fe ions acted as radical capturers during combustion via a gas-phase flame retardant mechanism. Additionally, $FeCl_3$ coordinated with the electronegative cyano groups of the PAN chains.

3.4 Electrospun polyvinylidene fluoride (PVDF)

Using the electrospinning method, Kim et al. [15] prepared thermally insulating, fire-retardant, smokeless and flexible PVDF nanofibers. The authors used silica aerogel as a filler to impart the fire resistance performance of the resultant PVDF nanofibers. While neat PDVF nanofibers burned completely after just 1 second of flame exposure, the aerogel-containing nanofibers did not burn even after 10 seconds of flame exposure. This behaviour was attributed to the synergy between PVDF and silica that improved the overall performance of the fibres.

Separators for lithium-ion batteries must possess certain features to be suitable for their purpose [19,37]. These materials need to have high wettability with electrolytes, high chemical or electrochemical stability, exceptional mechanical strength, appropriate porosity, and thermal stability. Thermal stability is crucial to ensure that the separators do not shrink during operation, as they are exposed to high temperatures which could lead to severe safety issues.

Although electrospinning technology offers highly porous materials with interconnected structures, the choice of material plays a significant role in meeting other essential criteria for qualifying as a separator in lithium-ion batteries [37]. PVDF (polyvinylidene fluoride) is highly polar, which provides a high dielectric constant, good chemical and thermal stability, and low surface energy. These properties make PVDF an excellent choice for anodic stability and electrolyte polarity, thus making it a suitable separator for lithium-ion batteries. For instance, PVDF was utilized as the shell and poly-m-phenyleneisophthalamide (PMIA) as the core in a co-axial method for creating a lithium-ion battery separator [37]. The resultant core-shell structure exhibited a tensile strength of approximately 19 MPa and an elongation of around 22%, which were superior to those of the neat PVDF nanofibrous membrane. This mechanical performance qualifies the core-shell structure for application in lithium-ion batteries from a mechanical safety perspective. The enhancement in mechanical properties was attributed to the synergy between

Materials Research Foundations 190 (2026) 268-294 https://doi.org/10.21741/9781644904077-10

PVDF and the high-strength PMIA, as well as the entanglement within the core-shell structure. Additionally, the membrane demonstrated no signs of shrinking when exposed to 140°C for 2 hours, indicating high thermal stability and, consequently, heat safety. Zhang and co-workers from Tianjin China utilized co-axial electrospinning to create a FR and anti-shrink separator for batteries [38]. The separator comprised a shell of poly(vinylidene fluoride-hexafluoropropylene) (PVDF-HFP) and a core of polyetherimide. Combustion tests revealed that the separator remained unchanged when heated to 190°C for 20 seconds, showing no signs of shrinkage. This stability was attributed to the decomposition of PEI into nitrogen and ammonia (NH_3), which diluted the available oxygen and prevented further combustion. Additionally, the formation of a carbon layer during combustion blocked the contact between oxygen molecules and combustibles, thereby reducing the spread of flames.

Similarly, sandwich-structured nanofibrous lithium-ion battery separators were investigated [16,19,20]. A sequential electrospinning technique was employed to produce materials with robust mechanical properties and high thermal stability [16,19,20]. PVDF/PMIA/PVDF sandwich structures were prepared using sequential electrospinning without any chemical or post-treatment, resulting in highly porous membranes (with a porosity of approximately 91%) [19]. These membranes exhibited high ionic conductivity, excellent mechanical strength (around 14 MPa), elongation of approximately 26%, and thermal stability, viz., showing no shrinkage after being exposed to 180°C for an hour.

3.5 Electrospun polyurethane (PU) nanofibers

Electrospun polyurethane (PU) nanofibers are extensively researched for applications like wound dressings, tissue engineering, and protective textiles due to their mechanical strength and porous structure [26]. The inclusion of additives into electrospun polyurethane was reported to be a suitable route to broaden its applications. For instance, different nanoparticles are often added in order to improve the antibacterial activity of PU to afford its application in textile. This is because of the microbes found on the skin of human including *S. Aureus* and *E. Coli* which often contribute to infections and odour. To further expand the application of electrospun PU in textile FRs are also included into the mixture to improve the fire resistance making them suitable for military, firefighting and industrial protective gear. In most cases the choice of the additive depends on the anticipated application for the resultant PU nanofibers. For instance, Wamtha et al. [26] incorporated both AgNPs and MgO into electrospun PU. The authors synthesized AgNPs and MgO separately using hydrothermal and solution combustion methods, respectively. The AgNPs, with sizes ranging between 51-76 nm, and spherical MgO ~70 nm in size were added to the PU solution followed by electrospinning using a horizontal technique. The resulting nanofibrous membranes were tested for flame-retardant properties using the UL 94 standard, recording char length and afterflame, and afterglow times. Adding 2-4% MgO to PU fibres improved the afterflame time from ~10.5 seconds to 26.5 seconds and reduced the afterglow time from ~8.5 seconds to ~7.7 seconds. The char length increased from ~6.5 cm to ~8.5 cm, resulting in a V1 UL 94 rating. When both AgNPs and MgO were added, the afterflame time increased from ~10.5 seconds to ~19.1 seconds, while the char length increased from ~5.5 cm to ~8.2 cm. The afterglow time decreased from ~8.5 seconds to ~5.1 seconds, maintaining a V1 UL 94 rating. The presence of Mg ions acted as catalysts, promoting crosslinking and dehydration reactions essential for char formation, which protects the underlying material. Additionally, the antibacterial activity was enhanced by the presence of both MgO and AgNPs, with inhibition zones reaching 21 mm and 30

 Materials Research Forum LLC
https://doi.org/10.21741/9781644904077-10

mm against *E. coli* and *S. aureus*, respectively. Similarly, Kim et al. [39] developed PU nanofibers with outstanding flame retardancy and antibacterial properties by incorporating boron-doped carbon nanotubes and tannic acid. They found that the tensile strength of the nanofibrous composite increased with higher filler content, achieving optimal performance at a filler content of 3%. The heat release rate and antibacterial properties also improved with increased filler content. This enhancement was attributed to the synergy between boron-doped CNTs and tannic acid. This study demonstrates that there is an optimal filler content at which the overall properties of the nanofibrous composite are maximized, beyond which the performance may decline.

Tannic acid-coated graphene was incorporated into PU followed by electrospinning to afford good flame retardant, antibacterial and mechanical properties [40]. The inclusion of 5 wt% led to a reduction in the peak heat release rate (PHRR) from approximately 341 W/g for neat PU to 235 W/g. Additionally, the heat release capacity (HRC) decreased from around 530 J/g·K to 357 J/g·K. This improvement was attributed to the excellent thermal oxidation stabilization, carbonization, and thermal shielding properties of reduced graphene oxide Furthermore, the tannic acid (TA) coating enabled the formation of a cohesive char layer during the condensation step, resulting in the suppression of second to third-order heat transfer. The antibacterial activity increased by about 181% compared to PU/TA-based nanofibers, due to RGO's ability to impose oxidative stress and mechanical stress through penetration of the cell membrane. Moreover, the tensile modulus and strength improved by 28% and 136%, respectively, compared to neat PU nanofibrous membranes. It is worth noting that PU is preferred for textile clothing due to its flexibility. However, the inclusion of FR additives enables the production of multifunctional materials, expanding its applications beyond textiles [41,42]. Yang et al. [41] developed a multifunctional nanofibrous membrane composed of barium titanate@polyurethane/polysulfonamide that can be used as a filter at higher temperatures. The nanofibrous membrane demonstrated excellent PM2.5 removal efficiency of 99.99%, a low-pressure drop of ~39 Pa, and a tensile strength of around 13 MPa. The membrane exhibited high thermal stability, maintaining its filtration efficiency when tested at temperatures between 50-250°C. Additionally, it showed chemical resistance in both alkali and acidic solutions for an hour, maintaining its structural integrity. In fire testing, the membrane displayed good fire resistance performance. Elsewhere, a blend of polyurethane (PU) and polyetherimide (PEI) was reinforced with SiO_2 nanoparticles to fabricate nanofibrous composites suitable for use as separators in lithium-ion batteries [12]. The authors electrospun the PU and PEI blend and then coated it with SiO_2 dispersed in acetone through a dip-coating process. The nanofibers were soaked and then squeezed for an hour under pressure, followed by drying at room temperature for an hour and vacuum drying at 70°C for 4 hours, resulting in a membrane with a thickness of approximately 35 μm.

The tensile strength increased from around 8.7 MPa to approximately 16.0 MPa due to the formation of strong bonds between adjacent layers during the coating process. Compared to Celgard (a commercially available separator), the PU-PEI blend composite membrane exhibited excellent flame retardancy. While Celgard ignited and was immediately engulfed in flames, both the blend and composite nanofibrous membranes showed excellent fire resistance due to the high flame retardancy of PEI (see Fig. 5). These fibrous materials only shrank, which is crucial from a fire safety perspective as it can prevent explosive burnings in lithium-ion batteries. The composite membrane had high porosity (~59%) compared to Celgard (~48%) and demonstrated high electrolyte uptake (~189%) versus 82% for Celgard. The composite membrane exhibited cycling stability with a retention capacity of approximately 99% after 50 cycles at a 0.2C rate, indicating its potential as a next-generation separator for lithium-ion batteries.

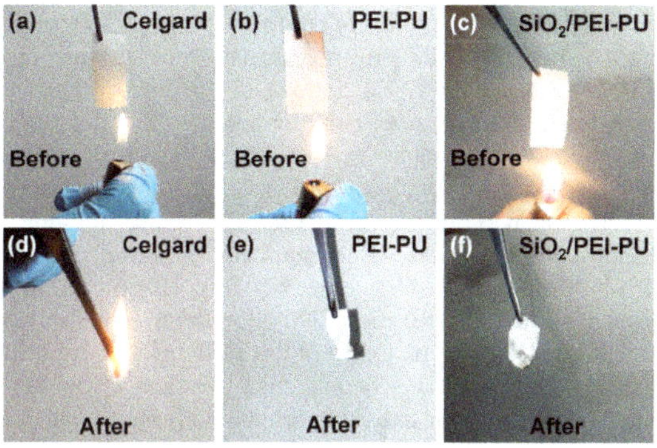

Figure 5. Fire testing of (a&d) Celgard, (b&e) PEI-PU and (c&f) SiO_2 coated PEI-PU nanofibrous membrane. Reproduced with permission from [12].

3.6 Electrospun cellulose nanofibers

Electrospun cellulose nanofibers are versatile materials used in various applications, including water filtration [43], biomedical fields [4,44,45], sensor development [13,46], electro-conductive materials [47] and active packaging [48,49]. Cellulose is a natural semicrystalline material with a polymer structure consisting of crystalline and amorphous regions [50-52]. The electrospinning of pure cellulose into cellulose nanofibers is challenging due to cellulose's tendency to aggregate and its limited solubility in common solvents [53]. The solubility of cellulose depends on the polydispersity and molecular weight [54]. Cellulose solvents are divided based on derivatising and non-derivatising solvents [55]. The non-derivatising solvents disrupt intermolecular hydrogen bonding, while the derivatising solvents aid in the dispersion of macromolecules [55]. Examples of non-derivatising solvents include aqueous solutions of inorganic salts such as cuprammonium hydroxide [56], N-methylmorpholine-N-oxide (NMMO) [57], Dimethyl sulfoxide/tetrabutylammonium fluoride trihydrate (DMSO/TBFA) [55], and lithium chloride/N, N-dimethylacetamide (LiCl/DMAc) [58]. Derivatising solvents include N, N-dimethylformamide/dinitrogen tetroxide (DMF/N_2O_4) [59] and trifluoroacetic acid (TFA) [60]. The dissolution of cellulose using ionic liquids (IL) such as 1-butyl-3-methylimidazolium chloride (BmimCl) and 1-ethyl-3-methylimidazolium acetate (EmimAc) has been widely reported [61-63]. The interest in ionic liquids is due to their low vapour pressure, high polar character and ease of regeneration. Ionic liquids solubilise cellulose by forming hydrogen bonds through a non-derivatising process [56,63]. Additionally, cellulose derivatives with improved solubility can be created by replacing the hydrogen atoms on the active hydroxyl groups in cellulose's microstructure with alternative functional groups [64]. Examples of cellulose derivatives are cellulose acetate (CA), methyl cellulose (MC), carboxymethyl cellulose (CMC), ethyl cellulose (EC), and hydroxypropyl cellulose (HPC) [64].

Materials Research Foundations 190 (2026) 268-294
https://doi.org/10.21741/9781644904077-10

The fabrication of electrospun cellulose nanofibers using plant cellulose, cellulose derivatives (e.g., cellulose acetate), and bacterial cellulose has been reported [57,59,65]. Shu et al. [66] prepared biodegradable electrospun cellulose nanofibers by designing a new drum-type electrospinning receiving device and optimising process parameters. The electrospinning solution was made following a two-step dissolution process. The first step involved the activation of cellulose powder in DMAC solution at 120°C, followed by adding LiCl and stirring at 100 °C for 2 hours. The complete dissolution was achieved after cooling the cellulose//LiCl DMAc to 25°C and stirring the solution for 8 hours. The novel drum-type receiving device handled high boiling-point LiCl/DMAC solvents more effectively and produced nanofibers with uniform diameters. The performance of the electrospun cellulose nanofibers in oil-water filtration revealed separation efficiencies ranging from 98% to 99%, whilst maintaining performance over 20 cycles. The nanofibers also fully biodegraded in soil within 12 days. Dizge et al.[67] demonstrated the fabrication of silica nanoparticle (SiNP) modified electrospun cellulose nanofiber membranes following a sol-gel method. Additional treatment of the SiNP layer on the cellulose nanofiber was conducted by applying a fluoroalkylsilane coating using chemical vapour deposition (CVD). The successful modification resulted in an increased hydrophobicity and oleophobicity. The fabricated modified electrospun nanofibers showed strong resistance to wetting by low surface tension liquids, with the highest contact angles for water, ethanol, and mineral oil. A more recent study by Ali et al. [53] explored the electrospinning of cellulose acetate solution (CA) filled with cellulose nanocrystals (CNC). A bisolvent solution of anhydrous acetone and acetic acid was used to dissolve the CA. The results from that study revealed that the morphology of the electrospun fibres was greatly affected by the amount of CNC added and the final solution viscosity. Similarly, He et al. [68] attributed the uniform morphology, improved tensile properties and thermal stability of their electrospun cellulose nanofibers to the reinforcing effect of CNC.

Cellulose nanofibers, with an oxygen limiting index [13] of 18%, are highly flammable, which limits their applications [69]. This has led to several research studies aimed at improving the thermal stability and flame retardancy of cellulose nanofibers. Jian et al. [70] reported improved flame-retardancy of modified electrospun cellulose acetate (CA) nanofibers doped with flower-shaped hydromagnesite (MgO-P) and magnesium oxide (MgO). The addition of MgO-P and MgO significantly improved the flame retardancy of CA nanofibers by serving as physical barriers. They promoted char formation and released non-flammable gases that diluted both oxygen and flammable volatiles, thereby suppressing combustion. As a result, the flames extinguished on their own once the ignition source was removed. Dong et al. [71] successfully fabricated degradable flame-retardant cellulose fibre films by crosslinking micro- and nanocellulose fibres with calcium alginate. This modification led to a limiting oxygen index of 38% and a significant improvement in the thermal stability of the modified cellulose nanofibers. Zheng et al. [72] developed a new method for preparing flame-retardant electrospun cellulose fibres by coating them with magnesium hydroxide ($Mg(OH)_2$) nanoparticles during dry-jet wet electrospinning. The authors reported that coating nanoparticles on the fibre surface significantly increased combustion onset and reduced mass loss compared to embedding them inside. Zhang et al. [73] produced flame-retardant cellulose nanofibrils by phosphorylating cellulose with phosphorus pentoxide (P_2O_5) using ball milling, followed by melamine treatment. The modification resulted in a limiting oxygen index of 30% and a 62.8% reduction in peak heat release rate.

3.7 Electrospun polyimide nanofibers

The thermal, chemical and mechanical stability of electrospun polyimide (PI) nanofibers has deemed them versatile materials for various applications including; making protective clothing, air purification media, energy storage/conversion and environmental protection [74-76]. Typically, PIs are classified into aliphatic, semi-aromatic, and aromatic groups, with the aromatic ones being the most important. Aromatic PIs are generally resistant to combustion and do not usually need to be mixed with flame retardant materials [68,74]. In the work done by Gallo et al. [77], they evaluated the flame retardancy properties of electrospun polyamic acid (PAA) nanofiber mats and PI nanofiber mats, when free-standing and upon laying them down on a PA66 substrate. Amongst other findings, they reported that the low thermal conductivity of electrospun nanofiber mats act not only as sacrificial layers but also as a protective surface that delay ignition, the effect of which is influenced by the fiber diameters. Moreover, they reported that low density and large surface-area-to-volume ratio of PI nanofibers result in a smoldering-like fire behavior. For the purpose of developing innovative nanofibers with higher thermal stability and flame-resistance properties, PI is often also incorporated with other additives or in polymer blends [78,79]. For instance, in the work done by Serbezeanu et al. [78], they electrospun non-woven fibrous membranes based on polymer blends with different ratios of a poly(ether-ether-ketone) (PEEK) and a phosphorus-containing PI. Their study showed that the surface morphology of the electrospun polymer blends is not affected by the thermal treatment, indicating that they can be used for a long time in harsh conditions, without losing their structural integrity. Moreover, a higher char yield was obtained in the polymer blends relative to PEEK alone, while membranes with high PI exhibited a non-burning behavior. Selvakumar et al. [80] developed a textile coated fabric system with superior FR from electrospun PI/boric acid (BA) nanocomposites. They evaluated the flame-retardant properties of the fabric systems coated with different nanoparticle contents. They used testing equipment manufactured on the basis of Consumer Product Safety Commission 16 Code of Federal Regulations part 1610 standard. They also utilized thermogravimetric analysis as complementary technique. They found that the 0.05% of BA nanocomposite fiber web in particular exhibited an increase in flame-spreading time of >80%, indicating superior fire protection. Qiao et al. [81] electrospun PI nanofibers containing organic montmorillonite (OMMT), which served as separators in lithium-ion batteries. This is because the fairly poor heat resistance and weak fire resistance of commercial polyolefin separators are a safety hazard and easily increase the risk of short circuit and thermal runaway. When compared to the commercially available Celgard 2500 separator, their prepared composite separator displayed exceptional thermal dimensional stability up to 250 °C and superior fire resistance. The ignition tests showed that the Celgard 2500 separator burns instantly upon ignition, with no flame retardancy, while the composite has excellent flame retardancy and self-extinguishing properties. Similarly, Yuan et al. [82] fabricated a composite polymer electrolyte by combining electrospun PI with succinonitrile-based solid composite electrolyte, to be used in solid-state batteries. When exposed to a flame, the composite did not catch fire, showing a flame-retardant property. They attributed this to the highly thermal stability of PI, as well as the fact that succinonitrile releases CO, CO_2, NO gases, etc., at high temperatures, which can serve as fire extinguishing agents. Therefore, the synergy in their fabricated composite resulted in a fireproof material that is promising for safe solid-state batteries. Wang et al. [83] fabricated polyimide/organic montmorillonite nanofibrous (PI/OMMT ESNF) membranes by electrospinning polyamic acid/organic montmorillonite precursors followed by a thermal imidization process. They found the optimum concentration of organic montmorillonite to be 3%,

Materials Research Foundations 190 (2026) 268-294 https://doi.org/10.21741/9781644904077-10

wherein the tensile strength increased by ~92% and tensile modulus by ~49%. Moreover, the LOI value rose from ~29 to ~30%. Their results also demonstrated that the incorporation of organic montmorillonite is beneficial for enhancing the mechanical properties and flame retardancy of PI nanofibrous membranes. The heat blocking effect of PI/OMMT membranes shows that cotton that was placed on the top of the prepared membrane (PI/M3) that suffered a flame with a temperature of ~520 only mild charred, while the membrane remained intact, indicating excellent heat-resistance performance for their PI/OMMT ESNF membrane.

Challenges ahead of flame-retarded electrospun nanofibers

The fragility of electrospun nanofibers remains a significant challenge for their adoption in a wide variety of applications. One potential solution to address this issue is the addition of fillers to the electrospun polymer solution, which can enhance the mechanical properties of the nanofibers. Additionally, the inherent flammability of polymers necessitates the inclusion of flame retardants (FRs) in the system. Polymers that are less flammable tend to be expensive and are often reserved for high-end engineering applications. It is important to note that the spinnability of a neat polymer does not guarantee the spinnability of its composite. Optimization is required to ensure the spinnability of composite solutions. Each laboratory employs its own electrospinning techniques, making it challenging to replicate conditions when attempting to advance technology based on existing research. Therefore, it is essential to optimize spinning conditions based on the available techniques. Advanced electrospinning techniques, especially those designed for high production, are often expensive, which can hinder the advancement of technologies that rely solely on electrospinning. Despite advancements in electrospinning, such as the development of co-axial and triaxial techniques to accommodate unspinnable materials, there remains a lack of scalability for these technologies. The mechanical performance of flame-retardant electrospun nanofibers is often not consistently reported, and when it is, the results are based on tests designed by individual laboratories. This inconsistency makes it challenging to compare results and establish a baseline for how these materials perform when flame retardants (FRs) are incorporated into the system.

To address this issue, standardized testing protocols should be developed and adopted across laboratories to ensure comparability of results. This would involve creating uniform guidelines for testing mechanical properties such as tensile strength, elasticity, and durability of flame-retardant electrospun nanofibers. By doing so, researchers can more accurately assess the impact of FRs on the mechanical performance of these materials.

Furthermore, collaboration between laboratories and sharing of data can help build a comprehensive understanding of the mechanical behaviour of flame-retardant electrospun nanofibers. This collective effort can lead to the development of more robust and reliable materials, ultimately facilitating their adoption in various applications, from protective clothing to advanced engineering fields.

State-of-the-art and future opportunities

Electrospinning technologies present a promising opportunity to develop novel, multifunctional materials that extend their applications beyond traditional uses. These technologies can be further advanced to incorporate various additives tailored to specific applications. Although scaling up these technologies remains challenging, there is significant potential for engineers and scientists

Materials Research Foundations 190 (2026) 268-294 https://doi.org/10.21741/9781644904077-10

to collaborate, ensuring high-quality and high-quantity production to facilitate the commercialization of functionalized novel materials. With the banning of most halogenated flame retardants (FRs) in various countries, natural FRs and other fillers have emerged as potential alternatives. Despite the abundance of studies on natural FRs as future flame-retardant materials, there is a notable lack of research on their use in electrospun polymers. Electrospinning can produce flame-retardant nanofibrous membranes that outperform traditional membranes due to their unique features, such as a large surface area and high porosity. This advancement has broadened the application of fibers beyond classic textile uses to more advanced fields like energy storage. Moreover, flame-retardant materials qualify these fibers for use in protective clothing for firefighting, military, and industrial protective gear. These materials can also be utilized in household items and high-tech applications. The integration of natural FRs into electrospun polymers not only enhances their flame-retardant properties but also opens up new avenues for innovation and application in various industries. Standardization and collaboration are essential for advancing the field and ensuring that flame-retardant electrospun nanofibers meet the necessary performance criteria for their intended applications. By establishing uniform testing protocols and fostering cooperation among researchers, this can create a robust framework for evaluating these materials.

These standardized protocols can be implemented to test both the fire resistance and mechanical performance of the nanofibers. This comprehensive approach will provide a clearer understanding of how flame retardants affect the overall properties of the materials. Additionally, collaboration between laboratories will facilitate the sharing of data and best practices, leading to more reliable and comparable results.

Ultimately, this concerted effort will help in developing high-quality flame-retardant electrospun nanofibers that can be confidently used in various applications, from protective clothing to advanced engineering solutions. By working together and adhering to standardized testing methods, the field can achieve significant advancements and ensure the materials meet the highest performance standards.

Conclusion and remarks

Over the past few decades, the electrospinning method has garnered significant interest due to its versatility in fabricating polymeric fibres with diameters ranging from nano- to micro-scale. These nanofibers hold potential for various applications, including smart textiles, tissue engineering, drug delivery, biosensors, biomedical uses, filtration, catalysis, and energy storage. However, the applications of electrospun flame-retardant nanofibers are still largely limited to textiles and energy storage. There is considerable work to be done to expand their applications beyond these traditional uses. To facilitate this expansion, it is crucial to standardize the fire testing and mechanical performance evaluation of flame-retardant nanofibrous membranes. Currently, the fire tests commonly used are those designed for textile clothing, with few tests extending beyond this scope. Establishing standardized testing protocols will help build a baseline for comparison and ensure consistent performance evaluation across different studies. Electrospinning technologies that can incorporate flame retardants, such as coaxial and triaxial electrospinning, present additional challenges. These techniques are difficult to upscale as they are still based on traditional single-needle spinning methods. Furthermore, there are very few studies that have explored other electrospinning technologies capable of producing large quantities of fibres suitable for

Materials Research Forum LLC
https://doi.org/10.21741/9781644904077-10

commercialization. More research using such scalable technologies is needed to ensure that these materials can be produced at a commercial scale. Advancing electrospinning technologies and developing new methods to incorporate flame retardants will be essential for broadening the applications of flame-retardant nanofibers. This will involve not only improving the scalability of these techniques but also ensuring that the resulting materials meet the necessary performance standards for a wide range of applications. By addressing these challenges, this can unlock the full potential of electrospun flame-retardant nanofibers and pave the way for their use in diverse and innovative fields.

Acknowledgements

The authors are thankful to the Department of Science, Technology and Innovation/Nanotechnology Innovation Centre Mintek for financial support.

References

[1] M.E. El-Naggar, A.R. Wassel, M.S. Hasanin, Chapter 31 - Functionalized nanofibers as fire-retardant fabrics, in: K. Deshmukh, S.K.K. Pasha, A. Barhoum, C. Mustansar Hussain (Eds.), Functionalized Nanofibers, Elsevier, 2023, pp. 893-921. https://doi.org/10.1016/B978-0-323-99461-3.00009-1

[2] J. Rodrigues, N.G. Shimpi, Nanostructured flame retardants: An overview, Nano-Struct. Nano-Objects. 39 (2024) 101253. https://doi.org/10.1016/j.nanoso.2024.101253

[3] M.E. Mngomezulu, M.J. John, V. Jacobs, A.S. Luyt, Review on flammability of biofibres and biocomposites, Carbohydr. Polym. 111 (2014) 149-182. https://doi.org/10.1016/j.carbpol.2014.03.071

[4] T.C. Mokhena, M.B. Chabalala, S. Mapukata, A. Mtibe, L. Hlekelele, Z. Cele, M.J. Mochane, B. Ntsendwana, T.A. Nhlapo, T.P. Mokoena, M.F. Bambo, K.P. Matabola, S.S. Ray, E.R. Sadiku, K. Shingange, Electrospun PCL-Based Materials for Health-Care Applications: An Overview, Macromol. Mater. Eng. 2300388.

[5] H.S. SalehHudin, M.E. Niza, M.W.N. Liza, A. and Muhammad Afifi, Multiple-jet electrospinning methods for nanofiber processing: A review, J. Manuf. Process. 33 (2018) 479-498. https://doi.org/10.1080/10426914.2017.1388523

[6] T.C. Mokhena, K.P. Matabola, T.H. Mokhothu, A. Mtibe, M.J. Mochane, G. Ndlovu, J.E. Andrew, Electrospun carbon nanofibres: Preparation, characterization and application for adsorption of pollutants from water and air, Sep. Purif. Technol. 288 (2022) 120666. https://doi.org/10.1016/j.seppur.2022.120666

[7] T.C. Mokhena, A.S. Luyt, Electrospun alginate nanofibres impregnated with silver nanoparticles: Preparation, morphology and antibacterial properties, Carbohydr. Polym. 165 (2017) 304-312. https://doi.org/10.1016/j.carbpol.2017.02.068

[8] T.C. Mokhena, A.S. Luyt, Development of multifunctional nano/ultrafiltration membrane based on a chitosan thin film on alginate electrospun nanofibres, J. Clean. Prod. 156 (2017) 470-479. https://doi.org/10.1016/j.jclepro.2017.04.073

[9] T.C. Mokhena, N.V. Jacobs, A.S. Luyt, Electrospun alginate nanofibres as potential bio-sorption agent of heavy metals in water treatment, Express Polym. Lett. 11 (2017) 652-663. https://doi.org/10.3144/expresspolymlett.2017.63

[10] O. Mapazi, P. Matabola, T. Mokhena, H. van der Walt, The catalytic degradation of dyes using polyethersulfone/poly(vinylidene fluoride)/copper oxide composite nanofibers, J. Appl. Polym. Sci. 140 (2023) e53679. https://doi.org/10.1002/app.53679

[11] T.C. Mokhena, E.R. Sadiku, M.J. Mochane, S.S. Ray, Mechanical properties of fire retardant wood-plastic composites: A review, Express Polym. Lett. 15 (2021) 744-780. https://doi.org/10.3144/expresspolymlett.2021.61

[12] Y. Zhai, K. Xiao, J. Yu, B. Ding, Fabrication of hierarchical structured SiO2/polyetherimide-polyurethane nanofibrous separators with high performance for lithium ion batteries, Electrochim. Acta. 154 (2015) 219-226. https://doi.org/10.1016/j.electacta.2014.12.102

[13] A. Petropoulou, S. Kralj, X. Karagiorgis, I. Savva, E. Loizides, M. Panagi, T. Krasia-Christoforou, C. Riziotis, Multifunctional gas and pH fluorescent sensors based on cellulose acetate electrospun fibers decorated with rhodamine B-functionalised core-shell ferrous nanoparticles, Sci. Rep. 10 (2020) 367. https://doi.org/10.1038/s41598-019-57291-0

[14] C. Liu, K. Xu, Y. Shi, J. Wang, S. Ma, Y. Feng, Y. Lv, F. Yang, M. Liu, P. Song, Fire-safe, mechanically strong and tough thermoplastic Polyurethane/MXene nanocomposites with exceptional smoke suppression, Mater. Today Phys. 22 (2022) 100607. https://doi.org/10.1016/j.mtphys.2022.100607

[15] Y.-G. Kim, H.S. Kim, S.M. Jo, S.Y. Kim, B.J. Yang, J. Cho, S. Lee, J.E. Cha, Thermally insulating, fire-retardant, smokeless and flexible polyvinylidene fluoride nanofibers filled with silica aerogels, Chem. Eng. J. 351 (2018) 473-481. https://doi.org/10.1016/j.cej.2018.06.102

[16] L. Wang, Z. Wang, Y. Sun, X. Liang, H. Xiang, Sb2O3 modified PVDF-CTFE electrospun fibrous membrane as a safe lithium-ion battery separator, J. Membr. Sci. 572 (2019) 512-519. https://doi.org/10.1016/j.memsci.2018.11.041

[17] Z.-S. Hang, L.-H. Tan, X.-M. Cao, F.-Y. Ju, S.-J. Ying, F.-M. Xu, Preparation of melamine microfibers by reaction electrospinning, Mater. Lett. 65 (2011) 1079-1081. https://doi.org/10.1016/j.matlet.2011.01.010

[18] J. He, W. Zhang, X. Jin, C. He, W. Li, Blow spinning mass production of melamine@polyacrylonitrile composite nanofibers with enhanced flame-retardant and strength, Prog. Nat. Sci.: Mater. Int. 33 (2023) 634-643. https://doi.org/10.1016/j.pnsc.2023.11.008

[19] Y. Zhai, N. Wang, X. Mao, Y. Si, J. Yu, S.S. Al-Deyab, M. El-Newehy, B. Ding, Sandwich-structured PVdF/PMIA/PVdF nanofibrous separators with robust mechanical strength and thermal stability for lithium ion batteries, J. Mater. Chem. A. 2 (2014) 14511-14518. https://doi.org/10.1039/C4TA02151G

[20] D. Wu, C. Shi, S. Huang, X. Qiu, H. Wang, Z. Zhan, P. Zhang, J. Zhao, D. Sun, L. Lin, Electrospun nanofibers for sandwiched polyimide/poly (vinylidene fluoride)/polyimide separators with the thermal shutdown function, Electrochim. Acta. 176 (2015) 727-734. https://doi.org/10.1016/j.electacta.2015.07.072

[21] E. Ismar, A.S. Sarac, Nonflammable pre-carbonized polyacrylonitrile nanofiber webs, SN Appl. Sci. 2 (2020) 589. https://doi.org/10.1007/s42452-020-2407-y

[22] W. Bai, L. Xiao, T. Long, Z. Wang, J. Wang, Y.K.K. Richard, Y. Lu, Fire-retardant and thermally conductive polyacrylonitrile-based separators enabling the safety of lithium-ion batteries, J. Colloid Interface Sci. 684 (2025) 377-387. https://doi.org/10.1016/j.jcis.2024.12.229

[23] A. Yusuf, V.S. Avvaru, M. Dirican, S. Changchun, D.-Y. Wang, Low heat yielding electrospun phosphenanthrene oxide loaded polyacrylonitrile composite separators for safer high energy density lithium-ion batteries, Appl. Mater. Today. 20 (2020) 100675. https://doi.org/10.1016/j.apmt.2020.100675

[24] S. Moon, T. Emrick, High flame-resistant and strong electrospun polyacrylonitrile-carbon nanotubes-ochre nanofibers, Polymer 54 (2013) 1813-1819. https://doi.org/10.1016/j.polymer.2013.01.053

[25] T. Zhao, D. Teng, Y. Xu, X. Zhang, Y. Li, Y. Zeng, Multi-functional air filters with excellent flame retardancy and fire-warning capability, J. Colloid Interface Sci. 617 (2022) 236-245. https://doi.org/10.1016/j.jcis.2022.03.003

[26] V. Mamtha, H.N.N. Murthy, V.P. Raj, P. Tejas, C.S. Puneet, A. Venugopal, S. AanMankunipoyil, C. Manjunatha, Electrospun PU/MgO/Ag nanofibers for antibacterial activity and flame retardancy, Cloth. Text. Res. J. 41 (2021) 236-253. https://doi.org/10.1177/0887302X211009479

[27] J. Wei, G. Sun, T. Li, H. Li, F. Xu, Y. Lu, M. Liang, G. Zhang, Preparing durable flame retardant and formaldehyde free cotton fabrics with crosslinked flame retardant system, Polym. Degrad. Stab. (2025) 111382. https://doi.org/10.1016/j.polymdegradstab.2025.111382

[28] Q. Wang, Y. Yu, J. Ma, N. Zhang, J. Zhang, Z. Liu, G. Cui, Electrospun melamine resin-based multifunctional nonwoven membrane for lithium ion batteries at the elevated temperatures, J. Power Sources. 327 (2016) 196-203. https://doi.org/10.1016/j.jpowsour.2016.07.063

[29] L. Xiao, L. Xu, Y. Yang, S. Zhang, Y. Huang, C.W. Bielawski, J. Geng, Core-shell structured polyamide 66 nanofibers with enhanced flame retardancy, ACS Omega. 2 (2017) 2665-2671. https://doi.org/10.1021/acsomega.7b00397

[30] M.D. Thum, M. Tighe, N.K. Weise, N. Hoffman, R. Mosurkal, J.A. Orlicki, J.G. Lundin, Flame-retardant properties and characterization of nylon/tannic acid electrospun fibers, Adv. Eng. Mater. 26 (2024) 2301333. https://doi.org/10.1002/adem.202301333

[31] H. Wu, M. Krifa, J.H. Koo, Flame retardant polyamide 6/nanoclay/intumescent nanocomposite fibers through electrospinning, Text. Res. J. 84 (2014) 1106-1118. https://doi.org/10.1177/0040517513515314

[32] S.Y. Gu, J. Ren, Q.L. Wu, Preparation and structures of electrospun PAN nanofibers as a precursor of carbon nanofibers, Synth. Met. 155 (2005) 157-161. https://doi.org/10.1016/j.synthmet.2005.07.340

[33] R.D. Andrei, A. Marinoiu, E. Marin, S. Enache, E. Carcadea, Carbon Nanofibers Production via the Electrospinning Process, Energies, 2020. https://doi.org/10.3390/en13113029

[34] E. Ismar, A.S. Sarac, Oxidation of polyacrylonitrile nanofiber webs as a precursor for carbon nanofiber: aligned and non-aligned nanofibers, Polym. Bull. 75 (2018) 485-499. https://doi.org/10.1007/s00289-017-2043-x

[35] Y. Kang, J. Chen, S. Feng, H. Zhou, F. Zhou, Z.-X. Low, Z. Zhong, W. Xing, Efficient removal of high-temperature particulate matters via a heat resistant and flame retardant thermally-oxidized PAN/PVP/SnO2 nanofiber membrane, J. Membr. Sci. 662 (2022) 120985. https://doi.org/10.1016/j.memsci.2022.120985

[36] Y. Cai, D. Gao, Q. Wei, H. Gu, S. Zhou, F. Huang, L. Song, Y. Hu, W. Gao, Effects of ferric chloride on structure, surface morphology and combustion property of electrospun polyacrylonitrile composite nanofibers, Fibers and Polym. 12 (2011) 145-150. https://doi.org/10.1007/s12221-011-0145-x

[37] L. Wang, N. Deng, J. Ju, G. Wang, B. Cheng, W. Kang, A novel core-shell structured poly-m-phenyleneisophthalamide@polyvinylidene fluoride nanofiber membrane for lithium ion batteries with high-safety and stable electrochemical performance, Electrochim. Acta 300 (2019) 263-273. https://doi.org/10.1016/j.electacta.2019.01.115

[38] Y. Zhang, R. He, Y. Li, H. Liu, H. Liu, X.-X. Zhang, Fabrication of coaxially electrospun PEI@PVDF-HFP fibrous membrane as flame-retardant and anti-shrink separator for lithium-ion battery, Mater. Lett. 361 (2024) 136063. https://doi.org/10.1016/j.matlet.2024.136063

[39] Y.N. Kim, J.Y. Jo, Y. Kim, Y.-M. Ha, H. Han, D.C. Lee, J. Kim, Y.C. Jung, Structure stability, flame retardancy, and antimicrobial properties of polyurethane composite nanofibers containing tannic acid and boron-doped carbon nanotubes, Macromol. Mater. Eng. 306 (2021) 2100455. https://doi.org/10.1002/mame.202100455

[40] Y.N. Kim, Y.-M. Ha, J.E. Park, Y.-O. Kim, J.Y. Jo, H. Han, D.C. Lee, J. Kim, Y.C. Jung, Flame retardant, antimicrobial, and mechanical properties of multifunctional polyurethane nanofibers containing tannic acid-coated reduced graphene oxide, Polym. Test. 93 (2021) 107006.. https://doi.org/10.1016/j.polymertesting.2020.107006

[41] X. Yang, Y. Pu, Y. Zhang, X. Liu, J. Li, D. Yuan, X. Ning, Multifunctional composite membrane based on BaTiO3@PU/PSA nanofibers for high-efficiency PM2.5 removal, J. Hazard. Mater. 391 (2020) 122254. https://doi.org/10.1016/j.jhazmat.2020.122254

[42] H. Shi, H. Liu, S. Luan, D. Shi, S. Yan, C. Liu, R.K.Y. Li, J. Yin, Antibacterial and biocompatible properties of polyurethane nanofiber composites with integrated antifouling and bactericidal components, Compos. Sci. Technol. 127 (2016) 28-35. https://doi.org/10.1016/j.compscitech.2016.02.031

[43] D.-N. Phan, H. Lee, B. Huang, Y. Mukai, I.-S. Kim, Fabrication of electrospun chitosan/cellulose nanofibers having adsorption property with enhanced mechanical property, Cellulose 26 (2019) 1781-1793. https://doi.org/10.1007/s10570-018-2169-5

[44] G. Ramanathan, L.S. Seleenmary Sobhanadhas, G.F. Sekar Jeyakumar, V. Devi, U.T. Sivagnanam, P. Fardim, Fabrication of biohybrid cellulose acetate-collagen bilayer matrices as nanofibrous spongy dressing material for wound-healing application, Biomacromolecules 21 (2020) 2512-2524. https://doi.org/10.1021/acs.biomac.0c00516

[45] S. Ahmadian, M. Ghorbani, F. Mahmoodzadeh, Silver sulfadiazine-loaded electrospun ethyl cellulose/polylactic acid/collagen nanofibrous mats with antibacterial properties for wound healing, Int. J. Biol. Macromol. 162 (2020) 1555-1565. https://doi.org/10.1016/j.ijbiomac.2020.08.059

[46] R. Jia, W. Tian, H. Bai, J. Zhang, S. Wang, J. Zhang, Amine-responsive cellulose-based ratiometric fluorescent materials for real-time and visual detection of shrimp and crab freshness, Nat. Commun. 10 (2019) 795. https://doi.org/10.1038/s41467-019-08675-3

[47] Q. Lyu, B. Peng, Z. Xie, S. Du, L. Zhang, J. Zhu, Moist-induced electricity generation by electrospun cellulose acetate membranes with optimized porous structures, ACS Appl. Mater. Interfaces. 12 (2020) 57373-57381. https://doi.org/10.1021/acsami.0c17931

[48] A. Zaitoon, L.-T. Lim, C. Scott-Dupree, Activated release of ethyl formate vapor from its precursor encapsulated in ethyl Cellulose/Poly (Ethylene oxide) electrospun nonwovens intended for active packaging of fresh produce, Food Hydrocoll. 112 (2021) 106313. https://doi.org/10.1016/j.foodhyd.2020.106313

[49] B. Niu, L. Zhan, P. Shao, N. Xiang, P. Sun, H. Chen, H. Gao, Electrospinning of zein-ethyl cellulose hybrid nanofibers with improved water resistance for food preservation, Int. J. Biol. Macromol. 142 (2020) 592-599. https://doi.org/10.1016/j.ijbiomac.2019.09.134

[50] S. Absar, M. Khan, K. Edwards, J. Neumann, Investigation of synthesis and processing of cellulose, cellulose acetate and poly (ethylene oxide) nanofibers incorporating anti-cancer/tumor drug cis-diammineplatinum (II) dichloride using electrospinning techniques, J. Polym. Eng. 35 (2015) 867-878. https://doi.org/10.1515/polyeng-2015-0057

[51] T.C. Mokhena, V. Jacobs, A.S. Luyt, A review on electrospun bio-based polymers for water treatment, Express Polym. Lett. 9 (2015) 839-880. https://doi.org/10.3144/expresspolymlett.2015.79

[52] T.C. Mokhena, N.V. Jacobs, A.S. Luyt, Nanofibrous alginate membrane coated with cellulose nanowhiskers for water purification, Cellulose 25 (2018) 417-427. https://doi.org/10.1007/s10570-017-1541-1

[53] A. Khodayari, S. Vats, G. Mertz, C.N. Schnell, C.F. Rojas, D. Seveno, Electrospinning of cellulose nanocrystals; procedure and optimization, Carbohydr. Polym. 347 (2025) 122698. https://doi.org/10.1016/j.carbpol.2024.122698

[54] D. Klemm, B. Philpp, T. Heinze, U. Heinze, W. Wagenknecht, Comprehensive cellulose chemistry. Volume 1: Fundamentals and analytical methods, 1998. https://doi.org/10.1002/3527601929

[55] T. Heinze, A. Koschella, Solvents applied in the field of cellulose chemistry: A mini review, Polímeros 15 (2005) 84-90. https://doi.org/10.1590/S0104-14282005000200005

[56] A. Kramar, F.J. González-Benito, Cellulose-based nanofibers processing techniques and methods based on bottom-up approach-a review, Polymers 14 (2022) 286. https://doi.org/10.3390/polym14020286

[57] H. Peng, G. Dai, S. Wang, H. Xu, The evolution behavior and dissolution mechanism of cellulose in aqueous solvent, J. Mol. Liq. 241 (2017) 959-966. https://doi.org/10.1016/j.molliq.2017.06.103

[58] X. He, L. Cheng, X. Zhang, Q. Xiao, W. Zhang, C. Lu, Tissue engineering scaffolds electrospun from cotton cellulose, Carbohydr. Polym. 115 (2015) 485-493. https://doi.org/10.1016/j.carbpol.2014.08.114

[59] Y. Jing, X. Ma, C. Xu, H.-l. Tian, S.-w. Chen, Repair of dural defects with electrospun bacterial cellulose membranes in a rabbit experimental model, Mater. Sci. Eng. C. 117 (2020) 111246. https://doi.org/10.1016/j.msec.2020.111246

[60] M.Á. Robles-García, C.L. Del-Toro-Sánchez, E. Márquez-Ríos, A. Barrera-Rodríguez, J. Aguilar, J.A. Aguilar, F.J. Reynoso-Marín, I. Ceja, R. Dórame-Miranda, F. Rodríguez-Félix, Nanofibers of cellulose bagasse from Agave tequilana Weber var. azul by electrospinning: preparation and characterization, Carbohydr. Polym. 192 (2018) 69-74. https://doi.org/10.1016/j.carbpol.2018.03.058

[61] S.-L. Quan, S.-G. Kang, I.-J. Chin, Characterization of cellulose fibers electrospun using ionic liquid, Cellulose 17 (2010) 223-230. https://doi.org/10.1007/s10570-009-9386-x

[62] Y. Ahn, D.-H. Hu, J.H. Hong, S.H. Lee, H.J. Kim, H. Kim, Effect of co-solvent on the spinnability and properties of electrospun cellulose nanofiber, Carbohydr. Polym. 89 (2012) 340-345. https://doi.org/10.1016/j.carbpol.2012.03.006

[63] R. Phadagi, S. Singh, H. Hashemi, S. Kaya, P. Venkatesu, D. Ramjugernath, E. Ebenso, I. Bahadur, Understanding the role of dimethylformamide as co-solvents in the dissolution of cellulose in ionic liquids: Experimental and theoretical approach, J. Mol. Liq. 328 (2021) 115392. https://doi.org/10.1016/j.molliq.2021.115392

[64] J. Kerwald, C.F. de Moura Junior, E.D. Freitas, J.d.D.P. de Moraes Segundo, R.S. Vieira, M.M. Beppu, Cellulose-based electrospun nanofibers: A review, Cellulose 29 (2022) 25-54. https://doi.org/10.1007/s10570-021-04303-w

[65] A. Kiper, A. Özyuguran, S. Yaman, Electrospun cellulose nanofibers from toilet paper, Journal of Material Cycles and Waste Management 22 (2020) 1999-2011. https://doi.org/10.1007/s10163-020-01085-4

[66] D. Shu, P. Xi, B. Cheng, Y. Wang, L. Yang, X. Wang, X. Yan, One-step electrospinning cellulose nanofibers with superhydrophilicity and superoleophobicity underwater for high-efficiency oil-water separation, Int. J. Biol. Macromol. 162 (2020) 1536-1545. https://doi.org/10.1016/j.ijbiomac.2020.07.175

[67] N. Dizge, E. Shaulsky, V. Karanikola, Electrospun cellulose nanofibers for superhydrophobic and oleophobic membranes, J. Membr. Sci. 590 (2019) 117271. https://doi.org/10.1016/j.memsci.2019.117271

[68] X. He, Q. Xiao, C. Lu, Y. Wang, X. Zhang, J. Zhao, W. Zhang, X. Zhang, Y. Deng, Uniaxially aligned electrospun all-cellulose nanocomposite nanofibers reinforced with cellulose nanocrystals: scaffold for tissue engineering, Biomacromolecules 15 (2014) 618-627. https://doi.org/10.1021/bm401656a

[69] W. Shao, X. Zhang, X. Liang, W. Tao, M. Ma, S. Chen, Y. Shi, H. He, Y. Zhu, X. Wang, Cellulose nanofiber-based nanocomposite films with efficient electromagnetic interference shielding and fire-resistant performance, ACS Appl. Mater. Interfaces. 16 (2024) 42674-42686. https://doi.org/10.1021/acsami.4c10660

[70] L. Jiang, K. Li, H. Yang, X. Liu, W. Xu, B. Deng, Significantly improved flame-retardancy of cellulose acetate nanofiber by Mg-based nano flaky petal, Cellulose 26 (2019) 5211-5226. https://doi.org/10.1007/s10570-019-02451-8

[71] J. Dong, C. Geng, Y. Liu, T. Wang, X. Liu, J. Duan, X. Long, Z. Nie, L. Tan, Y. Xia, Flame-retardant and degradable cellulose micro/nanofiber films for packaging materials, ACS Appl. Nano Mater. 7 (2024) 13799-13808. https://doi.org/10.1021/acsanm.4c02590

[72] Y. Zheng, J. Miao, N. Maeda, D. Frey, R.J. Linhardt, T.J. Simmons, Uniform nanoparticle coating of cellulose fibers during wet electrospinning, J. Mater. Chem. A. 2 (2014) 15029-15034. https://doi.org/10.1039/C4TA03221G

[73] T. Zhang, M. Wu, S. Kuga, C.M. Ewulonu, Y. Huang, Cellulose nanofibril-based flame retardant and its application to paper, ACS Sustain. Chem. Eng. 8 (2020) 10222-10229. https://doi.org/10.1021/acssuschemeng.0c02892

[74] Y. Ding, H. Hou, Y. Zhao, Z. Zhu, H. Fong, Electrospun polyimide nanofibers and their applications, Prog. Polym. Sci. 61 (2016) 67-103. https://doi.org/10.1016/j.progpolymsci.2016.06.006

[75] Y. Li, L.-B. Zhong, Q.-J. Zhang, C.-Y. Guo, B.-Q. He, M. Younas, Y.-M. Zheng, Highly breathable and durable waterproof polyimide electrospun nanofibrous membrane for potential reusable protective clothing application: Preparation, characterization and performance, J. Membr. Sci. 693 (2024) 122354. https://doi.org/10.1016/j.memsci.2023.122354

[76] F. Zhang, Y. Si, J. Yu, B. Ding, Electrospun porous engineered nanofiber materials: A versatile medium for energy and environmental applications, Chem. Eng. J. 456 (2023) 140989. https://doi.org/10.1016/j.cej.2022.140989

[77] E. Gallo, Z. Fan, B. Schartel, A. Greiner, Electrospun nanofiber mats coating-new route to flame retardancy, Polym. Adv. Technol. 22 (2011) 1205-1210. https://doi.org/10.1002/pat.1994

[78] D. Serbezeanu, C. Hamciuc, T. Vlad-Bubulac, M.-D. Onofrei, A. Bargan, D. Rusu, D.M. Suflet, G. Lisa, Electrospun Nanofibers Based on Polymer Blends with Tunable High-Performance Properties for Innovative Fire-Resistant Materials, Polymers, 2022. https://doi.org/10.3390/polym14245501

[79] Q. Zhang, L. Ma, T. Xue, J. Tian, W. Fan, T. Liu, Flame-retardant and thermal-protective polyimide-hydroxyapatite aerogel fiber-based composite textile for firefighting clothing, Compos. B: Eng. 248 (2023) 110377. https://doi.org/10.1016/j.compositesb.2022.110377

[80] N. Selvakumar, A. Azhagurajan, T.S. Natarajan, M. Mohideen Abdul Khadir, Flame-retardant fabric systems based on electrospun polyamide/boric acid nanocomposite fibers, J. Appl. Polym. Sci. 126 (2012) 614-619. https://doi.org/10.1002/app.36640

[81] M. Qiao, G. Zhang, J. Deng, J. Guo, J. Zhang, Electrospun polyimide@organic-montmorillonite composite separator with enhanced mechanical and thermal performances for high-safety lithium-ion battery, J. Mater. Sci. 57 (2022) 11796-11808. https://doi.org/10.1007/s10853-022-07343-0

[82] B. Yuan, B. Zhao, Z. Cong, Z. Cheng, Q. Wang, Y. Lu, X. Han, A Flexible, Fireproof, Composite Polymer Electrolyte Reinforced by Electrospun Polyimide for Room-Temperature Solid-State Batteries, Polymers, 2021. https://doi.org/10.3390/polym13213622

[83] L. Wang, L. Cui, J. Fan, Y. Liu, Electrospun polyimide/organic montmorillonite composite nanofibrous membranes with enhanced mechanical properties and flame retardancy, J. Ind. Text. 49 (2018) 875-888. https://doi.org/10.1177/1528083718801360

Materials Research Forum LLC
https://doi.org/10.21741/9781644904077-11

Chapter 11

Flame-Retardant Materials for Batteries

E. Quiroz-Pérez[1], D.C. López-Díaz[2], F. Castillo-Leonardo[1], M.C. González-López[1,2], A.R. Hernandez-Martinez[1,2], J.A. de Lira-Flores[1*]

[1]Facultad de Química, Universidad Autónoma de Querétaro, Centro Universitario, Cerro de las Campanas, 76010, Querétaro, México

[2]Facultad de Química, Universidad Autónoma de Querétaro, Campus Pedro Escobedo, Av. Panamericana No. 180 Int. 100, Centro, Pedro Escobedo, 76700, Querétaro, México

julio.lira@uaq.mx

Abstract

Lithium-ion and solid-state batteries are widely used in energy storage systems, yet their thermal instability poses significant safety risks, including thermal runaway and fire hazards. This chapter examines flame-retardant strategies that improve safety without compromising electrochemical performance. Halogenated and halogen-free additives (APPs, phosphazenes), intumescent systems, and nonflammable electrolytes are discussed. Innovations in separators, binders, and microcapsules are also reviewed, along with recent technological developments. Challenges in scalability, compatibility, and sustainability for the design of safe energy storage solutions are discussed.

Keywords

Flame-Retardants, Lithium-Ion Batteries, Solid-State Batteries, Thermal Runaway, Non-Flammable Electrolytes

Contents

Materials Research Foundations 190 (2026) 295-332 https://doi.org/10.21741/9781644904077-11

1. Introduction

Since their invention, Li-ion batteries have undergone rapid and sustained development as energy storage systems. This can be attributed to some interesting advantages exhibited by these systems, such as high capacity and voltage, a wide range of operating temperatures, a low self-discharge rate, and an outstanding cycling performance. In this sense, the rapid proliferation of portable electronic devices—ranging from cell-phones and tablets to personal computers—has been directly enabled by the advantages of Li-ion battery technology [1,2]. In addition, the application of these batteries in different industries has also increased during the last decades, mainly due to their capacity for large scale energy storage. Such applications span the automotive industry, telecommunications backup systems, renewable energy storage, aviation, and rail transport [3-5]. Despite this increasing growth, safety is still a determining factor for further development of Li-ion batteries, especially in industrial scale applications.

During their cycle life, Li-ion batteries may exhibit one or more of the following failures: overcharge, overdischarge, and internal and/or external short circuit. These failures often lead to

Materials Research Foundations 190 (2026) 295-332 https://doi.org/10.21741/9781644904077-11

thermal runaway, a phenomenon triggered by an excessive and uncontrollable heat generation inside the batteries. If the heat generated in the batteries is not dissipated, it is eventually released along with flammable gases [6-8]. As a matter of fact, thermal runaway in Li-ion batteries have been acknowledged as the primary cause of fire and explosions related to the use of these devices [9,10]. It is worth mentioning that these accidents have often led to different consequences, including the loss of life and money, as well as property damage [11,12].

In order to solve the problems caused by thermal runaway, different strategies have been proposed [13,14]. One of these consists of the use of flame-retardants, which are chemical agents that inhibit or reduce the process of combustion by modifying pyrolysis or oxidation reactions [15,16]. In this sense, the use of flame-retardants in Li-ion batteries has been proposed by different authors in order to assure a safer operation of these devices [17-20]. In this case, there is a growing interest in the development of more efficient flame-retardants for Li-ion batteries, especially those that do not cause a negative effect on the electrochemical performance of these devices.

Taking the aforementioned aspects into account, the objective of this chapter is to present a review of the application of flame-retardant materials in Li-ion batteries and their relevance to industrial safety. Our revision focuses on relevant subjects, which include characteristics, types, applications, and innovations on flame-retardants. Based on these points, we also present the challenges and future perspectives on the development of effective flame-retardants for Li-ion batteries, considering the safety normatives. Finally, the conclusions of our review are presented.

2. Properties of Lithium-Ion Batteries

2.1 Components and Operating Mechanism

Batteries are devices designed to store and deliver electrical energy. In lithium-ion batteries, electrical energy is stored within one or more electrochemical cells, where reversible redox reactions facilitate the transfer of lithium ions between electrodes [21]. Lithium-ion batteries are characterized by relevant characteristics such as rechargeability, high energy density, elevated discharge potential, compact design, and extended cycling stability compared to other battery technologies. These attributes have enabled their continuous development alongside advanced electronic devices that are essential to support growing technological demands. Fig. 1 illustrates the main components of a lithium-ion cell, which operate bi-directionally in charge and discharge cycles. The components are defined as follows:

- Anode: The negative electrode, where a loss of electrons occurs through an oxidation reaction during the charging process. Anode materials commonly include graphite [22,23] and metal alloys based on Silicon (Si), Titanium (Ti), Copper (Cu), Tin (Sn), and Germanium (Ge) [24,25]. In addition, alternative metal oxides such as niobates [26] and titanium-based compounds have also been explored [27]. Recent research has focused on the use of nanocarbon structures and carbon-based surface coatings to enhance electrochemical performance [28]. The choice of anode material plays a crucial role in improving cycle stability, increasing charging rate, enhancing thermal stability, and enabling higher charge storage capacity.
- Cathode: The positive electrode, where a reduction reaction involves a gain of electrons. The selection of cathode materials must ensure high operating voltage, elevated energy density,

Materials Research Foundations 190 (2026) 295-332 https://doi.org/10.21741/9781644904077-11

excellent ionic and electronic conductivity, as well as thermal stability [29]. Common commercial cathode materials include lithium-rich compounds such as $LiCoO_2$, $LiMn_2O_2$ (LMO), $LiFePO_2$ (LFP), and $LiNiCoAlO_2$ (NCA). For example, when high energy storage, long cycle life, and enhanced safety cathode are required, $LiFePO_2$ can be considered as an optimal selection [30,31].

- Electrolyte: A reaction medium that transports the lithium ions (Li+) within the electrochemical cells. Generally, it consists of a non-aqueous solution composed of lithium salts dissolved in organic solvents. However, to increase the ionic conductivity and reduce the electrolyte decomposition at high voltage, phosphate, nitrate, or borate-based additives are added [32]. A critical characteristic in electrolyte selection, particularly for battery design, is maintaining long-term cycling stability to prevent solid electrolyte interphase (SEI) formation. The SEI is a passive layer that gradually forms during the battery cycling, which reduces the transport of Li+ and compromises electrochemical performance in cells [33].
- Separator: A barrier that prevents direct contact between the electrodes but allows ion migration. It typically consists of a porous membrane (40-60%) with commercial pore size ranging from 0.03 to 0.12 μm [34]. These separators must maintain thermal stability for a normal operating temperature range between -20 °C and 60 °C [35]. Common materials used in these membranes include Polypropylene and Polyethylene [36] as well as inorganic based component on refractory ceramics such as alumina (Al_2O_3), zirconia (ZrO_2) or transition metals [37].

The components of electrochemical cells in lithium-ion batteries must generally meet strict conditions, such as high energy density, durability over large numbers of charge and discharge cycles, and extended thermal stability, because these batteries are widely used in electric vehicles, electronic devices, and renewable energy systems.

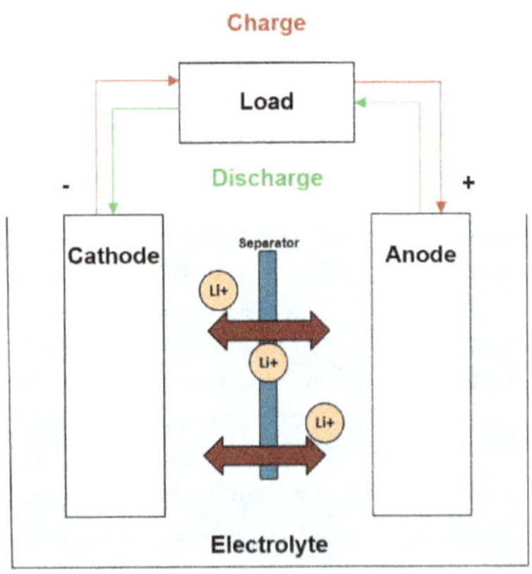

Figure 1. Schematic diagram of a lithium-ion battery.

Materials Research Foundations 190 (2026) 295-332 https://doi.org/10.21741/9781644904077-11

2.2 Charge mechanism

During this process, as previously mentioned in the component descriptions, a redox reaction describes the ion interaction between the electrodes. Assuming a graphite anode and a lithium-based cathode, the lithium–ion reduction can be described as follows:

$$C_6 + xLi^+ + xe^- \rightarrow Li_xC_6 \tag{1}$$

The cathode ion-reaction corresponds to oxidation:

$$LiCoO_2 \rightarrow Li_{1-x}CoO_2 + xLi^+ + xe^- \tag{2}$$

2.3 Discharge mechanism

The reverse redox reaction in the discharge process occurs. In this case, the anode is affected by an oxidation reaction:

$$Li_xC_6 \rightarrow xC_6 + xLi^+ + xe^- \tag{3}$$

The reduction reaction in the cathode is:

$$Li_{1-x}CoO_2 + xLi^+ + xe^- \rightarrow LiCoO_2 \tag{4}$$

If alternative materials are selected for the electrodes, the reaction mechanism is preserved. Finally, the generated electrons flow from the cathode to the anode through the external circuit, while lithium ions are transported across the separator via the electrolyte solution.

2.4 Risk of Lithium-Ion Batteries

The main risks associated with lithium-ion batteries are linked to materials used in internal cell components, particularly when these components are operated under prolonged and strenuous operating conditions, whether electrically, thermally, and/or mechanically. Among the most critical safety concerns are those related to fire and explosion hazards, such as thermal overloading, internal short circuits, or flammable electrolyte decomposition. These risks can result in effects during thermal runaway, such as gas accumulation, cell rupture, or high-temperature flame propagation, as presented in Fig. 2.

Ensuring battery safety is fundamental for design, operation, transport, and final disposal management. Adequate safety measures are essential to guarantee the reliable performance of the devices they power. The following section outlines the main fire and explosion risks associated with lithium-ion batteries, as well as the underlying causes of these effects.

Materials Research Foundations 190 (2026) 295-332 https://doi.org/10.21741/9781644904077-11

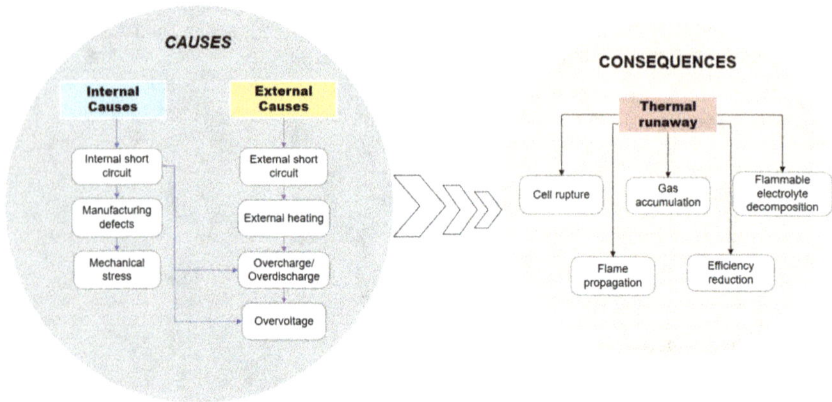

Figure 2. Causes and consequences related to Fire and explosion risks of Lithium-Ion Batteries.

2.4.1 The Causes

The causes of fire and explosion risk in lithium-ion batteries can be classified as internal and external factors. Internal causes are commonly associated with internal short circuits, while external causes occur in conditions outside the electrochemical cell, for example, external short circuits, external heating, overloading, and overvoltage.

In lithium-ion batteries, the causes that lead to internal short circuits can be classified as manufacturing or intrinsic defects (such as inefficient internal sealing, and improper electrolyte selection), extended and inadequate operating conditions [38] (including overcharging, deep discharging [39], or exposure to adverse weather conditions. An ISC is a risk that potentially generates toxic gases, excessive heat generation, or even explosions. However, this risk increases if the device is subjected to mechanical stress during transportation, as in vehicular applications [40,41].

To prevent safety incidents associated with internal short circuits in batteries, early identification of precursor conditions is critical to mitigate the risks. Detection methods involve the diagnosis of capacity balance and internal resistance variations [42]. Advanced approaches implement algorithms for electrochemical impedance spectroscopy data or frequency response analysis [43-45]. Accurate detection enables the identification of indicators that allow the adjustment of charge and discharge operations to prevent detrimental effects such as internal short circuits (ISC) [46].

2.4.2 The consequences

A common issue in lithium-ion batteries is the phenomenon known as thermal runaway, referring to uncontrolled internal overheating, which represents a significant hazard due to the possibility of producing side reactions in the electrodes that lead to fires. The charging and discharging processes are potential causes for thermal runaway. Therefore, some batteries are equipped with effective protection mechanisms for certain operating ranges [47]. This involves continuous monitoring of heat transfer or high-temperature traces for estimating the risk factor [48,49], as well as identifying the causes to implement thermal runaway suppression strategies [50].

During the thermal runaway process, the propagation of elevated temperatures within cell components also causes toxic gases and particle emissions, posing significant safety risks. Among the gas emissions are flammable gases such as hydrogen, ethane, and methane [51]. Their prolonged internal accumulation generates overpressure and ultimately causes cell rupture. Toxic particles originating from internal components may contain heavy metals, which may be released, posing serious health consequences to users [52].

3. Characteristics of Flame-Retardant Materials

Flame-retardant materials (FRMs) are essential for improving fire safety across a wide range of industries, including electronics, construction, and transportation. Traditionally, halogenated flame-retardants, especially brominated compounds, have dominated the market due to their high effectiveness and low cost [53]. However, increasing environmental and toxicological concerns have driven the development of safer alternatives, including halogen-free and phosphorus-based retardants.

3.1 Brominated Flame-retardants (BFRs)

Brominated flame-retardants, such as polybrominated diphenyl ethers (PBDEs), have historically accounted for approximately 30% of the global FRM market [54]. Their flame-retardant mechanism is based on the release of bromine radicals in the gas phase, which effectively scavenge the free radicals responsible for sustaining combustion, thereby disrupting chain reactions [54]. These compounds have been widely used in flexible polyurethane foams, epoxy coatings, and textiles, typically at concentrations ranging from 5 to 30 wt% [55]. Nevertheless, their environmental persistence, bioaccumulation, and toxicity have prompted the search for alternative flame-retardants with reduced environmental impact [55].

3.2 Ammonium Polyphosphate (APP)

In recent years, phosphorus-based flame-retardants have attracted increasing attention due to their high efficiency and lower environmental impact [56,57]. Among these, ammonium polyphosphate (APP) is a halogen-free inorganic flame-retardant that has emerged as an efficient and environmentally favorable alternative [58]. APP, its chemical formula $(NH_4,H)_{h+2}P_hO_{3h+1}$, is employed in polymer matrices such as polypropylene, polyamides, and EPDM rubbers, maintaining adequate mechanical properties (elastic modulus >2.5 MPa) while reducing the composite density (1.74–2.10 g/cm^3), demonstrating particular effectiveness in oxygen-deficient environments [56,57]. Furthermore, the incorporation of synergistic additives like phosphomolybdic acid (PMA) improves thermal stability and char formation, enhancing flame resistance in polyurethane foams [56,59]. For instance, the addition of 12% APP and 3% PMA increases char residue by 209.43% at 800°C and reduces total heat release (THR) by 24.62%, achieving UL-94 V-0 classification. The APP/borate combination has shown a reduction of up to 40% in heat release rate in epoxy matrices, attributed to the formation of thermally stable borophosphate glasses. These advances respond to increasing environmental regulations and the demand for multifunctional materials in the aerospace and wind energy sectors.

Materials Research Foundations 190 (2026) 295-332 https://doi.org/10.21741/9781644904077-11

3.3 Halogen-Free Flame-retardants

Halogen-free flame-retardants, including organic and inorganic phosphorus compounds, nitrogen-containing species, and certain metals, offer a sustainable alternative to BFRs. These materials mainly act in the condensed phase by promoting char formation and reducing the emission of toxic smoke during combustion. For instance, compounds such as red phosphorus and aluminum hypophosphite provide good thermal stability and flame retardancy (LOI > 32%) [58], although red phosphorus may release phosphine (PH_3), limiting its long-term storage.

Gutiérrez and Mina [60] reported the development of nine flame retardant masterbatch formulations with low environmental impact for incorporation into linear low-density polyethylene. Their study evaluated compounds based on N,N-ethylene bis(tetrabromoamide) mixed with antimony trioxide and zinc borate, aiming to produce crosslinked polyethylene with adequate flame-retardant properties for cable insulation. The XLPE-FR8 formulation, which consisted of the MBFR8 mixture that had the best results (45% N, N-Ethylene bis (tetrabromophthalamide) (EBTPI) and 1.25% dispersing agent (DA), mixed with antimony trioxide, zinc borate, linear low-density polyethylene, and polyethylene wax) was found to be feasible for use in the insulation of electrical cables that must comply with UL 44 (horizontal flame test). Similarly, Long et. al [58] have reported the synthesis of novel halogen-free flame retardants, such as phenylphosphinic aryl dimelamine ester (PDEPDM), which significantly increases the limiting oxygen index (LOI) while maintaining acceptable mechanical properties in polyethylene. They demonstrated that a 32 wt% addition of PDEPDM significantly improved the limiting oxygen index (LOI) to 29.8, while maintaining acceptable mechanical properties such as tensile strength of 10.06 MPa and impact resistance of 16.77 kJ/m^2.

Red phosphorus (density 2.34 g/cm^3) is effective in polycarbonates and polyamides through the formation of crosslinked aromatic structures [55]; however, its tendency to generate phosphine (PH_3) limits long-term storage applications. Aluminum hypophosphites have emerged as stable alternatives (density 1.85 g/cm^3) for thermoset composites, combining flame retardancy (LOI > 32%) with low smoke emission [5].

3.4 Advantages and Disadvantages of Different Types of Flame-Retardants

There is growing concern regarding the environmental impact of certain flame-retardants, particularly brominated flame-retardants (BFRs). While these compounds are effective and cost-efficient, they have been subjected to increasing restrictions by regulatory bodies such as the European Commission [61] and the United States Environmental Protection Agency (EPA) due to their potential to form brominated furans and dioxins compounds, recognized for their carcinogenic potential during combustion [54].

Brominated flame-retardants are highly lipophilic and bioaccumulative, raising concerns about their environmental persistence. Numerous studies have documented this persistence and accumulation, which has led to usage restrictions, especially for brominated diphenyl oxides, as regulated under the RoHS Directive [62]. Current trends favor hybrid systems such as APP/borate combinations, which integrate condensed-phase and gas-phase mechanisms of action and achieve UL-94 V-0 classifications at loadings below 15% [57]. These advances align with circular economy goals and aim to reduce the ecological footprint in strategic sectors such as electric mobility and aerospace [57]. Brominated flame-retardants are highly effective at low concentrations (1–5 wt%) and offer lower production costs compared to halogen-free alternatives.

Materials Research Foundations 190 (2026) 295-332 https://doi.org/10.21741/9781644904077-11

However, their toxicity, bioaccumulative behavior, and the release of hazardous compounds during combustion have led to regulatory restrictions and a shift toward safer alternatives.

On the other hand, halogen-free flame-retardants, especially phosphorus-based systems, offer significant advantages in terms of reducing toxic emissions and enhancing compatibility with recycling processes. Although they generally require higher loadings to achieve comparable flame retardancy, their development has advanced considerably, achieving UL-94 V-0 classifications at loadings below 15% through component synergism.

3.5 Mechanisms of Action: How Flame-Retardants Interfere with Combustion and Prevent Fire Propagation

Flame-retardant materials disrupt the self-sustaining combustion process by acting as a thermal barrier. They operate by simultaneously preventing ambient heat (from the gas phase) from reaching the flammable substrate, while also inhibiting the release of combustible gases (fuel) from the solid material back into the flame [57]. This dual capability explains why halogen-containing flame-retardants, for instance, are traditionally favored and considered highly efficient for applications in construction materials. These compounds decompose rapidly and absorb combustion heat, releasing non-flammable molecules, such as halogen hydrides, that effectively quench radical chain reactions [63]. In summary, brominated flame-retardants (BFRs) release bromine radicals in the gas phase that neutralize reactive combustion species, interrupting radical propagation chains [54], and often require synergists such as antimony trioxide to maximize their effectiveness [56].

Intumescent flame-retardant systems function by forming a charred carbon layer upon exposure to heat. This layer acts as a barrier, preventing oxygen from reaching the combustible material and thereby halting the combustion process. It also inhibits contact between the molten polymer and the flame, serving as a thermal insulator [62]. Phosphorus-based systems further promote molecular crosslinking, which enhances the thermal stability of the polymer matrix [56]. APP is a flame-retardant that operates in the condensed phase through multiple mechanisms: (i) Char layer formation, linked to thermal decomposition that produces phosphoric acids, which catalyze surface crosslinking [56]; (ii) Gas dilution, via the release of ammonia and water vapor that reduces the concentration of combustible gases [56]; Thermal barrier effect, through the formation of an intumescent carbonaceous layer that lowers heat transfer during combustion [53].

3.6 Commonly Used Materials

Research has increasingly focused on developing lower-toxicity alternatives, such as N,N-ethylene bis(tetrabromoacetamide) (EBTPI), which, according to environmental protection agencies in the United States and Denmark, does not exhibit the same level of adverse environmental or health effects as other brominated compounds, such as decabromodiphenyl ether (Deca-BDE) [54].

There is also growing interest in halogen-free flame-retardants, particularly those based on phosphorus chemistry. The study by Long et al. (2018) demonstrated that such compounds can provide excellent flame-retardant performance without generating toxic smoke during combustion [56].

4. Applications of Retardant Materials in Lithium-Ion Batteries

Table 1 shows implementations of flame-retardant materials by battery component. It is presented this way, with the intention of comparing approaches, mechanisms, performance, and the challenges for researchers, engineers, and other stakeholders. For example, the first line shows that the use of common phosphate additives, such as Triphenyl Phosphate (TPP), Trimethyl Phosphate (TMP), and Triethyl Phosphate (TEP), in the electrolyte is a low-cost and highly effective strategy [64]. Their main mechanism of action is in the gas phase; during a fire, they release phosphorus radicals (PO·) that capture and neutralize free radicals (H· and ·OH), effectively extinguishing the fire. However, it has a negative impact on battery performance. These additives increase the viscosity of the electrolyte and decrease its ionic conductivity, hindering the movement of ions. Furthermore, their instability is a critical aspect due to contact with the graphite anode, which causes degradation of the solid electrolyte interphase (SEI) protective layer, reducing the battery's lifespan and performance after several charge and discharge cycles. Therefore, there is a trade-off between sacrificing battery performance and cycle life for fire protection. However, microencapsulations and multifunctional materials improve safety with minimal impact on battery performance. Therefore, it is important to balance fire-retardant efficiency, electrochemical compatibility, and manufacturability.

Table 1. Applications of Fireproof Materials in Lithium-Ion Batteries

Component	Strategy and Specific Material	Main Mechanism of Action	Impact on Electrochemical Performance	Advantages and Challenges
Electrolyte	Additive: Phosphates (Triphenyl Phosphate, Trimethyl Phosphate, Triethyl Phosphate) [64]	Gas Phase: Capture of H· and ·OH radicals by radicals PO·(Phosphoryl Radical).	- Decreases ionic conductivity. - Increases viscosity. - Poor reductive stability in graphite anodes, degrading solid electrolyte interphase (SEI) and service life.	+ Low cost, high flame-retardant (FR) efficiency. - Significantly compromises cycle performance.
Electrolyte	Additive: Phosphonates (Dimethyl Methylphosphonate) [65]	Gaseous Phase: Like phosphates.	- Less negative impact than phosphates, but can still affect performance. - Can help form a Cathode Electrolyte (CEI) stable.	+ Good balance between safety and performance at low concentrations (<20%). - Long-term compatibility remains a challenge.
Electrolyte	Additive: Phosphazenes (Hexafluorocyclotriphosphazene, Ethoxy(pentafluoro)cyclotriphosphazene) [64]	Gas Phase: Release of P· and F· radicals. Condensed Phase: Promotes char formation.	- Relatively minor impact on performance. - May improve low-temperature performance and reduce polarization. - Does not co-intercalate with graphite.	+ High FR efficiency, multifunctional. - Synthesis cost and purity are challenges.
Electrolyte	Additive: Microencapsulated (HCP@UF, TPP@PVDF-HFP) [66]	Thermal release of FR agent (Gas/Condensed Phase) only in case of thermal abuse.	- Minimal impact on performance during normal operation. - Capacity retention is very high (e.g., 88.8% after 500 cycles with PAN-HCP@UF).	+ Decouples the safety function from normal operation. - Process complexity, cost, and long-term stability of the capsules.

Electrolyte	Solvent: Triacetin (GTA) [67]	High flash point (138 °C) and boiling point (258 °C) solvent, intrinsically less flammable.	- High viscosity and low ionic conductivity. - Requires cosolvents (EC, FEC) and film-forming additives (LiDFOB) for robust SEI/CEI.	+ Allows operation at high temperatures (>100°C). - Electrochemical performance does not yet match conventional carbonate electrolytes.
Separator	Covering: Aluminum Diethylphosphinate (ADEP) in Polyacrylonitrile (PAN) [68]	Gaseous and Condensed Phase: ADEP (phosphorus) decomposes to inhibit the flame.	- Higher ionic conductivity (0.83 mS/cm) and electrolyte absorption (137%) than commercial PP (0.65 mS/cm, 63%). - Better discharge capacity at high rates.	+ Excellent heat resistance and self-extinguishing. - The electrospinning and coating process adds complexity.
Separator	Intrinsic Polymer: Phosphorylated Cellulose [69]	Thermally stable cellulose structure. Phosphate groups act as FRs in the condensed phase (charring).	- Greater porosity and ionic conductivity than commercial separators. - Cycle performance comparable to that of commercial membranes.	+ Sustainable, biodegradable, low-cost, and multifunctional. - Mechanical strength may be lower than that of synthetic polymers.
Electrodes	Binder: Polyacrylic Acid (PAA)- Flame-retardant Epoxy Resin (FREP) for Si anode [70]	The FREP component (P and N) acts as FR in the condensed phase and gaseous.	- Dampens Si volume expansion. - Improves adhesion to the current collector. - Maintains excellent electrochemical performance.	+ Multifunctional solution (mechanical, adhesive, safety) for high-energy anodes. - The synthesis of the binder is more complex than that of PVDF or CMC.
Electrodes	Incorporation: en-TPP in cathode [71]	Thermal release of encapsulated TPP (Gas Phase).	No sacrifice in electrochemical performance, as the TPP is isolated from the electrolyte under normal conditions.	+ Place the flame-retardant directly on the heat source. - Increases the electrode's inactive mass; uniform dispersion is crucial.

In conclusion, the battery components can be modified including materials to provide greater safety, reducing the risk of fires without compromising electrochemical performance. The advantages and disadvantages of each material highlight the importance of appropriate selection of materials and strategies. Multifunctional materials are the most promising, applying inherent safety principles such as minimization or intensification. The following sections delve into the scientific and technological advances of these materials and strategies.

5. Innovations in Flame-Retardant Materials for Batteries - Part 1: Review of Scientific Literature

The development of flame-retardant materials in lithium batteries is evident in reported academic research, as well as in the development of applied technology. This section reviews the state-of-the-art scientific studies focused on electrolyte formulations, solid and hybrid systems, and structural materials. The reader is expected to identify fire-extinguishing mechanisms, electrochemical performance, and current design challenges. The goal of developing these materials is to improve thermal safety without sacrificing battery performance.

Various non-flammable electrolytes have been explored for lithium batteries, such as aqueous solutions, phosphates, ionic liquids, solid-state systems, fluorides, deep eutectic solvents, and fluorinated phosphazenes. Among these, organic phosphate-based liquids were the first and remain the most widely studied flame-retardant electrolytes for lithium-ion batteries. They provide inherent non-flammability, a wide operational temperature range, low cost, high lithium salt solubility, and low viscosity, making them suitable both as primary solvents and flame-retardant additives. This balance of safety, performance, and affordability has established phosphate-based systems as a standard reference for developing advanced and reliable lithium battery electrolytes [72].

One of the most widely used is trimethylphosphate (TMP), which has the characteristic of having a simple structure, although it has the disadvantage that it is necessary to increase its content to achieve non-flammability, and it also exhibits good oxidative stability but poor reductive stability, which leads to reductive decomposition [72]. Adding co-solvents can help reduce electrolyte decomposition, but too much can compromise non-flammability, so co-solvents with high boiling points and fewer hydrogen atoms are preferred. TMP interacts with lithium metal, forming a high-resistance surface film that indicates poor compatibility, while amorphous carbon anodes with disordered laminar structures can prevent TMP co-intercalation. The stability of lithium bis(perfluoroethylsulfonyl)imide ($LiN(SO_2C_2F_5)_2$) in TMP non-flammable electrolyte is enhanced by adding vinylene carbonate (VC), vinylethylene carbonate (VEC), and cyclohexane. Introducing a diluent in high-concentration electrolytes further improves lithium-ion solvation, electrochemical stability, and overall performance, making TMP-based systems suitable for high-performance lithium batteries while maintaining safety [73,74].

Introducing the diluent 1,1,2,2-tetrafluoroethyl-2,2,3,3-tetrafluoropropyl ether (TTE) into a LiFSI/TMP high-concentration electrolyte (HCE) allows the creation of a local high-concentration electrolyte (LHCE) combining LiFSI and TMP+TTE. This LHCE lowers viscosity, increases ionic conductivity, and maintains key properties such as non-flammability and a stable solvation structure, forming a solid electrolyte interface (SEI) from FSI anions that stabilizes the Gr anode. Electrolytes containing TMP have been utilized in lithium metal batteries (LMBs); however, lithium's high reactivity limits TMP's compatibility with the anodes, causing decreased coulombic efficiency and shorter cycle life. To overcome these limitations, a 4 M LiFSI/TMP HCE was developed, achieving non-flammability while suppressing lithium dendrite growth and enhancing overall battery stability [75,76]. This formulation demonstrated superior safety and stability, achieving a zero self-extinguishing time (SET) while also showing high oxidative resistance and strong chemical compatibility. The system's effectiveness was confirmed in symmetric Li||Li cells; testing in this configuration showed only a low overpotential and resulted in a uniform, dense lithium deposition. Also, when applied in full cell configurations against both LiFePO4 and Li4Ti5O12 cathodes, the electrolyte also enabled outstanding cycling stability. Furthermore, a highly concentrated 5 M LiFSI/TMP electrolyte was found to be compatible with graphite and effectively suppressed lithium dendrite growth by forming a LiF-rich solid electrolyte interphase (SEI) [77].

Another electrolyte explored in recent years is triethyl phosphate (TEP), which, similar to TMP, exhibits effective flame-retardant properties, high solubility in lithium salts along with reduced viscosity. Nevertheless, the compatibility it exhibits with Gr and lithium metal remains suboptimal [72]. Comparative studies revealed a critical difference between the solvents: TEP was found to avoid co-intercalation with lithium ions. Conversely, TMP was shown to be prone to this mechanism, a reaction which directly leads to the destructive exfoliation of the graphite layers [78]. On the other hand, replacing the methyl group with an ethyl group slowed the decomposition kinetics of PET compared to TMP, resulting in relatively good reductive stability. However, an excessive amount of PET had a pronounced negative impact on battery lifespan [79]. Studies have shown that the electrode–electrolyte interface in TEP-based systems can be stabilized by controlling the salt-to-solvent ratio. At a 1:2 proportion (≈2.2 M), solvent molecules remain strongly coordinated within the solvation shell, reducing side reactions with graphite and lithium metal. A LiFSI/TEP electrolyte with FEC and LiBOB displayed electrochemical behavior comparable to carbonate-based systems, enabling efficient lithium plating and stripping with >99% coulombic efficiency. Commercial 18650 cells using this formulation resisted nail penetration without ignition or explosion, confirming its safety. Moreover, the addition of a diluent to tune the solvation environment improved compatibility with high-voltage electrodes, ensuring both stability and flame-retardant characteristics [80].

A non-flammable LHCE consisting of 1.2 M LiFSI/TEP+EC+BTFE forms a thin SEI on graphite and a ~2 nm CEI on NCM811, enhancing electrode stability and minimizing transition metal dissolution. This enables reliable cycling of high-energy Gr||NCM811 cells. Additionally, a 1.2 M LiFSI electrolyte in TEP/FEC/BTFE has been investigated for silicon-anode batteries, confirming its potential for advanced high-performance systems [81,82].

In recent years, non-flammable TEP-based electrolytes have gained importance in LMBs, first serving as co-solvents in Li||S systems. In LiBOB/TEP+FEC blends, FEC functions as a film-former, while LiBOB reduces preferentially at the anode, producing a robust SEI that stabilizes TEP. At 30% FEC content, an optimal trade-off between electrochemical performance and flame resistance is achieved, with capacity retention of ≈91% after 500 cycles and ≈79% after 1000 [83].

Dimethyl methylphosphonate (DMMP) was first introduced in 2007 as a flame-retardant additive in LIB electrolytes, becoming one of the most studied phosphorus-based compounds. Its low viscosity, moderate dielectric constant, and enhanced flame resistance (stemming from a higher phosphorus content compared to TEP and TMP) make it a promising choice. Nevertheless, limited compatibility with graphite electrodes remains a drawback. Studies confirm that incorporating 10% DMMP into $LiPF_6$/EC+DEC electrolytes maintain adequate electrochemical performance while achieving non-flammability. In parallel, the inclusion of 10% FEC in $LiPF_6$/DMMP systems acts as a film-forming additive, enabling stable cycling in SiO||LFP cells. Furthermore, other phosphate flame retardants such as TPP, TAP, and TBP have also been reported, providing additional alternatives to improve both the safety and stability of liquid and polymer-based electrolytes [84-88].

Fluorinated electrolytes can maintain a low overpotential and facilitated density as flame-retardant agents, with FEC being the most commonly used. They promote the formation of a dense LiF-rich SEI, enabling uniform lithium deposition and high coulombic efficiency, which improves cycling performance in lithium metal batteries. Increasing the lithium salt concentration while using FEC as the sole solvent is one strategy to achieve a non-flammable electrolyte. Following this approach, Yu et al. [89] developed a specific flame-retardant HCE (high-concentration electrolyte) using a 6.5 M LiTFSI/FEC system, which demonstrated enhanced safety and stability for Li||S batteries. If you are interested in learning more about the physical characteristics of the most common flame retardants, please consult Zhang et al. [72].

Materials Research Foundations 190 (2026) 295-332 https://doi.org/10.21741/9781644904077-11

5.1 Polymer electrolyte with flame-retardant properties derived from PEO

PEO is the most widely used substrate in solid polymer electrolytes due to its ability to form complexes with various lithium salts and provide good solubility. PEO-based electrolytes are flexible, helping to accommodate electrode volume changes during charge and discharge cycles. However, its low dielectric constant limits lithium salt dissociation, and high crystallinity reduces ionic conductivity at room temperature. Above 65 °C, PEO exhibits high lithium-ion conductivity and low interfacial impedance, making it suitable for solid electrolytes without liquids. Its flammability requires the addition of lithium salts and flame-retardant agents for thermal stability. Common flame retardants include phosphorus compounds, nano-sized metal oxides, nitrogen-containing additives, bio-derived materials, and halogenated retardants. These additives not only improve flame resistance but also create Li^+ diffusion pathways, enhancing electrochemical performance and overall stability of PEO-based systems [90-92].

5.2 Phosphonium Flame-retardant

Similarly, there are other flame-retardants called nitrogen flame-retardants. They offer significant benefits in terms of flame resistance and environmental friendliness. The flame-retardant function of these is mainly based on the decomposition of heat and the generation of non-combustible gases to dilute combustible materials. Nitrogen within these flame-retardants can remain stable within a polymer matrix. Incorporating nitrogen-containing flame retardants into PEO polymer electrolytes can enhance thermal stability and reinforce mechanical characteristics [90].

Polybenzimidazole (PBI), an aromatic heterocyclic polymer, is highly valued as an additive for polymer electrolytes due to its combination of high thermal resistance, flame retardancy, mechanical robustness, and chemical stability. Capitalizing on these strengths, cost-effective PBI/PEO composite electrolytes have been developed, which can be prepared via a straightforward and scalable hybrid method [90,93]. Its imidazole groups interact well with polar electrolytes, suppressing PEO crystallization and facilitating Li^+ transport. Membranes incorporating PBI show enhanced thermal and electrochemical stability, significantly improving both safety and performance in lithium-ion batteries. For Electrochemical properties of flame-retardants for different polymeric solid electrolytes from recent years see Liao et al. [90,93].

One option to improve the safety of lithium-ion batteries (LIBs) is to use polymeric separators that are non-flammable and possess thermal and electrochemical stability to replace conventional polyethylene (PE) or polypropylene (PP) separators. Table 2 lists some relevant properties of non-flammable polymers commonly used in LIBs. Among them, polymers containing chlorine (Cl) or bromine (Br) can release halogen or halide radicals, which act by scavenging the free radicals responsible for maintaining the flame. Fluorinated polymers, on the other hand, are significantly more stable than the others because of their strong carbon-fluorine (C-F) bonds and because they do not generate halogen radicals. On the other hand, polymers with aromatic or heterocyclic groups and nitriles in their main chain show good flame retardancy, due to their tendency to cross-link and stable carbonisation. To assess the flammability of a polymer, the limiting oxygen index (LOI), a parameter which determines the lowest oxygen concentration in a mixture of oxygen and nitrogen that is necessary to maintain the combustion of the material, is used [94,95]. The acronyms in Table 2 correspond to the following characteristics: Melting points (MP), thermal degradation temperature (Tdt), ignition temperature (It) and tensile strength (Ts).

Materials Research Forum LLC
https://doi.org/10.21741/9781644904077-11

Reported advances demonstrate the existence of diverse methods and materials that enhance battery safety, including solvents, polymers, and electrolytes. However, there are challenges in scaling up production or practical implementation. For this reason, the following section presents a review of patents and technological implementations to understand the transition from laboratory to industrial scale.

Table 2. Key properties of the most common non-flammable polymers used in LIB [95].

Material	Mp [°C]	Tdt [°C]	LOI [%]	It[°C]	Ts [MPa]
Poly(tetrafluoroethylene) (PTFE)	327	543	95	630	19.6-35.2
Poly(acrylonitrile) (PAN)	320	293	18	460	2.2-3.8
Polyamide 6 (PA 6)	260	424	25	432	64.7
Poly(vinylidene fluoride) (PVDF)	160-170	438	43-65	643	35
Polyvinyl chloride (PVC)	189	273	50	395	52
Poly(vinylidene chloride) (PVDC)	195	225	42	-	-
Poly(ether-ether-ketone) (PEEK)	340	570	37	570	100
Polybutylene terephthalate (PBT)	237	382	20-24	382	34-60
Polybenzimidazoles (PBI)	-	584	42	-	160
Polyetherimide (PEI)	-	527	47	528	-85
Poly(mphenyleneisophthalamide) (NOMEX)	407	-	29	-	-
Polyphenylene sulfide (PPS)	287	504	44	575	45-172

6. Innovations in Flame-retardant Materials for Batteries - Part 2: Patent Review

This section presents a review of registered patents from 2020 to March 2025, with a focus on flame-retardant materials in batteries. The following types have been considered:

- Lithium-ion (Li-ion) batteries: the most widely used in electric vehicles, stationary storage, and portable electronic devices.
- Solid-state batteries: These share potential solutions with Li-ion batteries in the use of non-flammable electrolytes and advanced encapsulations.
- Sodium (Na-ion), nickel-metal hydride (NiMH), and lead-acid batteries have been excluded from this analysis due to their limited presence in the current landscape; however, some examples of technological developments are presented.

The patent search, conducted via the Espacenet database [96], yielded an extensive set of relevant CPC (Cooperative Patent Classification) codes. This is an international system that organizes patents according to their technical content. Table 3 shows an example of the basic structure of a CPC code, in this case H01M10/0565, which is divided into five elements: section (H), class (01),

Materials Research Foundations 190 (2026) 295-332 https://doi.org/10.21741/9781644904077-11

subclass (M), main group (10), and finally, the subgroup (/0565). The different codes returned by the search included chemical technologies, materials, and battery components. However, not all of them were considered relevant to this chapter. Priority was given to those related to battery thermal safety, the incorporation of flame-retardant materials, and the prevention of thermal runaway (Table 4). Among them, codes such as H01M10/0565 (active battery components, especially electrolytes), C08K3/22 (functional additives in polymers), A62C3/00 (flame-retardant chemical compositions), and H01M50/414 (cell structures and encapsulation) stand out, among others. Therefore, classes associated with technologies that are too general, such as adhesives (C09J11/207), decorative coatings (C09D5/234, C09D7/148), generic polymers (C08L23/163) or compounds with broad industrial application unrelated to batteries (C09K5/92, C08G18/108), which do not have a direct link with the objectives of this review, are omitted.

Table 3: Basic Structure of a CPC Code example: H01M10/445

CPC Code Segment	Meaning
H	Section – General technology area (e.g., Electricity)
01	Class – Broad technology field (e.g., Basic electric elements)
M	Subclass – Specific group (e.g., Devices for electrical energy storage)
10	Main Group – General category within the subclass (e.g., Electrochemical cells)
/ 0565	Subgroup – Highly specific subject matter (e.g., Gel or flame-retardant electrolytes)

Table 5 presents a functional mapping of CPC codes for flame-retardant materials for lithium-ion and solid-state batteries. Each row represents a CPC subclass, and each column denotes a functional component within the battery system. The relationship between each code and the application is shown; the degree of interaction is indicated by a solid box for high relevance, a hatched box for medium or indirect relevance, and an empty box for low or no direct relevance. The purpose of this representation is to identify innovation clusters, development hotspots, and potential technology gaps in fire protection solutions for energy storage systems. For example, classifications C08K5/10, C08K3/22, and C08L2205/04 appear with high relevance in columns such as Electrolytes, Separators, and Chemical Additives, evidencing strong technological activity in the formulation of polymeric compounds and additives that improve fire resistance from within the cells.

Materials Research Forum LLC
https://doi.org/10.21741/9781644904077-11

Table 4: Description of Relevant CPC Classifications for Flame-Retardant Battery Technologies

CPC Classification	Short Title	Relevance to the Topic	Type of Technological Focus
H01M10/0565	Gel or flame-retardant electrolytes	Core: flame-retardant gel electrolytes	Electrolyte formulations
H01M10/0567	Semi-solid / solid-state safety components	Core: critical for thermal stabilization	Advanced cell chemistries
H01M50/414	Flame-safe cell structures	Core: structural modules with thermal control	Battery casing and assembly
H01M50/44	Design for thermal dissipation	Medium: thermal mitigation by design	Cell-level heat management
C08K5/10	Phosphorus-based flame-retardant additives	Core: widely used flame inhibitors	Functional additives in polymers
C08K3/22	Polymer base formulations	Medium: polymeric compatibility with retardants	Material chemistry
C08L2205/04	Polymers with integrated fire properties	Core: key in gel and separator engineering	Polymer chemistry
C08L2203/04	Modified polymer electrolyte blends	Medium: relevant to safety and performance	Polymer design
A62C3/16	Fire suppression systems	Core: extinguishing designs in battery modules	Safety integration in transport/storage
A62C3/00	Flame-retardant compositions	Direct: foundational for material safety	Chemical additives and protective gels
Y02E60/10	Clean energy storage solutions	Core: thermal-safe battery innovations	Sustainable and secure energy storage
Y02P70/50	Eco-friendly flame-retardants	Medium: sustainability in flame-retardant design	Environmental material optimization

In contrast, the Sustainability approach is addressed in codes Y02E60/10 and Y02P70/50, showing that there is progress towards flame-retardant solutions with an environmental focus; this axis still represents an area of opportunity for sustainable innovation in the development of safe materials for batteries.

Below is a sample of the technological developments found during the patent review. These were classified into different topics; however, they are not exclusive; some of them cover a broad spectrum of topics. Therefore, the reader is encouraged to review the entire section to gain a clear overview of the progress in this area.

Materials Research Forum LLC
https://doi.org/10.21741/9781644904077-11

Table 5: Heatmap Matrix of CPC Classifications for Flame-Retardant Materials in Batteries

CPC Classification	Chemical Additives	Electrolytes	Polymers / Separators	Enclosure / Cell Housing	Thermal Safety	Sustainability
H01M10/0565	Medium	High	Medium		High	
H01M10/0567		Medium	High		High	
H01M50/414		Medium		High	Medium	
H01M50/44				High	Medium	
C08K5/10	High		High		Medium	
C08K3/22	High		Medium			
C08L2205/04	Medium	High	High		High	
C08L2203/04	Medium	Medium	High		Medium	
A62C3/16	High		Medium	High	High	
Y02E60/10		High	Medium	Medium	High	High
Y02P70/50	Medium	Medium	Medium	Medium	High	High

Legend: ■ High relevance, ▒ Medium relevance and □ Not directly relevant

6.1. Flame-retardant additives for electrolytes

This section presents flame-retardant additives in liquid, gel, or solid electrolytes, with the aim of mitigating or suppressing violent thermal reactions that can trigger thermal runaway events in lithium-ion batteries and their solid or semi-solid variants.

There are phosphate and halogenated additives for gelled systems. For example, Zhao et al. [97] proposed a gel electrolyte with a non-flammable phosphate solvent and a polymerization system (solid interfacial film, SIF). Moreover, Yang et al. [98] developed a preparation method for gel electrolyte batteries using tetrabromobisphenol A ester as a polymerizable monomer and retarder additive without compromising conductivity. Also, Wen et al. [99] used phosphate monomers that are activated by a thermal crosslinking reaction. These technologies thus facilitate efficient manufacturing and enhance thermal runaway resistance.

There are inventions related to aromatic additives and functional solvents. Xie et al. [100] developed a non-flammable electrolyte based on a mixture of lithium salts, phosphate ester, and an aromatic additive that reduces viscosity and improves conductivity. Also, Xu et al. [101] formulated a quasi-solid polyether-based electrolyte with a flame-retardant solvent that polymerizes by ring-opening. The resulting solution has good conductivity and fire resistance.

On the other hand, hybrid systems with organic or inorganic additives are being prepared. Shin et al. [102] developed a gel electrolyte based on a complex mixture of oligomers, non-aqueous solvents, lithium salts, initiators, and a retardant additive, which can be used with high-voltage batteries to improve overall thermal stability. Also, there are innovations with integrated functional polymers; for example, Teran et al. [103] patented functionalized fluoropolymers with high ionic conductivity for use in nonflammable electrolytes. Their molecular design allows for low volatility and high compatibility with lithium-ion batteries.

6.2 Quasi-solid and functional electrolytes with thermal shut-down.

This section addresses modified electrolyte technologies for lithium-ion batteries. These include polymer gels, quasi-solid formulations, and functionalized solids, with thermal shut-down mechanisms, self-crosslinking, or the release of quenching agents, to suppress thermal runaway events and improve operational stability.

Polymer-based gel electrolyte formulations are used to reinforce encapsulation. For example, Jang [104] describes a quasi-solid electrolyte composition with vinyl phosphonate monomers, which when polymerized generate a fire-resistant and thermally stable matrix. The invention [105] adds a gel electrolyte that combines lithium salts, flame-resistant monomers and Zr-based solid electrolytes, resulting in high thermal stability without affecting ionic conduction.

There are quasi-solid electrolyte formulations and retarding gels that combine lithium/sodium salts with polymerizable monomers. Zhou et al. [106] used monomers, salts and crosslinkers in the presence of the active materials. Liu et al. [107] proposed a heat-sealing gel based on a polymeric bilayer incorporating an ionic liquid and a self-sealing matrix. Other high-performance examples include: polyester-phosphate systems [108], perfluorohexane as a solvent [109], fire-retardant gel precursors based on methacrylate and phospholipids [110] and multifunctional phosphorus monomers [111]. Finally, Zhou et al. [106] designed an electrolyte that improves thermal stability and avoids the use of toxic solvents by means of a film that impregnates the electrodes.

On the other hand, functionalized solid electrolytes include additives capable of improving ionic conductivity and thermal safety. Yao et al. [112] proposed a solid electrolyte incorporating closed-type or carbonized flame-retardants (ammonium pyrophosphate, guazinyl pyrophosphate, ammonium polyphosphate and chitosan). Yuan et al. [113] designed a solid cell with double injection, one of polymerizable gel and another of extinguishing microcapsules. Zhou et al. [114] developed a garnet-type solid electrolyte-based gel with high thermal stability. Also, Wen et al. [99] developed pouch-type cells, and Chen et al. [115] used a continuous powder production system for solid electrolytes.

6.3 Non-flammable liquid electrolytes and hybrid formulations

In this area, nonflammable liquid electrolytes and hybrid formulations are also being developed. This category includes liquid solutions modified with flame-retardant solvents or molecular structures that reduce flammability. Song et al. [116] developed a liquid electrolyte with low flammability and the ability to operate at high voltages. Chen and Lyu [117] presented a system with polymer, additives, and sodium salts that form a gel upon overheating, preventing combustion. Lu et al. [118] proposed a liquid solvated with phosphorus-containing diluents and a low dielectric constant. Other examples include: an aromatic system that optimizes solvation [100] and a molten salt based on non-metallic organic cations [119].

6.4 Functional separators with thermal release or response

Functional separators and active protective structures have also been developed, using coated layers or internal chemical modifications. For example, a polymer with flame-retardant functional groups inserted into the separator matrix [120], a nanofiber separator with encapsulated flame-retardant release agents [121], and a chemically cross-linked separator activated by internal condensation in the cell [122]. Patent [123] details a separator with layers that expand above 80°C, blocking the passage of ions and halting the electrochemical reaction (shutdown). Furthermore, Li et al. [124] propose a fire suppression system composed of polymers that release inert agents when heated.

On the other hand, Pekala et al. [125] propose a ceramic coating on microporous substrates that increases mechanical and thermal resistance without compromising ionic conductivity. Similarly, Liu and Guo [126] introduce mesoporous polyimide thin films to suppress dendrite formation in alkali metal batteries. Honmoto [127] produces films with functional fillers that improve the separator's strength and structural integrity. Finally, Sun et al. [128] present membranes with optimized polymer blends.

6.5 Fire-resistant polymers, foams and structural coatings

Several developments involve advanced polymer formulations. For example, Chen [129] describes an aromatic polycarbonate resin with flame-retardant additives and inorganic reinforcements such as wollastonite. Similarly, Higashi et al. [130] and Yamaguchi and Kondo [131] explore polymer compositions with phosphorus-organic agents and hydrated minerals, which exhibit high thermal stability and minimal structural degradation. Also, there are foamed materials, such as [132], which incorporate silicone and hollow microbubbles, resulting in a lightweight and insulating material, ideal for heat-resistant encapsulations. Some inventions focus on the separator structure as a critical factor. Yen [133] and Kim et al. [134] introduce methods for manufacturing microporous separators made of cross-linked polyolefins with high porosity and dimensional stability.

6.6 Active materials or safe anodes

Inventions of active electrode materials or their coatings have been developed to reduce reactivity and prevent thermal expansion. Lin et al. [135] generated a coating on lithium metal anodes to reduce interfacial reactions, Kuribayashi et al. [136] changed the cathode collector to stainless steel to prevent ignition, and Tokizane et al. [137] applied selective coatings on active particles that control the lithium concentration and its controlled release.

6.7 Microcapsules, valves, thermal suppression systems

Onnerud et al. [138] present endothermic materials that release gases when reacting with heat, which helps cool and ventilate the system by increasing pressure; that is, it combines insulation, energy absorption, and controlled gas evacuation. Also, Ootsuki et al. [139] propose a fire-resistant resin (temperatures below 800°C) with an endothermic agent (300 J/g). Ju et al. [140] describe microcapsules that release an adsorbent core (between 110–200°C) that reduces the heat of combustion of the electrolyte. Liu et al. [141] developed a membrane with materials that, when exposed to heat, change phase. The microcapsules then release an initiator that polymerizes a monomer dissolved in the electrolyte, thermally blocking the cell. Also, Kwok et al. [121] developed a separator coated with polymeric nanofibers loaded with encapsulated retardants.

For its part, the invention [142] proposes a mechanical system with an internal air bag with a relief valve and a flame-retardant, which reduces the pressure of the device. In [143], a system with sensors to detect and control the surface temperature for energy storage stations is proposed. Likewise, the invention [144] presents an anti-explosion valve that prevents the entry of water

Materials Research Foundations 190 (2026) 295-332
https://doi.org/10.21741/9781644904077-11

vapor and controls the internal pressure of the battery pack. Also, Robin [145] and Jin [146] introduce a method and a protective system that combines sensors, microencapsulated materials and extinguishing mechanisms. Finally, He and Lu [144] developed an anti-explosion valve with a breathable membrane and water vapor absorbent material inside a reusable housing.

6.8 Thermal encapsulations, housings and structural barriers

Fire-retardant adhesive barriers, laminates, and coatings are external or internal structural solutions that act as passive thermal barriers, using composite materials or special adhesives. Liu and Meyer [147] proposed an adhesive tape with an inorganic fiber and silicone matrix, useful as a structural thermal barrier. Taniguchi et al. [148] formulated a phosphazenic epoxy resin with high heat and moisture resistance. Ootsuki et al. [149] developed a multilayer laminate with thermally expandable layers and flame-retardant additives.

At the module or system level, several proposals integrate retardant or ventilation layers that protect multiple cells in a coordinated manner. Qi et al. [150] propose a module with exhaust channels, a relief valve, and filters that capture harmful particles and gases.

Proposals aimed at structural encapsulation include [151], which proposes lithium cells that generate non-flammable smoke during thermal runaway events. The cell described in [152] incorporates a dual-electrolyte system, where a liquid phase contains flame-retardants and a polymeric phase acts as a structural barrier. In [153], a double shell with circulating coolant is used, accompanied by a vacuum unit and exhaust valves. Wang et al. [154] offer a spiral cell where the conventional separator is replaced by a solid electrolytic layer connected to a polymeric layer. Finally, the proposal of Lu et al. [155] combines a bipolar design with gels and self-supporting separators. This configuration allows high voltages to be reached with good thermal dissipation.

6.9 Sustainable solutions with low toxicity or curable in situ

Energy storage technologies must consider reducing environmental impact. This section presents a selection of inventions aimed at developing sustainable thermal solutions.

Zhang et al. [156] proposed a nitrile-based polymer gel and a highly polar and porous polybenzimidazole membrane, without requiring halogenated compounds, thus reducing the toxicity of the system. Xin et al. [157] introduced a gel electrolyte based on a cyclotriphosphazene additive as a retarder, avoiding persistent or bioaccumulative substances. Shin et al. [102] formulated a gel electrolyte that includes ionic liquids, initiators and retarding additives, eliminating the use of harmful solvents. Zhao et al. [158] proposed a porous polymeric material designed to selectively dissociate without releasing hazardous byproducts. Li et al. [159] proposed a gel electrolyte that prevents leaks. Yeh et al. [160] designed an electrolyte composition that integrates polymers with functional groups such as amide and epoxy, allowing their in situ crosslinking.

On the other hand, with an emphasis on recycling, Liang et al. [161] presented an electrolyte that thickens upon mechanical impact and returns to a liquid state at rest; that is, it reinforces the structure of the battery pack. Yuan et al. [113] developed a solid cell manufacturing process that combines monomer injection, thermal polymerization and retarding microcapsules, without adding external components that complicate recycling. Jang et al. [162] developed a separator composed of two polymers, whose architecture eliminates the use of adhesives or additional coatings, which minimizes toxic waste and maximizes recyclability.

These analysis patents demonstrate the trend toward developing multifunctional electrolytes and cells focused on ensuring thermal safety without reducing electrochemical performance. Examples of thermal protection include self-crosslinking polymer gels, hybrid solid electrolytes, nonflammable formulations, as well as separators and functionalized active materials. Furthermore, several technologies are being adapted to commercially compatible manufacturing processes, contributing to risk mitigation. Applications extend to electric vehicles, high-energy storage, and high-density electronics. Finally, innovation is opening up paths toward truly sustainable energy storage.

7. Challenges and Perspectives

Based on the aspects discussed above, flame retardants clearly offer significant advantages for the strategic containment of fires. However, it has been reported that these agents have some side effects, which may negatively affect the electrochemical performance of Li-ion batteries. In this sense, Shim et al. [163] studied the effect of the concentration of diphenyloctyl phosphate (DPOF) in the electrochemical cell performance. According to the authors, the battery's electrochemical performance suffered significantly at high concentrations of the additive; specifically, a 30 wt.% content of DPOF led to a deterioration in both rate capabilities and overall cycling stability. In addition to this, Chawla et al. [164] reported that the direct addition of flame-retardant additives to conventional $LiPF_6$-organocarbonate electrolytes causes a significant reduction in the electrochemical performance of Li-ion batteries.

These findings have driven the development of new flame-retardant agents designed to minimize impact on the electrochemical performance of batteries. For instance, Wang et al. [165] studied the use of a salt-concentrated organic electrolyte and trimethyl phosphate (TP), as a flame-retardant agent. According to the authors, this combination allows a stable charge-discharge cycling with carbon and graphite anodes for both Na-ion and Li-ion batteries. In this case, for more than 1000 cycles/year the observed degradation was negligible. The authors also performed flame tests in order to demonstrate the effectiveness of TP as a flame-retardant. In recent work, Kim et al. [166] indicated that phosphorous-based additives are better options as flame-retardant agents than fluorine-based additives. According to the authors, the former exhibit significant advantages, including high chemical and thermal stability, and both lower toxicity and smoke production. In addition, phosphorous-based additives have a minimum impact on the electrochemical performance of batteries.

Taking the aforementioned into account, it is necessary to continue with the development of new flame-retardant agents that offer a good compromise between safety, electrochemical performance and efficiency of Li-ion batteries.

Regarding normatives on the safe use of Li-ion batteries, many international organizations have developed safety standards for energy storage devices. The most relevant standards are outlined below:

International Electrotechnical Commission (IEC) - Standard IEC62133. Secondary cells and batteries containing alkaline or other non-acid electrolytes - Safety requirements for portable sealed secondary cells, and for batteries made from them, for use in portable applications [167].

Materials Research Foundations 190 (2026) 295-332 https://doi.org/10.21741/9781644904077-11

International Standardization Organization (ISO) - Standard ISO12405-1. Electrically propelled road vehicles - Test specification for lithium-ion traction battery packs and systems - Part 1: High-power applications [168].

International Standardization Organization (ISO) - Standard ISO12405-2. Electrically propelled road vehicles - Test specification for lithium-ion traction battery packs and systems - Part 2: High-energy applications [169].

Society of Automotive Engineers (SAE) - Standard J2464. Electric and hybrid electric vehicle rechargeable energy storage system (RESS) - Safety and abuse testing [170].

United Nations (UN) Manual of Tests and Criteria - Standard UN38.3. Transport of dangerous goods [171].

Underwriters Laboratories (UL) - Standard UL2580. Standard for safety batteries for use in electric vehicles [172].

The aforementioned standards cover a series of safety tests that batteries must pass in order to obtain a certification for their use in specific applications. These safety tests include overcharge tests, heating tests, short circuit tests and crush tests.

Internal generation and accumulation of heat have been recognized as the main causes of accidents related to the use of Li-ion batteries [173-175]. Since this work is based on the importance of flame-retardant materials in Li-ion batteries, focus will be given to safety tests that consider these mechanisms.

Overcharge/over-discharge tests. Overcharge poses safety risks during the use of Li-ion batteries, which include the Li plating on the anode surface, decomposition of both the electrolyte and the solid electrolyte interphase. These risks, in turn, can cause overheating and a subsequent thermal runaway. Safety standards include the guidelines of overcharge/over-discharge tests for the evaluation of the safety performance of batteries under overcharge conditions. During these tests, batteries are charged and/or discharged up to specific limits provided by manufacturers; based on these results, the condition of batteries is established according to different pass/fail criteria. Overcharge tests on Li-ion batteries have been reported in various studies [176-178].

Heating tests. Both external and internal factors may cause an increase in the temperature of batteries. Heating poses a risk to batteries since it can cause interfacial electrolyte-electrode reactions, electrolyte decomposition, and a subsequent heat release (explosion). Heating tests are performed in order to analyze the heat distribution and thermal stability of batteries. In this case, batteries are placed in a chamber at high temperature for a certain time to check if there is no leakage or combustion. Thermal stability of Li-ion batteries has been studied by several authors through heating tests [179-181].

External and internal short circuit tests. External short circuits represent a risk to Li-ion batteries, since these can lead to heat generation with a subsequent thermal runaway, and even a reduction in the isolation resistance. During external short circuit tests, a current is applied to a battery with an overcurrent protection device, which limits or reduces the current flow. These tests have been used in different studies [182-184]. Internal short circuits, on the other hand, pose safety risks such as venting and sparking, which can cause the ignition of the electrolyte vapors flowing from the batteries. It is worth mentioning that the internal short circuit tests are in development nowadays. In this sense, some authors have proposed internal short-circuit tests for Li-ion batteries [185-187].

Materials Research Foundations 190 (2026) 295-332
https://doi.org/10.21741/9781644904077-11

As mentioned before, safety standards have been proposed by different organizations. In this sense, these standards provide a point of reference to establish the level of safety of the Li-ion batteries used in different applications.

While all the aforementioned safety standards have been developed with a solid technical basis and adopted by different industries, there are important concerns that should be addressed in order to assure a higher confidence in these standards.

To start with, each one of the safety standards has some unique features, including test conditions, terminology and pass-fail criteria. These differences pose a serious concern for different industries, especially when choosing the right safety standard for a specific application. Therefore, it is necessary to develop strategies that allow for reaching a consensus among standards by unifying the aforementioned aspects.

Another major concern is that some of the test conditions suggested in the standards may not represent the field conditions, or even worst-case scenarios. In such cases, guidelines and testing procedures must be updated to reflect real-world operating conditions of both current and emerging technologies and processes where batteries are used.

Given the importance of flame retardants in the safe use of Li-ion batteries, current standards should include appropriate guidelines, tests and procedures that consider the incorporation of these agents into the batteries and their effect on the safety of different applications. Moreover, these standards must consider the growing incorporation of novel materials and chemical compounds increasingly used as flame retardant agents. Therefore, this subject represents a challenge that must be addressed during the stages of proposal, modification and development of safety standards.

8. Conclusions

The development of flame-resistant materials for batteries has significant implications for the safety of energy storage. Throughout this chapter, various formulations — including phosphorus-based compounds, intumescent systems, hybrid polymers, non-flammable electrolytes, and functional separators — have been identified as promising strategies for mitigating ignition risks without severely compromising electrochemical performance.

Nonetheless, critical challenges remain. The integration of these materials may adversely affect ionic conductivity, interfacial stability, and long-term cyclability, requiring careful evaluation of component compatibility. Furthermore, existing safety testing standards often fail to accurately reflect real-world operational scenarios, thereby limiting their industrial applicability.

Accordingly, there is a pressing need for regulatory frameworks that incorporate updated methodologies and account for the deployment of emerging flame-retardant systems. Achieving this goal requires an interdisciplinary approach involving researchers, manufacturers, and standards bodies to advance safe, efficient, and environmentally sustainable battery technologies.

In summary, the future of thermally stable energy storage will depend on the development of multifunctional materials that combine flame-retardant efficacy, electrochemical compatibility, and industrial feasibility.

References

[1] D.D. Friel, Electronic options for lithium-ion batteries, in: Lithium-ion batteries, Elsevier, 2014, pp. 361-386. https://doi.org/10.1016/B978-0-444-59513-3.00016-9

[2] H. Deng, K.E. Aifantis, Applications of lithium batteries, in: K.E. Aifantis, R.V. Kumar, P. Hu (Eds.) Rechargeable ion batteries: Materials, design and applications of Li-ion cells and beyond, WILEY-VCH, Germany, 2023, pp. 83-103. https://doi.org/10.1002/9783527836703.ch4

[3] M. Dubarry, N. Vuillaume, B.Y. Liaw, Origins and accommodation of cell variations in Li-ion battery pack modeling, Int. J. Energy Res., 34 (2010) 216-231. https://doi.org/10.1002/er.1668

[4] Y. Lavoie, F. Danet, B. Lombard, Lithium-ion batteries for industrial applications, in: 2017 Petroleum and Chemical Industry Technical Conference (PCIC), IEEE, 2017, pp. 283-290. https://doi.org/10.1109/PCICON.2017.8188747

[5] Y. Ding, Z.P. Cano, A. Yu, J. Lu, Z. Chen, Automotive Li-ion batteries: Current status and future perspectives, Electrochem. Energy Rev., 2 (2019) 1-28. https://doi.org/10.1007/s41918-018-0022-z

[6] K.O.A. Amano, S.-K. Hahn, N. Butt, P. Vorwerk, E. Gimadieva, R. Tschirschwitz, T. Rappsilber, U. Krause, Composition and explosibility of gas emissions from lithium-ion batteries undergoing thermal runaway, Batteries., 9 (2023) 300. https://doi.org/10.3390/batteries9060300

[7] O. Willstrand, M. Pushp, P. Andersson, D. Brandell, Impact of different Li-ion cell test conditions on thermal runaway characteristics and gas release measurements, J. Energy Storage., 68 (2023) 107785. https://doi.org/10.1016/j.est.2023.107785

[8] K. Masalkovaitė, P. Gasper, D.P. Finegan, Predicting the heat release variability of Li-ion cells under thermal runaway with few or no calorimetry data, Nat. Commun., 15 (2024) 8399. https://doi.org/10.1038/s41467-024-52653-3

[9] B. Xu, J. Lee, D. Kwon, L. Kong, M. Pecht, Mitigation strategies for Li-ion battery thermal runaway: A review, Renewable Sustainable Energy Rev., 150 (2021) 111437. https://doi.org/10.1016/j.rser.2021.111437

[10] G. Wei, R. Huang, G. Zhang, B. Jiang, J. Zhu, Y. Guo, G. Han, X. Wei, H. Dai, A comprehensive insight into the thermal runaway issues in the view of lithium-ion battery intrinsic safety performance and venting gas explosion hazards, Appl. Energy., 349 (2023) 121651. https://doi.org/10.1016/j.apenergy.2023.121651

[11] E.-y. Oh, D.S. Min, J.Y. Han, S. Jung, T.-s. Kang, Consequence analysis of toxic gases generated by fire of lithium ion batteries in electric vehicles, J. Korean Inst. Gas., 23 (2019) 54-61.

[12] B. Zhang, R.L. Bewley, T.R. Tanim, L.K. Walker, Electric vehicle post-crash recovery-Stranded energy issues and mitigation strategy, J. Power Sources., 552 (2022) 232239. https://doi.org/10.1016/j.jpowsour.2022.232239

[13] X. Feng, D. Ren, X. He, M. Ouyang, Mitigating thermal runaway of lithium-ion batteries, Joule, 4 (2020) 743-770. https://doi.org/10.1016/j.joule.2020.02.010

[14] M. Ghiji, V. Novozhilov, K. Moinuddin, P. Joseph, I. Burch, B. Suendermann, G. Gamble, A review of lithium-ion battery fire suppression, Energies, 13 (2020) 5117. https://doi.org/10.3390/en13195117

[15] T. Dagger, B.R. Rad, F.M. Schappacher, M. Winter, Comparative performance evaluation of flame retardant additives for lithium ion batteries-I. Safety, chemical and electrochemical stabilities, Energy Technol., 6 (2018) 2011-2022. https://doi.org/10.1002/ente.201800132

[16] M. Pigłowska, B. Kurc, Ł. Rymaniak, Starch as the flame retardant for electrolytes in lithium-ion cells, Materials, 15 (2022) 523. https://doi.org/10.3390/ma15020523

[17] D.G. Belov, D. Shieh, A study of tetrabromobisphenol A (TBBA) as a flame retardant additive for Li-ion battery electrolytes, J. Power Sources., 247 (2014) 865-875. https://doi.org/10.1016/j.jpowsour.2013.08.143

[18] W. Wang, C. Liao, L. Liu, W. Cai, Y. Yuan, Y. Hou, W. Guo, X. Zhou, S. Qiu, L. Song, Comparable investigation of tervalent and pentavalent phosphorus based flame retardants on improving the safety and capacity of lithium-ion batteries, J. Power Sources., 420 (2019) 143-151. https://doi.org/10.1016/j.jpowsour.2019.02.037

[19] X. Tian, Y. Yi, B. Fang, P. Yang, T. Wang, P. Liu, L. Qu, M. Li, S. Zhang, Design strategies of safe electrolytes for preventing thermal runaway in lithium ion batteries, Chem. Mater., 32 (2020) 9821-9848. https://doi.org/10.1021/acs.chemmater.0c02428

[20] Z. Gao, S. Rao, T. Zhang, W. Li, X. Yang, Y. Chen, Y. Zheng, Y. Ding, T. Dong, S. Li, Design strategies of flame-retardant additives for lithium ion electrolyte, J. Electrochem. Energy Convers. Storage., 19 (2022) 030910. https://doi.org/10.1115/1.4053968

[21] B.R. Thomas, Linden's Handbook of Batteries, Fourth Edition, Fourth Edition ed., McGraw-Hill Education, New York, 2011.

[22] K. Xiong, T. Qi, X. Zhang, Advancements in Graphite Anodes for Lithium-Ion and Sodium-Ion Batteries: A Review, Electroanalysis, 37 (2025) e202400318. https://doi.org/10.1002/elan.202400318

[23] Y. Wen, K. He, Y. Zhu, F. Han, Y. Xu, I. Matsuda, Y. Ishii, J. Cumings, C. Wang, Expanded graphite as superior anode for sodium-ion batteries, Nature Communications, 5 (2014) 4033. https://doi.org/10.1038/ncomms5033

[24] J. Saddique, H. Shen, J. Ge, X. Huo, N. Rahman, A.A.A. Ahmadi, M. Mushtaq, Nanostructure Sn/C Composite High-Performance Negative Electrode for Lithium Storage, Molecules, 27 (2022) 4083. https://doi.org/10.3390/molecules27134083

[25] J. Liu, K. Song, C. Zhu, C.-C. Chen, P.A. van Aken, J. Maier, Y. Yu, Ge/C Nanowires as High-Capacity and Long-Life Anode Materials for Li-Ion Batteries, ACS Nano, 8 (2014) 7051-7059. https://doi.org/10.1021/nn501945f

[26] H. Ding, Z. Song, H. Zhang, H. Zhang, X. Li, Niobium-based oxide anodes toward fast and safe energy storage: A review, Materials Today Nano, 11 (2020) 100082. https://doi.org/10.1016/j.mtnano.2020.100082

[27] A. Balasankar, S.E. Arthiya, S. Ramasundaram, P. Sumathi, S. Arokiyaraj, T. Oh, K. Aruchamy, G. Sriram, M.D. Kurkuri, Recent Advances in the Preparation and Performance of Porous Titanium-Based Anode Materials for Sodium-Ion Batteries, Energies, 15 (2022) 9495. https://doi.org/10.3390/en15249495

[28] S. Yuan, Q. Lai, X. Duan, Q. Wang, Carbon-based materials as anode materials for lithium-ion batteries and lithium-ion capacitors: A review, Journal of Energy Storage, 61 (2023) 106716. https://doi.org/10.1016/j.est.2023.106716

[29] A.K. Koech, G. Mwandila, F. Mulolani, P. Mwaanga, Lithium-ion battery fundamentals and exploration of cathode materials: A review, South African Journal of Chemical Engineering, 50 (2024) 321-339. https://doi.org/10.1016/j.sajce.2024.09.008

[30] G. Ma, X. Luo, M. Cheng, D. Ju, Effect of impurities in FePO4 raw materials on the performance of LiFePO4 cathode materials, Scientific Reports, 15 (2025) 15122. https://doi.org/10.1038/s41598-025-99729-8

[31] S. Evro, A. Ajumobi, D. Mayon, O.S. Tomomewo, Navigating battery choices: A comparative study of lithium iron phosphate and nickel manganese cobalt battery technologies, Future Batteries, 4 (2024) 100007. https://doi.org/10.1016/j.fub.2024.100007

[32] R.I. Pushparaj, A.R. Kumar, G. Xu, Enhancing safety in lithium-ion batteries with additive-based liquid electrolytes: A critical review, Journal of Energy Storage, 72 (2023) 108493. https://doi.org/10.1016/j.est.2023.108493

[33] S. Kainat, J. Anwer, A. Hamid, N. Gull, S.M. Khan, Electrolytes in Lithium-Ion Batteries: Advancements in the Era of Twenties (2020's), Materials Chemistry and Physics, 313 (2024) 128796. https://doi.org/10.1016/j.matchemphys.2023.128796

[34] H. Zhang, J. Liu, M. Guan, Z. Shang, Y. Sun, Z. Lu, H. Li, X. An, H. Liu, Nanofibrillated Cellulose (NFC) as a Pore Size Mediator in the Preparation of Thermally Resistant Separators for Lithium Ion Batteries, ACS Sustainable Chemistry & Engineering, 6 (2018) 4838-4844. https://doi.org/10.1021/acssuschemeng.7b04203

[35] S. Ma, M. Jiang, P. Tao, C. Song, J. Wu, J. Wang, T. Deng, W. Shang, Temperature effect and thermal impact in lithium-ion batteries: A review, Progress in Natural Science: Materials International, 28 (2018) 653-666. https://doi.org/10.1016/j.pnsc.2018.11.002

[36] J. He, X. Li, Recent materials development for Li-ion and Li-S battery separators, Journal of Energy Storage, 112 (2025) 115541. https://doi.org/10.1016/j.est.2025.115541

[37] Y. Wang, W. Zhuang, H. Bian, B. Yang, D. Bin, J. Shi, H. Lu, Lithium-ion battery separator using inorganic materials: A mini review, Journal of Power Sources, 643 (2025) 237015. https://doi.org/10.1016/j.jpowsour.2025.237015

[38] Z. An, T. Shi, X. Du, X. An, D. Zhang, J. Bai, Experimental study on the internal short circuit and failure mechanism of lithium-ion batteries under mechanical abuse conditions, Journal of Energy Storage, 89 (2024) 111819. https://doi.org/10.1016/j.est.2024.111819

[39] G.-E. Lee, Y.-J. Min, H.-C. Shin, Assessment of internal short-circuit risks in overdischarged lithium batteries under various charging conditions, Electrochimica Acta, 512 (2025) 145449. https://doi.org/10.1016/j.electacta.2024.145449

[40] M. Cortada-Torbellino, D.G. Elvira, A. El Aroudi, H. Valderrama-Blavi, Review of Lithium-Ion Battery Internal Changes Due to Mechanical Loading, Batteries, 10 (2024) 258. https://doi.org/10.3390/batteries10070258

[41] H. Bąkowski, I. Przytuła, W. Cebulska, D. Hadryś, J. Ćwiek, The Impact of Mechanical Failure of 18650 Batteries on the Safety of Electric Transport Operations, Energies, 17 (2024) 5980. https://doi.org/10.3390/en17235980

[42] X. Lai, Y. Ke, X. Tang, Y. Zheng, Balanced capacity-based quantitative method for detecting internal short circuits in Lithium-ion battery modules, Journal of Energy Storage, 123 (2025) 116622. https://doi.org/10.1016/j.est.2025.116622

[43] Q. Tao, Y. Wang, E. Tian, H. Tao, H. Chen, Y. Chen, Internal short circuit fault diagnosis for automotive lithium-ion batteries using an adaptive parameter identification model, Measurement, 253 (2025) 117664. https://doi.org/10.1016/j.measurement.2025.117664

[44] N. Harting, N. Wolff, U. Krewer, Identification of Lithium Plating in Lithium-Ion Batteries using Nonlinear Frequency Response Analysis (NFRA), Electrochimica Acta, 281 (2018) 378-385. https://doi.org/10.1016/j.electacta.2018.05.139

[45] J. Ulrich, A. Lindner, T. Brake, M. Winter, S. Wiemers-Meyer, A. Weber, U. Krewer, Early detection of lithium plating during fast charging of lithium-ion batteries using nonlinear frequency response analysis, Journal of Power Sources, 647 (2025) 237358. https://doi.org/10.1016/j.jpowsour.2025.237358

[46] S. Park, Y. Song, S.W. Kim, Simultaneous diagnosis of cell aging and internal short circuit faults in lithium-ion batteries using average leakage interval, Energy, 290 (2024) 130220. https://doi.org/10.1016/j.energy.2023.130220

[47] W.H. Yan, W.X. Huang, Y. Yang, Z.W. Wei, H.S. Zhen, Y. Lin, Research on overcharge mitigations and thermal runaway risk of 18650 lithium-ion batteries, Journal of Energy Storage, 120 (2025) 116372. https://doi.org/10.1016/j.est.2025.116372

[48] K. Maher, A. Boumaiza, R. Amin, Thermal management challenges in lithium-ion batteries: understanding heat generation mechanisms, Chemical Communications, 61 (2025) 3740-3743. https://doi.org/10.1039/D4CC05619A

[49] X. He, C. Zhao, Z. Hu, F. Restuccia, F. Richter, Q. Wang, G. Rein, Heat transfer effects on accelerating rate calorimetry of the thermal runaway of Lithium-ion batteries, Process Safety and Environmental Protection, 162 (2022) 684-693. https://doi.org/10.1016/j.psep.2022.04.028

[50] H. Sun, L. Zhang, Q. Duan, S. Wang, S. Sun, J. Sun, Q. Wang, Experimental study on suppressing thermal runaway propagation of lithium-ion batteries in confined space by various fire extinguishing agents, Process Safety and Environmental Protection, 167 (2022) 299-307. https://doi.org/10.1016/j.psep.2022.09.016

[51] P. Wang, C. Xu, J. Huang, M. Zhang, F. Jiang, X. Feng, Experimental and simulation study on internal thermal runaway development drives venting and flammable gas risk evaluate of Lithium-ion battery, Applied Energy, 385 (2025) 125545. https://doi.org/10.1016/j.apenergy.2025.125545

[52] M. Meister, S. Sharma, X. He, P.S. Chepaitis, T. Waddey, M. Wilson, V. Premnath, J. Jeevarajan, M. Black, C. Wright, Evaluating inhalation risks and toxicological impacts of lithium-ion battery thermal runaway emissions, Environment International, 199 (2025) 109466. https://doi.org/10.1016/j.envint.2025.109466

[53] M. Chen, Q. Guo, Y. Yuan, A. Li, B. Lin, Y. Xiao, L. Xu, W. Wang, Recent Advancements of Bio-Derived Flame Retardants for Polymeric Materials, Polymers, 17 (2025) 249. https://doi.org/10.3390/polym17020249

[54] B. Liu, L. Ding, L. Lv, Y. Yu, W. Dong, Organophosphate esters (OPEs) and novel brominated flame retardants (NBFRs) in indoor dust: A systematic review on concentration, spatial distribution, sources, and human exposure, Chemosphere, 345 (2023) 140560. https://doi.org/10.1016/j.chemosphere.2023.140560

[55] S. Birnbaum Linda, F. Staskal Daniele, Brominated flame retardants: cause for concern?, Environmental Health Perspectives, 112 (2004) 9-17. https://doi.org/10.1289/ehp.6559

[56] G. Vahidi, D.S. Bajwa, J. Shojaeiarani, N. Stark, A. Darabi, Advancements in traditional and nanosized flame retardants for polymers-A review, Journal of Applied Polymer Science, 138 (2021) 50050. https://doi.org/10.1002/app.50050

[57] Y. Liu, L. Yuan, G. Liang, A. Gu, Developing intrinsic halogen-free and phosphorus-free flame retardant biobased benzoxazine resins with superior thermal stability and high strength, European Polymer Journal, 180 (2022) 111581. https://doi.org/10.1016/j.eurpolymj.2022.111581

[58] J. Long, S. Li, B. Liang, Synthesis and properties of a new halogen-free flame retardant for polyethylene, Pigment & Resin Technology, 47 (2018) 208-215. https://doi.org/10.1108/PRT-04-2016-0048

[59] Z. Xu, Z. Xu, R. Tao, L. Mao, J. Zhan, J. Xiao, T. Yu, Combustion and Thermal Properties of Flame Retardant Polyurethane Foam With Ammonium Polyphosphate Synergized by Phosphomolybdic Acid, Frontiers in Materials, Volume 9 - 2022 (2022). https://doi.org/10.3389/fmats.2022.944368

[60] L. Gutiérrez, J.H. Mina, Desarrollo de compuestos XLPE con retardantes de llama de baja toxicidad para aislamiento de cables, Tecnológicas, 22 (2019) 73-90. https://doi.org/10.22430/22565337.1104

[61] C. European, Directive - 2011/65 - EN - rohs 2 - EUR-Lex, in, Official Journal of the European Union, 2011.

[62] G.P. Sánchez, Estudio de la incorporación de arcilla de paligorskita, recubierta con poliurea en estructuras núcleo, coraza, y su efecto sobre las propiedades fisicomecánicas y de inflamabilidad en sistema PLA/EVA, (2022).

[63] M. Zhu, Z. Ma, L. Liu, J. Zhang, S. Huo, P. Song, Recent advances in fire-retardant rigid polyurethane foam, Journal of Materials Science & Technology, 112 (2022) 315-328. https://doi.org/10.1016/j.jmst.2021.09.062

[64] Z. Chen, Y. Chao, W. Li, G.G. Wallace, T. Bussell, J. Ding, C. Wang, Abuse-tolerant electrolytes for lithium-ion batteries, Advanced Science, 8 (2021) 2003694. https://doi.org/10.1002/advs.202003694

[65] J. Deng, Z. Hu, J. Chen, J. Zhao, Z. Bai, Safety Methods for Mitigating Thermal Runaway of Lithium-Ion Batteries-A Review, Fire, 8 (2025) 223. https://doi.org/10.3390/fire8060223

[66] C. Zhang, C. Li, K. Zhang, S. Zhang, J. Liu, M. Wang, L. Wang, Building Flame-Retardant Polymer Electrolytes via Microcapsule Technology for Stable Lithium Batteries, ACS Applied Materials & Interfaces, 16 (2024) 27470-27480. https://doi.org/10.1021/acsami.4c04154

[67] X. Wu, T. Liu, Y.-G. Lee, J.F. Whitacre, Glycerol Triacetate-Based Flame Retardant High-Temperature Electrolyte for the Lithium-Ion Battery, ACS Applied Materials & Interfaces, 16 (2024) 24590-24600. https://doi.org/10.1021/acsami.4c02323

[68] S.H. Kang, H.Y. Jeong, T.H. Kim, J.Y. Lee, S.K. Hong, Y.T. Hong, J. Choi, S. So, S.J. Yoon, D.M. Yu, Aluminum Diethylphosphinate-Incorporated Flame-Retardant Polyacrylonitrile Separators for Safety of Lithium-Ion Batteries, Polymers, 14 (2022) 1649. https://doi.org/10.3390/polym14091649

[69] T.C. Turossi, H.L.O. Júnior, F.M. Monticeli, O.T. Dias, A.J. Zattera, Cellulose-Derived Battery Separators: A Minireview on Advances Towards Environmental Sustainability, Polymers, 17 (2025) 456. https://doi.org/10.3390/polym17040456

[70] H. Liu, T. Chen, Z. Xu, Z. Liu, J. Yang, J. Chen, High-Safety and Long-Life Silicon-Based Lithium-Ion Batteries via a Multifunctional Binder, ACS Applied Materials & Interfaces, 12 (2020) 54842-54850. https://doi.org/10.1021/acsami.0c17563

[71] P.-H. Huang, S.-J. Chang, C.-C. Li, Encapsulation of flame retardants for application in lithium-ion batteries, Journal of Power Sources, 338 (2017) 82-90. https://doi.org/10.1016/j.jpowsour.2016.11.026

[72] S. Zhang, S. Li, Y. Lu, Designing safer lithium-based batteries with nonflammable electrolytes: A review, eScience, 1 (2021) 163-177. https://doi.org/10.1016/j.esci.2021.12.003

[73] X. Wang, E. Yasukawa, S.J.J.o.T.E.S. Kasuya, Nonflammable trimethyl phosphate solvent-containing electrolytes for lithium-ion batteries: II. The use of an amorphous carbon anode, 148 (2001) A1066. https://doi.org/10.1149/1.1397774

[74] X. Wang, C. Yamada, H. Naito, G. Segami, K.J.J.o.T.E.S. Kibe, High-concentration trimethyl phosphate-based nonflammable electrolytes with improved charge-discharge performance of a graphite anode for lithium-ion cells, 153 (2005) A135. https://doi.org/10.1149/1.2136078

[75] K. Takada, Y. Yamada, A.J.A.a.m. Yamada, interfaces, Optimized nonflammable concentrated electrolytes by introducing a low-dielectric diluent, 11 (2019) 35770-35776. https://doi.org/10.1021/acsami.9b12709

[76] T. Zhang, Y. Li, N. Chen, Z. Wen, Y. Shang, Y. Zhao, M. Yan, M. Guan, F. Wu, R.J.A.A.M. Chen, Interfaces, Regulating the solvation structure of nonflammable electrolyte for dendrite-free Li-metal batteries, 13 (2021) 681-687. https://doi.org/10.1021/acsami.0c19075

[77] P. Shi, H. Zheng, X. Liang, Y. Sun, S. Cheng, C. Chen, H.J.C.c. Xiang, A highly concentrated phosphate-based electrolyte for high-safety rechargeable lithium batteries, 54 (2018) 4453-4456. https://doi.org/10.1039/C8CC00994E

[78] H. Nakagawa, M. Ochida, Y. Domi, T. Doi, S. Tsubouchi, T. Yamanaka, T. Abe, Z.J.J.o.P.S. Ogumi, Electrochemical Raman study of edge plane graphite negative-electrodes in electrolytes containing trialkyl phosphoric ester, 212 (2012) 148-153. https://doi.org/10.1016/j.jpowsour.2012.04.013

[79] H. Li, L. Peng, D. Wu, J. Wu, Y.J. Zhu, X.J.A.E.M. Hu, Ultrahigh-capacity and fire-resistant LiFePO4-based composite cathodes for advanced lithium-ion batteries, 9 (2019) 1802930. https://doi.org/10.1002/aenm.201802930

[80] Z. Zeng, V. Murugesan, K.S. Han, X. Jiang, Y. Cao, L. Xiao, X. Ai, H. Yang, J.-G. Zhang, M.L.J.N.E. Sushko, Non-flammable electrolytes with high salt-to-solvent ratios for Li-ion and Li-metal batteries, 3 (2018) 674-681. https://doi.org/10.1038/s41560-018-0196-y

[81] X. Cao, Y. Xu, L. Zhang, M.H. Engelhard, L. Zhong, X. Ren, H. Jia, B. Liu, C. Niu, B.E.J.A.E.L. Matthews, Nonflammable electrolytes for lithium ion batteries enabled by ultraconformal passivation interphases, 4 (2019) 2529-2534. https://doi.org/10.1021/acsenergylett.9b01926

[82] H. Jia, L. Zou, P. Gao, X. Cao, W. Zhao, Y. He, M.H. Engelhard, S.D. Burton, H. Wang, X.J.A.E.M. Ren, High-performance silicon anodes enabled by nonflammable localized high-concentration electrolytes, 9 (2019) 1900784. https://doi.org/10.1002/aenm.201900784

[83] H. Yang, Q. Li, C. Guo, A. Naveed, J. Yang, Y. Nuli, J.J.C.C. Wang, Safer lithium-sulfur battery based on nonflammable electrolyte with sulfur composite cathode, 54 (2018) 4132-4135. https://doi.org/10.1039/C7CC09942H

[84] J. Feng, P. Ma, H. Yang, L.J.E.A. Lu, Understanding the interactions of phosphonate-based flame-retarding additives with graphitic anode for lithium ion batteries, 114 (2013) 688-692. https://doi.org/10.1016/j.electacta.2013.10.104

[85] H. Xiang, H. Xu, Z. Wang, C.J.J.o.P.S. Chen, Dimethyl methylphosphonate (DMMP) as an efficient flame retardant additive for the lithium-ion battery electrolytes, 173 (2007) 562-564. https://doi.org/10.1016/j.jpowsour.2007.05.001

[86] Y.E. Hyung, D.R. Vissers, K.J.J.o.p.s. Amine, Flame-retardant additives for lithium-ion batteries, 119 (2003) 383-387. https://doi.org/10.1016/S0378-7753(03)00225-8

[87] X. Liu, X. Shen, F. Zhong, X. Feng, W. Chen, X. Ai, H. Yang, Y.J.C.C. Cao, Enabling electrochemical compatibility of non-flammable phosphate electrolytes for lithium-ion batteries by tuning their molar ratios of salt to solvent, 56 (2020) 6559-6562. https://doi.org/10.1039/D0CC02940H

[88] Z. Zeng, B. Wu, L. Xiao, X. Jiang, Y. Chen, X. Ai, H. Yang, Y.J.J.o.P.S. Cao, Safer lithium ion batteries based on nonflammable electrolyte, 279 (2015) 6-12. https://doi.org/10.1016/j.jpowsour.2014.12.150

[89] Z. Yu, J. Zhang, C. Wang, R. Hu, X. Du, B. Tang, H. Qu, H. Wu, X. Liu, X.J.J.o.E.C. Zhou, Flame-retardant concentrated electrolyte enabling a LiF-rich solid electrolyte interface to improve cycle performance of wide-temperature lithium-sulfur batteries, 51 (2020) 154-160. https://doi.org/10.1016/j.jechem.2020.03.034

[90] Y. Liao, X. Xu, X. Luo, S. Ji, J. Zhao, J. Liu, Y. Huo, Recent Progress in Flame-Retardant Polymer Electrolytes for Solid-State Lithium Metal Batteries, Batteries, 9 (2023). https://doi.org/10.3390/batteries9090439

[91] C. Shi, T. Hamann, S. Takeuchi, G.V. Alexander, A.M. Nolan, M. Limpert, Z. Fu, J. O'Neill, G. Godbey, J.A.J.A.A.M. Dura, Interfaces, 3D asymmetric bilayer garnet-hybridized high-energy-density lithium-sulfur batteries, 15 (2022) 751-760. https://doi.org/10.1021/acsami.2c14087

[92] D. Zhang, L. Li, X. Wu, J. Wang, Q. Li, K. Pan, J.J.F.i.E.R. He, Research progress and application of PEO-based solid state polymer composite electrolytes, 9 (2021) 726738. https://doi.org/10.3389/fenrg.2021.726738

[93] Q. Zhang, H. Huang, T. Liu, Y. Wang, J. Yu, Z.J.P. Hu, Molecular composite electrolytes of polybenzimidazole/polyethylene oxide with enhanced safety and comprehensive performance for all-solid-state lithium ion batteries, 239 (2022) 124450. https://doi.org/10.1016/j.polymer.2021.124450

[94] S.V. Levchik, Introduction to flame retardancy and polymer flammability, Flame retardant polymer nanocomposites, (2007) 1-29. https://doi.org/10.1002/9780470109038.ch1

[95] X. Zhang, Q. Sun, C. Zhen, Y. Niu, Y. Han, G. Zeng, D. Chen, C. Feng, N. Chen, W. Lv, W. He, Recent progress in flame-retardant separators for safe lithium-ion batteries, Energy Storage Materials, 37 (2021) 628-647. https://doi.org/10.1016/j.ensm.2021.02.042

[96] Espacenet - patent search.

[97] D. Zhao, W. Lin, H. Song, H. Shang, Q. Tong, Gel polymer electrolyte based on flame-retardant local high-concentration electrolyte as well as preparation method and application of gel polymer electrolyte, in, Shenzhen Jintang New Energy Tech Co Ltd, 2025.

[98] S. Yang, W. Wang, H. Yue, S. Cheng, L. Guan, C. Hu, X. Meng, W. Cui, X. Yan, G. Yuan, X. Wang, H. Guo, Preparation method of flame-retardant gel electrolyte lithium ion battery and lithium ion battery, in, Xinxiang Huarui Lithium New Energy Co Ltd; the Battery Res Institute of Henan Co Ltd; Univ Henan Normal, 2021.

[99] Y. Wen, X. Yu, J. Qiu, H. Ming, G. Cao, M. Li, Y. Wang, Y. Zhang, Phosphate flame-retardant gel electrolyte and preparation method thereof, in, Chemical Res Institute Academy of Military Sciences Pla China, 2023.

[100] J. Xie, B. Zhai, X. Li, M. Qin, Z. Zeng, High-safety electrolyte based on aromatic hydrocarbon modification and application thereof, in, State Grid Xinjiang Electric Power Co Ltd Electric Power Res Inst; Univ Huazhong Science Tech; State Grid Corp China, 2023.

[101] Y. Xu, J. Zhao, M. Li, Polyether quasi-solid electrolyte and preparation method and application thereof, in, Univ Tianjin, 2023.

[102] W.K. Shin, K.H. Ahn, C.H. Lee, J.H. Lee, Gel Polymer Electrolyte Composition and Lithium Secondary Battery Comprising Same, in, Lg Chemical Ltd, 2020.

[103] A. Teran, B. Rupert, E. Nasybulin, J. Burdynska, Functionalized Fluoropolymers and Electrolyte Compositions, in, Blue Current Inc, 2016.

[104] B.Z. Jang, Flame-Resistant Electrolyte Compositions from Phosphonate Vinyl Monomers, Quasi-Solid and Solid-State Electrolytes, and Lithium Batteries, in, Global Graphene Group Inc, 2022.

[105] C. Wang, J. Teng, X. Ouyang, Z. Yang, D. Yuan, X. Chen, J. Wei, B. Shi, Flame-retardant gel polymer electrolyte and preparation method thereof, in, Guizhou Meiling Power Supply Co Ltd, 2023.

[106] G. Zhou, C.t.S. Kwok, J. Yu, F. Ciucci, Synchronous Polymerized Non-Flammable Quasi-Solid-State Electrolyte for Solid-State Batteries, in, Univ Hong Kong Science & Tech, 2022.

[107] L. Liu, B. Xie, Flame-retardant polymer electrolyte with thermal shutdown function and preparation method of flame-retardant polymer electrolyte, in, Univ Beijing Technology & Business, 2021.

[108] B.Z. Jang, Flame-resistant electrolyte compositions, quasi-solid and solid-state electrolytes, and lithium batteries, in, Global Graphene Group Inc, 2023.

[109] Y. Li, J. Zhong, J. Tian, Y. Yi, J. Du, Gel polymer electrolyte, solid-state lithium ion battery, preparation method and application, in, Shenzhen Power Supply Bureau, 2025.

[110] C. Li, Z. Lu, J. Chen, Z. Yang, J. Yi, G. Liu, C. Yang, Non-Flammable Gel Electrolyte Precursor, Modified Solid-State Electrolyte, Lithium Battery, and Preparation Method Therefor, in, Shenzhen Inst Adv Tech, 2022.

[111] L. Liu, J. Zheng, C. Wang, Flame-retardant gel polymer electrolyte suitable for wide-temperature-range sodium-ion battery and application of flame-retardant gel polymer electrolyte, in, Anhui Changhuai New Mat Co Ltd, 2023.

[112] Y. Yao, X. Lai, X. Sun, J. Zhou, Solid electrolyte, preparation method thereof and lithium ion battery, in, Huizhou Liwei Electronic Tech Co Ltd, 2023.

[113] X. Yuan, C. Wu, T. Gao, F. Yan, W. Wu, Preparation method of solid-state battery cell and solid-state battery cell, in, Wuhan Zhongke Advanced Material Tech Co Ltd, 2024.

[114] L. Zhou, Z. Geng, S. Li, W. Liao, Gel polymer electrolyte, solid-state battery, preparation method of gel polymer electrolyte and application, in, Shanghai Electric Group Co Ltd, 2021.

[115] X. Chen, Solid electrolyte preparation equipment of solid lithium battery, in, Evergrande New Energy Technology Shenzhen Co Ltd, 2021.

[116] S.W. Song, K.H. an, Flame-retardant or non-flammable electrolyte and lithium secondary battery containing the same, in, Iac in Nat Univ Chungnam, 2022.

[117] L. Chen, J. Lyu, Sodium ion electrolyte for preventing thermal runaway, preparation method, battery and electric device, in, Wuxi Pangu New Energy Co Ltd; Shenzhen Pangu Sodium & Power Co Ltd, 2024.

[118] Y. Lu, Z. Li, Q. Su, Non-Flammable Solvate Ionic Liquid Electrolyte with Diluters, in, Gm Global Tech Operations Llc, 2022.

[119] H. Nakagawa, S. Katayama, T. Onda, Non-Aqueous Electrolyte Battery, in, Gs Yuasa Corp, 2007.

Materials Research Foundations 190 (2026) 295-332 https://doi.org/10.21741/9781644904077-11

[120] K.a. Kwon, K.S. Lee, Flame Retardant Group-Containing Polymer, Separator of Non-Aqueous Electrolyte Battery Comprising the Same, and Non-Aqueous Electrolyte Battery, in, Lg Energy Solution Ltd, 2023.

[121] C.H. Kwok, K.I. Lee, C. Liu, I.K.Y. Lau, Fire-Retardant Based Nanofiber Coated Separators for Li-Ion Batteries and Producing Method Thereof, in, Nano & Advanced Materials Inst Ltd, 2020.

[122] X. Zhang, Y. Akita, H. Kobayashi, Y. Fukunaga, Composite Single-Layer Chemically Cross-Linked Separator, in, Asahi Chemical Ind, 2022.

[123] Y. Shi, Z. Yan, J. Shi, F. Luo, Composite diaphragm for reducing thermal runaway risk of lithium battery, preparation method and lithium battery, in, Liyang Tianmu Pilot Battery Material Tech Co Ltd, 2020.

[124] M. Li, T. Tang, L. Dong, F. Liu, X. Zhao, X. Shi, Integrated fire suppression device of electric motor coach, in, Yantai Chungway New Energy Tech Co Ltd, 2024.

[125] R. Pekala, W. Wood, D. Spitz, S. Butler, Coating Method and Resultant Ceramic-Modified Separators, in, Amtek Res International Llc, 2024.

[126] G. Liu, D. Guo, Mesoporous Polyimide Thin Films as Dendrite-Suppressing Separators for Alkali Metal Batteries, in, Virginia Tech Intellectual Properties Inc; Virginia Polytechnic Institute and State Univ, 2023.

[127] H. Honmoto, Method for Producing Composite Film, in, Teijin Ltd, 2016.

[128] Y. Sun, F. Tao, C. Ohm, Porous Membranes and Polymer Blend Made Therewith, in, Celanese Int Corp; Sun Yingying; Tao Fangfang, 2025.

[129] X. Chen, Flame-Retardant Aromatic Polycarbonate Resin Composition, in, 2009.

[130] K. Higashi, K. Oda, A. Sumita, K. Minagawa, Flame Retardant Polycarbonate Resin Composition and Molded Article Thereof, in, Teijin Ltd, 2024.

[131] T. Yamaguchi, T. Kondo, Reinforced Flame Retardant Resin Composition and Molded Article, in, Yamaguchi Toru; Kondo Tomohiro; Asahi Kasei Chemicals Corp, 2012.

[132] X. Yao, M. Hu, X. Tai, L. Zou, Y. Guo, Silicone Foam Material for Battery Fire Protection, in, Dow Silicones Corp; Yao Xuesi; Hu Minbiao; Tai Xiangyang; Zou Lu; Guo Yi, 2024.

[133] W.W. Yen, Microporous sheet product and methods for making and using the same, in, Yen William Winchin, 2020.

[134] B.H. Kim, P.Y. Park, K.J. Cho, Y.H. Choi, J.H. Lee, K.H. Choi, Method for Manufacturing Separator, in, W Scope Korea Co Ltd; W Scope Chungju Plant Co Ltd, 2024.

[135] Z. Lin, D. Xiao, B. Zhang, S. Zhang, M. Zhang, Modified lithium metal negative electrode, preparation method thereof and application of modified lithium metal negative electrode in all-solid-state lithium metal battery, in, Univ Guangdong Technology, 2023.

[136] I. Kuribayashi, High Capacity Lithium Ion Secondary Battery, in, Kee Kk, 2024.

[137] M. Tokizane, J. Imaizumi, M. Imazaki, T. Kikuchi, Lithium-Ion Secondary Battery, and Manufacturing Method Lithium-Ion Secondary Battery, in, Kaneka Corp, 2024.

[138] T.P.J. Onnerud, J.J. Shi, R.E. Bowersock, G.E. Gayman, J.P. Street, Lithium ion battery with thermal runaway protection, in, Cadenza Innovation Inc, 2020.

[139] K. Ootsuki, A. Dohi, M. Shimamoto, Fire-Resistant Resin Composition, Fire-Resistant Sheet, Fire-Resistant Laminate, and Battery, in, Sekisui Chemical Co Ltd, 2020.

[140] S. Ju, J. Ma, M. Wang, L. Wu, Z. Yan, H. Lin, X. Hao, Thermally excited active safety mechanism flame-retardant microcapsule, preparation method thereof and lithium ion battery, in, Dongying Hi Tech Spring Chemical Ind Co Ltd, 2021.

[141] J. Liu, B. Chen, C. Wu, L. Chen, K. Chen, Lithium ion battery capable of blocking thermal runaway stage by stage and preparation method of lithium ion battery, in, State Grid Hunan Electric Power Co Ltd; State Grid Hunan Electric Power Co Ltd Disaster Prevention and Reduction Center; State Grid Corp China, 2023.

[142] Y. Zhang, H. Wang, M. Ouyang, W. Li, C. Li, J. Li, L. Lu, X. Han, J. Du, Safety battery, in, Univ Tsinghua, 2019.

[143] H. Chen, T. Zhang, Integrated equipment for preventing and controlling thermal runaway disaster of lithium ion battery energy storage power station, in, Guangzhou Liurui Firefighting Tech Co Ltd, 2024.

[144] M. He, Y. Lu, Anti-explosion valve capable of preventing water vapor from entering and battery pack with anti-explosion valve, in, Zhejiang Forever New Energy Tech Co Ltd, 2023.

[145] M.L. Robin, Thermal Protection of Lithium Ion Batteries, in, Chemours Co Fc Llc, 2023.

[146] X. Jin, Battery protector of power battery safety basic device, in, Suzhou Haiyun Coating Tech Co Ltd, 2022.

[147] J.J. Liu, S.R. Meyer, Composite Adhesive Fire Barrier and Method of Making and Using the Same, in, 3m Innovative Properties Company, 2024.

[148] M. Taniguchi, Y. Tada, Y. Nishioka, Flame-retardant epoxy resin composition, molded object thereof, and electronic part, in, Taniguchi Masatoshi; Tada Yuji; Nishioka Yoichi, 2003.

[149] K. Ootsuki, A. Dohi, M. Shimamoto, Fire-Resistant Laminate and Battery, in, Sekisui Chemical Co Ltd, 2021.

[150] P. Qi, B. Yi, Y. Zhang, Power battery safety fault processing system and vehicle, in, Deepal Automobile Technology Co Ltd, 2024.

[151] Z. Zhang, Safer Batteries and Separators for Batteries; Solutions on Li-Ion Edv/Ess Battery Safety; Solutions on Anode/Electrolyte/Cathode Reduction and/or Oxidation Issues; and, Improved Batteries, Anodes, and/or Separators, in, Celgard Llc, 2024.

[152] H. He, A. Zhamu, B.Z. Jang, Flame-Resistant Quasi-Solid Hybrid Electrolyte for Safe Lithium Batteries and Production Method, in, Global Graphene Group Inc, 2022.

[153] G. Ji, Anti-overheating, anti-spontaneous combustion and anti-spontaneous explosion method for battery pack, in, Ji Guojun, 2019.

[154] Y. Wang, K. Chen, Y. Feng, Z. Li, H. He, Spiral storage battery cell and battery, in, Suzhou Qingtao New Energy Tech Co Ltd, 2024.

[155] Y. Lu, Z. Li, M. Hou, Q. Su, M. Wu, X. Que, H. Liu, S. Chen, Designs of High-Power Gel-Assisted Bipolar Solid-State Battery, in, Gm Global Tech Operations Llc, 2023.

[156] W. Zhang, L. Zou, Z. Yang, J. Lu, Flame-retardant nitrile-based gel polymer electrolyte based on strong-polarity porous diaphragm and preparation method of flame-retardant nitrile-based gel polymer electrolyte, in, Univ Hefei Technology, 2024.

[157] S. Xin, Q. Feng, S. Tan, W. Wang, Y. Guo, Gel-state flame-retardant polymer electrolyte for in-situ polymerization of sulfur-based battery, in, Inst Chemistry Cas, 2023.

[158] X. Zhao, Y. Cui, C. Liu, M. Hou, H. Cao, Porous buffer material, secondary battery and preparation method thereof, in, Ruipu Lanjun Energy Co Ltd; Shanghai Ruipu Qingchuang New Energy Co Ltd, 2024.

[159] Q. Li, T. Hou, X. Chen, D. Zeng, C. Wang, C. Wang, Gel polymer electrolyte, preparation method of gel polymer electrolyte, and battery, in, Huizhou Eve Energy Co Ltd, 2019.

[160] T.-J. Yeh, C.-M. Chang, T.-H. Wang, C.-Z. Lu, C.-E. Liu, S.-C. Liao, Electrolyte, composition for electrolyte and lithium battery employing the same, in, Ind Tech Res Inst, 2019.

[161] X. Liang, Q. Huang, B. Liu, K. Hu, H. Deng, X. Gong, L. Yang, N. Zhou, X. Jin, D. Liang, Shear thickening electrolyte based on liquid-solid-liquid conversion mechanism, preparation method and secondary battery thereof, in, Univ Hefei, 2023.

[162] B.Z. Jang, Heat/Flame-Resistant Polymer Composite-Based Solid Electrolyte Separator, Lithium Secondary Battery, and Manufacturing Method, in, Global Graphene Group Inc, 2023.

[163] E.G. Shim, T.H. Nam, J.G. Kim, H.S. Kim, S.I. Moon, Effect of the concentration of diphenyloctyl phosphate as a flame-retarding additive on the electrochemical performance of lithium-ion batteries, Electrochim. Acta., 54 (2009) 2276-2283. https://doi.org/10.1016/j.electacta.2008.10.037

[164] N. Chawla, N. Bharti, S. Singh, Recent advances in non-flammable electrolytes for safer lithium-ion batteries, Batteries., 5 (2019) 19. https://doi.org/10.3390/batteries5010019

[165] J. Wang, Y. Yamada, K. Sodeyama, E. Watanabe, K. Takada, Y. Tateyama, A. Yamada, Fire-extinguishing organic electrolytes for safe batteries, Nat. Energy., 3 (2018) 22-29. https://doi.org/10.1038/s41560-017-0033-8

[166] J.H. Kim, J.H. Hyun, S. Kim, W.H. Park, S.H. Yu, Phosphorus-Based Flame-Retardant Electrolytes for Lithium Batteries, Adv. Energy Mater., (2025) 2500587. https://doi.org/10.1002/aenm.202500587

[167] C. Millsaps, IEC 62133: the standard for secondary cells and batteries containing alkaline or other non-acid electrolytes is in its final review cycle, Battery Power., 16 (2012) 16-18.

[168] ISO, ISO 12405-1: Electrically propelled road vehicles - Test specification for lithium-ion traction battery packs and systems - Part 1: High-power applications, in, 2011.

[169] ISO, ISO 12405-2: Electrically propelled road vehicles - Test specification for lithium-ion traction battery packs and systems - Part 2: High-energy applications, in, 2012.

[170] D.H. Doughty, SAE J2464: EV & HEV rechargeable energy storage system (RESS) safety and abuse testing procedure, in, SAE, 2010. https://doi.org/10.4271/2010-01-1077

[171] UN, Recommendations on the Transport of Dangerous Goods: Model Regulations - Twenty-third Revised Edition, United Nations, 1995.

[172] UL, UL 2580: Standard for batteries for use in electric vehicles, in, 2011.

[173] Y.W. Wang, C.M. Shu, Hazard characterizations of Li-ion batteries: thermal runaway evaluation by calorimetry methodology, in: Z. Zhang, S.S. Zhang (Eds.) Rechargeable Batteries: Materials, Technologies and New Trends, Springer International Publishing, Switzerland, 2015, pp. 419-454. https://doi.org/10.1007/978-3-319-15458-9_15

[174] Z. Rao, P. Lyu, P. Du, D. He, Y. Huo, C. Liu, Thermal safety and thermal management of batteries, Battery Energy., 1 (2022) 20210019. https://doi.org/10.1002/bte2.20210019

[175] S.S. Madani, C. Ziebert, M. Marzband, Thermal characteristics and safety aspects of Lithium-ion batteries: An in-depth review, Symmetry., 15 (2023) 1925. https://doi.org/10.3390/sym15101925

[176] D. Ren, X. Feng, L. Lu, X. He, M. Ouyang, Overcharge behaviors and failure mechanism of lithium-ion batteries under different test conditions, Appl. Energy., 250 (2019) 323-332. https://doi.org/10.1016/j.apenergy.2019.05.015

[177] D. Ren, X. Feng, L. Lu, J. Li, M. Ouyang, Comparison of the overcharge behaviors of lithium-ion batteries under different test conditions, Energy Procedia., 158 (2019) 4921-4926. https://doi.org/10.1016/j.egypro.2019.01.699

[178] A. Stein, D. Kehl, C. Jackmann, S. Essmann, F. Lienesch, M. Kurrat, Thermal electrical tests for battery safety standardization, Energies., 15 (2022) 7930. https://doi.org/10.3390/en15217930

[179] L. Zhang, W. Fan, Z. Wang, W. Li, D.U. Sauer, Battery heating for lithium-ion batteries based on multi-stage alternative currents, J. Energy Storage., 32 (2020) 101885. https://doi.org/10.1016/j.est.2020.101885

[180] W. Tang, W.C. Tam, L. Yuan, T. Dubaniewicz, R. Thomas, J. Soles, Estimation of the critical external heat leading to the failure of lithium-ion batteries, Appl. Therm. Eng., 179 (2020) 115665. https://doi.org/10.1016/j.applthermaleng.2020.115665

[181] C. Zhao, T. Wang, Z. Huang, J. Wu, H. Zhou, M. Ma, J. Xu, Z. Wang, H. Li, J. Sun, Experimental study on thermal runaway of fully charged and overcharged lithium-ion batteries under adiabatic and side-heating test, J. Energy Storage., 38 (2021) 102519. https://doi.org/10.1016/j.est.2021.102519

[182] H. Ji, Y.-H. Chung, X.-H. Pan, M. Hua, C.-M. Shu, L.-J. Zhang, Study of lithium-ion battery module's external short circuit under different temperatures, J. Therm. Anal. Calorim., 144 (2021) 1065-1072. https://doi.org/10.1007/s10973-020-09506-0

[183] J.-B. Jung, M.-G. Lim, J.-Y. Kim, B.-G. Han, B. Kim, D.-S. Rho, Safety assessment for external short circuit of Li-ion battery in ESS application based on operation and environment factors, Energies., 15 (2022) 5052. https://doi.org/10.3390/en15145052

[184] Z. An, T. Shi, Y. Zhao, Q. Gong, D. Zhang, J. Bai, X. Du, Study on aging and external short circuit mechanisms of Li-ion cells with different electrode thicknesses, Appl. Energy., 350 (2023) 121796. https://doi.org/10.1016/j.apenergy.2023.121796

[185] W. Cai, H. Wang, H. Maleki, J. Howard, E. Lara-Curzio, Experimental simulation of internal short circuit in Li-ion and Li-ion-polymer cells, J. Power Sources., 196 (2011) 7779-7783. https://doi.org/10.1016/j.jpowsour.2011.04.024

[186] W. Fang, P. Ramadass, Z.J. Zhang, Study of internal short in a Li-ion cell-II. Numerical investigation using a 3D electrochemical-thermal model, J. Power Sources., 248 (2014) 1090-1098. https://doi.org/10.1016/j.jpowsour.2013.10.004

[187] R. Zhao, J. Liu, J. Gu, Simulation and experimental study on lithium ion battery short circuit, Appl. Energy., 173 (2016) 29-39. https://doi.org/10.1016/j.apenergy.2016.04.016

Materials Research Foundations 190 (2026) 333-359 | https://doi.org/10.21741/9781644904077-12

Chapter 12

Flame Retardant Materials: Applications in Microwave-Based Sensors

Dnyandeo Pawar[1]*, Ajay Kumar[2], Rajesh Kanawade[2]

[1]Centre for Materials for Electronics Technology (C-MET), Thrissur, Kerala, India

[2]Physical and Materials Chemistry Division, CSIR-National Chemical Laboratory, Pune, Maharashtra, India

pawar.dnyandeo@cmet.gov.in

Abstract

The flame-retardant (FR) materials are the key interest for researchers in several applications especially in the microwave domain due to their high heat resistance, high thermal stability, low loss, easy fabrication, and low-cost fabrication. In this chapter, the flame-retardant materials along with their types, coatings, properties of FR materials, types of FR substrates, interaction of microwave frequency with microwave dielectric substrates have been discussed. This chapter emphasized the microwave-based applications of FR materials in communication, chemical, biological, energy harvesting, material characterization, and other applications along with their advantages and disadvantages in detail. The benefits and weaknesses of flame-retardant materials have also been discussed. The conclusion and outlook are also presented.

Keywords

Flame Retardant, Fire-Resistant, Microwave Substrate, Microwave Sensor, Antenna

Contents

Materials Research Foundations 190 (2026) 333-359 https://doi.org/10.21741/9781644904077-12

1. Introduction

Flame-retardant materials are types of materials which are harder to burn due to thermo-oxidative decomposition. Flame retardant properties could be easily attained by chemical or physical treatment. To achieve the retardancy, certain chemicals like alum, vinegar, clay, $FeSO_4$, gypsum, NH_4Cl, $(NH_4)_3PO_4$, and several acids could be added [1–3].

Normally, the polymer gets burning by thermo-oxidation degradation process. In this, the hydroperoxyl groups get formed depending upon the elements or chemical groups attached to carbon atoms. Flame retardant also commonly denoted as FR which typically indicates that it will not catch and spread the fire easily when exposed to flames. To improve the fire-resistant property, the materials are blended organically (addition of halogens like Br, I, and Cl) or inorganically (mixing with aluminum trihydrate (ATH)) with flame retardants [4].

2. Flame Retardant Coatings

Flame retardant materials need to be coated for many applications. It protects the substrate from the flame and thermal damage due to overheating. It slows the spread of a flame, decreases the magnitude, and reduces the smoke production. The coating might have number of functions such as adhesion, catalytic, magnetic, protective, electrical, and decorative. The formulation of coatings can be done through additive, solvent, binder, and filler. The coatings can be done via various processes such as chemical vapor, physical vapor, electro-chemical, spray coating etc. After this process, the coating can be analyzed by various analysis techniques and methods [5].

The coating should have following characteristics such as a lower thermal conductivity, strong adhesion, highly durable, nonflammable, and cheap. The coating has three types such as flame-resistant polymeric materials, thermal barriers, and intumescent coatings.

The flame-resistant coatings are classified as nonintumescent coatings and intumescent coatings. The nonintumescent coatings typically slow the flame spread and smoke. Based on type of nature, it is classified into 3 types such as class A, B, and C. In this, the substrate and film thickness play a crucial role. In other category, the intumescent coatings get swell when it comes in contact with heat or flame [6].

3. Properties of FR material

The dielectric permittivity of FR material particularly fire retardant (FR4) and number '4' specifies the woven glass reinforced epoxy resin varies in 3.8-4.8 dependent on type and composite. Typically, for high frequency applications, a lower dielectric constant material is preferred. The FR material exhibits various interesting properties useful for industrial applications. The FR materials are self-extinguishable and therefore, cannot sustain the combustion. Hence, the FR material is ideal for electronic components and devices. After interaction with heat, it does not emit toxic fumes and therefore, not harmful to human beings. It shows a high thermal stability in the range of 50 °C to 115 °C. It exhibits a high tensile strength normally in the range from 300 MPa to 500 MPa. It displays a high bending strength i.e. resist deformation under bending loads normally in 350 MPa to 600 MPa. It minimizes the signal distortion, dispersion, and attenuation and thus, good for signal transmission. It shows a low dissipation factor or tangent loss ($\tan\delta$) typically in the range of 0.015- 0.025 depending on the type of material. It controls the impedance and thus, allows efficient power transfer. The FR material has relatively a low thermal conductivity

Materials Research Foundations 190 (2026) 333-359 https://doi.org/10.21741/9781644904077-12

(~0.3 - 0.5 W/m·K). The glass transition temperature of FR4 is in the range of 130 °C - 180 °C. The FR materials hold a low moisture absorption property i.e. ~0.10%.

Among the various FR material, the one type i.e. FR4 is widely used in microwave and other electronic applications. The FR4 is a compound material made using two compositions such as fiber-glass cloth and epoxy resin (shown in Fig.1). In this the 'FR' stands for fire retardant and number '4' specifies the woven glass reinforced epoxy resin. In this composite structure, the fiberglass is mechanically strong and stable whereas, the epoxy resin having good thermal resistance and electrically resistant [7,8]. The ratio of composite can be tuned differently but normally the wt% ratio is 60% fiberglass and 40% epoxy resin. Considering the use of FR4 substrate in microwave and electronic applications, various advantages and disadvantages have been discussed below.

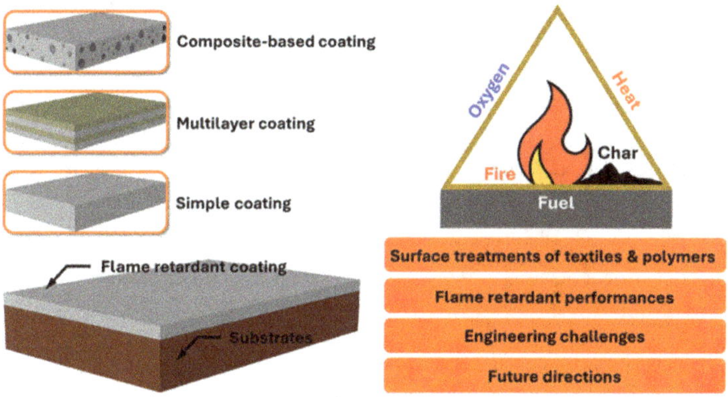

Figure 1. The flame-retardant materials coating and combustion process based on obtainability of heat, fuel, and oxygen as oxidizing agent. [Reproduced with permission from ref. [1]].

3.1 Advantages of FR4 materials

The FR4 exhibits superior properties and therefore, could be used in numerous applications. It has a low cost and hence, alternative to other materials. It has high mechanical strength, lightweight, and available in different thicknesses. It is also resistant to different environments.

3.2 Disadvantages of FR4 material

FR4 has lower thermal conductivity than ceramics or metal-based substrates. This will eventually create challenges in heat dissipation and therefore, limited in power applications. However, few techniques will be utilized to improve the performance such as thermal vias, thick copper cladding, hybrid composite formation etc.

As per the regulatory bodies such as the Restriction of Hazardous Substances (RoHS) and Waste Electrical and Electronic Equipment (WEEE) directions, an environmental compliance could also

Materials Research Foundations 190 (2026) 333-359 https://doi.org/10.21741/9781644904077-12

be raised due to use of halogenated components such as bromine resins [9]. Thus, the halogen-free resins need to be fabricated.

4. Types of FR4

The FR4 materials show the following types depending on the materials, fabrication process, and properties as follows.

(a) Standard FR4: It is a basic type of material widely used in electronic applications. Its glass transition temperature varies from 130 °C to 140 °C and therefore, suitable for moderate thermal-based applications.

(b) High T_g FR4: This material exhibits a higher T_g in the range from 170 °C to 180 °C. This type of materials could be used in power electronics and auto-mobile applications.

(c) High microwave frequency FR4: This material exhibits a lower dielectric constant. It provides applications in satellite and telecommunication applications.

(d) FR4 G10: It shows a high mechanical strength, low dissipation factor, and good dielectric properties.

The FR4 is called as printed circuit board (PCB). Different types of PCBs have been classified as follows.

1. Additive based retardants
2. Reactive flame retardants
3. Halogen-based retardants
4. Non-halogen-based retardants
5. Inorganic retardants
6. Organic or bio-based flame retardants

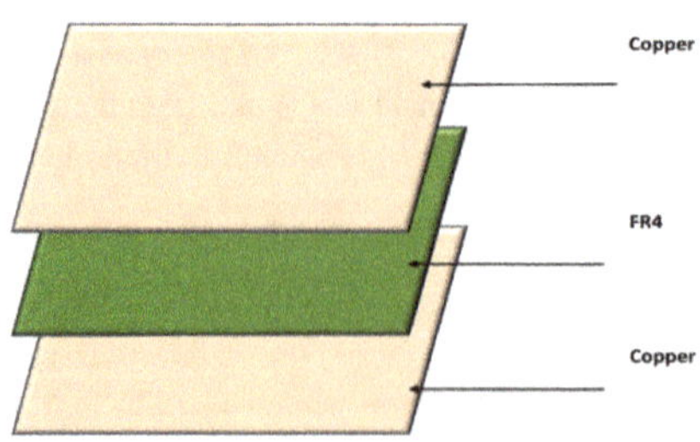

Figure 2. A schematic of a flame retardant (FR4 type) sheet.

Materials Research Foundations 190 (2026) 333-359 https://doi.org/10.21741/9781644904077-12

5. Interaction of microwave radiation with microwave substrates

When microwave interacts with the FR4 substrate, the microwave energy gets transferred to the substrates through interaction with electrons within the material and generates heat inside it. The electromagnetic field interaction with FR4 is mainly defined by skin effect in which the EM field interacts with metal sheet mostly copper, causing an electrical current to pass on the surface of the conductor rather flowing in the entire section of the material. The dielectric loss occurred because of the interaction of energy flowing and transfer of electrons of the FR4 or PCB which get converted into heat.

FR-4 substrate is a woven fiber glass with epoxy resin composite material copper cladded on both sides. When EM wave interacts with the electrons of FR4 material at micro level, there will be both transmission and refraction. There are other parameters such as frequency of incident wave, materials temperature, applied pressure, electric and magnetic field orientation also need to be considered. During an interaction of EM with electrons, materials get polarized due to polarization of electrons and photons lost their energy by relating with the polarized materials. Due to this, the speed of the photons gets slower, and beam changes its direction. The amount of absorbed photons are dependent on the thickness of laminate and materials structure [10,11]. The light transmission is given by the simple equation: $R + A + T = 1$, where R is reflection, A is absorption, and T is transmission.

6. Applications of flame retardants in microwave domain

This section discusses different applications of FR4 substrate in microwave domain. Based on the type of application, it can be used in communication, chemical sensing, resonator applications, energy harvesting, etc.

The sub-6 GHz 5G antennas for communication purpose were fabricated on FR4 substrates [12]. The dual-band monopole antennas on FR4 laminate on a size of 40 mm x 12 mm x 1.6 mm were fabricated. The antennas were designed in ANSYS HFSS software. The inverse C-shaped microstrip line radiated from 4.8 GHz to 5.7 GHz and the L-type microstrip line radiated in 2.3 GHz-2.6 GHz. This antenna demonstrated an enhanced reflection coefficient as <-10 dB, peak gain (2.7@2.4 GHz and 1.8@5.2 GHz), and radiation efficiency (95% at 2.4 GHz and 93% at 5.2 GHz), respectively. A circularly patch antenna at 2.4 GHz for Wi-Fi and other WLAN uses were studied [13]. The 25 mm x 20 mm x 1.5 mm size antenna made-up on FR4 substrate with dielectric permittivity of 4.3 and tangent loss of 0.025 shown a return loss of -52.89 dB which could be useful for wireless communication applications. Different microwave sensors were fabricated on various materials like Polyimide, Rogers RO3010, Rogers RO4350, Rogers RT5880, and FR-4 as shown in Fig. 3 (a, b, c). Based on FR4 substrate, the sensors response characteristics was studied for three different thicknesses [14]. It observed that the response of FR4 substrate was better than other substrates. The response and Q-factor for 2.88 GHz and 3.5 GHz were 0.66% & 0.19%, and 1413.29 & 1140.16, respectively.

Dielectric resonators have also been extensively used for communication applications. Recently, FR4 substrate was used by placing a $Ba_3V_2O_8$ ceramic material dielectric resonator which resonated at 8.06 GHz as shown in Fig. 3(d) [15]. The FR4 epoxy with size of 25 mm x 25 mm was used as a substrate to fabricate the resonator. With reference to change in the resonator placement on microstrip fabricated on FR4 substrate, the resonance frequency and return loss also

Materials Research Foundations 190 (2026) 333-359 https://doi.org/10.21741/9781644904077-12

got changed. The S11 and VSWR results indicated that the good impedance matching, and bandwidth were achieved on the FR4 substrate. However, changes in temperature and humidity could impact the performance of antenna.

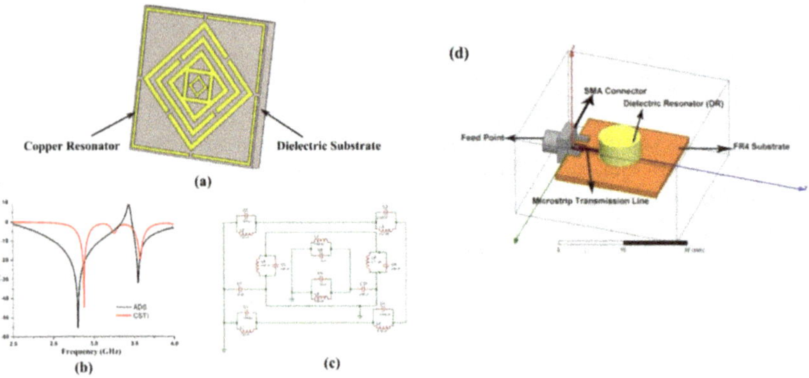

Figure 3. (a) A schematic of metamaterial sensor, (b) Reflection coefficient resulting from the CST and the ADS circuit, (c) Equivalent electrical circuit. [Reproduced with permission from ref. [14]]; and (d) Another sensor fabricated using a cylindrical dielectric resonator placed on FR4 substrate [Reproduced with permission from ref.[15]].

Another FR4 substrate used resonator antenna was fabricated [16]. The FR-4 laminate (dielectric constant = 4.4 and tanδ= 0.02) was chosen because of its cheap and quick availability. The 50 Ω rectangular strip was used for power transmission to the antenna. The developed resonator showed a reflection coefficient < 10 dB with bandwidth of 430 MHz at resonating frequency of 4.32 GHz. The sensor showed a gain of 6.28 dBi with an efficiency of 88.2%. FR4 laminate (dielectric permittivity of 4.3) was used for resonator-based application [17]. The $Ba(Zn_{1/3}Nb_{2/3})O_3$ (BZN) resonator of dimensions in 12 mm diameter and 6 mm thickness was used. The sensor showed a bandwidth of 590 MHz with radiation efficiency more than 90% at 7.08 GHz. However, the humidity and temperature response on sensing performance was not studied. In another research work, a triple band dielectric resonator was designed and fabricated on FR4 substrate [18]. The antenna array dimension was 50 × 60 × 1.6 mm^3. The antenna operated in various frequencies like 1.5 Gigahertz, 2.4 Gigahertz, and 5.8 Gigahertz and showed a gain= 2.8 dBi, 3.4 dBi, and 5.8 dBi, respectively. An antenna exhibited an efficiency and VSWR of > 75% and 1.5, respectively. A small alteration in the simulated and experimental results were observed because of glue layer attached with the resonator and FR4 substrate. Also, the system should be mechanically stable for better results. Another flame retardant FR4 laminate with Dk of 4.4 and height of 1.6 mm was explored for wireless communication application [19]. The total dimension of antenna was 29.5 × 22 × 1.6 mm^3. An antenna geometry developed on FR4 laminate was able to resonate multiple bands at S-band, C, 5G NR, X, Ku, and K bands corresponding to 3.5, 6.95, 8.89, 10.75, 14.74, 15.68, and 19.78 GHz, respectively. The sensor showed a gain and efficiency of 4.23 dBi and 86.23 %, respectively. However, some limitations such as the FR4 substrate provide a maximum loss at Ku-band and K-bands, which might impact the antenna characteristics in higher frequency range. Additionally, some insertion loss was created due to use of PIN diode. A circular ring microstrip antenna based on FR4 substrate was used for the X-band applications [20]. The FR4 board with dielectric permittivity $\varepsilon_r = 4.4$, h = 1.6 mm, and loss $\delta = 0.02$ was preferred for sensing because of its good mechanical properties and economical. The sensor geometry was fabricated in HFSS software. The sensor showed an efficiency of 83% at 9.7 GHz and 69% at 11.4 GHz. It could be used for RF switches applications. However, the difference in the simulation and experimental was observed due to limitation of the feeding element, material, and soldering. A FR4-based microwave sensor for crack detection was fabricated [21]. The FR4 substrate had a size of 40 × 40 mm^2 and thickness of 3.2 mm. The sensor was capable of detect a crack of size length of 17 mm, width of 2.4 mm, and angle of inclination of 40.5° to the y-axis. However, this sensor has required a characterization to use for other dielectric applications.

Experimental and threotical study of utilization of three different substrates such as FR-4, Rogers RO4350B, Rogers, and RT5880 for thickness monitoring was performed [22]. The FR4 with three different thicknesses were used. The quality factor (Q-factor) was 435. The resonance frequency shifted about 80 MHz to changed FR4 thickness of 1.6 mm and 1.3 mm and 110 MHz shift for thickness for 1.6 and 1 mm thick FR4 substrate. It also noted that the sensor performance was better in lower frequency than higher frequency.

A complementary split ring resonator (CSRR) sensor was designed for solid material detection. Four types of samples under tests like air, Roger 5880, Roger 4350, FR4, and their complete performance was analyzed at 2.5 GHz [23]. It noted that maximum frequency shift of 296 MHz was observed for FR4 material. This type of sensor could be useful in characterization of different types of material. The thickness measurement using a nondestructive testing based on various microwave sensors fabricated using FR4 substrate was also conducted [24]. In this study, different microwave structures such as microstrip line (MS-line), meander microstrip line, complementary split-ring resonator, ring resonator, and microstrip antenna developed on FR4 of height 1.45 mm and dielectric permittivity of 4.3 were used to detect the thickness of dielectric material. The results were based on dual scale frequency and amplitude and therefore, could be potentially utilized in thickness monitoring applications.

A star shaped microwave sensor operated at C (4.46 GHz) and X band (11.48 GHz) were optimized for sensing solid and liquid samples such as solids (FR-4, RO-3035, RO-5880, and RO-6202), and liquids (milk, oil, water, and energy drinks) [25]. The study indicated how these substrates were performed differently for solid and liquid sensing.

The FR4 based CSRR sensor was used for dielectric characterization of various materials as shown in Fig. 4. [26]. The dual band sensor was developed on FR4 substrate with dielectric permittivity of 4.4 and tangent loss of 0.02. The height of FR4 was changed from 0.1 mm to 1.5 mm. A comparative study of some commonly available dielectric material such as HDPE (23), PVC (27), Rubber (3), Plexiglass (34), Polyimide-Quartz (4), Glass (55), TaconicRF-60(7), Marble (83), Roger TMM10 (92), Sapphire (10), and Silicon (119) were used as materials under test. The simulation and measured results also compared for better performance.

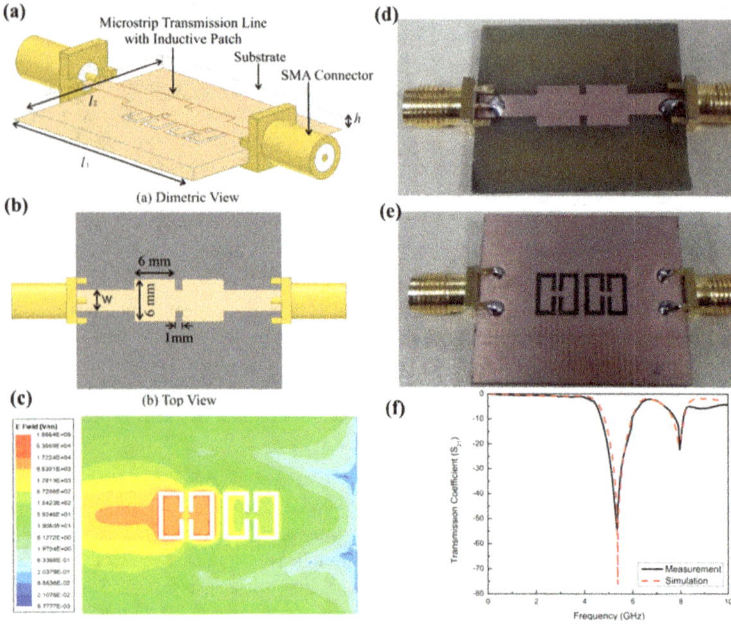

Figure 4. (a) & (b) A schematic of FR4-based set up, (c) Distribution of electric field, (d) & (e) Fabricated prototype structure, and (f) Comparison of simulated and measured results. [Reproduced with permission from ref. [26]].

Two sensors using microstrip line and split ring resonator fabricated using copper cladded FR4 with dielectric permittivity of 4.4, tangent loss of 0.02, and height of 1.6 were explored to study adulteration in vegetable oils [27]. It was detected that the center frequency changed to lower frequency with increased in dielectric constant of the test samples. However, the sensor recovery needs to be evaluated. The microwave antenna T-based resonator developed on FR4 for characterizing solid material was also utilized [28]. The FR-4 with dielectric permittivity= 4.3, tangent loss= 0.0265, and thickness= 1.6 mm was used for measurement. The sensor exhibited two frequencies at 1.81 GHz and 2.34 GHz and showed a sensitivity of 1.15% and 0.16%, respectively. This FR4-based geometry could be used for noncontact material characterization. Some other sensors [29,30] for dielectric characterization of low-permittivity of materials have been displayed in Figure 5.

Materials Research Forum LLC
https://doi.org/10.21741/9781644904077-12

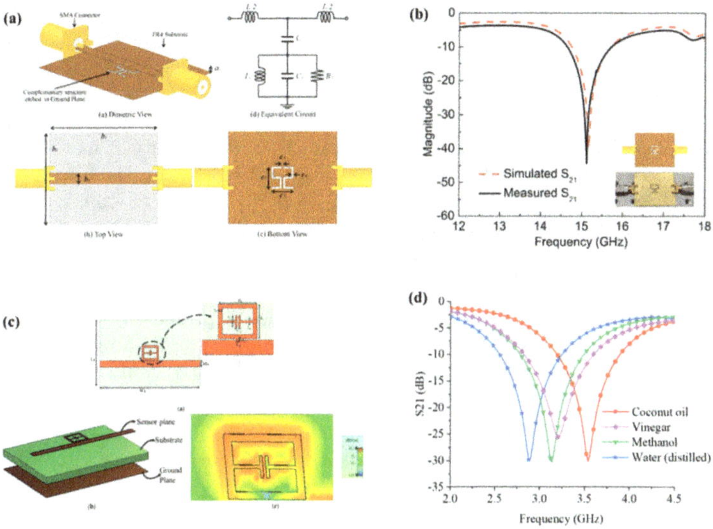

Figure 5. (a) A sensor geometry based on FR4 platform along with the dimensions, (b) The simulated and measured results of sensor at 15.17 GHz with notch depth −39.84 dB [Reproduced with permission from ref. [29]]; (c) Another sensor fabricated with inductively coupled capacitively loaded ring resonator used for liquid measurement, and (d) Measured S21 vs. frequency graph with various sample liquids such as coconut oil, vinegar, methanol, and water. [Reproduced with permission from ref. [30]].

The different types of commercially available films like polyethylene terephthalate (PET), polyvinyl chloride (PVC), and FR4 for coating thickness monitoring were used [31]. The dielectric constants of these films were 2.87, 2.84, and 4.3, respectively. The sensing mechanism showed a linear relation between the coating thickness and frequency shift of the cavity sensor. The accuracy for thickness up to 2 mm was obtained. In another study, the sensor size of 19.9 × 19.9 × 1.6 mm^3 of FR-4 laminate was explored for material properties and thickness of coated material [32]. This sensor showed absorption of 99.9%, 99.9%, and 98.6% at 1.93 GHz, 3.55 GHz, and 6.35 GHz, respectively. The sensor performance was measured with varied thickness and changed dielectric constant of oil. The sensor exhibited a sensitivity of 0.454 %, 0.061 %, and 1.55 %, respectively.

The FR4 substrate material for non-destruction crack detection was utilized [33]. The FR4 material was selected for sensing purposes because of its small contraction and expansion due to low thermal expansion. Two sensors operated at 418 MHz and 4.59 GHz were compared. The study showed a power shift of 23% for microwave region and around 8% for radio-wave frequency. This study is beneficial for range of crack studies.

The FR4-based sensor also used in chemical sensing. The transmission line-based sensor was fabricated on FR4 laminate (dielectric permittivity of 4.3, tangent loss of 0.02, height of 1.6 mm, and copper-clad of 0.035 mm) [34]. FR4 was preferred because of minimum loss, mechanically

Materials Research Foundations 190 (2026) 333-359 https://doi.org/10.21741/9781644904077-12

good, and cheap. The sensor was used for detecting various types of branded and unbranded diesel. The total bandwidth of 60 MHz at 10 GHz was observed for blended and unblended diesel. However, the sensors repeatability, temperature and humidity impact, and performance under small change in blending concentration need to be performed. The FR4 or PCB board-based microwave sensors have been utilized in food industry. In this context, the quality of milk was checked by measuring fat content present in the milk. The CSRR sensor with total dimensions of 20 mm x 20 mm was designed on FR4 substrate operated at 2.4 GHz [35]. The sensor was capable to distinguish fat content in milk varied from 0.94% to 3.17% in 5 ml sample liquids. With reference to change in fat content, the S21 parameter was changed and got reflected on VNA. This is a non-invasive study and hence, it could be potentially used in food industry. However, further study on sensors reliability and adaptability under varied operating conditions is required.

A CSRR structure based on FR4 with height 0.158 cm, dielectric permittivity 4.4, and tangent loss 0.02 for soil water content determination was fabricated [36]. The sensor showed a frequency shift of 165 MHz under soil water content variation of 0% to 15%. These types of studies could be used to study magnetic permeability depending on the soil type.

The CSRR based antenna operated at 2.4 GHz and made-up of FR4 with dielectric constant 4.4, tangent loss 0.02, and height 3.0 mm for water-ethanol mixture sensing was performed [37]. The complex permittivity of the samples was detected by this sensor. The water-ethanol mixture concentration varied from 0% to 100% and corresponding resonant frequency shift was measured. It noted that as the water proportion in the mixture increased the corresponding resonant frequency was decreased. The structure was simple and low-cost. This sensor could be used to test other mixtures also.

Real time glucose sensing was performed by using FR4 laminate with height of 1.57 mm, dielectric permittivity of 4.7, tangent loss of 0.014, and single side copper clad thickness of 0.0175 mm [38]. The interdigital structure is fixed in co-planar transmission line. The sensing area was confined by the PDMS cavity and exposed to varied glucose level from 0 g/mL to 1 g/mL. The device showed a sensitivity of 235.32 MHz/g/mL. In comparison with other sensors, this sensor was easy to operate, cheap, and simple fabrication. A honey shaped CSRR microwave prototype antenna on FR4 substrate operated at resonant frequency of 2.4 for aqueous glucose sensing in the range of 70–120 mg/dL GHz was demonstrated [39]. The sensor had parameters as dielectric constant= 4.4, tangent loss= 0.02, length= 66 mm, width= 20 mm, and thickness h = 0.8 mm. The FR4-based prototype system was compact and could be used for diabetic monitoring. Additional sensors [40,41] for glucose and sweat monitoring have been shown in Figure 6.

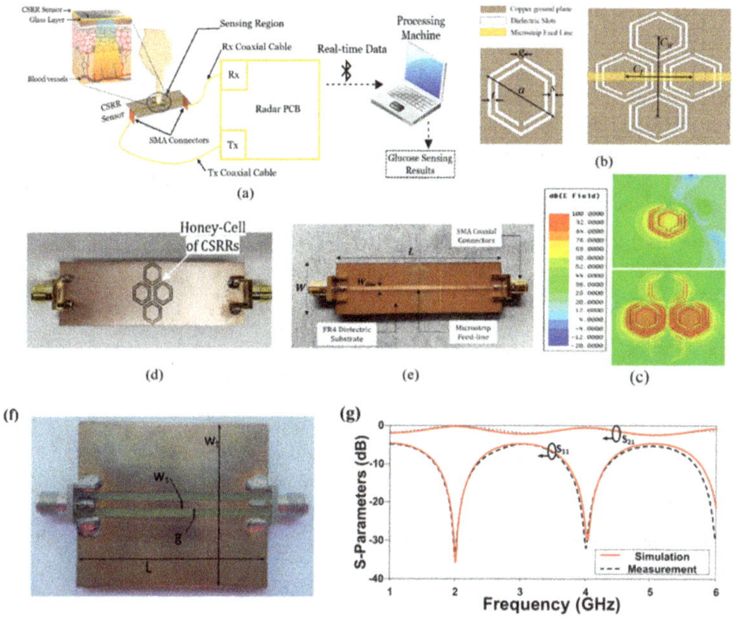

Figure 6. (a) A honey-cell structure-based CSRR sensor using FR4 substrate for blood glucose sensing, (b) Configuration of structure, (c) Electric field distribution, (c) & (e) Fabricated prototype [Reproduced with permission from ref. [40]]; (f) Another sensor on FR4 substrate-based coplanar waveguide for sweat monitoring, and (g) Measured and simulated S-parameters of the coplanar waveguide-based sensor. [Reproduced with permission from ref. [41]].

Microwave sensor fabricated using FR4 substrate was also used in biomedical application [42]. The CSRR biosensor for leukemia cancer detection fabricated using dielectric FR-4 epoxy substrate was performed. For sensor design fabrication, the FR4 substrate with dielectric permittivity of 4.4 and substrate height of 1.6 mm was designed at resonance frequency 2.45 GHz. Sensor performance was monitored on a vector network analyzer that provided high accuracy and real time testing.

The FR4 based microwave interdigital capacitor sensor was also used in gas sensing [43]. The surface was functionalized with titanium dioxide layer on the electrodes and explored towards O_2 gas. The O_2 gas detection was performed ranging from 0% to 100%. The resonant frequency changed as the O_2 concentration increased from 0% to 100%. The sensor performance was measured in S-parameter.

The chemical sensing was performed by using FR4 substrate. The sensor developed on a 10 mm height copper cladded FR4 board [44]. The sensor was able to detect transformer oil, corn, cotton and olive oils, branded and unbranded diesels, and aniline doped ethyl-alcohol, and benzene doped carbon tetrachloride. The sensor exhibited a corresponding frequency change of 250 MHz, 200

 Materials Research Forum LLC
https://doi.org/10.21741/9781644904077-12

MHz, 250 MHz, 150 MHz, and 50 MHz, respectively. The sensor could be used in chemical and medical field.

The 10 × 10 mm^2 size FR4 board was used for oil sensing [45]. The FR4 board was used because of its cheap, mechanically stable, and low lossy. The FR4 had a tangent loss of 0.02, dielectric constant of 4.3, and height of 1.6 mm. The double sided copper cladded substrate had a thickness of 0.035 mm with conductivity of 5 × 10^8 S/m. Sensor showed an absorption of 91.46%, 99.72%, 96.89%, and 98.45% to 3.64 GHz, 5.99 GHz, 7.47 GHz, and 10.01 GHz, respectively. Additionally, the sensors Q-factor variation might impact the sensitivity.

The 2.5 mm thick FR4 dielectric substrate was used for microwave absorber applications [46]. The FR4 with Dk = 5 and tanδ = 0.02 was used as a substrate. The FR4 was coated with 0.035 mm and conductivity of 5.8 10^7 S/m thick copper on both sides. In this study, an absorption of FR4 under change in dielectric constant of material was studied. It is observed that as the permittivity increased an absorption level was also increased. The FR4 board of 1.6 mm thick with total dimensions of 13 × 13 × 0.035 mm^3 was used to fabricate a split ring resonator (SRR) and an octagonal split ring resonator (OSRR) structure [47]. These structures were operated in C-band and could be used for communication, defense, and security applications. The sensors showed a very strong absorption of 99.99% and 99.40% at resonant frequencies of 6.15 and 7.36 GHz, respectively. However, some distorted output signal was observed due to radiation from surroundings, calibration error, and distortion in metallic copper. Another metamaterial sensor [48] for 5G applications and sub-6 GHz 5G band, i.e., 3.5 GHz has been reported as shown in Fig. 7.

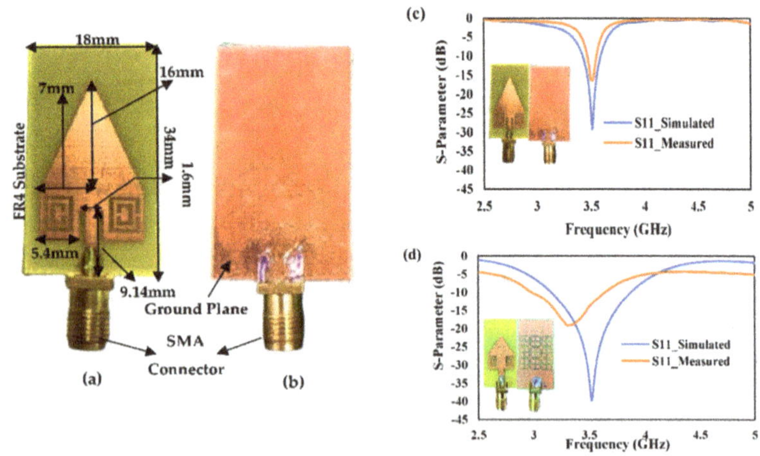

Figure 7. (a) & (b) Fabricated prototype structure of antenna using FR4 substrate with top and bottom layer, Simulated and measured S-parameter for (c) A single arrow antenna, and (d) For 3 × 3 array structure. [Reproduced with permission from ref. [48]].

Flame retardant FR4 dielectric with permittivity of 4.3, permeability of 1, height of 1.6 mm, metallic copper cladded on both sides with electrical conductivity 4.8×10^7 S/m, and thermal

Materials Research Forum LLC
https://doi.org/10.21741/9781644904077-12

conductivity of 401 W/mK was utilized as a microwave absorbing material in S, C, and X-band applications [49]. Simulation study of different substrates like RT 5870, RT 5880, RT 6002, and FR4 were also considered. The results showed that the FR4 exhibited a better performance than others. The absorption for FR4 was 99.86%, 97.58%, and 97.46% at resonating frequencies of 3.02, 5.71 and 8.36 GHz, respectively. However, the size and structure of absorbers are important for absorbance-based studies.

The FR4-based sensors are also utilized for biochemical sensing applications. Recently, a rectangular patch antenna was used for the detection of COVID-19 affected lungs [50]. The antenna designed on FR4 laminate with height of 1.62 mm and further drop-cast and spin coated with multiwalled carbon nanotubes. The antenna operated at 6.63, 7.291, 7.29, and 7.22 GHz. The highest specific absorption rate was 1.77 W/kg for 10 g of tissues. The antenna could be used on-body for early detection of COVID-19-affetced lungs.

A temperature sensor was also fabricated using FR4 dielectric board with dielectric permittivity= 4.4, thickness= 1.57 mm, and $\tan\delta$= 0.002 [51]. FR4 was selected due to its high resistance to heat and chemical, and mechanically stable. The sensor performance was measured at resonance frequency of 2.21 GHz under temperature range from 20 ^{0}C to 120 ^{0}C and exhibited a sensitivity of 0.640 MHz/^{0}C. The future work could be based on the real time characterization using machine learning algorithm. The FR4 is a low-loss and less temperature dependent low-cost material and therefore, could be used in cryogenic applications. A coplanar waveguide structure fabricated on PCB and its resonance peak was monitored under cryogenic temperature [52]. It was observed that the 9% reduction in dielectric constant and 70% reduction in tangent loss of FR4 substrate under variation of cryogenic temperature varied from 300 K to 4 K. This study provided futuristic approach for studying any type of substrate under cryogenic conditions.

FR4 based substrates were also used as a solid platform for energy harvesting applications. A rectenna was fabricated using a FR4 laminate with thickness= 1.6 mm and Dk= 4.4. This rectenna showed a broadband frequency signal of 1.8 to 2.8 GHz, gain of 5 to 6.5, measured bandwidth of 19.6%, radiation efficiency of 93%, and maximum efficiency reached to 65% [53]. This rectenna could be used for low-power IoT applications. However, more compact antennas with a high gain is needed for IoT applications.

An omnidirectional rectenna array fabricated using a FR4 substrate was also utilized for energy harvesting applications. The 3×3 network was able to distribute varied RF energy received by three antennas [54]. The height of FR4 board was 1.5 mm. The rectenna exhibited a single band frequency of 2.15 to 2.56, gain of 7.4, and maximum efficiency of 73% under input power of -10 to 20 dBm. The following Table 1 provides a summary of various microwave sensors fabricated using flame retardant materials.

Table 1. Comparative performance of various microwave sensors.

Sr. No.	Sensor type	Substrate	Center frequency (GHz)	Analyte	Detection range	Response	Ref.
1.	Complementary split ring resonator	FR-4	15.12	Dielectric characteristics of low permittivity materials	Relative permittivity varies from 2.1– 3	Relative sensitivity = 6.7%	[29]
2.	Half-mode substrate integrated waveguide (HMSIW) resonators	-	2.45	Material characterization	RO5880, F4B, RO4350, FR–4, TP1, and TP2	258.8 mV/ε'_r (Vmag)	[55]
3.	Octagonal DS-CSRR		2.4	Permittivity measurement of liquid samples	Dielectric constant: 1–111	Normalized sensitivity of 1.52%	[56]
4.	Complementary split ring resonator	FR-4	2.45	Material characterization	Dielectric constant varies from 1–10	4.12% in solid and 0.78% in liquid	[57]
5.	Asymmetric Electric Split-Ring Resonator (AESRR)	FR-4	11.575	Material characterization	Liquid sample permittivity changes from 1 to 7; Solid substrates permittivity changes from 1 to 9	425 MHz for liquid; 3315 MHz for liquid	[58]
6.	Inductively coupled capacitively loaded ring resonator (IC-CLRR)	FR-4	3.56	Liquid characterization	Permittivity varies from 2 to 80	-	[30]
7.	Passive layered permittivity sensor using a resonant absorbing surface	FR-4	9.91	Permittivity measurement	1-2.1	1610 MHz	[59]
8.	Asymmetric single split resonator (ASSR)	FR-4	1.27	Permittivity of both solid and liquid samples	-	0.910	[60]
9.	Complementary split-ring resonator	FR-4	2.45	Water- ethanol mixture	0–100%	273 MHZ	[61]
10.	Planar Ring Resonator	FR-4	2.72	Ethanol-water mixture	10–90%	0.726%	[62]

11.	Inkjet-printed coupled ring resonators	FR-4	2.2–2.6	Ethanol concentrations in water	0–100%	188 kHz/% ethanol	[63]
12.	Planar semicircular resonator	FR-4	2.85	Purity of liquid ethanol-water mixture	0.9–10	0.52%	[64]
13.	Complementary split-ring resonator	FR-4	2.6	Distilled water, fresh water, sea water	-	Reflection coefficient (dB): -16.2086, -20.1407, -22.7447	[65]
14.	Circular split ring resonators	FR-4	1.57	Palm kernel oil, coconut oil, and castor oil	-	0.069%	[66]
15.	Complementary split ring resonator	FR-4	5.2	Oil adulteration detection	10% to 30% adulteration in the olive oil	530 MHz	[67]
16.	Planar antenna	FR-4	0.2 to 10	Sediment density	1.54 g/cm^3 to 1.73 g/cm^3	820 MHz g^{-1}.cm^3	[68]
17.	Complementary split-ring resonator	FR-4	2.45	Blood glucose level	40–140 mg/dL	0.94 MHz/ (mg/dL)	[40]
18.	Nested split-ring resonator (NSRR) microwave sensor integrated with phase-locked loop (PLL)	FR-4	0.743	Glucose	1.2 microlitre	17.2 dB/mg /mL	[69]
19.	Planar Yagi–Uda antenna	FR-4	5.5	Glucose monitoring	0–500 mg/dL	26 MHz	[70]
20.	Complementary split ring resonator	FR-4	1.65	Glucose monitoring	3-1444 5-1444	1.607 (DI) 1.317 (Saline)	[71]
21.	Molecularly imprinted polymer (MIP)-coated	FR-4	7	Glucose	50 mg/dLto400 mg/dL	1.3 MHz/(mg/dL)	[72]
22.	Double-spiral transmission line	FR-4	1.44	Glucose and isopropyl alcohol	0–80, 0–100	1.186 %ml/mg, 1 %ml/mg	[73]
23.	Microstrip parallel coupled line	FR-4	2.45	Soil moisture	0–100%	200 MHz	[74]
24.	Hollow microwave resonator	FR-4	4.05	Humidity sensing	19% ~ 79% RH	40.75 kHz/RH	[75]
25.	Microstrip line	FR-4	3.3	Humidity	20–80% RH	1.75 MHz/RH%	[76]

26.	Moore fractal geometry	FR-4	0.8	Water content in crude oil	0% to 100%	150 MHz	[77]
27.	Optical-microwave moore fractal structure	FR-4	0.76, 1.2	Water content in crude oil	0–100%	Q-factor range: 72-20	[78]
28.	Complementary split-ring resonator	FR-4	38.59	Oil detection	Permittivity range studied: 1–3	1.54 GHz/DCU	[79]
29.	Dumbbell shaped structure loaded modified circular ring resonator	FR-4	3.26 11.6, 17.13	Absorber with liquid sensing	S, X, Ku band	Peak absorption (%): 93.8, 96.74, 99.95	[80]
30.	Archimedes'-spiral-shaped ultrathin metamaterial absorbers	FR-4	8.49, 8.88, 9.3, 9.87, 10.65, 11.58, 12.24	Refractive index sensing	-	Absorption rates: 98.5%, 77.1%, 88.7%, 98.2%, 99.7%, 83.4%, and 99.9%	[81]
31.	Wideband microwave patch antenna	FR-4	3.50, 5.80, and 9.30,	Salinity and sugar content detection	0–80%	Salinity measurement @5.80 GHZ sensitivity: 0.1875 dB/(mg/mL); Sugar content measurement @5.80 GHZ sensitivity: 0.216 dB/(mg/mL)	[82]
32.	Triple circle split ring resonator	FR-4	8–12	Benzene and carbon tetrachloride	-	Shift of resonance frequency: 90	[83]
32.	Parallel coupled microstrip lines	FR-4	~1.5	Assessment of pickled fish samples with varying salt concentration	0–90%	Response measured in S-parameter	[84]
34.	Complementary ring structure	FR-4	2.45	Aspergillus Niger detection	-	S_{21}: -24.840	[85]
35.	Complementary split-ring resonator	FR-4	~1.8	Urea in saline	0–250	Sensitivity: 72.7 MHz/M	[86]
36.	Hilbert's fractal architecture	FR-4	0.46	Urine	0–100%	Frequency shift: 18 MHz	[87]
37.	Complementary split ring resonator	FR-4	1.41	Saccharides	10 mg/ml to 50 mg/ml	-	[88]

Materials Research Foundations 190 (2026) 333-359 https://doi.org/10.21741/9781644904077-12

38.	Dual-band dual-polarized annular ring antenna	FR-4	2.5	pH sensing	4–10	4 MHz/pH	[89]
39.	Coplanar waveguide (CPW) transmission line	FR-4	2	Sweat monitoring	NaCl concentration: 0.05, 0.35, and 0.7 mol/L	6.71 dB/g/L	[41]
40.	Hairpin resonator	FR-4	6	Nitrogenous fertilizers in soil and water	In soil: 0–200 ppm, In water: 0–400 ppm	5%, 83.58%	[90]
41.	Complementary split ring resonators	FR-4	2.18	L-Lysine	1200 microlitre	0.242%	[91]
42.	H-slot antenna	FR-4	4.2	Termites	-	-	[92]
43.	Split ring resonators	FR-4	8–15	Hyperthermia for skin cancer	-	-	[93]
44.	Electromagnetic band gap (EBG) based ground plane with double sided rings of C shaped	FR-4	3.81	Deep brain temperature measurement	SAR/ volume loss density, electric field	25 w/kg, 2.16 w/m^3, 33.69 v/m	[94]
45.	Metamaterial (MTM) based 1 × 4 linear antenna arrays	FR-4	3.5	Communication	Main beam angle: $-45^\circ < \theta < +45^\circ$	Peak gain: 4.8–6.1 dB	[95]
46.	Complementary split-ring resonator metamaterial	FR-4	3.5	Fifth generation (5G) indoor distributed antenna systems	-	Gain/bandwidth: 100 MHz/2.6 dBi	[48]
47.	Octagonal split ring resonator	FR-4	3.31, 8.6, 12.99	S-, X- and Ku-band applications	-	Effective medium ratio (EMR): 10.07	[96]
48.	Frequency modulated continuous wave (FMCW) radar sensor	FR-4	480	Communication	-	Tuning range: 49 GHz	[97]
49.	Monopole antenna	FR-4	0.868, 0.915	Wireless communication	-	Gain: 1.9, 2	[98]
50.	Planar antenna	FR-4	2.54–5.84	Wireless communication	-	On-body radiation efficiency exceeds 70% after 4 GHz	[99]
51.	H-shaped antenna	FR-4	2.5–4.5	Wireless communication	-	Gain is 8.2 dBi (86%)	[100]

52.	Dual-polarized differential-feed fractal slotted ground antenna	FR-4	1.8–2.6	Energy harvesting	Input power: -20–5 dBm	Maximum efficiency: 50%	[101]
53.	Two-port differential rectifier integrated with hybrid coupler	FR-4	1.55–2.6	Energy harvesting	Input power: -20–10 dBm	Maximum efficiency: >50%	[102]
54.	Complementary split-ring resonator	FR-4	5.8	Angular displacement	−60◦ to +90◦	1.32◦/ ◦	[103]
55.	Complementary split ring resonator	FR-4	~4.5	Angular displacement	-80° to +80°	10.65°/°	[104]
56.	Micro-loop-gap resonator (μ-LGR)	FR-4	6.83	Chip-scale atomic clock (CSAC)	-	High field orientation factor (FOF): 0.75	[105]
57.	Microwave array	FR-4	2 to 6	Distance	0–1.1 mm	161 MHz/mm	[106]
58.	Complementary split-ring resonator	FR-4	418 MHz and 4.59	Crack detection	0.18 mm to 0.56 mm	Level of detection: 0.18 mm	[33]
59.	Microstrip patch antenna	FR-4	1.3, 2.1, 2.9, 3.6, and 4.1	Partial discharge detection	-	Peak-to-peak voltage (Vpp): 248 mV	[107]
60.	Planar microstrip	FR-4	2.4	Power line fault detection	-	Average error of 0.15 m (5.04%)	[108]
61.	Rat-race coupler structure	FR-4	450 MHZ	Reader for reflective-mode microwave sensor	-	-	[109]

7. Conclusion and outlook

In this chapter, the recent development regarding flame retardant materials, fabrication process, and applications of these substrates in microwave regime is presented. The mechanism related to FR materials, properties, different types of FR substrates, interaction of microwave with FR substrate, and their applications are deduced in detail.

The FR4 material-based substrate has many applications in microwave region. However, there is a tremendous need to prepare other types of FR materials for microwave applications. In this scenario, the cost and availability of raw materials also need to be considered for better utilization. Certain low-cost materials such as metal oxides-based composites, metal coated, natural textile, nanomaterials, and porous FR materials could be potentially used as a substrate for microwave applications. The materials like porous polymer flame-retardant materials have several applications in microwave domain such as microwave absorbers, energy harvesting, satellite communication etc.

An extensive use of bromine and chlorinated FR substrates could be a serious threat to health and environment. Therefore, development of halogen-free FR is expected to gain momentum soon. The development of smart cities and infrastructure could also boost the requirement of FR materials.

Several challenges also need to be considered in certain areas like FR synthesis, manufacturing, and applications. Therefore, the future prospect will be focused on better dispersion of filler material into polymer matrix to improve the flame retardancy, interfacial compatibility, functionalization strategies, advanced manufacturing processes, cost, scalability, and environmental approach.

References

[1] O.Y. Wen, M.Z.M. Tohir, T.C.S. Yeaw, M.A. Razak, H.S. Zainuddin, M.R.A. Hamid, Fire-resistant and flame-retardant surface finishing of polymers and textiles: A state-of-the-art review, Prog. Org. Coatings 175 (2023) 1-27. https://doi.org/10.1016/j.porgcoat.2022.107330

[2] S. Sun, Q. Yu, B. Yu, F. Zhou, New progress in the application of flame-retardant modified epoxy resins and fire-retardant coatings, Coatings 13 (2023) 1-23. https://doi.org/10.3390/coatings13101663

[3] Y.F. Lv, W. Thomas, R. Chalk, S. Singamneni, Flame retardant polymeric materials for additive manufacturing, Mater. Today Proc. 33 (2020) 5720-5724. https://doi.org/10.1016/j.matpr.2020.05.081

[4] M.S. Murad, A.K. Hamzat, E. Asmatulu, R. Asmatulu, Flame-retardant fiber composites: synergistic effects of additives on mechanical, thermal, chemical, and structural properties, Adv. Compos. Hybrid Mater. 8 (2024) 1-43. https://doi.org/10.1007/s42114-024-01111-1

[5] M.M. Mohd Sabee, Z. Itam, S. Beddu, N.M. Zahari, N.L. Mohd Kamal, D. Mohamad, N.A. Zulkepli, M.D. Shafiq, Z.A. Abdul Hamid, Flame retardant coatings: Additives, binders, and fillers, Polymers (Basel). 14 (2022) 1-33. https://doi.org/10.3390/polym14142911

[6] T. Mariappan, Fire Retardant Coatings, New technologies in protective coatings, C. Giudice, G. Canosa (Eds.), IntechOpen, Rijeka, 2017. https://doi.org/10.5772/67675

[7] R. Sanapala, B. Sood, D. Das, M. Pecht, Effect of lead-free soldering on key material properties of FR-4 printed circuit board laminates, IEEE Trans. Electron. Packag. Manuf. 32 (2009) 272-280. https://doi.org/10.1109/TEPM.2009.2029566

[8] A. Honarbari, P. Cataldi, A. Zych, D. Merino, N. Paknezhad, L. Ceseracciu, G. Perotto, M. Crepaldi, A. Athanassiou, A green conformable thermoformed printed circuit board sourced from renewable materials, ACS Appl. Electron. Mater. 5 (2023) 5050-5060. https://doi.org/10.1021/acsaelm.3c00799

[9] P. Cusack, T. Perrett, The EU RoHS Directive and its implications for the plastics industry, Plast. Addit. Compd. 8 (2006) 46-49. https://doi.org/10.1016/S1464-391X(06)70584-0

[10] R.R. Mishra, A.K. Sharma, Microwave-material interaction phenomena: Heating mechanisms, challenges and opportunities in material processing, Compos. Part A Appl. Sci. Manuf. 81 (2016) 78-97. https://doi.org/10.1016/j.compositesa.2015.10.035

[11] R. Zamorano Ulloa, M.G. Hernández, V.L. Villegas Rueda, The interaction of microwaves with materials of different properties, electromagnetic fields and waves K.H. Yeap, K. Hirasawa (Eds.), IntechOpen, Rijeka, 2019. https://doi.org/10.5772/intechopen.83675

[12] P. Pradeep, J.S. Kottareddygari, C.S. Paidimarry, Design of a dual-band monopole antenna for internet of things and sub-6 GHz 5G applications, in: 2024 IEEE Wirel. Antenna Microw. Symp., 2024: pp. 1-4. https://doi.org/10.1109/WAMS59642.2024.10528093

[13] K.P. Lad, K. Mhapsekar, S. Baudha, A compact circular patch antenna for 2.4 GHz Wi-Fi and other WLAN applications, in: 2022 IEEE Microwaves, Antennas, Propag. Conf., 2022: pp. 716-718. https://doi.org/10.1109/MAPCON56011.2022.10046876

[14] M.A. Khalil, W.H. Yong, M.T. Islam, A. Hoque, M.S. Islam, C.C. leei, M.S. Soliman, Double-negative metamaterial square enclosed Q.S.S.R for microwave sensing application in S-band with high sensitivity and Q-factor, Sci. Rep. 13 (2023) 1-17. https://doi.org/10.1038/s41598-023-34514-z

[15] M.A. Dinesh, R. Kumar, V. Rani Gupta, V. Dayal, Synthesis, characterization, and application of Ba3V2O8 ceramics in cylindrical dielectric resonator antenna: Simulation and experimental study, Mater. Sci. Eng. B 311 (2025) 1-11. https://doi.org/10.1016/j.mseb.2024.117791

[16] S. Mohapatra, R. Barman, T.K. Das, T. Badapanda, Y. Huang, J. Xiao, S.N. Tripathy, Structure property correlation of (1-x)MgTiO3 - xSrTiO3 microwave dielectric ceramics for dielectric resonator antenna applications, Ceram. Int. 50 (2024) 31792-31808. https://doi.org/10.1016/j.ceramint.2024.05.390

[17] H. Li, F. Liu, L. Cheng, J. Huang, S. Liu, Microstructure and microwave properties of complex perovskite-structured Ba(Zn1/3Nb2/3)O3 ceramics and dielectric resonator antenna by digital light processing technology, J. Eur. Ceram. Soc. 45 (2025) 1-8. https://doi.org/10.1016/j.jeurceramsoc.2025.117492

[18] A. Vahora, K. Pandya, A miniaturized cylindrical dielectric resonator antenna array development for GPS / Wi-Fi / wireless LAN applications, E-Prime - Adv. Electr. Eng. Electron. Energy 2 (2022) 1-8. https://doi.org/10.1016/j.prime.2022.100044

[19] R. Parasher, D. Yadav, A. Saharia, Design and analysis of hybrid fractal frequency-band reconfigurable metamaterial antenna for multi-standard wireless applications, Results Eng. 26 (2025) 1-16. https://doi.org/10.1016/j.rineng.2025.104696

[20] L. Magthelin Therase, J. Thangappan, A novel microstrip antenna using circular ring defected ground structure for X band applications, Measurement 183 (2021) 1-8. https://doi.org/10.1016/j.measurement.2021.109768

[21] V.R. Samsingh, M. Kanagasabai, E.F. Sundarsingh, A novel metamaterial enhanced microwave testing system for bare PCB substrates using image rendering approach, IEEE Trans. Compon., Packag. Manuf. Technol. 7 (2017) 285-291. https://doi.org/10.1109/TCPMT.2016.2633974

[22] M. Rashedul Islam, M. Tariqul Islam, M.S. Soliman, B. Bais, M. Jit Singh, H. Alsaif, M. Shabiul Islam, Metamaterial sensor based on reflected mirror rectangular split ring resonator for the application of microwave sensing, Measurement 198 (2022) 1-11. https://doi.org/10.1016/j.measurement.2022.111416

[23] A.J. Al-Gburi, Z. Zakaria, N. Abd Rahman, S. Alam, M.A. Said, A compact and low-profile curve-feed complementary split-ring resonator microwave sensor for solid material detection, Micromachines 14 (2023) 1-21. https://doi.org/10.3390/mi14020384

[24] R.U. Tariq, M. Ye, Y. He, Comparative study of various microwave sensors in determining the thickness of dielectric materials, in: 2020 Int. Conf. Microw. Millim. Wave Technol., 2020: pp. 1-3. https://doi.org/10.1109/ICMMT49418.2020.9386628

[25] M.Z.B. Chowdhury, M.T. Islam, A. Alzamil, M.S. Soliman, M. Samsuzzaman, A tunable star-shaped highly sensitive microwave sensor for solid and liquid sensing, Alexandria Eng. J. 86 (2024) 644-662. https://doi.org/10.1016/j.aej.2023.12.001

[26] A. Armghan, T.M. Alanazi, A. Altaf, T. Haq, Characterization of dielectric substrates using dual-band microwave sensor, IEEE Access 9 (2021) 62779-62787. https://doi.org/10.1109/ACCESS.2021.3075246

[27] A. Raveendran, S. Raman, Low-cost multifunctional planar RF sensors for dielectric characterization and quality monitoring, IEEE Sens. J. 21 (2021) 24056-24065. https://doi.org/10.1109/JSEN.2021.3114257

[28] S. Alam, Z. Zakaria, I. Surjati, N.A. Shairi, M. Alaydrus, T. Firmansyah, Integrated microwave sensor and antenna sensor based on dual T-shaped resonator structures for contact and noncontact characterization of solid material, IEEE Sens. J. 23 (2023) 13010-13018. https://doi.org/10.1109/JSEN.2023.3273008

[29] T. Haq, C. Ruan, X. Zhang, S. Ullah, A.K. Fahad, W. He, Extremely sensitive microwave sensor for evaluation of dielectric characteristics of low-permittivity materials, Sensors 20 (2020) 1-17. https://doi.org/10.3390/s20071916

[30] S. Agarwal, M.C. Garg, Design of an accurate, planar, resonant microwave sensor for testing a wide range of liquid samples, Electronics 13 (2024) 1-10. https://doi.org/10.3390/electronics13224510

[31] Z. Li, Z. Meng, F. Fei, Z. Chen, Microwave cavity resonator sensor with an octagonal cross section for thickness measurement of coatings on carbon fibre composites, NDT E Int. 130 (2022) 1-9. https://doi.org/10.1016/j.ndteint.2022.102680

[32] S.M.A. Haque, M.T. Islam, A.G. Alharbi, A. Miah, S. Al Zafir, M. Samsuzzaman, Double stripe square enclosed SRR based triple band metamaterial absorber for materials variation and thickness sensing applications, Opt. Commun. 583 (2025) 1-12. https://doi.org/10.1016/j.optcom.2025.131740

[33] R. Srivastava, A. Vaishnav, S.N. Kale, Non-destructive methodology for crack detection using machine learning-assisted resonant sensor, Measurement 229 (2024) 1-8. https://doi.org/10.1016/j.measurement.2024.114429

[34] A. Tamer, F. Karadağ, E. Ünal, Y.I. Abdulkarim, L., Deng, O., Altintas, M., Bakır, M., & Karaaslan, M. (2023). Metamaterial-based sensor integrating a transmission line for the detection of branded and unbranded diesel fuel. Chem. Phys. Lett. 742 (2020) 1-7. https://doi.org/10.1016/j.cplett.2020.137169

[35] K. Karthikeyan, P.S.P. Anandh, M. Ramesh, V. Saran, Design and parameter optimization of microwave sensor for fat content detection in milk, AIP Conf. Proc. 3221 (2024) 1-10. https://doi.org/10.1063/5.0236788

[36] J.G.D. Oliveira, E.N.M.G. Pinto, V.P. Silva Neto, A.G. D'Assunção, CSRR-based microwave sensor for dielectric materials characterization applied to soil water content determination, Sensors 20 (2020) 1-16. https://doi.org/10.3390/s20010255

[37] E.L. Chuma, Y. Iano, G. Fontgalland, L.L.B. Roger, H. Loschi, PCB-integrated non-destructive microwave sensor for liquid dielectric spectroscopy based on planar metamaterial resonator, Sensors Actuators A Phys. 312 (2020) 1-6. https://doi.org/10.1016/j.sna.2020.112112

[38] Z. Abedeen, P. Agarwal, Microwave sensing technique based label-free and real-time planar glucose analyzer fabricated on FR4, Sensors Actuators A Phys. 279 (2018) 132-139. https://doi.org/10.1016/j.sna.2018.06.011

[39] A.E. Omer, G. Shaker, S. Safavi-Naeini, Portable radar-driven microwave sensor for intermittent glucose levels monitoring, IEEE Sensors Lett. 4 (2020) 1-4. https://doi.org/10.1109/LSENS.2020.2986208

[40] A.E. Omer, G. Shaker, S. Safavi-Naeini, H. Kokabi, G. Alquié, F. Deshours, R.M. Shubair, Low-cost portable microwave sensor for non-invasive monitoring of blood glucose level: Novel design utilizing a four-cell CSRR hexagonal configuration, Sci. Rep. 10 (2020) 1-20. https://doi.org/10.1038/s41598-020-72114-3

[41] M.M.Y.R. Riad, A.R. Eldamak, Coplanar waveguide based sensor using paper superstrate for non-invasive sweat monitoring, IEEE Access 8 (2020) 177757-177766. https://doi.org/10.1109/ACCESS.2020.3027033

[42] P.P. Kumar, S.R.P. Eedala, A. Yammanuru, A.N. Grace, A.C. Josephine Malathi, Non-invasive leukemia detection via CSRR sensor, J. Hazard. Mater. Adv. 18 (2025) 1-8. https://doi.org/10.1016/j.hazadv.2025.100654

[43] G. Gugliandolo, M. Latino, G. Campobello, Z. Marinkovic, G. Crupi, N. Donato, On the gas sensing properties of microwave transducers, in: 2020 55th Int. Sci. Conf. Information, Commun. Energy Syst. Technol., 2020: pp. 161-194. https://doi.org/10.1109/ICEST49890.2020.9232765

[44] Y.I. Abdulkarim, L. Deng, H. Luo, S. Huang, M. Karaaslan, O. Altıntaş, M. Bakır, F.F. Muhammadsharif, H. N.Awl, C. Sabah, K.S.L. Al-badri, Design and study of a metamaterial based sensor for the application of liquid chemicals detection, J. Mater. Res. Technol. 9 (2020) 10291-10304. https://doi.org/10.1016/j.jmrt.2020.07.034

[45] M.G. Rabbani, M.T. Islam, A. Hoque, B. Bais, S. Albadran, M.S. Islam, M.S. Soliman, Orthogonal centre ring field optimization triple-band metamaterial absorber with sensing application, Eng. Sci. Technol. an Int. J. 49 (2024) 1-16. https://doi.org/10.1016/j.jestch.2023.101588

[46] K.S.L. Al-badri, Electromagnetic broad band absorber based on metamaterial and lumped resistance, J. King Saud Univ. - Sci. 32 (2020) 501-506. https://doi.org/10.1016/j.jksus.2018.07.013

[47] M. Salah Uddin Afsar, M. Rashed Iqbal Faruque, M. Bellal Hossain, M. Tariqul Islam, SRR inspired modified psi shaped perfect metamaterial absorber for C-band application, J. Magn. Magn. Mater. 582 (2023) 1-12. https://doi.org/10.1016/j.jmmm.2023.171010

[48] A. Karimbu Vallappil, B.A. Khawaja, M.K.A. Rahim, M.N. Iqbal, H.T. Chattha, Metamaterial-inspired electrically compact triangular antennas loaded with CSRR and 3 × 3 cross-slots for 5G indoor distributed antenna systems, Micromachines 13 (2022) 1-16. https://doi.org/10.3390/mi13020198

[49] M.S.U. Afsar, M.R.I. Faruque, S. Abdullah, M.J. Hossain, Rotational symmetric solar system shaped triple band perfect metamaterial absorber for S-, C-, and X-band application, Sensors Actuators A Phys. 365 (2024) 1-9. https://doi.org/10.1016/j.sna.2023.114839

[50] R.R. Hasan, A.M. Saleque, A.B. Anwar, M.A. Rahman, Y.H. Tsang, Multiwalled carbon nanotube-based on-body patch antenna for detecting COVID-19-affected lungs, ACS Omega 7 (2022) 28265-28274. https://doi.org/10.1021/acsomega.2c02550

[51] M.J. Duarte, J.G.D. Júnior, V.P. da Silva Neto, A novel ultra-miniaturized 2.5-D frequency selective surface temperature sensor with linear frequency-temperature behavior, Sensors Actuators A Phys. 378 (2024) 1-8. https://doi.org/10.1016/j.sna.2024.115763

[52] A. Paghi, G. Trupiano, C. Puglia, H. Burgaud, G. De Simoni, A. Greco, F. Giazotto, Estimation of the FR4 microwave dielectric properties at cryogenic temperature for quantum-chip-interface PCBs design, IEEE Trans. Instrum. Meas. 73 (2024) 1-7. https://doi.org/10.1109/TIM.2024.3372217

[53] S.K. Bairappaka, A. Ghosh, O. Kaiwartya, M. Aljaidi, Y. Cao, R. Kharel, A novel design of broadband circularly polarized rectenna with enhanced gain for energy harvesting, IEEE Access 12 (2024) 65583-65594. https://doi.org/10.1109/ACCESS.2024.3397016

[54] H. Sun, J. Huang, Y. Wang, An omnidirectional rectenna array with an enhanced RF power distributing strategy for RF energy harvesting, IEEE Trans. Antennas Propag. 70 (2022) 4931-4936. https://doi.org/10.1109/TAP.2021.3138542

[55] H. Sun, S. Xiao, T. Li, G. Du, F. Zhang, H. Huang, G. Tian, An integrated microwave sensor system for defect evaluation of materials, IEEE Sens. J. 24 (2024) 41885-41893. https://doi.org/10.1109/JSEN.2024.3485104

[56] Y. Beria, G.S. Das, A. Buragohain, B.B. Chamuah, Highly sensitive miniaturized octagonal DS-CSRR sensor for permittivity measurement of liquid samples, IEEE Trans. Instrum. Meas. 72 (2023) 1-9. https://doi.org/10.1109/TIM.2023.3307172

[57] X. Han, K. Liu, S. Zhang, P. Peng, C. Fu, L. Qiao, Z. Ma, CSRR metamaterial microwave sensor for measuring dielectric constants of solids and liquids, IEEE Sens. J. 24 (2024) 14167-14176. https://doi.org/10.1109/JSEN.2024.3373755

[58] Y. Cao, C. Ruan, K. Chen, X. Zhang, Research on a high-sensitivity asymmetric metamaterial structure and its application as a microwave sensor, Sci. Rep. 12 (2022) 1-20. https://doi.org/10.1038/s41598-022-05255-2

[59] A.M. Gargari, L. Markley, Wireless high sensitivity low-profile permittivity sensor using resonant absorbing surfaces, in: 2020 IEEE Int. Symp. Antennas Propag. North Am. Radio Sci. Meet., 2020: pp. 897-898. https://doi.org/10.1109/IEEECONF35879.2020.9330490

[60] A.P. Viswanathan, R. Moolat, M. Mani, S. VA, M. Pezholil, A simple electrically small microwave sensor based on complementary asymmetric single split resonator for dielectric characterization of solids and liquids, Int. J. RF Microw. Comput. Eng. 30 (2020) 1-13. https://doi.org/10.1002/mmce.22462

[61] A. Javed, A. Arif, M. Zubair, M.Q. Mehmood, K. Riaz, A low-cost multiple complementary split-ring resonator-based microwave sensor for contactless dielectric characterization of liquids, IEEE Sens. J. 20 (2020) 11326-11334. https://doi.org/10.1109/JSEN.2020.2998004

[62] F. Jafari, M.R. Malekshahi, A Low-Cost Microwave Microfluidic Sensor Based on Planar Ring Resonator, IEEE Sens. J. 23 (2023) 21070-21077. https://doi.org/10.1109/JSEN.2023.3301813

[63] G. Gugliandolo, G. Vermiglio, G. Cutroneo, G. Campobello, G. Crupi, N. Donato, Development, characterization, and circuit modeling of inkjet-printed coupled ring resonators for application in biological samples, IEEE Trans. Instrum. Meas. 72 (2023) 1-10. https://doi.org/10.1109/TIM.2023.3271010

[64] F. Jafari, M.R. Malekshahi, A new technique of using a neural network classifier in microwave microfluidic sensors for purity detection of liquids, IEEE Sens. J. 24 (2024) 25662-25669. https://doi.org/10.1109/JSEN.2024.3418871

[65] K. K., R. S., B. S., Design and analysis of non-invasive rf microwave based sensor for water quality monitoring, in: 2025 Int. Conf. Vis. Anal. Data Vis., 2025: pp. 476-479. https://doi.org/10.1109/ICVADV63329.2025.10961883

[66] S.M.A. Haque, M.T. Islam, I. Hossain, F. Alorifi, Z.A. Shamsan, K. Almuhanna, S.K.A. Rahim, M. Samsuzzaman, Left-handed metamaterial based on circular split ring (CSRR) resonator for microwave sensing Application, Opt. Mater. 152 (2024) 1-13. https://doi.org/10.1016/j.optmat.2024.115480

[67] M.H. Bhatti, M.A. Jabbar, M.A. Khan, Y. Massoud, Low-cost microwave sensor for characterization and adulteration detection in edible oil, Appl. Sci. 12 (2022) 1-9. https://doi.org/10.3390/app12178665

[68] R. Mansour, S. Rioual, B. Lescop, P. Talbot, M. Abboud, W. Farah, G. Tanné, Development of a resonant microwave sensor for sediment density characterization, Sensors 20 (2020) 1-13. https://doi.org/10.3390/s20041058

[69] X. Fu, J. Wu, X. Wang, X. Gu, C. Wang, Y. Wu, J. Liang, NSRR microwave sensor based on pll technology for glucose detection, IEEE Trans. Microw. Theory Tech. 71 (2023) 763-772. https://doi.org/10.1109/TMTT.2022.3193696

[70] A. Nella, R.W. Aldhaheri, J.B. Kamili, N.M. Sobahi, A non-invasive method of glucose monitoring using FR4 material-based microwave antenna sensor, 30 (2023) 1-13. https://doi.org/10.1515/secm-2022-0187

[71] S. Hosseinzadeh, M. Pakvarjouy, Microwave-based single port glucose sensor arranging interdigital and complementary split ring resonator, Measurement 246 (2025) 1-8. https://doi.org/10.1016/j.measurement.2025.116695

[72] A.H. Omidvar, A. Amanati Shahri, A.L. Serrano, J. Gruber, G. Pamplona Rehder, A highly sensitive molecularly imprinted polymer (MIP)-coated microwave glucose sensor, Sensors 22 (2022) 1-15. https://doi.org/10.3390/s22228648

[73] M.I.O. Souza, A.F.D. Mota, V.M. Pepino, J.P. Carmo, B.-H. V Borges, Multi-purpose microwave biosensor based on signal encoding technique and microfluidics for improved sensitivity, IEEE Sens. J. 21 (2021) 4571-4581. https://doi.org/10.1109/JSEN.2020.3033970

[74] W. Karasaeng, J. Nualkham, C. Summatta, S. Sonasang, Measurement of soil moisture using microwave sensors based on bsf coupled lines, Eng. Proc. 58 (2023) 1-7. https://doi.org/10.3390/ecsa-10-16029

[75] K. Zhang, B. Gao, J. Wen, X. Gong, P. Wang, L. Tong, Design and experiment of a hollow structure microwave humidity sensor, in: 2021 IEEE Int. Geosci. Remote Sens. Symp. IGARSS, 2021: pp. 8245-8248. https://doi.org/10.1109/IGARSS47720.2021.9554520

[76] J. Bourely, L. De Sousa, N. Fumeaux, O. Vorobyov, C. Beyer, D. Briand, Biodegradable materials as sensitive coatings for humidity sensing in S-band microwave frequencies, Micro Nano Eng. 19 (2023) 1-8. https://doi.org/10.1016/j.mne.2023.100185

[77] R.K. Abdulsattar, T.A. Elwi, Z.A. Abdul Hassain, A new microwave sensor based on the moore fractal structure to detect water content in crude oil, Sensors 21 (2021) 1-18. https://doi.org/10.3390/s21217143

[78] R.K. Abdulsattar, M. Alibakhshikenari, B.S. Virdee, R. Sharma, T.A. Elwi, L. Kouhalvandi, Z.A. Abdul Hassain, S. Mansoor Ali, N.T. Tokan, P. Livreri, F. Falcone, E. Limiti, Optical-microwave sensor for real-time measurement of water contamination in oil derivatives, AEU - Int. J. Electron. Commun. 170 (2023) 1-10. https://doi.org/10.1016/j.aeue.2023.154798

[79] Q. Zhang, Z. Kou, J. Yang, Y. Li, Z. Yin, G. Deng, Highly sensitive and robust metasurface-inspired microfluidic sensor for oil detection, IEEE Sens. J. 24 (2024) 40847-40854. https://doi.org/10.1109/JSEN.2024.3488812

[80] M.G. Rabbani, M.T. Islam, M. Moniruzzaman, S. Alamri, A.A.M. Rahman, A.M. Moubark, M.S. Islam, M.S. Soliman, Dumbbell shaped structure loaded modified circular ring resonator based perfect metamaterial absorber for S, X and Ku band microwave sensing applications, Sci. Rep. 14 (2024) 1-21. https://doi.org/10.1038/s41598-024-56251-7

[81] O.S. Lateef, M. Al-Badri, K.S.L. Al-badri, S.A. Mohammed, Polarization-insensitive Archimedes'-spiral-shaped ultrathin metamaterial absorbers for microwave sensing application, Sci. Rep. 13 (2023) 1-9. https://doi.org/10.1038/s41598-023-46363-x

[82] M.N. Rahman, S.M.T. Islam, M. Maniruzzaman, M.T. Islam, Salinity and sugar content detection by wideband microwave-based patch antenna sensor, J. Electr. Comput. Eng. 2025 (2025) 1-10. https://doi.org/10.1155/jece/4813865

[83] M.R. Islam, M.T. Islam, M.S. M., B. Bais, S.H.A. Almalki, H. Alsaif, M.S. Islam, Metamaterial sensor based on rectangular enclosed adjacent triple circle split ring resonator with good quality factor for microwave sensing application, Sci. Rep. 12 (2022) 1-18. https://doi.org/10.1038/s41598-022-10729-4

[84] W. Karasaeng, J. Nualkham, C. Chomchalao, T. Pongthavornkamol, D. Wongpatsa, S. Sonasang, Development of microwave sensor using coupled lines for pickled fish assessment, in: 2024 8th Int. Conf. Inf. Technol., 2024: pp. 520-524. https://doi.org/10.1109/InCIT63192.2024.10810631

[85] R. Kumari, P.N. Patel, R. Yadav, An eng-inspired microwave sensor and functional technique for label-free detection of aspergillus niger, IEEE Sens. J. 18 (2018) 3932-3939. https://doi.org/10.1109/JSEN.2018.2822306

[86] S. Hosseinzadeh, M. Yousefi, A permittivity and conductivity sensor based on microstrip transmission line with defective ground for detection of urea in saline, IEEE Trans. Instrum. Meas. 73 (2024) 1-11. https://doi.org/10.1109/TIM.2024.3385840

[87] R.K. Abdulsattar, M.T.S. Al-Kaltakchi, I.A. Mocanu, A.A. Al-Behadili, Z.A. Abdu Hassain, Study on sensing urine concentrations in water using a microwave sensor based on hilbert structure, Sensors 24 (2024) 1-19. https://doi.org/10.3390/s24113528

[88] K. Wadhwani, S. Hussaini, A. Syed, Real-time qualitative and quantitative analysis of Saccharides using CSRR based RF sensor, in: 2022 IEEE Sensors, 2022: pp. 1-4. https://doi.org/10.1109/SENSORS52175.2022.9967068

[89] R. Raju, G.E. Bridges, A compact wireless passive harmonic sensor for packaged food quality monitoring, IEEE Trans. Microw. Theory Tech. 70 (2022) 2389-2397. https://doi.org/10.1109/TMTT.2022.3142063

[90] A. Kundu, S.M.R.H. Shawon, S. Kapali, A. Helal, K. Ali, Hairpin resonator-based microwave sensor for detection of nitrogenous fertilizers in soil and water, IEEE Sens. J. 24 (2024) 27436-27445. https://doi.org/10.1109/JSEN.2024.3426559

[91] K. Wadhwani, S. Hussaini, A. Mazumder, A. Syed, Solvent-based optimization of CSRR and IDC RF bio-sensors, IEEE Sens. J. 22 (2022) 5651-5661. https://doi.org/10.1109/JSEN.2022.3148349

[92] S.A. Elbendera, A. Allam, A. Magdy, R.K. Pokharel, A.B. Abdel-Rahman, H-slot sensing antenna system for termites detection, IEEE Sens. J. 25 (2025) 4494-4502. https://doi.org/10.1109/JSEN.2024.3512359

[93] Y. Gu, D. Kendig, M. Shakouri, D.W. van der Weide, Dual mode split ring resonator sensing and hyperthermia array for skin, IEEE J. Electromagn. RF Microwaves Med. Biol. 8 (2024) 206-212. https://doi.org/10.1109/JERM.2024.3373537

[94] T. Singh, P.K. Mishra, A. Pal, P.K. Shukla, P. Sharma, V.S. Tripathi, Defected ground circular shaped non-invasive microwave sensor for deep brain temperature measurements, in: 2023 IEEE Microwaves, Antennas, Propag. Conf., 2023: pp. 1-6. https://doi.org/10.1109/MAPCON58678.2023.10463875

[95] A.K. Vallappil, M.K.A. Rahim, B.A. Khawaja, M.N. Iqbal, A miniaturized metamaterial-loaded switched-beam antenna array system with enhanced bandwidth for 5G applications, IEEE Access 12 (2024) 6684-6697. https://doi.org/10.1109/ACCESS.2024.3351475

[96] Ismatul N Idrus, Mohammad R I Faruque, Sabirin Abdullah, Mohammad T Islam, Mayeen U Khandaker, Jamel Nebhen, An octagonal split ring resonator-based double negative metamaterial for S-, X- and Ku-band applications, Proc. Inst. Mech. Eng. Part L J. Mater. Des. Appl. 236 (2021) 2269-2280. https://doi.org/10.1177/14644207211017155

[97] F. Vogelsang, J. Bott, D. Starke, M. Hamme, B. Sievert, H. Rücker, N. Pohl, A fully integrated 0.48 THz FMCW radar sensor in a SiGe technology, IEEE J. Microwaves 5 (2025) 572-582. https://doi.org/10.1109/JMW.2025.3553681

[98] M.S. Yahya, S. Soeung, S.K.A. Rahim, U. Musa, S.S.B. Hashwan, M.A. Haque, Machine learning-optimized compact frequency reconfigurable antenna with rssi enhancement for long-range applications, IEEE Access 12 (2024) 10970-10987. https://doi.org/10.1109/ACCESS.2024.3355145

[99] S. Aksimsek, E.A. Ozek, H. Ozpinar, Laser ablation cutting-based metal patterning technique enabling 3D-printed broadband antennas for sub-6 GHz wireless communications applications, IEEE Trans. Components, Packag. Manuf. Technol. 11 (2021) 1506-1513. https://doi.org/10.1109/TCPMT.2021.3100713

[100] S Kannadhasan, R Nagarajan, Development of an H-shaped antenna with FR4 for 1-10 GHz wireless communications, Text. Res. J. 91 (2021) 1687-1697. https://doi.org/10.1177/00405175211003167

[101] H. Mahfoudi, M. Tellache, H. Takhedmit, A wideband rectifier array on dual-polarized differential-feed fractal slotted ground antenna for RF energy harvesting, Int. J. RF Microw. Comput. Eng. 29 (2019) 1-9. https://doi.org/10.1002/mmce.21775

[102] C. Sandhya, K.K. Kumar, A.M. Jaleel, Broadband integrated rectenna using differential rectifier and hybrid coupler, IET Microwaves, Antennas Propag. 14 (2020) 1384-1395. https://doi.org/10.1049/iet-map.2019.1127

[103] A.K. Jha, A. Lamecki, M. Mrozowski, M. Bozzi, A highly sensitive planar microwave sensor for detecting direction and angle of rotation, IEEE Trans. Microw. Theory Tech. 68 (2020) 1598-1609. https://doi.org/10.1109/TMTT.2019.2957369

[104] Y. Beria, A. Borah, G.S. Das, A. Buragohain, P.P. Kalita, T. Doloi, Design and study of a highly sensitive CSRR-based microwave angular displacement sensor, in: 2023 IEEE Silchar Subsect. Conf., 2023: pp. 1-4. https://doi.org/10.1109/SILCON59133.2023.10404848

[105] Y. Su, M. Pellaton, C. Affolderbach, G. Mileti, A.K. Skrivervik, Mode suppression and homogeneous field bandwidth enhancement of a tuning-free micro-loop-gap resonator using FR4 for chip-scale rubidium clock, IEEE Trans. Microw. Theory Tech. 72 (2024) 3711-3721. https://doi.org/10.1109/TMTT.2023.3326482

[106] J.-K. Wu, W. Yue, K. Gao, S. von Gratowski, X.-F. Gu, L. Pan, N.-Y. Kim, J.-G. Liang, Microwave-sensor array for decoupling detection of distance, shape, dielectric, and morphology, IEEE Trans. Instrum. Meas. 72 (2023) 1-10. https://doi.org/10.1109/TIM.2023.3300411

[107] J.P. Uwiringiyimana, U. Khayam, Suwarno, G.C. Montanari, Design and implementation of ultra-wide band antenna for partial discharge detection in high voltage power equipment, IEEE Access 10 (2022) 10983-10994. https://doi.org/10.1109/ACCESS.2022.3144416

[108] S. V Yadav, A. Chittora, A noncontact planar microstrip sensor for power line fault detection, IEEE Sensors Lett. 7 (2023) 1-4. https://doi.org/10.1109/LSENS.2023.3266160

[109] M. Elgeziry, F. Costa, P. Vélez, F. Paredes, F. Martín, S. Genovesi, Toward a low-cost portable reader for reflective-mode microwave sensors, IEEE Trans. Instrum. Meas. 73 (2024) 1-9. https://doi.org/10.1109/TIM.2023.3338651

Chapter 13

Inorganic Flame Retardants and Mineral Compounds

T.S. Motsoeneng[1,*], S. Khumalo[1], S.M. Mohomane[1], T.C. Mokhena[2]

[1]Department of Chemistry, University of Zululand, 1 Main Road, Vulindlela, KwaDlangwezwa, 9886, South Africa

[2]DSTI/Nanotechnology Innovation Centre (NIC), Advanced Materials Division, Mintek, Randburg, South Africa

motsoenengts@gmail.com

Abstract

Flame- or fire-retardant materials are crucial and valuable when enhancing the flame-retardant properties of materials or compositions that are prone to fire and/or easily combustible. Due to their nature, flame retardants are available in various forms and compositions spanning from organic to inorganic compounds. These compounds can be used individually or in combination as hybrid materials to enhance the flame-retardancy of other materials that are physically and chemically susceptible to fire. When used as hybrid materials, these materials exhibit significantly improved flame-retardant properties. Flame-retardants are widely used in various fields to enhance fire resistance across a broad range of materials and applications. Fire-resistant materials are primarily used in synthetic and/or natural polymers to produce composites with enhanced flame resistance. There is a plethora of characterisation techniques and tests that can be used to determine the fire-resistance properties of a composition incorporating a fire-retardant material. This chapter covers various aspects of flame-resistant materials, including their chemical and physical properties, recent industrial applications, and testing methodologies.

Keywords

Fire Retardants, Endothermic Decomposition, Inorganic Retardants, Mineral Fillers, Layered Silicates, Flammability

Contents

Materials Research Foundations 190 (2026) 360-373 https://doi.org/10.21741/9781644904077-13

1. Introduction

Mineral chemicals and inorganic flame retardants are essential for enhancing the fire resistance of polymeric materials used in textiles, packaging, electronics, transportation, and construction. These chemicals reduce flammability, delay ignition, and limit smoke production. Aluminium hydroxide (ATH), magnesium hydroxide (MDH), ammonium polyphosphate, antimony trioxide, and various mineral fillers, such as clay, talc, and intumescent compounds, are examples of typical inorganic flame retardants [1]. These materials are used in a wide range of industries that are mandated to adhere to strict fire safety standards and to address environmental and health problems associated with conventional halogenated fire retardants [2].

The usage of flame retardants may be traced back to ancient Egypt, where alum was used to make textiles less flammable. The demand for efficient flame-retardant systems increased in the 20th century as synthetic polymers gained popularity. Initially, halogenated organic compounds were primarily considered in industries due to their efficacy; however, growing concerns about their toxicity and environmental persistence prompted further investigation into safer alternatives [3]. Inorganic flame retardants have gained popularity since the 1970s as environmentally benign, non-toxic, and thermally stable alternatives. The functional potential of mineral-based additions has been further enhanced by breakthroughs in nanotechnology and materials science [4].

Due to their natural flammability, polymers pose significant fire risks. Inorganic flame retardants reduce these hazards by absorbing heat, diluting flammable gases, or encouraging char formation. MDH and ATH decompose endothermically, thereby reducing the combustion rate by releasing water and cooling the material surface [5]. Additionally, these chemicals decrease harmful gas emissions and improve smoke suppression. The inclusion of inorganic flame retardants is now necessary for the manufacture of safer, more environmentally friendly polymer-based products as global fire safety regulations get stricter [6].

Materials Research Foundations 190 (2026) 360-373
https://doi.org/10.21741/9781644904077-13

2. Flame-retardant mechanisms of inorganic compounds

Inorganic flame retardants and mineral-based additives use a variety of physical and chemical processes to slow down or prevent the combustion of polymeric materials. Endothermic decomposition, char formation, gas-phase dilution, and the physical barrier effect are the main processes. These processes typically operate in concert to enhance the fire resistance of materials without generating hazardous byproducts. Figure 1 illustrates various retardant mechanisms that typically occur during the burning of flame-retardant-containing materials.

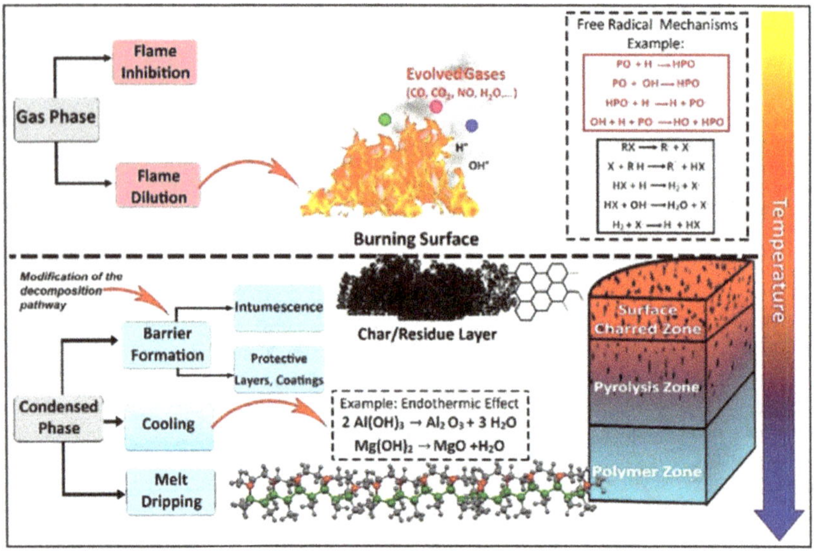

Figure 1. Different potential mechanisms for the protection of flammable materials when embedded with flame retardants [3]

Endothermic decomposition

Endothermic breakdown, in which the additive absorbs heat from the environment during thermal breakdown, is one of the most prevalent processes in inorganic flame retardants. When subjected to high temperatures (over 200°C for ATH and 300°C for MDH), chemicals like aluminium hydroxide (ATH) and magnesium hydroxide (MDH) break down, absorbing heat and releasing water vapour [7].

$2Al(OH)_3 \rightarrow Al_2O_3 + 3H_2O$ (at ~200°C)

$Mg(OH)_2 \rightarrow MgO + H_2O$ (at ~300°C)

This endothermic process involves a reduction in the temperature of the burning material, which slows the pyrolysis of the polymer and delays ignition and flame propagation.

Materials Research Foundations 190 (2026) 360-373 https://doi.org/10.21741/9781644904077-13

Char formation

The formation of char involves the creation of a carbonaceous layer on the surface polymer as it burns. Certain mineral-based retardants, especially those used in intumescent systems, act as thermal insulators, shielding the underlying material from further heat exposure and inhibiting the release of combustible volatiles. Char-forming substances strengthen the structure and reduce the amount of fuel available to the fire [8].

Gas-phase dilution

Many inorganic flame retardants produce inert gases during decomposition, such as water vapour, ammonia, or nitrogen, which reduces the concentration of combustible gases in the combustion zone. By decreasing the partial pressure of oxygen and combustible vapours and disrupting the chain reactions involved in flame propagation, this gas-phase dilution reduces the likelihood of maintaining combustion. The release of H_2O from ATH or NH_3 from ammonium polyphosphate dilutes flammable pyrolysis products and cools the flame [9].

The effect of a physical barrier

Some mineral fillers, such as clays and talc, and intumescent coatings create a protective barrier layer when heated. This barrier prevents volatile degradation products from flowing, thereby preventing oxygen and heat from entering the polymer. It enhances the integrity of the surface structure. Layered silicates, such as montmorillonite clay, may also produce a "labyrinth effect", which slows the diffusion of gases and increases thermal stability [10].

Compared to halogenated substances, inorganic flame retardants are less harmful and more environmentally friendly because they function mostly through physical processes. The fire resistance of polymeric materials is greatly enhanced by the integration of endothermic cooling, protective char layers, gas dilution and physical barriers. These methods are crucial for ensuring compliance with current fire safety standards while also fostering the development of environmentally friendly materials [11].

3. Classification of inorganic flame retardants

Metal hydroxides are compounds that have metal cations that are bound with hydroxide (OH^-) groups. They do not dissolve in water and are effective at neutralising acids, as they predominantly exhibit a basic nature and are used in flame retardants. Metal oxides form when a metal reacts with oxygen; they have high melting points and do not dissolve in water. They are also used as catalysts and pigments. Metal borates are compounds that comprise a metal cation and a borate anion, exhibiting very high thermal stability and serving as flame retardants. Clay is a natural hydrous aluminium silicate with a layered structure. Additionally, it has a very high surface area and cation-exchange capacity, and is used as a filler in plastics and rubbers. A synergistic combination of two or more substances that work together to produce a result greater than the total of their individual results, which is observed in flame retardant systems, poly-additives and in formulations. Figure 2 illustrates the different commonly used flame retardants in various applications.

Materials Research Foundations 190 (2026) 360-373 https://doi.org/10.21741/9781644904077-13

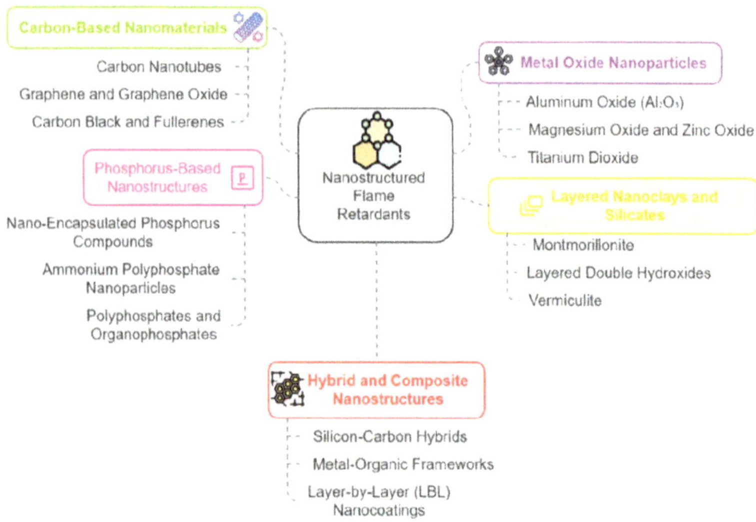

Figure 2. Classification of nanostructured flame-retardants and their major types [3]

3.1 Aluminium trihydroxide (ATH)

Aluminium trihydroxide (ATH) is the most used FR filler in commodity polymers. Under the FR mechanism, ATH undergoes a single-step endothermic thermal breakdown at 200°C, producing water vapour and leaving a stable aluminium oxide (Al_2O_3) residue. During polymer combustion, water vapour plays an important role in cooling the system by acting as a diluent, thereby forming an aluminium oxide residue that serves as a protective layer [13,14].

The following reaction illustrates the equation for ATH decomposition:

$$2Al(OH)_3 \rightarrow Al_2O_3 + 3H_2O$$

$Al(OH)_3$ decomposes to alumina (Al_2O_3) and water vapour, with the released water serving as a gaseous byproduct that absorbs heat (endothermic cooling). An alumina barrier layer acts as a protective barrier, being chemically inert and thermally stable. Alumina can retard the decomposition of material beneath, in a flame thrower; it also suppresses smoke and flame spread. Water vapour can decrease the excess oxygen supply to the combustible substrate, which is why alumina is crucial [15,16]. Moreover, thermal stability is crucial for ATH, as it determines its performance when used as a polymer additive. ATH provides an earlier decomposition temperature for ideal polymers, particularly for thermoplastics that require high processing temperatures and for thermally stable thermosets [13,17].

3.2 Magnesium hydroxide (MH)

The endothermic degradation, which starts at around 330°C for $Mg(OH)_2$ (as compared to around 230°C for ATH), extracts heat from the substrate, retarding the rate of thermal degradation of the

Materials Research Foundations 190 (2026) 360-373 https://doi.org/10.21741/9781644904077-13

plastic, thus being particularly well-suited for use as a flame retardant in higher temperature polymer systems [18,19]. The MH decomposition process occurs as per the reaction below:

$Mg(OH)_2 \rightarrow MgO + H_2O$

The relatively high heat capacities of both magnesium hydroxide and the decomposition products formed upon its decomposition reduce the thermal energy available to degrade the plastic. The decomposition products provide increased insulation of the plastic from the heat source by forming char [20]. The high filler content typically associated with magnesium hydroxide-treated plastics serves as a solid-phase diluent.

3.3 Antimony trioxide (AT)

Antimony trioxide acts as a synergist with halogenated flame retardants, significantly enhancing their performance by promoting char formation and interfering with the free-radical chemistry of the flame. Its role as synergist is when, combined with halogenated flame retardant, antimony trioxide forms volatile antimony halides during combustion according to the following decomposition reaction [6,21]:

$Sb_2O_3 + 6HX \rightarrow 2SbX_3 + 3H_2O$, where X can be Cl or Br.

The halides act in the gas phase to chemically suppress the flame, enhancing the flame-inhibiting action of halogens. A char formation promotes charring in some matrices, such as polyvinyl chloride (PVC), helping to retain structural integrity during fire, slow heat, and mass transfer, and also insulating the underlying material [22]. A free radical capture during the combustion, flame propagation relies heavily on hydrogen (H•) and hydroxyl (OH•) radicals, reaction chain:

$SbX_3 + H\bullet \rightarrow HX + SbX_2$

This reduces heat feedback to the material, lowering the burning rate and the halogen is recycled in the process, increasing efficiency.

3.4 Zinc compounds (zinc borate, zinc hydroxystannate)

Zinc borate is widely used as an intumescent flame retardant in a variety of materials. Zinc borate will thermally degrade to a char-like, insulating protective layer when exposed to high heat. Intumescence char is a critical component of its intumescent nature [23]. Thermal decomposition upon heating zinc borate breaks down into water and boric oxide (B_2O_3). Boric oxide then combines with the other ingredients in the material to form a char layer. The released water vapour helps cool the substrate by carrying away heat, and boron trioxide helps form a solid, protective char [7,24]. This char is foamy and expands when exposed to further heat, forming a thick, insulating layer. This process significantly increases the volume of the material while reducing its thermal conductivity. This expanded char provides a barrier between the heat source and the underlying material, effectively slowing down the rate of temperature increase and thus delaying combustion [7,25]. The resultant char layer is low in thermal conductivity and provides a physical barrier to the spread of flames and heat. This barrier helps protect the underlying material from further degradation and combustion. The compound zinc hydroxy stannate, $ZnSn(OH)_6$, is commonly used in flame-retardant formulations, particularly for suppressing smoke during combustion [26]. It does so predominantly by regulating the formation of smoke while reducing the generation of toxic fumes. Zinc hydroxy stannate can scavenge free radicals (such as the hydroxyl radical, OH) generated during combustion. These radicals lead to the production of

Materials Research Foundations 190 (2026) 360-373 https://doi.org/10.21741/9781644904077-13

gaseous byproducts and smoke. During thermal degradation, zinc hydroxystannate releases zinc oxide (ZnO) and stannic oxide (SnO_2). These oxides can help trap and neutralise toxic gases, such as carbon monoxide (CO) and hydrogen cyanide (HCN), which are common byproducts of combustion, thereby further reducing smoke [27,28]. Unlike halogenated fire-suppressant chemicals, zinc hydroxystannate does not create harmful gases. This makes it a safer alternative to conventional smoke-suppression agents. In addition, its presence can minimise the formation of thick smoke plumes that can obscure visibility [11].

3.5 Clay and layered silicates

Organo-modified montmorillonite is a smectite-type clay composed of layered silicates, modified with organic cations to enhance its compatibility with organic polymers. Barrier effects greatly enhance the moisture, gas, and chemical barrier properties in polymer matrices, likely due to reduced oxygen and water vapour transmission rates, which are essential for coatings and protective films. To enhance mechanical properties, the high aspect ratio and stiffness of the silicate layers improve tensile strength, modulus, and impact resistance. Organo-modified montmorillonite may serve as a nucleating agent that affects the crystallinity of the polymer, thereby improving its mechanical behaviour, resulting in a material that is stronger, stiffer, and tougher [29,30].

3.6 Emerging mineral fillers (expandable graphite, nano-TiO_2)

The advancement of sophisticated flame-retardant systems has increasingly shifted towards environmentally sustainable, halogen-free solutions that not only mitigate flammability but also improve mechanical, thermal, and barrier characteristics. Mineral-based nanofillers, such as expandable graphite and nanotitanium dioxide, have attracted significant attention due to their multifunctional properties in polymer composites [31,32]. Graphite, when treated with an intercalating agent, forms an intumescent char upon exposure to heat, which expands. Nano-titanium dioxide is primarily utilised in organic nanoparticles, which are recognised for their thermal stability, photocatalytic properties, and UV resistance. When combined with traditional flame retardants, graphite and titanium dioxide provide synergistic enhancements that contribute to mechanical reinforcement, thermal insulation, smoke suppression, and char integrity. The integration of expandable graphite and nano-titanium dioxide presents a promising avenue for developing high-performance, multifunctional flame-retardant composites [7,9].

4. Synergistic and composite systems

The use of synergistic flame-retardant systems has gained popularity in contemporary fire safety engineering, enabling the achievement of demanding fire performance criteria while preserving material properties. These systems combine mineral flame retardants (such as aluminium hydroxide, magnesium hydroxide, and various clays) with other flame-retardant families, including silicones, organophosphorus compounds, and intumescent additives, to provide greater fire resistance through complementary mechanisms [33].

4.1 Organophosphorus compounds combined with mineral flame retardants

Organophosphorus flame retardants are highly effective at preventing flame propagation in both the gaseous and condensed phases. When combined with mineral-based retardants, phosphorus synergistically promotes char formation, while minerals (such as ATH or MDH) cool the system

Materials Research Foundations 190 (2026) 360-373 | https://doi.org/10.21741/9781644904077-13

via endothermic breakdown. Dehydration processes are catalysed by acids produced by phosphorus, which improves the stability and integrity of the char. The release of inert gas from minerals helps dilute flammable volatiles [34]. Organophosphorus compounds can be used in polyolefins, polyamides, and polyesters for applications such as electronic housings, automotive parts, and construction panels.

4.2 Silicone compounds plus mineral flame retardants

Silicone-based additives (such as polysiloxanes or silsesquioxanes) are valued for their thermal stability, low toxicity and barrier-forming capabilities. At high temperatures, silicones create a ceramic-like silica coating on the surface, serving as a physical barrier. The structural integrity of the char is improved, and the rate of heat release is slowed by mineral fillers such as ATH or nanoclays [35]. Due to their hydrophobicity, silicones are more compatible with polymers and are used in a variety of applications, including transportation components that require high thermal resistance, elastomers, and wire and cable insulation.

4.3 Intumescent systems plus mineral flame retardants

Intumescent flame retardants (IFRs) are composed of a carbon source (such as pentaerythritol), an acid source (such as ammonium polyphosphate) and a blowing agent (such as melamine).

4.4 Synergistic impacts

The IFR creates a larger carbonaceous foam that insulates the substance. The intumescent mechanical strength char is increased by mineral fillers. Endothermic minerals like MDH help reduce surface temperatures and suppress smoke. Combining mineral flame retardants with silicone, organophosphorus compounds, or intumescent systems yields synergistic effects that include enhanced char strength and thermal insulation, lower peak heat release rates (PHRR), reduced smoke density and toxic gas emissions, and maintenance or improvement of mechanical and processing properties [36]. These beneficial properties resulting from hybrid materials make the IFR suitable for use in coatings, textiles, cable jackets, and foams for fire and thermal protection.

5. Applications in polymers and composites

In various polymers and composite systems, inorganic flame retardants are often utilised to enhance fire safety without compromising structural integrity. The following is an overview of their use in rubber, epoxy resins, and polyethylene, along with the major advantages and disadvantages of each system.

5.1 Polyethylenes (PEs)

PEs, due to their efficiency and non-toxicity, aluminium hydroxide (ATH) and magnesium hydroxide (MDH) are the most widely used flame retardants for polyethylenes [37].

Improvements to the property

Improved fire resistance during burning, combined with high loadings (up to 60 wt%), significantly reduces the peak heat release rate and smoke production. No dripping works well with synergists, such as organophosphorus compounds or nanoclays, to increase efficiency and reduce loading levels [38].

Materials Research Foundations 190 (2026) 360-373 https://doi.org/10.21741/9781644904077-13

Trade-offs in performance

Increasing the filler content often results in a decline in mechanical properties, such as tensile strength and elongation at break. Due to higher viscosity and low melt flow, processability may be impacted. Wire and cable insulation, pipes, films and geomembranes are among the applications [39].

5.2 Epoxy resins

The use of intumescent systems in conjunction with mineral additions, such as ATH, MDH, or phosphorus-containing nanoclays, is beneficial for epoxies. Property improvements, intumescent synergy, enhance char creation and barrier effects. Increased thermal stability and decreased flammability, frequently earning UL-94 V-0 ratings. The heat release rate can be reduced while maintaining mechanical integrity by adding phosphorus- or silicone-modified fillers [40].

Performance trade-offs

The inclusion of fillers can lead to brittleness and reduced transparency. To preserve the homogeneity and mechanical properties of the resin, thorough dispersion methods are necessary. These resins can be applied in flame-retardant coatings, structural composites, electrical encapsulation, and printed circuit boards [41].

5.3 Rubber

Clay-based retardants, ATH, and MDH are frequently employed in rubber systems, sometimes in conjunction with phosphorus or silicone chemicals.

Improvements to the property

Enhance thermal stability and flame resistance while preserving flexibility. Silicone and intumescent additives produce a ceramic-like char, which enhances high-temperature performance. Reduction of toxic gases and smoke suppression, both of which are especially important in the transportation and construction industries.

Trade-offs in performance

Excessive mineral filler loading might compromise tear strength, elastic recovery, and abrasion resistance. Without surface treatment, achieving compatibility and dispersion can be difficult. Automotive seals, gaskets, flame-retardant foams, and cable sheathing are examples of uses.

By utilising physical and chemical processes, the integration of inorganic and synergistic flame-retardant systems into polyethylene, epoxy resins, and rubber enhances fire performance. However, material optimisation, compatibilisers, or nanocomposite design techniques must be used to address compromises such as lower mechanical strength or processability. These systems pave the way for the safe application of polymers in fire-sensitive applications while advancing toward more environmentally friendly, halogen-free flame-retardant solutions [42].

6. Characterisation and testing methods

Flame-retardant-containing composites are characterised using several techniques and tested using different methods to evaluate their flame resistance and establish the decomposition mechanism of the underlying materials. Thermal gravimetric analysis (TGA) is commonly used to determine the char residue produced when a material burns. Additionally, in TGA, degradation kinetics can

Materials Research Foundations 190 (2026) 360-373 https://doi.org/10.21741/9781644904077-13

be evaluated under various conditions, including different temperatures, heating rates, and atmospheres. Moreover, differential scanning calorimetry (DSC) and dynamic mechanical analysis (DMA) can be used to determine the thermal properties of fire-retardant-containing species under external stress, providing a comprehensive understanding of their behaviour during dynamic combustion. The limiting oxygen index (LOI) and the horizontal or vertical burning test (UL94 HB) are the most commonly used methods for evaluating the flammability of materials [43-45].

7. Conclusion and future perspectives

This chapter provides a summary of the inorganic and mineral-based FRs, which can have a greatly negative impact on the fire hazard properties of polymer-based materials, concurrently due to their different flame-retardant actions (including endothermic decomposition, char formation, gas-phase dilution, and physical barrier), but contribute to the synergism effect of the fire retardancy of polymer-based materials. It methodically groups the various classes of inhibitors, such as metal hydroxides (ATH, MDH), metal oxides and borates, and organo-modified clays, as well as covers their individual thermal behaviours, efficacy, and decomposition behaviour. Environmentally problematic halogenated materials can be avoided by using synergistic compositions, including organophosphorus, silicone, and intumescent systems, which would improve char integrity and smoke suppression and reduce peak heat release rates. Applications across polyethylenes, epoxies, and rubber composites are examined, highlighting both fire-safety gains and mechanical or processability trade-offs that inform selection and optimisation strategies.

Considering the prospects, this field would be characterised by advancements in multifaceted materials comprising bio-sourced and halogen-free synergists. The synthesis of nanoscale mineral fillers with improved dispersion and interfacial bonding must be considered. This would result in the development of multifunctional composites that balance flame retardancy with mechanical performance and material recyclability. Innovation in characterisation, particularly at the nano- and micro-scales, will be critical to establishing clear structure–property relationships, enabling stringent fire-safety regulations to be met while minimising environmental impact.

References

[1] C.A. Wilkie, A.R. Horrocks, P. Hornsby, Chapter 6 - Flame retardant fillers, in fire retardancy of polymeric materials, in: A.B. Morgan and C.A. Wilkie (Eds.), Fire retardancy of polymeric materials, Taylor & Francis, 2024, pp.97-115. https://doi.org/10.1201/9781003380689-6

[2] T. Nazrun, M.K. Hassan, M.R. Hasnat, M.D. Hossain, B. Ahmed, S. Saha, A comprehensive review on intumescent coatings: formulation, manufacturing methods, research development, and issues, Fire 8 (2025) 1-39. https://doi.org/10.3390/fire8040155

[3] R. Gao, H. Ye, J. Chen, S. Chen, X. Fang, M. Zhou, Q. Xue, J. Chai, B. Xing, N. Chen, Recent advances in layered nanomaterial-based flame retardants for polymer composites, J. Appl. Polym. Sci. (2025) 1-18. https://doi.org/10.1002/app.58069

[4] M.K. Singh, S.M. Rangappa, M. Misra, A.K. Mohanty, S. Siengchin, Recent advancements in nanostructured flame-retardants: Types, mechanisms, and applications in polymer composites, J. Nanostruct. 42 (2025) 101-468. https://doi.org/10.1016/j.nanoso.2025.101468

[5] A.l. Mosawi, Flame retardants, their beginning, types, and environmental impact: a review, JSBCM 74 (2022) 2-8. https://doi.org/10.14382/epitoanyag-jsbcm.2022.1

[6] Y. Hu, Y. Ye, J. Wang, T. Zhang, S. Jiang, X. Han, Functionalization of chitosan and its application in flame retardant: a review, Int. J. Biol. Macromol. 295 (2025) 1-25. https://doi.org/10.1016/j.ijbiomac.2025.139615

[7] R. Gupta, M.K. Singh, S.M. Rangappa, S. Siengchin, H.N. Dhakal, S. Zafar, 2024. Recent progress in additive inorganic flame retardant polymer composites: Degradation mechanisms, modeling, and applications, Heliyon 10 (2024) 1-23. https://doi.org/10.1016/j.heliyon.2024.e39662

[8] M.M.S. Mohd Sabee, Z. Itam, S. Beddu, N.M. Zahari, N.L. Mohd Kamal, D. Mohamad, N.A. Zulkepli, M.D. Shafiq, Z.A. Abdul Hamid, Flame retardant coatings: additives, binders, and fillers, Polym. 14 (2022) 1-23. https://doi.org/10.3390/polym14142911

[9] R. Giri, L. Nayak, M. Rahaman, Flame and fire retardancy of polymer-based composites, Mater. Res. Innov. 25 (2021) 104-132. https://doi.org/10.1080/14328917.2020.1728073

[10] M. Liu, J. Qiao, X. Zhang, Z. Guo, X. Liu, F. Lin, M. Yang, J. Fan, X. Wu, Z. Huang, Flame retardant strategies and applications of organic phase change materials: a review, Adv. Funct. Mater. 35 (2025) 2412492. https://doi.org/10.1002/adfm.202412492

[11] A.J. Andrew, M. Sain, S. Ramakrishna, M. Jawaid, H.N. Dhakal, Environmentally friendly fire retardant natural fibre composites: a review, Int. Mater. Rev. 69 (2024) 267-308. https://doi.org/10.1177/09506608241266302

[12] M.N.A.M. Taib, P. Antov, V. Savov, W. Fatriasari, E.W. Madyaratri, R. Wirawan, L.M. Osvaldová, L.S. Hua, M.A.A. Ghani, S.S.A.O. Al Edrus, L.W. Chen, Current progress of biopolymer-based flame retardant, Polym. Degrad. Stab. 205 (2022) 1-28. https://doi.org/10.1016/j.polymdegradstab.2022.110153

[13] J. Rodrigues, N.G. Shimpi, Nanostructured flame retardants: an overview, J. Nanostruct. 39 (2024) 1-14. https://doi.org/10.1016/j.nanoso.2024.101253

[14] T. Maake, J.K. Asante, W. Mhike, B. Mwakikunga, Fire-retardant wood polymer composite to be used as building materials for South African formal and informal dwellings-a review, Fire 8 (2025) 1-30. https://doi.org/10.3390/fire8020081

[15] T. Kashiwagi, Polymer combustion and flammability-role of the condensed

phase, Symp. (Int.) Combust. 25 (1994) 1423-1437. https://doi.org/10.1016/S0082-0784(06)80786-1

[16] R. Fatima, S. Naseer, M.R.H.S. Gilani, M. Aamir, J. Akhtar, chapter 3 - Metal hydroxides, in: Inamuddin, T. Altalhi, S.A. Mohammed (Eds.), Sustainable materials for electrochemical capacitors, Wiley, 2023, pp.33-64. https://doi.org/10.1002/9781394167104.ch3

[17] P. Selvam, P. Saravanan, S. Palanisamy, L. Umate, R. Saminathan, Improving insulation performance with woven flax/PP hybrid composite reinforced by synthetic alumina and aluminium trihydrate, Interactions 245 (2024) 1-16. https://doi.org/10.1007/s10751-024-02121-7

[18] A. Bîrcă, O. Gherasim, V. Grumezescu, A.M. Grumezescu, Chapter 1 - Introduction to thermoplastic and thermosetting polymers, in: V. Grumezescu, A.M. Grumezescu (Eds.), Materials for biomedical engineering, Elsevier, 2019, pp.1-28. https://doi.org/10.1016/B978-0-12-816874-5.00001-3

[19] P.J. Anderson, R.F. Horlock, Thermal decomposition of magnesium hydroxide, Trans. Faraday Soc. 58 (1962) 1993-2004. https://doi.org/10.1039/tf9625801993

[20] Q. Chen, S. Xu, R. Li, B. Sun, H. Fang, Q. Zhang, Synthesis of highly dispersed magnesium hydroxide and its application in flame-retardant EVA composites, RSC Adv. 15 (2025) 12854-12865. https://doi.org/10.1039/D5RA01067E

[21] S. Kodani, S. Iwasaki, L. Favergeon, N. Koga, Revealing the effect of water vapor pressure on the kinetics of thermal decomposition of magnesium hydroxide, Phys. Chem. Chem. Phys. 22 (2020) 13637-13649. https://doi.org/10.1039/D0CP00446D

[22] Z. Zhou, H. Yu, G. Wang, M. Li, K. Shi, High antimony resistance strain Enterobacter sp. Z1 mediates biomineralization of antimony trioxide, Environ. Int. 195 (2025) 1-11. https://doi.org/10.1016/j.envint.2024.109237

[23] Y. Hu, Y. Zhang, Mechanisms and modes of action in flame retardancy of polymers, in: Y. Hu, X. Wang (1st Ed.), Flame Retardant Polymer Materials, Taylor & Francis, 2019, pp.13-34. https://doi.org/10.1201/b22345-2

[24] M. Kılınç, Production and characterization of boron-based additives and the effect of flame retardant additives on pet-based composites, Middle East Technical University (Turkey), Thesis (2009).

[25] M.J. Rozo, S. Sunder, W. Tabaka, B. Klaffke, H. Ruckdaeschel, B. Schartel, Unveiling aluminum diethyl phosphinate's dual identity: transfer from epoxy resins to glass fiber-reinforced composites, Polym. Compos. 46 (2025) 12981-12999. https://doi.org/10.1002/pc.29911

[26] X. Hui, D. Fang, G. Wang, F. Ran, X. Zhang, Z. Fu, O. Ruzimuradov, The flame retardant and smoke suppressant effect of transition metal hydroxystannate (MSn(OH)6, M = Fe, Co, Mn) for epoxy resin, J. Nanopart. Res. 27 (2025) 1-15. https://doi.org/10.1007/s11051-025-06223-3

[27] W.H. Pan, W.J. Yang, C.X. Wei, L.Y. Hao, H.D. Lu, W. Yang, Recent advances in zinc hydroxystannate-based flame retardant polymer blends, Polym. 14 (2022) 1-15. https://doi.org/10.3390/polym14112175

[28] N. Gobile, D.C. Onwudiwe, O.A. Oyewo, Zinc stannate modified graphitic carbon nitride for the photocatalytic reduction of Cr (VI) to Cr (III), Northwest University (South Africa), Dissertation (2022).

[29] S. Lin, Y. Cao, Z. Liu, Y. Li, J. Qian, K. Zheng, J. Jiang, X. Chu, Experimental and mechanistic study on the flame propagation behaviors of corn starch deflagration inhibited by different powder inhibitors, Fire Mater. 49 (2025) 1205-1216. https://doi.org/10.1002/fam.70013

[30] Y.X. Guo, J.H. Liu, W.P. Gates, C.H. Zhou, 2020. Organo-modification of montmorillonite, Clays Clay Miner. 68 (2020) 601-622. https://doi.org/10.1007/s42860-020-00098-2

[31] A.B. Mapossa, N. Mohammad Mehdipour, U. Sundararaj, Thermal, mechanical, flammability properties, and morphologies of polypropylene/calcium carbonate-based on carbon nanomaterials and layered silicates composites: A future perspective of new sheet material for flooring application, J. Vinyl Addit. Technol. 31 (2025) 146-181. https://doi.org/10.1002/vnl.22162

[32] Y. Lei, X. Zhao, L. Xu, H. Li, J. Liang, G.H. Yeoh, W. Wang, Natural flame retardant minerals for advanced epoxy composites, Fire 7 (2024) 1-20. https://doi.org/10.3390/fire7090308

[33] R. Yadav, A.K. Sharma, S. Sharma, Advanced development in natural graphite material and its applications: a review, Min. Metall. Explor. 42 (2025) 361-385. https://doi.org/10.1007/s42461-024-01167-z

[34] G. Beyer, Chapter 1 - Introduction to flame retardant systems, in Flame retardant nanocomposites, in: T. Sabu, V. Henri, S. Lakshmipriya (Eds.), Flame retardant nanocomposites, Woodhead Publishing, 2024, pp.1-22. https://doi.org/10.1016/B978-0-443-15421-8.00011-2

[35] T.L. Mhlabeni, E.B. Chemere, T. Jamiru, W. Mhike, Flame retardancy of biopolymers enhanced by bio-based flame retardants: a review, J. Fire Sci. 43 (2025) 79-114. https://doi.org/10.1177/07349041241309308

[36] E. Głowińska, P. Wiśniewska, E. Movahedifar, M.E. Penoff, O. Das, H. Vahabi, M.R. Saeb, Cone calorimetry ranking of unsaturated polyester thermoset resins using flame retardancy index (FRI), J. Therm. Anal. Calorim. 150 (2025) 7489-7506. https://doi.org/10.1007/s10973-025-14149-0

[37] R. Sauerwein, Chapter 3 - Mineral filler flame retardants, in: A.B. Morgan, Non-halogenated flame retardant handbook (2nd Ed.), Scrivener Publishing LLC, 2022, pp.101-168. https://doi.org/10.1002/9781119752240.ch3

[38] S. Sajan, D.P. Selvaraj, A review on polymer matrix composite materials and their applications, Mater. Today Proc. 47 (2021) 5493-5498. https://doi.org/10.1016/j.matpr.2021.08.034

[39] A. Kumar, S. Dixit, S. Singh, S. Sreenivasa, P.S. Bains, R. Sharma, Recent developments in the mechanical properties and recycling of fiber-reinforced polymer composites, Polym. Compos. 46 (2025) 3883-3908. https://doi.org/10.1002/pc.29261

[40] Y.H. Kim, S. Kumar, X. Li, S.Y. Kim, D.H. Shin, Temperature-dependent mechanical properties and material modifications of carbon fiber composites for optimized structures in high-end industrial applications, Compos. B Eng. 303 (2025) 1-25. https://doi.org/10.1016/j.compositesb.2025.112602

[41] W. Jilani, A. Bouzidi, M. Al-Dossari, I.S. Yahia, Impact of tributyl phosphate on epoxy polymer-based composites: Investigation of their composites' structural, optical, and dielectric properties for optoelectronic applications, Inorg. Chem. Commun. 174 (2025) 1-15. https://doi.org/10.1016/j.inoche.2025.114036

[42] J.J. Fekiač, M. Krbata, M. Kohutiar, R. Janík, L. Kakošová, A. Breznická, M. Eckert, P. Mikuš, Comprehensive review: optimization of epoxy composites, mechanical properties, & technological trends, Polym. 17 (2025) 1-39. https://doi.org/10.3390/polym17030271

[43] A. Salehi, M. Kheradmandkeysomi, R. Rahmati, S.S. Rahman, M. Fashandi, C.B. Park, Tailoring interfacial, rheological, and energy absorption properties of in situ nanofibrillated ethylene propylene diene monomer rubber for enhanced toughness in polypropylene composites, ACS Appl. Polym. Mater. 7 (2025) 2250-2264. https://doi.org/10.1021/acsapm.4c03204

[44] S. Hong, J. Yang, S. Ahn, Y. Mun, G. Lee, Flame retardant performance of various UL94 classified materials exposed to external ignition sources, Fire Mater. 28 (2004) 25-31. https://doi.org/10.1002/fam.840

[45] T.R. Hull, Chapter 11 - Challenges in fire testing: reaction to fire tests and assessment of fire toxicity, in: A.R. Horrocks, D. Price (Eds.), Advances in fire retardant materials, Woodhead Publishing, 2008, pp.255-290. https://doi.org/10.1533/9781845694701.2.255

Materials Research Foundations 190 (2026) 374-387
https://doi.org/10.21741/9781644904077-14

Chapter 14

Fire Safety Standards for Furniture, Textiles, Electronics, and Building Products

Ankit Singh[1]*, Arnab Kumar De[1], Hridoy Kumer Das[1,4], Rubel Khan[1,2], Wazed Ali[1], Shantanu Basak[3], Sohel Rana[1]

[1]Indian Institute of Technology, Delhi, Hauz Khas, New Delhi, Delhi 110016

[2]Bangladesh University of Textiles (BUTEX), 92 Shaheed Tajuddin Ahmed Ave, Dhaka 1208, Bangladesh

[3]National Institute of Natural Fiber Engineering & Technology, 12 Regent Park, Kolkata 700040, India

[4]University of Tennessee, Knoxville, Tennessee 37996, USA

ankitlkw@gmail.com

Abstract

Fire safety standards regulate the flammability and fire-resistance of furniture, textiles, electronics, and building materials to protect customers and the built environment. This chapter emphasizes the National Building Code of India (NBC), which sets fire safety standards for building materials and construction practices, and the Bureau of Indian Standards (BIS), which regulates product flammability and fire protection. This content also focuses on furniture and fabric standards such as IS 15741 and IS 15742. The electronics industry considers IS 13252 and IS 616, which are equivalent to IEC 60695, and provide guidelines for fire safety in electrical and electronic equipment. The chapter details textile and furniture regulations, including the UK Furniture and Furnishings (Fire Safety) and California Technical Bulletin 117.

Keywords

Fire Safety Standards, Fire-Resistant Materials, Building Code Compliance, Flame Retardants, Fire Risk Mitigation, Product Safety Regulations

Contents

Materials Research Foundations 190 (2026) 374-387 https://doi.org/10.21741/9781644904077-14

1. Introduction

Consumers, manufacturers, and government agencies are paying closer attention to the fire risks associated with synthetic materials, as they are inherently flammable. To meet the needs of a wide range of applications, flame retardants are added to materials to reduce their flammability and limit the release of smoke or harmful fumes when they catch fire. In many fields, the demand for flame-resistant materials, particularly flame-resistant plastics, has increased rapidly. This demand is primarily driven by four main areas: furniture, textiles, electronics, and building materials [1]. According to Dodson et al. (2017), fire was one of the most critical discoveries humans have ever made, but it also killed about 4,000 people a year in the US and 5,000 in Europe while costing the economy about 0.3% of its GDP. A study examining the causes of various accidents found that making plastic, cloth, computer, and building materials more fire-resistant could help mitigate the effects of inevitable mishaps [2]. Significant property and human damage have resulted from recent fire occurrences. Therefore, to effectively prevent these incidents, fire safety regulations and flame-retardant materials must be in place.

Construction engineering has experienced substantial growth recently, with a growing focus on flame-retardant polymer materials, categorized as intrinsic or doped based on their chemical composition [1]. Fire hazards provide significant dangers to individuals, assets, and the ecosystem. Consequently, fire prevention is crucial across several sectors, including construction, textiles, electronics, and transportation. To mitigate these risks, it is essential to use standardized assessments and adhere to established protocols. A systematic and scientific method exists to evaluate the behavior of materials and products in fire scenarios via standardized fire testing. Simultaneously, adherence to national and international safety regulations ensures that only products certified as fire-resistant are accessible.

This chapter provides a comprehensive overview of fire safety standards relevant to furniture, textiles, electronics, and building materials, with a primary focus on India-centric regulations and testing protocols. It examines key standards from the Bureau of Indian Standards (BIS), the National Building Code (NBC) of India, and fire safety guidelines from regulatory authorities. To offer a broader perspective, the chapter also compares these with prominent international standards, including NFPA (USA), BS (UK), ISO, and EN (Europe) frameworks. Through this comparative approach, the chapter highlights the alignment, gaps, and opportunities for harmonization between Indian and global fire safety practices.

2. Purpose and Principle of Fire Safety Standard

In recent years, fires have caused significant damage to people and property. Due to this, it is crucial to develop flame-retardant materials to prevent such fires from occurring. Polymers like polyethylene (PE) and polypropylene (PP), for instance, have a limited oxygen index (LOI) number of 17.0 to 18.0%, which means they can catch fire [3-5]. Insulating cables or wires with these won't stop the fire from spreading. Phenolic foam, polystyrene foam, and polyurethane foam insulate building walls well, but they don't prevent fires [6]. Thus, raising polymers' FR is a serious issue for their safety and reliability in many applications.

Over the past several decades, construction engineering has undergone significant advancements. A large number of researchers have therefore devoted their efforts and resources to developing superior flame-resistant polymer materials for commercial use. Such polymers are classified into two categories: "intrinsic" and "doped," based on the components used in their production. Flame retardants (FRs) exemplify the challenge of balancing the need to mitigate fire hazards with the potential adverse effects of these compounds. Manufacturers establish and maintain flammability standards for safety, often necessitating the use of flame retardants (FRs) to ensure compliance with these standards. Numerous governmental regulations assess the safety of each flame retardant to ensure public health is safeguarded. These include the Bureau of Indian Standards (BIS), the Toxic Substances Control Act (TSCA) in the United States, the Chemical Management Plan (CMP) in Canada, and the Registration, Evaluation, Authorisation of Chemicals (REACH) in the European Union.

Adding FRs to foam, fabrics, or other parts of furniture has usually been enough to meet furniture flammability regulations [7]. Flame retardants (FRs) make items less likely to catch fire, delay the spread of flames, and test the boundaries of each functional standard. The types and amounts of FRs used are also affected by the qualities of the textile materials. While prioritizing safety, clear rules support the advancement of study and technology in fire-resistant building materials and methods. Standards for fire safety are based on scientific tests that demonstrate how materials respond in a fire. Standardized tests from organizations like NFPA, ASTM, or ISO are conducted on materials in regulated laboratories. These tests examine factors such as flame spread, burning time, heat release, and smoke production. If a material meets the set safety standards, it is recognized as compatible. It is tested regularly to make sure that all production batches are safe. As the field of fire science grows, these standards are changed to reflect new information and tools [8].

3. Regulatory Framework and Key Authorities

3.1 Indian Regulatory Framework

The National Building Code (NBC), issued by the Bureau of Indian Standards (BIS), is a comprehensive set of rules that architects, engineers, and developers can follow to ensure buildings are safe during the planning and construction process. In Part 4 of the NBC, which is called "Fire and Life Safety." There are basic rules for preventing fires, protecting people inside, and evacuating the building safely in the event of a fire. Buildings are categorized into different usage

Materials Research Foundations 190 (2026) 374-387 https://doi.org/10.21741/9781644904077-14

types by the code, including residential, commercial, and public gathering. This makes it possible to do accurate estimates of the fire risk. It also listed the fire protection ratings for different parts of the building, like doors, walls, and beams. These ratings are based on standard time-temperature graphs. It also highlights the importance of properly planning exits, including minimum exit widths, maximum trip lengths, pressurizing stairwells, and ensuring the presence of emergency lighting to facilitate a safe and quick escape [9].

The Bureau of Indian Standards (BIS) holds paramount responsibility for developing and maintaining Indian Standards (IS codes) that describe the requirements for fire-safe materials and goods in various areas, including technology, furniture, building, and textiles. One of the primary responsibilities of the BIS is to ensure that all fire test methods are standardized. Some crucial standards include IS 3808, which tests the fire resistance of building parts, IS 11871, which assesses the flammability of materials, and IS 15061, which evaluates the fire resistance of building materials and components. BIS also utilizes labeling and licensing programs to ensure that fire-resistant items, such as electrical cords, cushioned furniture, and plastic panels, display valid ISI marks indicating compliance with relevant standards. To maintain high-quality and safety standards, BIS requires that materials used in public infrastructure, transportation systems, and high-rise buildings be regularly tested and comply with the relevant regulations [10].

Several crucial international standardization groups have established fire safety rules that are used worldwide, including in India, ensuring consistency across the globe. A group in the United States, called ASTM International, develops widely used fire testing methods, such as ASTM E84, which evaluates how building materials burn on their surface. The International Electro-technical Commission (IEC) sets standards for technology and electrical equipment, with one key set of standards, the IEC 60695 series, which checks electrical devices for fire hazards. The European Committee for Standardization (CEN) puts out the EN series. For example, EN 13501 is used for building materials related to fire safety, and EN 1021 assesses the flammability of cushioned furniture. The British Standards Institution (BSI) also has specific standards, such as BS 5852, that are often used to determine the fire resistance of padded seats. These global standards enable India to develop comprehensive and internationally aligned fire safety regulations.

3.2. International Regulatory Framework

Summary of International fire testing standards, such as:

ISO 5660-1 – Testing for fire resistance: cone calorimeter technique for heat release, smoke generation, and mass loss rate

ISO 9705 – Full-scale room fire test for surface products

ASTM E84 – Standard Test Method for Surface Burning Characteristics of Building Materials (Steiner Tunnel Test)

ASTM D6413 – Standard Test Method for Flame Resistance of Textiles (Vertical Test)

EN 13501-1 – Classification of building components and goods according to fire safety standards

EN 13823 – Single Burning Item (SBI) test for wall and ceiling products

Materials Research Forum LLC
https://doi.org/10.21741/9781644904077-14

4. Fire Safety Standard by sector

4.1 Furniture and Textile Materials

India

The Bureau of Indian Standards (BIS) has established several key standards for textiles used in furniture, interiors, and safety clothing in public places to enhance their fire safety. The IS 15768:2008 standard specifies safety rules for sofa fabrics used in the manufacture of furniture, such as couches, chairs, and seats, in public places and on public transit. To determine how likely they are to catch fire, textiles undergo several tests, including the burning cigarette test, the match flame test, and the crib test. The rule specifies the duration of fire resistance, the dangers of smoke, and its impact on vision. Each item must have a sticker indicating it is safe to use and won't catch fire. A Quality Control Order [11] from the Indian government makes this standard mandatory.

The IS 15741:2007 standard outlines the use of curtains and panels in public places such as hospitals, restaurants, theatres, and workspaces. Curtains and drapes that meet this standard are not likely to catch fire and have been tested for their ability to catch fire, their longevity, their smoke-related toxins, and their ability to be seen during burning. Small sparks or electricity problems can quickly start a fire, so this is important. IS 15612 (Part 1) categorizes linens into two classes based on their flammability: Class 1 and Class 2. These groups and tests make sure that only safe and appropriate textiles are used in high-risk public places [12].

IS 15742:2007 covers protective clothing made from materials that do not readily spread flames. It's designed for individuals who work near fires and heat, such as firefighters and welders. This rule ensures that clothes can withstand fire, direct heat, and rising heat and remain functional after exposure. The testing carefully checks for longevity, heat retention, and the ability to stop fires from spreading. Manufacturers must follow this standard and obtain BIS approval to demonstrate that their safety gear is safe and functions properly [13]. Together, the guidelines reduce the likelihood of fires, making them safer for everyone and ensuring that fabrics can be used safely in public buildings and high-risk workplaces.

International

This ensures that furniture and seat materials are safe in the event of a fire. Many countries have their own regulations regarding flammability. These standards help ensure that foams and fabrics are safe for use in public and private settings, and that they do not ignite or burn quickly and easily. We also check how they burn. Foreign standards, such as California TB117-2013 (USA), BS 5852 (UK), and EN 1021 (Europe), are essential and widely used.

United States, California: TB117-2013 - This flammability standard is primarily used in the U.S., particularly in California. It replaced the 1975 version and mainly focused on the resistance of furniture parts, such as fabrics, foams, obstacles, and railing materials, to catching fire. Instead of open flame tests, the new standard uses smoldering cigarette tests, similar to those in UFAC/NFPA 260. It permits companies to use fewer or no flame-retardant chemicals, which makes furniture safer and healthier for workers and customers. Compliance became required in 2015, but some makers had already done it. Furniture that isn't flame-safe must have a sticker indicating this, and customers can ask the store for more information [14].

BS 5852 (UK) - The British Standard BS 5852 is a widely recognized fire safety test used globally to assess the flammability of upholstered furniture. It's designed to resemble a real fire by utilizing

Materials Research Foundations 190 (2026) 374-387
https://doi.org/10.21741/9781644904077-14

various types of ignition. For example, Source 0 mimics a burning cigarette left on a chair, Source 1 models a small flame similar to that from a match or lighter, and Crib 5 recreates a more intense fire using stacked wooden sticks that burn more vigorously. These tests help people who make things figure out what happens to a substance when it's heated up. To pass, the fabric or foam must not catch fire, and if it does, it must extinguish the fire quickly without allowing it to spread. This makes the furniture much safer in places like homes, hotels, and offices, where it is used daily [15].

CE Marks EN 1021-1 and EN 1021-2 - The European Standard EN 1021 is widely used across the EU to check how well padded furniture materials avoid catching fire. A chair-shaped model, made of foam and fabric, is used in the test setup to simulate real furniture. The test consists mainly of two parts. In EN 1021-1, a lit cigarette is put in the corner where the backrest and seat meet. After an hour, there should be no burning or flames left. In EN 1021-2, a small spark, similar to that of a match, is applied for 15 seconds in the same spot. Once the light is removed, the cloth shouldn't continue to burn after two minutes. These tests are essential for public places, such as restaurants, schools, and hospitals, where fire safety is of utmost importance. They help furniture makers ensure their goods are significantly less likely to catch fire or spread flames [16].

4.2 Electronics and Electrical Devices

India

Under the Compulsory Registration Scheme, the Bureau of Indian Standards (BIS) has established essential safety standards to protect consumers from everyday electronics that may pose risks. IS 13252 (Part 1):2010 and IS 616:2017 are crucial standards for ensuring that Indian IT and audiovisual equipment is safe around electricity and fire. Laptops, computers, tablets, printers, and other IT gear are all covered by IS 13252 (Part 1):2010. Its main goal is to make these kinds of things electrically safe, so that they are less likely to cause shocks, fires, burns, and other risks while in use. Before selling these things in India, manufacturers must get BIS approval. This standard is regularly updated to reflect new safety concerns and advancements in technology. This ensures that approved gadgets meet India's safety requirements [17].

IS 616:2017 also covers audio, video, and devices that work with them, such as TVs, game consoles, Bluetooth speakers, and entertainment systems. It ensures that users are safe from mechanical, electrical, radiation, and fire hazards. The standard outlines the product's structure, insulation, protection against electrical hazards, and safety features for users. Like IS 13252, IS 616:2017 must be followed, and products approved by ISI are marked with a symbol to indicate that they are safe and of good quality [18]. All of these rules help make homes, companies, and public places safer from fire and electricity hazards. They keep users safe and help companies ensure their products are reliable, which makes the entire tech industry in the country safer.

International

The International Electrotechnical Commission (IEC) has established several global safety guidelines to ensure the safe operation of electrical equipment. People can receive electric shocks, experience burns, and even catch fire in computers, printers, and servers, according to IEC 60950-1. It sets rules for building, electricity, and insulation. To protect against electrical and mechanical risks, IEC 60065 covers TVs, speakers, and amplifiers. The IEC 62368-1 standard, which is newer and more unified, works better with modern equipment than these two older ones. You should

Materials Research Foundations 190 (2026) 374-387 https://doi.org/10.21741/9781644904077-14

also be aware of IEC 60695, a fire safety standard for electronic technology. It checks how the parts of the gadget react to fires or heat, observing whether they catch fire, the amount of smoke they produce, and whether the smoke is hazardous. Although some older standards are being replaced, IEC 60695 is still frequently used to help manufacturers produce fire-safe goods. These rules make everyday electronics safer to use at home, at work, and in public places [19].

4.3 Building and Construction Materials

India

Some fire safety rules, such as IS 3808:1979 and IS 12459:1988, were established by the Bureau of Indian Standards (BIS). These standards check how well materials resist fire and lower the risk of fire when installing electrical cables. The Indian Standard IS 3808:1979 tests building materials to make sure they don't catch fire. At high temperatures, it checks to see if a substance can catch fire, burn, or give off explosive vapours. This test identifies materials that won't catch fire and can be used for building parts, ceilings, walls, and padding. Controlled heating is used on a wide range of covered and bare building goods to measure flame output and heat release. The 2021 update to IS 3808 enables architects, engineers, and safety experts to select fire-safe building materials that reduce the likelihood of fires occurring in a building and limit the spread of flames [20].

A code of practice, IS 12459:1988, known as the "Code of Practice for Fire Safety in Cable Runs," protects cable lines in buildings and factories. This standard isn't about naming materials or types of fire; instead, it's about how to make ditches, caves, and passageways less likely to catch fire when electrical cables are present. It advises separating cable lines into compartments to prevent fires from spreading and using fire-resistant wires in key areas to maintain circuit functionality during emergencies. To help firefighters get to the scene of the fire quickly, the standard also emphasizes the use of early warning devices, such as fire detectors and sirens. To support a combined electricity system fire protection plan [21], IS 12459 also outlines other Indian fire safety regulations.

IS 3808 and IS 12459 enhance building fire safety at all levels, from the materials used to the entire system. They also help make the planning, construction, and upkeep of modern infrastructure safer.

International

Standardization is necessary for testing the fire behavior of building materials to ensure the safety of people and compliance with regulations. In North America, ASTM E84 is widely used. In the EU, on the other hand, EN 13501 is the standard for fire rating. Both standards assess how well materials resist fire and how easily they can spread smoke and flames, but they do so in different ways and employ distinct classification systems. There is a test called ASTM E84 that checks how building materials burn on the surface. It checks the Flame Spread Index (FSI) and the Smoke Developed Index. In a Steiner tunnel, the piece of material is held horizontally in front of a controlled flame for 10 minutes. The FSI records the speed of the flame, while the SDI uses light to measure the amount of smoke that is released. Surface burning is a method used to distinguish between building materials on the inside, such as insulation, coatings, wall panels, and ceiling tiles [22].

EN 13501 rates the fire protection of building materials and construction items. The EU's Construction Products Regulation (CPR) supports CE marking and ensures more consistent fire safety rules. EN 13501-1 examines how materials react to fire in the initial stages. Euro classes range from A1 (not flammable) to F (very flammable), with smoke output ranging from s1 to s3 and flame drops or particles ranging from d0 to d2. B-s1, d0 means the fire is working well with little smoke and no flaming drops. The primary test for categorizing items into groups is the Single Burning Item (SBI) test, which ranks items based on how quickly they catch fire. Wall and ceiling coverings, floor coverings, pipe insulation, and roofing goods are all covered by EN 13501. For outside fire risk, EN 13501-5 is also included [23].

The standards ASTM E84 and EN 13501 help builders choose safe materials and develop plans by evaluating and comparing the effectiveness of fire safety in various locations.

5. Core Technical Parameters in Fire Testing

5.1 Stages of Fire Development

When examining for fires, focusing on specific steps, the current conditions must be taken into account. The majority of fires come from small sources. Before a burning spark, there may be an induction stage with shouldering. This is followed by a rise in temperature until ventilation-controlled combustion occurs (often between 800 and 1000 °C), and finally a fall when the fuel is exhausted.

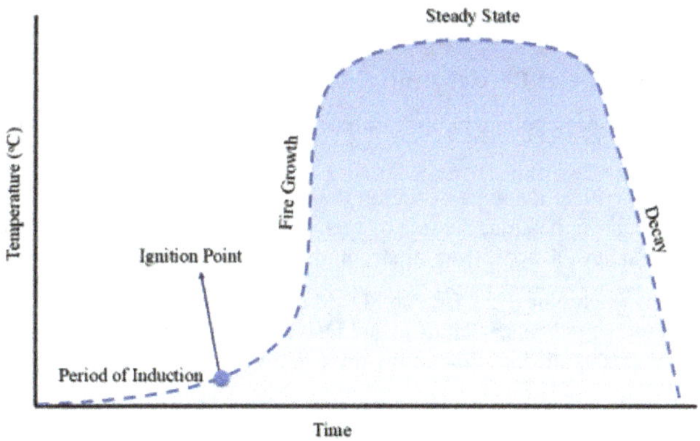

Fig. 1 - Stages of fire

To accurately replicate the effects of fire, it is essential to heat materials and products under conditions that closely resemble real-life scenarios.

Ignition: For tested ignition to happen, there must be a source (like a fire, cigarette, or glow wire), a little sample size (between 1 and 10 cm), and the surface temperature of the sample must be at ignition temperatures (between 300 and 400 °C). The area must also be well-ventilated.

Developing fire: As a fire grows, it creates an external heat flux of about 20 to 60 kW, which means that samples need to be bigger (10 cm to 1 m), the temperature outside needs to be higher than the starting temperature (400 to 600 °C), and there needs to be enough air flow.

Materials Research Foundations 190 (2026) 374-387 https://doi.org/10.21741/9781644904077-14

Fully developed fire: During the penultimate phase of fire development, external heat levels exceed 50 kW, sample sizes range from 1 to 5 meters, ambient temperatures surpass the ignition threshold of 600 °C, and airflow is minimal.

Ultimately, fire science and testing prioritize specific safety goals over others for valid reasons. Some common protection goals include stopping a fire from spreading, preventing it from reigniting, or acting as a barrier against fire. Most fire tests simulate a real fire situation and monitor a specific threat from a single example, rather than examining the material's properties. Additionally, how a subject reacts to a fire or fire test can have a significant impact on the overall fire situation. There are three main points we can make:

1. In different fire tests, it's hard to compare how fire acts. Predictions may not be accurate because the properties of materials can alter their effectiveness in various situations. However, imprecise correlations or correlations that are specific to the topic have been effective.

2. The size of the sample is significant in fire science, which makes moving up and down hard. Empirical methods often struggle to predict how a fire will behave, particularly when its intensity is exceptionally high. Forecasts of how fires will behave are getting more accurate thanks to better prediction systems.

3. The relationship between the properties of the parts and the properties of the material itself is complicated and can change.

5.2 Factors Affecting Fire Development

These factors influence how polymers burn in real flames and should be considered in a well-designed test.

- When there are critical amounts of fuel in the gas phase, it can catch fire and burn during production. During heating, the rate of fuel production is crucial, but the airflow around the sample can have a significant impact on the burning temperature.
- Stable radicals, including Cl.., Br., or PO. (Acting as inhibitors or diluents), reduce the quantity of active radicals, such as H and OH, required in the flame zone. The effect is most pronounced at the initiation of the fire and diminishes as it progresses.
- The flow properties of polymers that are breaking down—some depolymerise, which lowers viscosity. This makes it easier for heat to spread out and for things to move away from the heat source. Drops of harmless dripping or burning oil may fall below the flame zone. Nanofillers and other additives increase the viscosity, which prevents leaks and accelerates the rise in surface temperature. The time it takes to light will be cut down. Free radical initiators may be added to the fuel to cause it to drip, allowing it to be removed from the heat source.
- *Charring* - Charring on polymer surfaces makes it harder for heat to move to the samples. As they bubble up, intumescent chars make a stronger wall. In fire tests, the growth usually faces the heat source, which sends more heat to the sample.
- *Absorption of radiation* - Polymers need to be able to take in radiation from fires or solar screens. Adding absorbent centres, like conjugated double bonds or black dyes, could help

Materials Research Foundations 190 (2026) 374-387 https://doi.org/10.21741/9781644904077-14

heat stay where it's needed. On the other hand, in some tests, an extremely reflective surface may cause the time it takes for light to reach its destination to last much longer.

- *Making smoke* - Smoke can either radiate or restrict the flame's energy to the polymer.

6. Challenges of Fire Testing

To quantify the flammability of something, it's necessary to use specific testing methods to obtain numerical measurements. Since large-scale fire tests are costly, it is expected that these trials can be connected to actual fires that we want to stop. Since resources for testing are scarce, most tests are conducted on a small scale. When compared to full-scale tests, small-scale fire tests are much easier to repeat and show better results. Many tests are flawed because they can't mimic the strong effects of heat found in big fires, which could lead to false results when used in the wrong situation [24].

There are two different schools of thought in the field of fire science. One group suggests that lowering the peak heat release rate is crucial for preventing fires from spreading, as it controls how the initial flame propagates from the object that has caught fire to other objects. On the other hand, some argue that higher guided ignition temperatures or longer wait times before ignition indicate that ignition resistance is the primary factor affecting fire spread. For thermally thick materials, the time to fire is equal to the square of the ignition temperature. For thermally thin materials, it is equal to the ignition temperature. Usually, a flame spreads over a flat surface, and the rate at which the fire grows is related to the square of the time after ignition. This shows that the rate of growth depends on both the time of burning and the temperature. What effect will be most noticeable will depend on the fire situation [25].

7. Emerging Challenges and Global Trends

Developing sustainable flame retardants (FRs) is a significant challenge worldwide, as traditional halogenated chemicals are causing increasing health and environmental problems. The primary focus is on developing non-toxic, recyclable flame retardants that are more effective against fires while being safer for people and the environment. To truly be sustainable, it's essential to consider the entire life cycle of flame retardants, as demonstrated by a new study. Machine learning-assisted molecular design is a promising method that utilizes AI models based on known structure-property databases, such as Heat Release Capacity (HRC) values, to predict molecular structures that are effective at suppressing flames. Adding input from fire performance tests, persistence, bioaccumulation, and toxicity (PBT) studies, and life cycle analysis (LCA) into the AI loop can accelerate the process of identifying the next generation of environmentally friendly flame retardants (FRs).

People are interested in flame retardants derived from natural sources, but the fact that they need to be modified with petrochemicals before use detracts from their environmental benefits. Bio-refining methods need to get better so that biomolecules can be made in a way that is both cost-effective and useful. At the same time, prioritizing green synthesis, which includes processes that don't use solvents, atom economy, and reusing waste products, is crucial for making molecules more sustainable. Although improvements have been made, a lack of PBT and toxicology data remains, particularly for halogen-free flame retardants and the products that contain them. The way these products are currently evaluated doesn't accurately reflect what will happen to the world

Materials Research Foundations 190 (2026) 374-387 https://doi.org/10.21741/9781644904077-14

in the long run. For comprehensive chemical risk studies, scientists, toxicologists, ecologists, and AI experts must collaborate across disciplines. Implementing these ideas into machine learning models can aid in the development of long-lasting and effective flame retardants [26].

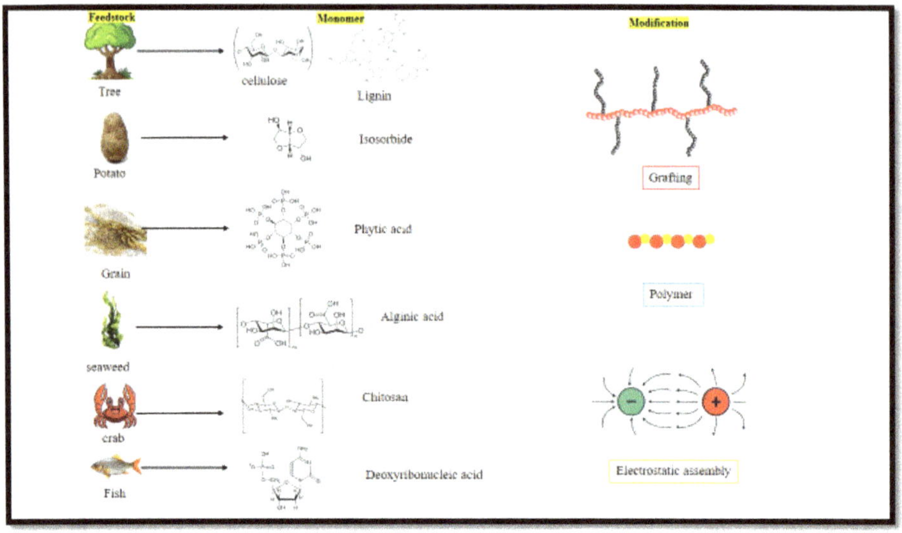

Fig 2. Natural feedstock as raw materials for sustainable flame retardants [26]

8. Case Studies and Industrial Implementation

Public places in India need to adhere to fire safety regulations. In Hyderabad, a hospital and school with multiple floors had significant problems with adhering to the fire safety rules outlined in IS 15768 and NBC 2005 (Part IV). A reputable public limited company ran the hospital construction site. Some of the rules were followed, including storing LPG and acetylene, using double-insulated cables, and maintaining basic fire extinguishers. However, there were no complete fire protection systems, vulnerability assessments, or scientifically designed safety measures. Even fewer fire extinguishers and buckets were in the school building, which a smaller private company made. These results clearly indicate that stricter regulation and regular inspections are necessary to ensure that public buildings, such as airports, hospitals, and schools, comply with the rules [27]. India's fire testing labs and governing bodies lack unified monitoring tools and insufficient knowledge of compliance. It is common for BIS standards, CPWD contracts, and local laws, such as those in the Greater Hyderabad Municipal Corporation, not to work well together. CPWD's contract framework is one of many that don't require fire safety rules during construction, especially for temporary buildings. This puts workers and places at risk.

The change in California's rules affected the flame-resistant properties of polyurethane foam used in furniture worldwide. By getting rid of required chemical FRs and switching to performance-based flammability testing (Technical Bulletin 117-2013), California made consumers safer and

Materials Research Foundations 190 (2026) 374-387 https://doi.org/10.21741/9781644904077-14

reduced their exposure to chemicals. The world example shows how laws based on science can lead to better materials and changes in industry standards [28].

9. Conclusion

The International Forum of Fire Research Directors' wants to make a significant change to how flammability is measured to keep up with the changing needs of fire safety. A variety of tests, ranging from small Bunsen burner doses to full-scale 15-meter fire models, have been used worldwide in the past to determine the flammability of materials. Because these test results often depend on the exact setup and conditions used, they can't be used to predict how something will work in real life. The Forum states that current small- and large-scale tests are insufficient to accurately predict the results of each other's tests, let alone model new or complex fire situations. They want a change to fundamentally based, small-scale tests of material properties that give true, first-principles computer models accurate inputs. With these models, it would be possible to conduct accurate studies on how fires function for various purposes. Although this method is more complicated and requires additional resources, it has been used before, can be used again, and is supported by scientific evidence. It opens the door to more accurate and useful fire safety plans.

In a world that is becoming more urbanised, there is a growing need for fire safety measures that work together and last. Due to new building materials, global trade, and shifting safety risks, outdated and disjointed testing methods require reevaluation. Science-based, integrated assessment systems that combine material innovation, computer modeling, and international regulatory convergence are likely to be crucial for the future of fire safety. This will ensure that fire safety keeps pace with technological advancements and social progress.

References

[1] S. Agnihotri, J.N. Sheikh, S.P. Singh, B.K. Behera, Flame-retardant textile structural composites for construction applications: a review, J. Mater. Sci. 59 (2024) 1788-1818. https://doi.org/10.1007/s10853-023-09312-7

[2] R.E. Dodson, K.M. Rodgers, G. Carey, J.G.C. Laurent, A. Covaci, G. Poma, G. Malarvannan, J.D. Spengler, R.A. Rudel, J.G. Allen, Flame retardant chemicals in college dormitories: flammability standards influence dust concentrations, Environ. Sci. Technol. 51 (2017) 4860-4869. https://doi.org/10.1021/acs.est.7b00429

[3] W. Guo, Y. Zhao, X. Wang, W. Cai, J. Wang, L. Song, Y. Hu, Multifunctional epoxy composites with highly flame-retardant and effective electromagnetic interference shielding performances, Compos. Part B Eng. 192 (2020) 107990. https://doi.org/10.1016/j.compositesb.2020.107990

[4] H. Huang, K. Zhang, J. Jiang, J. Li, Y. Liu, Highly dispersed melamine cyanurate flame-retardant epoxy resin composites, Polym. Int. 66 (2016) 85-91. https://doi.org/10.1002/pi.5244

[5] H. Wu, R. Ortiz, R.D.A. Correa, M. Krifa, J.H. Koo, Self-extinguishing and non-drip flame-retardant polyamide six nanocomposite: mechanical, thermal, and combustion behavior, J. Fire Retard. Therm. Stab. Mater. 1 (2018) 1-13. https://doi.org/10.1515/flret-2018-0001

[6] L. Liu, Y. Xu, S. Li, M. Xu, Y. He, Z. Shi, B. Li, A novel strategy for improving fire safety, water resistance and compatibility of thermoplastic polyurethane composites via biomimetic intumescent flame-retardant system, Compos. Part B Eng. 176 (2019) 107218. https://doi.org/10.1016/j.compositesb.2019.107218

[7] M.H. Stapleton, S. Sharma, G. Getzinger, P.L. Ferguson, M. Gabriel, T.F. Webster, A. Blum, Novel and high-volume-use flame retardants in U.S. couches reflective of the 2005 PentaBDE phase-out, Environ. Sci. Technol. 46 (2012) 13432-13439. https://doi.org/10.1021/es303471d

[8] V. Babrauskas, A. Blum, R. Daley, L. Birnbaum, Flame retardants in furniture foam: benefits and risks, Fire Saf. Sci. 10 (2011) 265-278. https://doi.org/10.3801/IAFSS.FSS.10-265

[9] Bureau of Indian Standards, National Building Code of India, Part 4: Fire and Life Safety, BIS, New Delhi, 2005.

[10] Bureau of Indian Standards, Fire Protection of Electronic Data Processing Installation - Code of Practice, IS 11871, BIS, New Delhi, 2004.

[11] Bureau of Indian Standards, IS 15768: Textiles - Resistance to Ignition of Upholstered Composites Used for Non-Domestic Furniture, BIS, New Delhi, 2008.

[12] Bureau of Indian Standards, IS 15741: Textiles - Resistance to Ignition of Curtains and Drapes, BIS, New Delhi, 2007.

[13] Bureau of Indian Standards, IS 15742: Textiles - Requirements for Clothing made of Limited Flame-Spread Materials and Material Assemblies Affording Protection Against Heat and Flame, BIS, New Delhi, 2007.

[14] California Bureau of Home Furnishings, Technical Bulletin 117-2013: Requirements, Test Procedure and Apparatus for Testing Smolder Resistance of Materials Used in Upholstered Furniture, 2013.

[15] British Standards Institution, BS 5852 - Methods of Test for Assessment of the Ignitability of Upholstered Seating, BSI, London, 2006.

[16] European Committee for Standardization, EN 1021-1/2: Furniture - Assessment of Ignitability of Upholstered Furniture, CEN, Brussels, 2006.

[17] Bureau of Indian Standards, IS 13252-1: Information Technology Equipment - Safety, Part 1: General Requirements, BIS, New Delhi, 2010.

[18] Bureau of Indian Standards, IS 616: Audio, Video and Similar Electronic Apparatus - Safety Requirements, BIS, New Delhi, 2010.

[19] M. Lin, Impact analysis on test items between IEC 60065, IEC 60950-1 and IEC 62368-1, IEEE Symp. Prod. Compliance Eng. (2011) 1-6. https://doi.org/10.1109/PSES.2011.6088253

[20] R. Kumar, A.A. Ansari, A. Kumar, Fire-safe interior surface finishes: a selection guide, AMC Indian J. Civ. Eng. 4 (2021) 29-33. https://doi.org/10.17010/ijce/2021/v4i1/160808

[21] Bureau of Indian Standards, IS 12459: Code of Practice for Fire Safety in Cable Runs, BIS, New Delhi, 1989.

[22] W.J. Parker, Investigation of the fire environment in the ASTM E84 tunnel test, Natl. Bur. Stand. Rep. 13 (1977). https://doi.org/10.6028/NBS.TN.945

[23] H. Yoshioka, T. Noguchi, K. Kobayashi, M. Kanematsu, T. Ando, S. Pareek, T. Hayakawa, Development of a new intermediate-scale box test on sandwich panel products compared with ISO 13784-1 room test and EN 13501-1 classification system, Fire Mater. 45 (2021) 1025-1034. https://doi.org/10.1002/fam.2984

[24] C.A. Wilkie, A.B. Morgan, Fire retardancy of polymeric materials, CRC Press, 2024. https://doi.org/10.1201/9781003380689

[25] D. Drysdale, An introduction to fire dynamics, Wiley, 2011. https://doi.org/10.1002/9781119975465

[26] J. Feng, L. Liu, Y. Zhang, Q. Wang, H. Liang, H. Wang, P. Song, Rethinking the pathway to sustainable fire retardants, Exploration 3 (2023) 1-20. https://doi.org/10.1002/EXP.20220088

[27] M.I. Faheem, S.S. Basha, A.K.M. Ali, A study on construction safety practices in buildings, SAGE J. 25 (2010) 5-24. https://doi.org/10.1177/2977657020100101

[28] J. Hinze, The construction safety record since 1971, in: Civil Engineers Influencing Public Policy, ASCE, 1996, pp. 113-120.

Keyword Index

About the Editors

Dr. Inamuddin is working as an Assistant Professor at the Department of Applied Chemistry, Aligarh Muslim University, Aligarh, India. He obtained a Master of Science degree in Organic Chemistry from Chaudhary Charan Singh (CCS) University, Meerut, India, in 2002. He received his Master of Philosophy and Doctor of Philosophy degrees in Applied Chemistry from Aligarh Muslim University (AMU), India, in 2004 and 2007, respectively. He has extensive research experience in multidisciplinary fields of Analytical Chemistry, Materials Chemistry, and Electrochemistry and, more specifically, Renewable Energy and Environment. He has worked on different research projects as a project fellow and senior research fellow funded by the University Grants Commission (UGC), Government of India, and the Council of Scientific and Industrial Research (CSIR), Government of India. He has received the Fast Track Young Scientist Award from the Department of Science and Technology, India, to work in the area of bending actuators and artificial muscles. He has also received the Sir Syed Young Researcher of the Year Award 2020 from Aligarh Muslim University. He has completed four major research projects sanctioned by the University Grant Commission, Department of Science and Technology, Council of Scientific and Industrial Research, and Council of Science and Technology, India. He has published 216 research articles in international journals of repute and nineteen book chapters in knowledge-based book editions published by renowned international publishers. He has published 200 edited books with Springer (U.K.), Elsevier, Nova Science Publishers, Inc. (U.S.A.), CRC Press Taylor & Francis Asia Pacific, Trans Tech Publications Ltd. (Switzerland), IntechOpen Limited (U.K.), Wiley-Scrivener, (U.S.A.) and Materials Research Forum LLC (U.S.A). He is a member of various journals' editorial boards. He has served as Associate Editor for journals (Environmental Chemistry Letter, Applied Water Science and Euro-Mediterranean Journal for Environmental Integration, Springer-Nature), Frontiers Section Editor (Current Analytical Chemistry, Bentham Science Publishers), Editorial Board Member (Scientific Reports-Nature)and Review Editor (Frontiers in Chemistry, Frontiers, U.K.) He has also guest-edited various special thematic issues for the journals of Elsevier, Bentham Science Publishers, and John Wiley & Sons, Inc. He has attended as well as chaired sessions at various international and national conferences. He has worked as a Postdoctoral Fellow, leading a research team at the Creative Research Initiative Center for Bio-Artificial Muscle, Hanyang University, South Korea, in the field of renewable energy, especially biofuel cells. He has also worked as a Postdoctoral Fellow at the Center of Research Excellence in Renewable Energy, King Fahd University of Petroleum and Minerals, Saudi Arabia, in the field of polymer electrolyte membrane fuel cells and computational fluid dynamics of polymer electrolyte membrane fuel cells. He is a life member of the Journal of the Indian Chemical Society. His research interest includes ion exchange materials, a sensor for heavy metal ions, biofuel cells, supercapacitors and bending actuators.

Dr. Arwa Alrooqi is an Assistant Professor in the Department of Chemistry, Faculty of Science, at Al Baha University, Saudi Arabia. She has published numerous research articles in internationally recognized journals. Dr. Alrooqi is actively engaged in multidisciplinary research, focusing on the synthesis and engineering of organic semiconductors and nanomaterials, their characterization, and their applications in smart window technologies. She has published several high-impact research articles and has co-edited numerous books with leading international publishers.

Dr. Tariq Altalhi is a Professor of Chemistry at Taif University, Saudi Arabia, where he has served as Head of the Chemistry Department and Vice Dean of the College of Science. He earned his Ph.D. from the University of Adelaide, Australia, receiving the Dean's Commendation for Doctoral Thesis Excellence. In 2015, his research was nominated among the top ten entries for the Green Tech Awards, Germany. His research interests include nanomaterial synthesis, membrane technologies, molecular separation, desalination, drug delivery, and biosensing. He published numerous research articles and has co-edited several scientific books and maintains strong collaborations with major industrial partners in Saudi Arabia.

Eng. Ayham Safklou

Eng. Ayham Safklou began his career in medical laboratories, specializing in clinical blood analysis, which established a strong foundation in analytical chemistry and quality control. He later led major environmental and waste-treatment projects for Saudi Aramco and other oil and gas companies, progressing to roles as Chief Operating Officer and General Manager. He also served as an Assistant Professor and currently contributes to advanced environmental R & D initiatives.

www.ingramcontent.com/pod-product-compliance
Lightning Source LLC
LaVergne TN
LVHW052223200926
848481LV00001B/190
9781644904060